Conservation Paleobiology

Conservation Paleobiology

Science and Practice

Edited by
Gregory P. Dietl and Karl W. Flessa

The University of Chicago Press

Chicago and London

The University of Chicago Press, Chicago 60637
The University of Chicago Press, Ltd., London
© 2017 by The Paleontological Society
Published 2017
Printed in the United States of America

26 25 24 23 22 21 20 19 18 17   1 2 3 4 5

ISBN-13: 978-0-226-50669-2   (cloth)
ISBN-13: 978-0-226-50672-2   (paper)
ISBN-13: 978-0-226-50686-9   (e-book)
DOI: 10.7208/chicago/9780226506869.001.0001

LIBRARY OF CONGRESS CATALOGING-IN-PUBLICATION DATA

Names: Dietl, Gregory P., editor. | Flessa, Karl W., editor.
Title: Conservation paleobiology : science and practice / edited by Gregory P. Dietl and Karl W. Flessa.
Description: Chicago : The University of Chicago Press, 2017. | "This book is an expanded reprint of the proceedings of a short course entitled "Conservation Paleobiology: Using the Past to Manage for the Future" that. . . was convened at the annual meeting of the Geological Society of America in Portland, Oregon, in 2009"—ECIP information. | Includes index.
Identifiers: LCCN 2017013691 | ISBN 9780226506692 (cloth : alk. paper) | ISBN 9780226506722 (pbk. : alk. paper) | ISBN 9780226506869 (e-book)
Subjects: LCSH: Paleobiology. | Paleoecology. | Conservation biology.
Classification: LCC QE719.8 .C66 2017 | DDC 560—dc23 LC record available at https://lccn.loc.gov/2017013691

♾ This paper meets the requirements of ANSI/NISO Z39.48-1992 (Permanence of Paper).

Contents

Foreword

Thomas E. Lovejoy

FOR THE bicentennial of the United States, the Smithsonian Institution coined the phrase "Look back, lest you fail to mark the path ahead." That is the essence of this volume—namely, to put a spotlight on how paleobiology can guide us in managing (and managing to avoid some) environmental change. The past is not the perfect predictor of the future by any means, but it can greatly enrich our understanding of what futures could be and what the choices before us are. Just the knowledge that sea levels were four to six meters higher during the last time the planet was two degrees warmer tells us a lot about how we should be managing the planet. It is also critical that we communicate such insights to the world of policy.

Paleontologists concern themselves with the history of life over an enormous period of time. Relatively recent paleontology provides an ecological perspective on how biotic systems came to be and how they respond primarily ecologically to environmental change. That area of study, of course, concerns itself with conditions and biology very akin to the present—the last brief moments before today compared to the full four billion years of life on Earth. Studies in deeper time give us a sense of evolutionary response to environmental change and may help us understand, for example, how corals responded to atmospheric carbon dioxide concentrations that were far in excess of current projections. This volume gives us insights into the value to conservation of near-time and deep-time paleontology.

During the past ten thousand years humanity and ecosystems have adjusted to a stable climate. In this world, the changes we have wrought have been primarily to the biological fabric of the planet (through direct harvest of particular resources, conversion of landscapes, and impacts on aquatic systems, including, of course, fisheries). Those kinds of changes are now occurring at a greater scale and at a faster pace. Beyond the addition of manmade chemicals, there are major distortions of the great global cycles of carbon and nitrogen as well as other gases that are triggering climate change and the acidification of the oceans. The total of all interacting changes is referred to as "global change." Humanity and its aspirations now face a world in ongoing and accelerating flux even if most do not recognize that.

Most people in the world of policy do not have the technical background to wade through—let alone understand—the kinds of publications and documents that are our habitual fare. So the responsibility falls to us to give them the essence of the science and its implications in plain simple language. This very volume is a first step in that direction. Yet it can be a real challenge even for someone who has had to do it on a regular basis for a long time. How, for example, does one explain ocean acidification or sequestering great amounts of carbon by restoring terrestrial ecosystems in a sound bite?

Nonetheless, the ability to sum up a complex situation is extremely valuable. The global climate system is complicated. It is hard to imagine anyone other than a specialist reading the enormous reports replete with ifs, ands, and buts that are produced periodically by the Intergovernmental Panel on Climate Change (IPCC). If we, as scientists who understand both the science and its implications, don't communicate clearly, it is hard to imagine how any of this global change can be taken as seriously as it patently deserves to be.

Communicating with scientific colleagues is one thing, but it is important to think how something might be interpreted or misinterpreted publicly. For example, a 2009 symposium at the National Museum of Natural History inadvertently gave rise to media stories that stated second growth in tropical forest regions was increasing and concerns about old-growth forest were overblown. An echo of that occurred recently in a back-and-forth in *Nature* about the importance of tropical primary forests versus secondary forests, versus sustainable logging. That would have been very confusing to the world of policy when the reality is we not only need all three but also need to increase the total tropical forest area with reforestation. It is another step to go beyond science and suggest something be done about an environmental trend, or even suggest what in particular might be done about it. Some colleagues are reluctant to cross the line—some even militantly so. I think the rest of us have a real responsibility to be engaged. If we aren't, how can we expect society to take an environmental trend seriously—or, for that matter, to adopt a policy that would actually work? I think it would be irresponsible *not* to offer our insights as scientists.

Having said that, I think it is important not to seem like rigid and proscriptive high priests and priestesses of science. Among other things, we must open the door to creative contributions from non-scientists and engage others in thinking them through. Pragmatic policy recommendations are needed. Those tend to come from engaging non-scientific stakeholders and their viewpoints, but of course should never be at the expense of weakening or ignoring what the basic science conveys about the problem at hand.

An interesting example relates to the changes taking place rapidly in the Arctic. The dramatic retreat of the sea ice along with other climate-driven influences is changing the Arctic Ocean ecosystem at the same time that the ocean is becoming accessible to fishing interests. Is it the role of science to communicate that the ecosystem is changing, or is there a responsibility to go beyond that?

What is clear is that allowing harvest in a changing ecosystem will make it impossible to understand how it is changing and understand the separate roles of climate change and harvesting—essentially we will have a marine ecosystem version of the Heisenberg uncertainty principle. So the rational decision would be to establish a moratorium on fishing for whatever length of time necessary to understand the changing ecosystem and how rational fishery management might be designed. For the area of the Arctic Ocean over which it has jurisdiction, the United States has established such a moratorium.

The paleobiological perspective this volume represents is of enormous practical importance in this age of global change. Indeed, the very knowledge of the relatively stable climate of the past ten thousand years should confer the need for caution with respect to forcing further change in the planet's systems.

Paleobiology can help us understand how past life on Earth responded to changing conditions. What can it suggest about the response of coral reefs not only to increasing temperature but also to increasing acidity? It is critical to integrate that perspective with the other effects of human activity on the planet. For example, the conversion of natural terrestrial landscapes to suit various human uses creates barriers to dispersal as plants and animals attempt to track their required climatic conditions and has driven populations of many species to dangerously low levels.

I have personally found paleobiology of central value in developing what should be a global policy to reduce potential climate change. To start with, climate change is already driving the ecosystem failure in tropical coral reefs (bleaching) and the vast coniferous tree mortality in western North America (because warmer winters and longer summers tip the balance in favor of native bark beetles). System change is manifest in ocean acidification. In the Amazon, the combined effects of fire, deforestation, and climate change suggest a tipping point is near that could result in dieback in the southern and eastern parts of that great forest. While CO_2 fertilization might ameliorate that because trees might retain more moisture in such a circumstance, we in fact do not know whether it would run into the nutrient limitation that was demonstrated in a temperate forest experiment. So I find myself driven to the conclusion that two degrees is too much for current ecosystems and that with something like 1.5 degrees coral reefs and the Amazon might be able to muddle through. The problem is that emissions have grown so fast that if the world chooses to stop at the (too high) two degrees, global emissions have to peak almost immediately. So the question arises: Is there a way to pull some of the excess CO_2 out of the atmosphere?

This is where paleobiology rides to the rescue. Twice in the history of life on Earth enormously high CO_2 levels were brought down to levels comparable to preindustrial times. The first was with the advent of plants on land, and the second was with the arrival of modern flowering plants. That took tens of millions of years—time we do not have.

It turns out, however, that paleobiology can suggest a modern policy. Close to half of current excess CO_2 comes from three centuries of ecosystem degradation and destruction. Were there to be a planetary scale program in ecosystem restoration, something like 50 parts per million of CO_2 could be recaptured over fifty years. That is not enough, but it is significant—it is enough to make the difference between current levels (ca. 400 parts per million) and what would be safe for coral reefs (350 parts per million).

As the climate change challenge becomes clearer and more ominous, our role as scientists will become even more important. It is virtually certain that a spate of geo-engineering solutions will be offered. Indeed, almost all the ones we currently know about address the symptom (temperature) and not the cause (CO_2). The National Academy's reports on the topic concluded that it was inappropriate to use the term "engineering" because engineering is not possible when a system is not actually well understood. More appropriate "geo-engineering" schemes will be ones that remove CO_2 and turn it into something inert.

The real point is that the planet does not work as just a physical system. It works as a combined biological and physical system, one that is now being distorted by the human enterprise. At some point humanity and the planet will certainly come to some sort of equilibrium, but biology and paleobiology will be needed for the outcome to be salutary.

Introduction

THIS BOOK is an expanded reprint of the proceedings of a short course entitled *Conservation Paleobiology: Using the Past to Manage for the Future* that was sponsored by the Paleontological Society. The short course was convened at the annual meeting of the Geological Society of America in Portland, Oregon, in 2009. The goals for the short course were (1) to review the many potential applications of geohistorical data[1] in conservation biology, (2) to stimulate new disciplinary research and cross-disciplinary opportunities for collaboration and synthesis, and (3) to increase paleontologists' understanding of how they might translate their research results into effective policies and management practices. The resulting short-course proceedings were thus intended to provide the reader with primers about the various techniques and source materials that can be used to collect and analyze physical, chemical, and biological information derived from geohistorical records—and how those records can be applied to address conservation issues.

The proceedings of the short course have been out of print for several years but have remained in high demand. While research has progressed in many of the areas covered here, this overview of the still rapidly developing field of conservation paleobiology retains its value for students, academics, practitioners, and anyone who has an interest in applying historical data to solve conservation problems. The short course and its proceedings marked a significant stage in the development of the field of conservation paleobiology. It brought together paleontologists—with expertise in a diversity of plant and animal groups, aquatic and terrestrial settings, and the perspectives of "near time" and "deep time"—to address a shared problem: the conservation (and restoration) of biodiversity, habitats, and ecosystem services. Instead of working in isolation, paleontologists who shared an interest in conservation issues were now learning from each other. Although the breadth of the proceedings does not encompass the full range of conservation paleobiology—particularly the rich literature of zooarchaeology and conservation biology (e.g., Lyman and Cannon, 2004; Rick and Erlandson, 2008; Wolverton and Lyman, 2012)—the short course remains the most comprehensive coverage of the discipline. There is continuing demand for the proceedings because the content remains timely, not least because conservation paleobiology is now at work on real-world problems.

We have expanded the original proceedings by adding this new introduction and by providing a perspective before each of the book's three sections. We have also added a discussion by five conservation paleobiologists who are applying the discipline's methods and an epilogue where we reflect on future directions in the field. We have not provided a review of developments in the field since the 2009 short course. Recent reviews (Dietl et al., 2015; Kidwell, 2015; Barnosky et al., 2017) make such a contribution unnecessary, and each of the three section introductions provides citations to other developments since the short course.

The book is organized into three sections. The first two sections recognize that research in conservation paleobiology utilizes geohistorical records at a variety of temporal scales.

In section 1, *Conservation Paleobiology in Near Time*, the chapters focus on the most recent part of the geologic record, generally within the last two million years. This approach uses the relatively familiar past to evaluate the quality of the record, develop proxy indicators of biotic and environmental conditions, and provide a geohistorical context for present-day conditions, typically using geohistorical records of extant species.

Section 2, *Conservation Paleobiology in Deep Time*, focuses on the older part of the geologic record. This approach uses the even longer geohistorical record of the distant past as an archive of repeated natural experiments

---

1. "[T]he individual stratigraphic sections, sediment or ice cores, tree ring series, fossil collections, specimens, or archaeological remains that provide temporal, environmental, and biologic information about a particular place" (NRC, 2005, p. 14).

to investigate biotic responses to system perturbations of diverse kinds and magnitudes and permit the testing of biotic responses to ecosystem disruptions under an array of conditions broader than those available in more recent times.

The distinction between near time and deep time is an arbitrary convenience. Each informs the other in an effort to understand biological vulnerability and resilience to environmental stressors.

Section 3, *Conservation Paleobiology at Work*, focuses attention on the ultimate test of conservation paleobiology: the application of its data and theories in actual management and policy contexts. One application is discussed in some detail while many others are mentioned in the course of a roundtable discussion among conservation paleobiologists now working in the "real world" (see Soulé, 1986) of management, policy, and political compromise. Success stories in this applied part of conservation paleobiology are rapidly accumulating.

We do not claim that this volume is the first word, the only word, or the last word on this subject. We think conservation paleobiology is a good idea, and like any good idea its roots are even deeper and broader than they may first appear. The neologism dates from 2002, but the approach predates the name and has gone by many other names, especially "historical ecology" (e.g., Martin, 1970; Delcourt and Delcourt, 1998; Swetnam et al., 1999; Burnham, 2001; Kowalewski, 2001; NRC, 2005; Burney and Burney, 2007; Willis and Birks, 2010; Louys, 2012; Rick and Lockwood, 2013; Gillson, 2015). Indeed, our original subtitle borrowed from an important paper in historical ecology (Swetnam et al., 1999). What is new about conservation paleobiology is that it provides a sense of community for paleontologists of all sorts who share a desire to play an active role in doing something about today's biodiversity crisis. We envisioned that the short course would serve as a catalyst for such a grassroots effort.

Even though several years have passed since the original short course and proceedings, it is likely still too soon to assess or isolate its impact. We could share anecdotes about increases in interest among undergraduate and graduate students, new courses and graduate programs, symposia at professional meetings, and, perhaps most tellingly, acceptance among conservation practitioners. The short course also gave rise to a report to the National Science Foundation (CPW, 2012) that explored the research opportunities for the earth sciences in conservation paleobiology. All of these changes are encouraging signs that a critical mass of people now agree that they are conservation paleobiologists.

The future looks bright. We are pleased to share the excitement evident in these proceedings.

REFERENCES

BARNOSKY, A. D., E. A. HADLEY, P. GONZALEZ, J. HEAD, P. D. POLLY, A. M. LAWING, J. T. ERONEN, D. D. ACKERLY, K. ALEX, E. BIBER, J. BLOIS, J. BRASHARES, G. CEBALLOS, E. DAVIS, G. P. DIETL, R. DIRZO, H. DOREMUS, M. FORTELIUS, H. W. GREENE, J. HELLMANN, T. HICKLER, S. T. JACKSON, M. KEMP, P. L. KOCH, C. KREMEN, E. L. LINDSEY, C. LOOY, C. R. MARSHALL, C. MENDENHALL, A. MULCH, A. M. MYCHAJLIW, C. NOWAK, U. RAMAKRISHNAN, J. SCHNITZLER, K. D. SHRESTHA, K. SOLARI, L. STEGNER, M. A. STEGNER, N. C. STENSETH, M. H. WAKE, AND Z. ZHANG. 2017. Merging paleobiology with conservation biology to guide the future of terrestrial ecosystems. Science, 355 (10 February 2017), doi:10.1126/science.aah4787.

BURNEY, D. A., AND L. P. BURNEY. 2007. Paleoecology and "inter-situ" restoration on Kaua`i, Hawai`i. Frontiers in Ecology and the Environment, 5:483–490.

BURNHAM, R. 2001. Is conservation biology a paleontological pursuit? Palaios, 16:423–424.

CPW (CONSERVATION PALEOBIOLOGY WORKSHOP). 2012. Conservation Paleobiology: Opportunities for the Earth Sciences. Report to the Division of Earth Sciences, National Science Foundation. Paleontological Research Institution, Ithaca, New York, 32 p.

DELCOURT, P. A., AND H. R. DELCOURT. 1998. Paleoecological insights on conservation of biodiversity: A focus on species, ecosystems, and landscapes. Ecological Applications, 8:921–934.

DIETL, G. P., S. M. KIDWELL, M. BRENNER, D. A. BURNEY, K. W. FLESSA, S. T. JACKSON, AND P. L. KOCH. 2015. Conservation Paleobiology: Leveraging knowledge of the past to inform conservation and restoration. Annual Reviews of Earth and Planetary Sciences, 43:79–104.

GILLSON, L. 2015. Biodiversity Conservation and Environmental Change: Using Palaeoecology to Manage Dynamic Landscapes in the Anthropocene. Oxford University Press, Oxford, 215 p.

KIDWELL, S. M. 2015. Biology in the Anthropocene: Challenges and insights from young fossil records. Proceedings of the National Academy of Sciences of the United States of America, 112:4922–4929.

KOWALEWSKI, M. 2001. Applied marine paleoecology: An oxymoron or reality? Palaios, 16:309–310.

LOUYS, J. 2012. Paleontology in Ecology and Conservation. Springer, Berlin, 273 p.

LYMAN, R. L., AND K. P. CANNON. 2004. Zooarchaeology and Conservation Biology. University of Utah, Salt Lake City, 288 p.

MARTIN, P. S. 1970. Pleistocene niches for alien animals. Bioscience, 20:218–221.

NRC (NATIONAL RESEARCH COUNCIL). 2005. The Geologic Record of Ecological Dynamics. Understanding the Biotic Effects of Future Environmental Change. The National Academies Press, Washington, D.C., 200 p.

RICK, T. C., AND J. M. ERLANDSON. 2008. Human Impacts on Ancient Marine Ecosystems: A Global Perspective. University of California Press, Berkeley, 336 p.

RICK, T. C., AND R. LOCKWOOD. 2013. Integrating paleobiology, archeology, and history to inform biological conservation. Conservation Biology, 27:45–54.

SOULÉ, M. E. 1986. Conservation biology and the "real world," p. 1–12. In M. E. Soulé (ed.), Conservation Biology: The Science of Scarcity and Diversity. Sinauer, Sunderland, Massachusetts.

SWETNAM, T. W., C. D. ALLEN, AND J. L. BETANCOURT. 1999. Applied historical ecology: Using the past to manage for the future. Ecological Applications, 9:1189–1206.

WILLIS, K. J., AND H. J. B. BIRKS. 2010. What is natural? The need for a long-term perspective in biodiversity conservation. Science, 314:1261–1265.

WOLVERTON, S., AND R. L. LYMAN. 2012. Conservation Biology and Applied Zooarchaeology. University of Arizona, Tucson, 256 p.

Conservation Paleobiology in Near Time

THE EIGHT chapters in this section explore the approaches of conservation paleobiology in the parts of the geohistorical record where the species and habitats are most like those of today. We use the arbitrary dividing line of approximately 2.5 million years: shortly before the onset of the climatic fluctuations that define the Pleistocene. Continents and oceans were largely in their present locations; sea level, though in fluctuation during the glacial cycles, was for the most part similar to what it is today; and human activities began to shape the earth's surface and climate. This time interval includes the Pleistocene, the Holocene, and what has come to be called, informally, the Anthropocene—the interval of geologic time in which human activities exert a significant effect on the Earth (Zalasiewicz et al., 2011; Lewis and Maslin, 2015).

The near-time approach takes great advantage of the wealth and quality of geohistorical data available from accumulating sediments, biotic remains, ice cores, tree rings, and archaeological sites, among other archives (see NRC, 2005; Dietl and Flessa, 2011). Many species alive today can be traced back through near time, and although the near-time record contains extinct species, their biotic roles and relationships are relatively easy to decipher. Near time also has the great advantage of the availability of high-precision dating techniques such as dendrochronology, radiocarbon, amino acid racemization, and uranium-thorium dating. Knowing the age of a specimen or its enclosing sediment is vital to understanding the times and rates of environmental and biotic change.

The geohistorical record of near time is rich and diverse. Michał Kowalewski (Chapter 1) and Susan Kidwell (Chapter 7) discuss the marine environment's abundance of shelled invertebrates. The terrestrial realm also yields abundant information, from the vertebrate remains in caves, packrat middens and raptor accumulations (Hadly and Barnosky, Chapter 3), or archaeological sites (Jackson and McClenachan, Chapter 5; Koch,

Fox-Dobbs, and Newsome, Chapter 6) to the plant remains represented by pollen, tree rings, and macrofossils (Jackson, Gray, and Shuman, Chapter 4). The sedimentary record of lakes (Smol, Chapter 2) has long been a rich source of information on environmental and biotic change in near time. Indeed, the paleolimnological record of the effects of acid rain is certainly one of conservation paleobiology's biggest success stories (see also Smol, 2008; Blais et al., 2015). Even the aerial realm is available for study, as illustrated by the case study of diets in California condors (Koch, Fox-Dobbs, and Newsome, Chapter 6).

Fossils and other organic remains are—or contain—proxy indicators. They "stand in" to indicate the presence, abundance, and distribution of species and the time during which they accumulated, their biotic interactions, and the environments in which they lived. Dead remains are, under the right circumstances, excellent proxies for the original live community's composition and relative abundance (Kidwell, Chapter 7; see also Kidwell, 2013), changing environmental conditions (Kowalewski, Chapter 1; Smol, Chapter 2; Hadly and Barnosky, Chapter 3; Jackson, Gray, and Shuman, Chapter 4; Lyons and Wagner, Chapter 8), and an organism's life history and behavior (Kowalewski, Chapter 1; Koch, Fox-Dobbs, and Newsome, Chapter 6). Biotic interactions and preservational processes can also leave their marks on the skeletal remains of organisms (Kowalewski, Chapter 1). And, even genetic variability can be tracked through time as ancient DNA becomes more easily acquired and analyzed (Hadly and Barnosky, Chapter 3).

Tools for conservation paleobiologists continue to be developed. Some are well-established and are in wide use as proxy indicators (see above). Others are new and are, as yet, not in wide use. Kidwell (Chapter 7) devises several measures of the difference between members of the living molluscan community and their adjacent remains. These metrics for mismatch may be useful proxies for not only the degree of impact, but also its cause. S. Kathleen Lyons and Peter Wagner (Chapter 8) borrow macroecology's approaches for the analysis of geographic ranges, body size distributions, diversity, and relative abundance to argue for their use in conservation biology.

The near-time fossil record abounds with examples of past biotic responses to environmental—including climatic—change. Some environmental change can be attributed to direct human impact (examples in Smol, Chapter 2; Jackson, Gray, and Shuman, Chapter 4; Jackson and McClenachan, Chapter 5; Koch, Fox-Dobbs, and Newsome, Chapter 6; Kidwell, Chapter 7), whereas other changes may have resulted from more natural disturbances, including climate change (examples in Smol, Chapter 2; Hadly and Barnosky, Chapter 3; Jackson, Gray, and Shuman, Chapter 4). Indeed, the ecological dynamics of the past provide an opportunity to use the temporal perspective to develop a predictive community ecology (Jackson, Gray, and Shuman, Chapter 4—see also Jackson and Blois, 2015).

REFERENCES

BLAIS, J. M., M. ROSEN, AND J. P. SMOL. 2015. Environmental Contaminants: Using Natural Archives to Track Sources and Long-term Trends of Pollution. Springer, Dordrecht, 509 p.

DIETL, G. P., AND K. W. FLESSA. 2011. Conservation paleobiology: Putting the dead to work. Trends in Ecology and Evolution, 26:30–37.

JACKSON, S. T., AND J. L. BLOIS. 2015. Community ecology in a changing environment: Perspectives from the Quaternary. Proceedings of the National Academy of Sciences of the United States of America, 112:4915–4921.

KIDWELL, S. M. 2013. Time-averaging and fidelity of modern death assemblages: Building a taphonomic foundation for conservation palaeobiology. Palaeontology, 56:487–522.

LEWIS, S. L., AND M. A. MASLIN. 2015. Defining the Anthropocene. Nature, 519:171–180.

NRC (NATIONAL RESEARCH COUNCIL) 2005. The Geological Record of Ecological Dynamics: Understanding the Biotic Effects of Future Environmental Change. The National Academies Press, Washington, D.C., 200 p.

SMOL, J. P. 2008. Pollution of Lakes and Rivers: A Paleoenvironmental Perspective—2nd Edition. Blackwell Publishing, Oxford, 383 p.

ZALASIEWICZ, J., M. WILLIAMS, A. HAYWOOD, AND M. ELLIS. 2011. The Anthropocene: A new epoch of geological time? Philosophical Transactions of the Royal Society London A, 369:835–841.

CHAPTER ONE

THE YOUNGEST FOSSIL RECORD AND CONSERVATION BIOLOGY: HOLOCENE SHELLS AS ECO-ENVIRONMENTAL RECORDERS

MICHAŁ KOWALEWSKI

Department of Geosciences, Virginia Tech, Blacksburg, VA 24061 USA

ABSTRACT.—This chapter reviews eco-environmental information that can be extracted from the youngest (surficial) fossil record. The main focus of this review are shell-producing macro-invertebrates, which are abundant at (or near) the sediment surface in many aquatic and terrestrial habitats. As demonstrated by geochronologic and sclerochronologic analyses, such Holocene shell accumulations provide direct, often continuous, records of the most recent centuries and millennia. They offer us a broad spectrum of data, including multiple taxonomic, ecologic, taphonomic, geochronologic, and geochemical parameters that can inform us about the recent history of organisms, communities, ecosystems, and environments. Eco-environmental analyses of shells and other bio-remains provide us thus with a long-term historical perspective; a quantitative baseline for understanding pre-industrial ecosystems, evaluating anthropogenic changes, and assessing restoration efforts. This research strategy, sometimes referred to as "conservation paleobiology," is often based on dead remains only, and consequently, represents a biologically non-invasive approach to conservation issues. The recently dead—the youngest fossil record of environmental and ecological processes at local, regional, and global scales—can help us to protect the living and should allow us to manage the future of our biosphere more effectively.

INTRODUCTION

SKELETAL REMAINS of dead organisms are copious in many modern environments: trillions of mollusk shells contour ocean shorelines, coastal dunes are streaked with dense layers of land snail conchs, and countless mussel valves litter streams and rivers. Even dolphin bones—a valuable collector's item and a non-exportable national commodity—may occur in large numbers in more remote areas (e.g., Liebig et al., 2003) yet undiscovered by beachcombers.

A few examples of surficial accumulations highlighted here on Figure 1 illustrate their exceptional qualities. Among others, these accumulations represent biomineralized remains that are easy to access and do not require expensive sampling machinery (Fig. 1.1, 1.2, 1.4, 1.5, and 1.6). Also, they often provide us with copious quantities of specimens that can be sampled from a single spot (e.g., Fig. 1.2, 1.3, 1.6, and 1.7), and the material is typically preserved exceedingly well; for example, articulated skeletons of vertebrates (Fig. 1.1) or ligament-preserving shells of freshwater mussels (Fig. 1.4) are not uncommon. Finally, these accumulations can be found in a broad spectrum of environmental and depositional settings—from eolian dunes (Fig. 1.5 and 1.6), through river banks (Fig. 1.4), down to various marine habitats, from supratidal mud flats

(Fig. 1.1) to open-shelf subtidal seafloors (Fig. 1.7).

To be sure, the wanton abundance of bioskeletal remains has been appreciated since the dawn of science. Starting with Aristotle and da Vinci, biologists and geologists have paid continuing attention to those surficial graveyards. However, only recently, have we begun to appreciate in full the wealth of historical data contained within shells, tests, bones, and other bio-remains scattered on the surface of our planet. This is for two reasons. First, geochronologic and sclerochronologic efforts of the last decades have demonstrated, repeatedly, that the youngest fossils yield continuous, multi-centennial to multi-millennia records of the most recent geological past (e.g., Flessa et al., 1993; Lazareth et al., 2000; Kowalewski et al., 2000; Carroll et al., 2003; Kidwell et al., 2005; Schöne et al., 2005; Kosnik et al., 2007; Yanes et al., 2007). Second, rapid advances in analytical instrumentation have provided us with increasingly diverse, sophisticated, and affordable techniques for dating, analyzing, and interpreting bio-remains left behind by dead organisms (e.g., Dettman and Lohmann, 1995; Kaufman and Manley, 1998; Riciputi et al., 1998; Lazareth et al., 2000; Rodland et al., 2003; Fiebig et al., 2005; Carroll et al., 2006; Fox et al., 2007; Schiffbauer and Xiao, in press). These and related analytical and conceptual advancements make it now possible to use the dead to help the living. That

---

In *Conservation Paleobiology: Using the Past to Manage for the Future, Paleontological Society Short Course, October 17th, 2009. The Paleontological Society Papers, Volume 15*, Gregory P. Dietl and Karl W. Flessa (eds.). Copyright © 2009 *The Paleontological Society.*

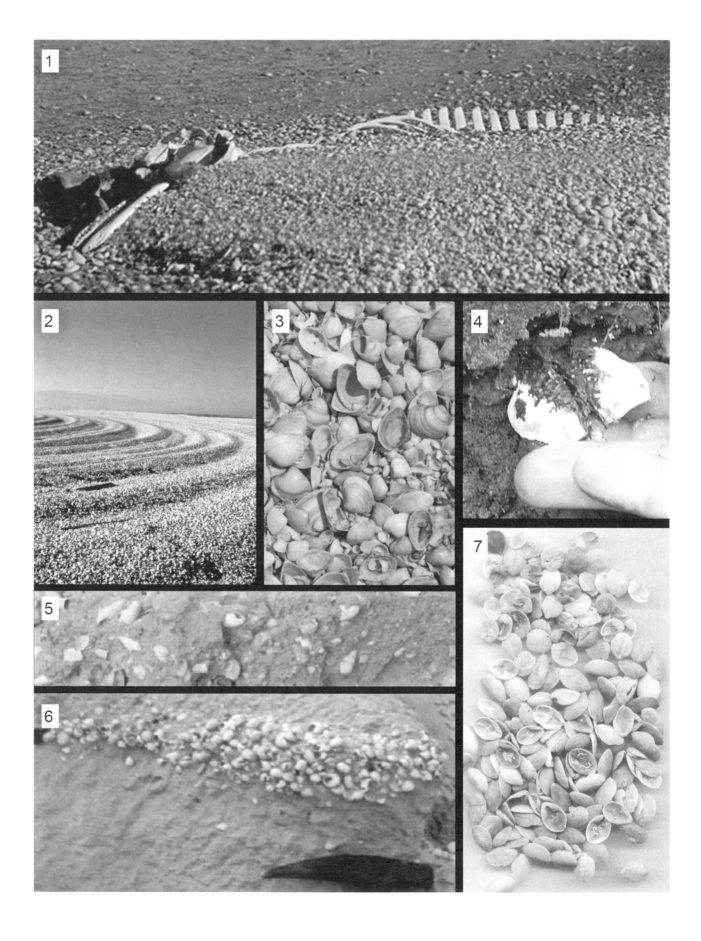

is, skeletal remains offer a quantifiable reference baseline that can help us to understand the living communities, assess the efficacy of restoration efforts, and, ultimately, protect their future. Just as important, this "Conservation Paleobiology" approach—as some refer to it nowadays (e.g., Flessa, 2002)—has a virtue of being partly or completely non-invasive. This is because conservation paleobiology involves sampling efforts that often do not require killing, or even disturbing notably, living biota.

Here, I review briefly eco-environmental information that can be extracted from the youngest (surficial) fossil record, with particular focus on shell-producing macro-invertebrates, although other types of macroscopic remains (bones, plants, etc.) will be discussed briefly when relevant. However, micropaleontological approaches to conservation paleobiology (subfossil pollen, test of plankton, etc.) are not discussed in this chapter. In addition, multiple case studies will be used to illustrate ways in which skeletal remains can be employed to augment our understanding of eco-environmental changes over centennial-to-millennial time scales. The main goal here is to highlight the unique historical potential of the data that can be provided by the youngest fossil record (e.g., Kowalewski et al., 2000; Jackson et al., 2001; Brown et al., 2005; Kidwell, 2007). After all, the historical perspective that reaches back in time far beyond the industrial revolution is necessary for the realistic assessment and effective mitigation of widespread anthropogenic changes occurring on our planet today (see especially Jackson et al., 2001; Pandolfi et al., 2003; Flessa et al., 2005; Flessa and Jackson, 2005).

A comment on citations cited in this chapter

Scientists care a great deal about fair and compre-hensive accounting of previous work, as attested by the increasing importance of the Science Citation Index and the ever increasing emphasis that peer-review journals place on adequate referencing. Therefore, it is appropriate to comment on how cited references were selected here. Many of the thematic areas discussed below (e.g., sclerochronology, amino-acid dating, taphonomy, and stable isotopes) represent active research directions with dozens, or hundreds, of new publications appearing every year. It would be arguably suitable to cite tens of papers in each sentence of the text below. Regrettably, this is not feasible. The citations used here are, therefore, an inevitably biased, subjective, and unrepresentative sample of the existing literature. This caveat is trivial, but necessary, given goals of short course publications such as this one. An effort has been made to select references that either provide synthetic and well-referenced overviews of topics, or are directly pertinent to the theme of this chapter, or represent up-to-date examples of some relevance. Nevertheless, the reader should appreciate the fact that the references cited here are neither free of the author's idiosyncratic parochialisms (e.g., ignorance, subjective preferences, inherent self-centeredness, etc.) nor satisfactorily representative of the existing literature. Those who use short course volumes such as this one to augment their research should bear in mind that this caveat likely applies to most chapters of the type presented here and in other such publications.

Fossils, "fossils", or subfossils?

It is fair to notice that Holocene fossils are not really fossils *sensu stricto,* being from the epoch that is geologically synonymous with the Recent. Some authors refer to such bioskeletal remains as "subfossils", some use "fossils" with quotation marks, and some use

←

FIGURE 1.—Copious skeletal remains left behind by various organisms found in a wide spectrum of present-day depositional environments; *1,* An articulated skeleton of a dolphin on the surface of macrotidal flats of the Lower Colorado River Delta, Mexico. Note also abundant mollusk shells on the surface; *2,* Multiple beach ridges made nearly exclusively of disarticulated valves of the bivalve mollusk *Mulinia coloradoensis* (the Lower Colorado River Delta); *3,* A close-up of one of the beach ridges; *4,* An articulated shell of a freshwater mussel protruding from a river bank (North Fork Holston River, SW Virginia); *5,* Abundant shells of terrestrial snails in a Late Quaternary paleosol, Furteventura Island, Canary Islands; *6,* A layer made nearly exclusively of shells of terrestrial snails embedded within a late Quaternary dune, Furteventura Island, Canary Islands; *7,* Calcitic shells of the brachiopod *Bouchardia rosea* dredged from the modern surface of the southern Brazilian Shelf. Photographs courtesy of, and copyrighted to, Karl Flessa, University of Arizona (Fig. *1.1-1.3*); Claudio de Francesco (University of La Plata) and John W. Huntley (University of Kentucky) (Fig. *1.4*); Yurena Yanes (Southern Methodist University and University of Granada) (Fig. *1.5-1.6*); and Marcello G. Simões (Sao Paulo State University) (Fig. *1.7*).

the term "fossil" without quotation marks. In my view, from all practical standpoints (taphonomic, logistic, methodological, etc.), Holocene bioskeletal remains do not differ from older fossils. Moreover, it makes little sense to call a specimen dated at 9,000 yrs BP a subfossil or "fossil", while calling another specimen from the same sample a true fossil, just because its radiocarbon date came back from the lab dated at 12,500 yrs BP. Here, for the sake of expedience and clarity, I will follow the simplest approach and refer to all specimens (Holocene or not) as fossils (without quotation marks). Terminological purists have every right to vehemently disagree with this terminological flippancy.

MAJOR TYPES OF DATA RECOVERABLE FROM SKELETAL REMAINS

With the right set of tools, diverse data can be obtained from skeletal remains. In this section, 9 types of data retrievable from shells, bones, and other biomineral debris are summarized briefly. This list is certainly not exhaustive, but rather serves to highlight the remarkable plethora of parameters and variables preserved in the youngest fossil record. These variables can be organized as follows (although this scheme is arbitrary as numerous ways of organizing those data types may be possible).

Intrinsic biological variables

These include shell parameters that relate directly to biological characteristics of skeletal remains. The four variables most commonly extracted include the following:

1. Taxonomic identity.—With few exceptions, geologically recent (late Pleistocene to Holocene) skeletal remains represent extant genera and species. This makes species identification possible for most specimens. The bulk sampling of skeletal remains can therefore provide quantitative datasets about taxonomic diversity and relative abundance of species. From the conservation biology perspective, this type of data can be used to carry out a fidelity analysis, in which skeletal remains ("death assemblage") are compared to sympatric living communities ("life assemblage"). This approach is used increasingly to detect recent anthropogenic changes in local communities (e.g., Greenstein et al., 1998; Simões et al., 2005; Kidwell, 2007; Terry, 2007; Ferguson and Miller, 2007; Ferguson, 2008).

2. Ecological characteristics.—Because most skeletal remains belong to organisms that are still around, ecological characteristics of those dead organisms can also be inferred with high accuracy. Thus, skeletal remains can provide information about relative abundance of various ecological groups, with specimens grouped by mode of life, feeding type, or even preferred bathymetry expressed in meters (e.g., Scarponi and Kowalewski, 2004). Absolute abundance of skeletal remains of a given species may also carry ecological information about population dynamics, bioproductivity, or intensity of trophic interactions in pre-industrial times (e.g., Walbran et al., 1989a, b; Kowalewski et al., 2000; see also later in the text).

3. Size/age/biomass.—Skeletal remains (unless too fragmented) can also often provide reasonably accurate estimates of size at the time of death. In many cases, this information can be used to estimate the ontogenetic age of the specimen, its biovolume, or its soft-tissue biomass (e.g., Staff et al., 1985; Zuschin et al., 1999; Novack-Gottshall, 2008; see also Kosnik et al., 2006; Krause et al., 2007).

4. Sclerochronology and growth patterns.—In organisms that secrete skeletal elements via incremental accretionary processes, growth rings or other comparable time-sequential structures can be often recognized and analyzed. Because rates of growth are influenced by environmental variables such as temperature, nutrient levels, and other parameters (e.g., Kennish and Olsson, 1975; Sato, 1997; Schöne et al., 2003a, 2005; and numerous references therein), statistical analyses of growth increments can not only provide geochronological estimates, but also inform us about past environmental conditions. This approach is particularly powerful when used in conjunction with geochemical data (e.g., Goodwin et al., 2001; Ivany et al., 2003; multiple papers in Schöne and Surge, 2005). In some organisms, growth lines can offer daily or even sub-daily resolution (e.g., Berry and Barker, 1968; see also Rodland et al., 2006a), which is particularly remarkable when considering that some long-lived shellfish, including both marine and freshwater forms (e.g., Schöne et al., 2004, 2005; Dunca et al., 2005) can live (and grow intermittently or continuously) for decades or even centuries (see also below).

Secondary biological variables

These are mostly represented by trace fossils and

skeletonized encrusters that provide direct or indirect records left on skeletal remains of shell-producing organisms. Two major types of records can be distinguished:

1. Syn-vivo records.—These are various traces and structures left by other organisms that interacted with the original owner of a given skeletal remain. Repair scars on skeletal remains recording failed predatory attacks (e.g., Vermeij et al., 1981; Vermeij, 1987; Cadeé et al., 1997; Alexander and Dietl, 2003; Voight and Sigward, 2007), drill holes made by carnivorous snails in shells of their prey (e.g., Carriker, 1981; Kitchell et al., 1981; Vermeij, 1983, 1987; Kitchell, 1986; Kelley and Hansen, 1993, 2003; Kowalewski, 2002; Leighton, 2002; Alexander and Dietl, 2001; Dietl and Herbert, 2005; Huntley and Kowalewski, 2007; Casey and Chattopadhyay, 2008; Schiffbauer et al., 2008), or syn-vivo domicile borings left by polychaetes (e.g., Rodrigues, 2007; Rodrigues et al., 2008; see also Martinell and Domènech, 2009) are all examples of ecological interactions recorded on skeletal remains of prey/host organisms. These records make it possible, for example, to assess trophic importance of prey organisms that produced those skeletal remains (e.g., Cintra-Buenostro et al., 2005; Stempien, 2005, 2007), estimate seasonal changes in intensity of predator-prey interactions (e.g., Kowalewski and Flessa, 2000), or evaluate predation pressures across environmental gradients (e.g., Dietl and Alexander, 2009).

2. Post-mortem records.—Organisms that colonize or mine skeletal remains of other organisms can produce various traces and biomineral structures that can be identified and interpreted. They can occur post-mortem (i.e., after the substrate-producing shell owner died), although some may also represent syn-vivo interactions. These records include diverse variety of encrustation structures left by biomineralizing organisms that colonize skeletons of other organisms (e.g., Lescinsky, 1997; Taylor and Wilson, 2003; Rodland et al., 2004, 2006b) and macroscopic and microscopic traces made by bioeroding organisms (e.g., Bromley et al., 1990; Cutler, 1995; Bromley, 1996; Walker et al., 1998; Taylor and Wilson, 2003). Encrusting and bioeroding organisms are sensitive to various environmental parameters and, as some argue, may be useful for assessing bioproductivity and eutrophication (e.g., Lescinsky, 1997; Holmes et al., 2000; Lescinsky et al., 2002).

Extrinsic variables

These include various physico-chemical parameters that can be extracted from skeletal remains. The three sets of parameters commonly extracted from biomineral solids, that are most relevant in the context of this review, include the following:

1. Taphonomic grades.—These variables have been used widely to quantify post-mortem alteration of subfossil (Holocene) specimens (e.g., Fürsich and Flessa, 1987; Davies et al., 1989; Meldahl and Flessa, 1990; Kowalewski et al., 1995; Best and Kidwell, 2000; Kidwell et al., 2001; Henderson et al., 2002; Tomašových and Rothfus, 2005; Lockwood and Work, 2006; Yanes et al., 2008b; Powell et al., 2008; and numerous references therein). Taphonomic data, typically expressed in terms of semi-qualitative taphonomic grades, include physicochemical features such as fragmentation and dissolution as well as biological parameters discussed above (bioerosion and encrustation). Whereas taphonomic analyses were initially motivated by actuopaleontological and deep-time paleoenvironmental questions (see especially taphofacies concept; Brett and Baird, 1986; Speyer and Brett, 1986), researchers have increasingly recognized in recent years that intensity of taphonomic alterations can provide various clues as to the health of local communities and the severity of anthropogenic changes (e.g., Holmes et al., 2000; Wapnick et al., 2004; Brown et al., 2005).

2. Geochronological estimates.—Many skeletal remains can be dated using various techniques, including radiocarbon (e.g., Flessa et al., 1993; Flessa and Kowalewski, 1994; Martin et al., 1996; Meldahl et al., 1997; Flessa, 1998); stable isotopes (e.g., Lazareth et al., 2000; Fig. 2), Uranium Thorium decay series (e.g., McCulloch and Mortimer, 2008; Zhao et al., 2009), free amino-acids (Powell et al., 1989), uncalibrated or partly calibrated amino-acid racemization [AAR] rates (e.g., Powell and Davies, 1990; Wehmiller et al., 1995), and ^{14}C-calibrated AAR rates (e.g., Goodfriend, 1989; Goodfriend et al., 2000; Wehmiller and Miller, 2000). The last approach, in particular, has been used with an increasing success to date numerous specimens of marine shellfish (e.g., Kowalewski et al., 1998; Carroll et al., 2003; Kidwell et al., 2005; Kosnik et al., 2007; Yanes et al., 2007; Krause et al., in press). Whereas AAR dating is not without limitations and assumptions (see Goodfriend et al., 1997, 2000; Barbour Wood et al., 2006; Kosnik et al., 2008; Kosnik and Kaufman,

2008; and references therein), the approach offers the most reliable and affordable way for assembling data-intensive time series of dated shells (see also later in the text).

 3. Geochemical data.—A great variety of geochemical signatures can be extracted from shells and other biominerals, including numerous stable isotope ratios and trace element concentrations. Stable isotopes of oxygen and carbon have been among the most common signatures extracted from Holocene and older shells (e.g., Rodriguez et al., 2001; Dietl et al., 2002; Schöne et al., 2003b; Metref et al., 2003; Dettman et al., 2004; Rohling et al., 2004; Carrè et al., 2005; Chauvaud et al., 2005; Rowell et al., 2005; Yanes et al., 2008a; and hundreds of other papers). Depending on the environment and analyzed organism, such isotopic signatures can provide insights into temperature, salinity, environment, ecosystem productivity, growth rates, diet, and other parameters of direct relevance for assessing environmental and ecological questions. However, interpretations of stable isotopic signatures are not always unambiguous. Isotope ratios for other elements such as, for example, hydrogen (e.g., Carroll et al., 2006), strontium (e.g., Briot, 2008), nitrogen (e.g., Kurata et al., 2001; Carmichael et al., 2008), or neodymium (Pomiès et al., 2002) can also be extracted either from the biomineral phase of the skeleton, or associated organic matrix, or soft tissues (in case of live-collected specimens). These isotopes have also shown promise for interpreting various eco-environmental parameters, including anthropogenic changes.

 Trace elements, often expressed as ratios standardized against Ca, are also routinely extracted from shells. For example, studies on present-day aragonitic hermatypic corals *Pavona clavus* (Lea et al., 1989; Shen et al., 1992), present-day and Pleistocene bivalves *Mercenaria* and *Spisula* (Stecher et al., 1996), and present-day *Mytilus edulis* (Vander Putten et al., 2000) all suggest that the Ba/Ca ratios in biogenic carbonates can provide a reliable proxy for high productivity. Another example is offered by Sr/Ca ratios that also can be extracted from biominerals, and have been used, especially for scleractinian corals, as a paleo-thermometer of sea-surface temperatures (Weber, 1973; McCulloch et al., 1994). However, numerous factors complicate trace element analyses. For example, Sr/Ca ratios may be affected by calcification/growth rates (De Villiers et al., 1994, 1995; Stecher et al., 1996) and differential

depletion of Ca vs. Sr in surface waters (Tsunogai and Watanabe, 1981; Bernstein et al., 1987, 1992). Similarly, numerous factors may complicate the application of Mg/Ca ratios as paleotemperature proxy (e.g., Stephenson et al., 2008). Examples of recent papers on various trace elements extracted from shells include Brown et al. (2005), Gillikin et al. (2006), Carroll and Romanek (2008), Barats et al. (2009) and numerous references therein.

 It is noteworthy that there are literally hundreds of papers published every year on geochemical signatures of biominerals—interested readers should browse for relevant citations in the sample of papers cited above and may also wish to consult recent books on stable isotope geochemistry and related topics (e.g., Norris and Corefield, 1998; Sharp, 2006; Fry, 2007; Hoefs, 2009; note that only some sections of those books will be of direct relevance here).

GEOCHRONOLOGY OF BIOSKELETAL REMAINS

 Bioskeletal remains can provide at least three distinct types of long-term records that can extend back over centennial to millennial timescales: (1) single-specimen sclerochronological records provided by individual specimens; (2) multi-specimen sclerochronological records assembled by correlating multiple, temporally overlapping specimens; and (3) assemblage-level geochronology based on direct dating of skeletal remains from time-averaged accumulations. Here, I will focus primarily on macrobenthic shellfish, whereas other types of records such as tree-ring dendrochronology or time-averaged bone accumulations will be noted only briefly.

Single-specimen sclerochronological records

 Aquatic invertebrates with accretionary growth strategies include many long-lived taxa that can produce skeletons representing decades or even centuries of growth. Such skeletal remains can provide high-resolution, long-term records of environmental conditions and may also be useful in retroactive monitoring of pollution (e.g., to evaluate onset of pollution by reconstructing long-term changes in concentration of trace elements incorporated by ancient organisms into biomineral solids). These types of records have been documented by now from a wide range of latitudes,

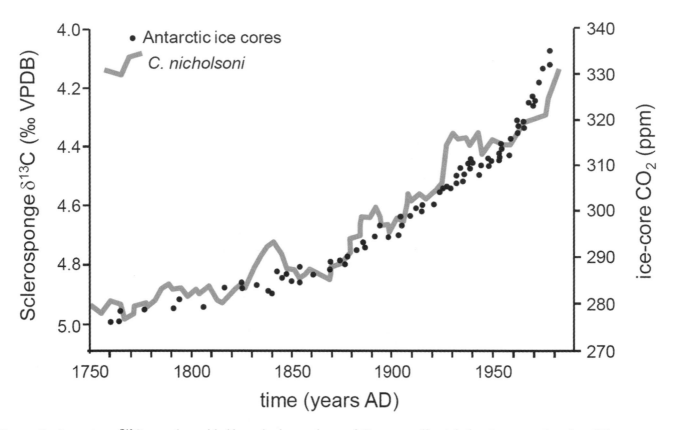

FIGURE 2.—Long-term $\delta^{13}C$ record provided by a single-specimen of *Ceratoporella nicholsoni* compared against CO_2 content estimated from ice-core air bubbles. Thick continuous line—*C. nicholsoni* three-point running-average ($\delta^{13}C$ profile in delta notation relative to Vienna Peedee belemnite (VPDB) standard) (data after Lazareth et al. 2000); open diamonds—CO_2 records from Antarctic ice cores (data from Etheridge et al., 1998; Neftel et al., 1985; Friedli et al., 1986). Plot redrafted based on a figure of Lazareth et al. (2000).

environments, and taxonomic groups. For example, Lazareth et al. (2000), using high-resolution micro-sampling techniques, demonstrated that a single scle-rosponge (collected from a Bahamian reef), despite being only 14 cm in diameter, represented over 223 years of growth (Fig. 2). The parallel time series of Pb concentrations extracted from the same specimen yielded proxy patterns remarkably consistent with the lead record extracted from Greenland ice cores (Mu-rozumi et al., 1969). In contrast to this shallow-water tropical example, Schöne et al. (2005) documented a single specimen of the bivalve *Arctica islandica* (live-collected in 1868 off the coast of Iceland) represent-ing an even more impressive record, which totaled 343 years of growth (Fig. 3). Such long-lived mollusks are not restricted to marine habitats. Freshwater mussel shells (especially at high latitudes) represent multiple decades of growth, and individuals representing more than 100 years of growth have been reported (Bauer,

1992; Schöne et al., 2004). Time-rich sclerochrono-logical records can be also extracted from remains of terrestrial organisms, including vertebrate records such as mammoth tusks (e.g., Fox et al., 2007) and vascular plant data based on tree rings; the latter being by far the most extensively studied type of sclerochronological records (e.g., Schweingruber, 1996).

Multi-specimen sclerochronological records

By combining multiple specimens that overlap in time, accretionary growth records can be cross-cor-related to form master chronologies that cover much longer time intervals than lifespans of individual or-ganisms. The validity of such an approach has been first established for tree rings, where time series (based on widths of individual growth increments) obtained from trees overlapping in time were assembled togeth-er into long-term tree-ring dendrochronologies. Such tree-ring master-chronologies can provide precise and

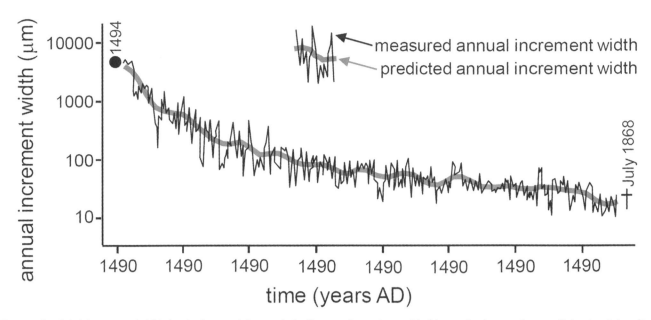

FIGURE 3.—Multi-centennial biological record (annual shell growth rate) provided by a single specimen of *Arctica islandica* live-collected in 1868 near Iceland (Schöne et al., 2005). Note that growth rates decrease with increasing ontogenetic age. Plot redrafted based on a figure presented in Schöne et al. (2005).

continuous records extending back for multiple millennia (e.g., Briffa et al., 1990, 1998; Becker et al., 1991; Scuderi, 1993).

The application of this approach to marine invertebrates is, by comparison, much less advanced. However, efforts focused on corals (e.g., Dunbar and Cole, 1993; Swart et al., 1996) and mollusk bivalves (e.g., Jones et al., 1989; Marchitto et al., 2000; Schöne, 2003; Schöne et al., 2004) indicate that such research strategies can be developed successfully for many taxa with accretionary growth patterns.

Time-averaged assemblage records

Dating efforts focused on surficial shell assemblages from nearshore and shelfal marine environments have demonstrated that skeletal remains of benthic organisms tend to represent a time-rich mixture of specimens that vary in age from recently dead to those that date back to hundreds or even thousands of years ago (see reviews by Behrensmeyer et al., 2000; Kowalewski and Bambach, 2003).

As mentioned above, dating efforts based on ^{14}C-calibrated amino-acid racemization ratios have been particularly successful in developing quantitative estimates of time-averaging and exploring geochronological age structures of bioskeletal accumulations.

These efforts have been particularly intense in recent years for marine shellfish. By now, several major age distributions (Fig. 4) have been assembled for mollusk bivalves and brachiopods from a broad spectrum of depositional and latitudinal settings (e.g., Kowalewski et al., 1998; Carroll et al., 2003; Kidwell et al., 2005; Kosnik et al., 2007; Krause et al., in press). These time series all represent multi-centennial to multi-millennial records. Some of those case studies provide a continuous, bioskeletal documentation for most of the Holocene (Fig. 4). The realistic resolution of these time-series, controlled by accuracy and precision of ^{14}C-calibrated AAR rates, is on the order of 50-150 years for most recent centuries and 200-500 years for early Holocene. Thus, while we cannot place dated shells in a temporal context as precise temporally as is possible for sclerochronological time series, we can group shells into 100 or 200 or 500 year bins and extract various parameters from shells to assemble multi-millennial time series. Such data can allow us to address many research questions that are otherwise inaccessible (e.g., Goodfriend and Gould, 1996; Kowalewski et al., 2000; Huntley et al., 2008).

It is also noteworthy that dating of terrestrial and lacustrine gastropod accumulations yielded estimates of temporal mixing comparable to those shown on

Figure 4; hundreds to thousands of years (e.g., Cohen, 1989; Goodfriend, 1987, 1989; Yanes et al., 2007). Similarly, terrestrial vertebrate bones often represent time-averaged accumulations that mix specimens over multi-centennial to multi-millennial time scales (e.g., Hadly, 1999; Behrensmeyer et al., 2000; Auler et al., 2006; Terry, 2008). Radiocarbon dating (Edinger et al., 2007) suggests that even in the case of biogenic *in situ* preserved structures such as coral reefs some time-averaging exists, although such age-mixing is much less extensive and of different genesis than post-mortem temporal mixing that typifies shell assemblages (Edinger et al., 2007).

Finally, it is noteworthy that even without applying direct dating, time-averaged records can provide useful information about long-term ecological and taphonomic patterns, as suggested both by theoretical modeling (e.g., Olszewski, 1999, 2004; Terry, 2008) as well as empirical data, which suggest that some age distributions of dated shells appear to provide continuous, time-weighted records of local communities ("uniform time-averaging"; see Kowalewski et al., 1998).

RESEARCH DIRECTIONS AND CASE STUDIES

As is clear from the overview provided in the two sections above (1) skeletal remains found at (or near) the surface provide biomineral solids from which diverse data of direct relevance to conservation biology can be extracted; and (2) these biomineral records can be assembled into multi-centennial to multi-millennial time series. In this section, specific case studies will be used to highlight how these qualities of the youngest fossil record can be used (separately or jointly) to augment efforts of conservation biology and historical ecology.

Fidelity

There is an increasing realization that the discordance between living communities and death assemblages does not need to represent taphonomic biases such as differential preservation of taxa or confounding effects of time-averaging. Instead, such disagreements in composition of dead versus living faunas may measure severity of anthropogenic changes. Recently, Kidwell (2007) has assembled an array of comparative (live-dead) datasets for marine benthic shellfish and showed that highly discordant systems, where death assemblages differ notably from living communities, typify ecosystems affected by severe anthropogenic changes. In contrast, relatively pristine habitats show much better concordance between dead and living. At a more local scale, such approaches have been used successfully for studying reefs (Greenstein et al., 1998), nutrient pollution (Ferguson, 2008), and anthropogenic changes in lake ecosystems (e.g., Alin and Cohen, 2004). Bone accumulations in various regions of the world can also offer important insights into the recent history of local ecosystems (e.g., Terry, 2008; Behrensmeyer and Western, 2009). This promising approach of conservation paleobiology is discussed in detail elsewhere in this volume (Kidwell, 2009; see also Kidwell, 2008; Tomašových and Kidwell, 2009a, b).

Admittedly, not all discrepancies between living communities and death assemblages reflect ecosystem changes (anthropogenic or natural). Organisms that differ notably in preservational potential may be affected by severe taphonomic biases, so dead-live disagreements in the composition, and especially relative abundance, of fauna may be driven by taphonomic processes (e.g., Schopf, 1976; Kowalewski, 1996; Stempien, 2005). Post-mortem transport may also move shells a notable distance, but without placing them outside their ecological range, so these shells may still appear to us as viably in place even though they may have traveled a long distance for a long time (see especially Flessa, 1998 for a compelling example). Thus, while discordances between dead and live may record anthropogenic changes, as shown conclusively by Kidwell (2007), taphonomic biases need to be accounted for on a case by case basis.

Absolute abundance

Absolute abundance (or density) of bioskeletal remains can also be a useful tool for fidelity analysis and can help to evaluate if intensity of ecological processes (e.g., bioproductivity, tropic flow, etc.) have been altered by anthropogenic changes. For example, Walbran et al. (1989a) used absolute abundance of skeletal elements of the crown-of-thorns starfish *Acanthaster planci* to assess if the population outbreak of this voracious predator, observed around the Great Barrier Reef since the late 1960s, was a historically unique (possibly anthropogenic) event. Using both surface sediments and subsurface cores material, multiple spikes in abundance of skeletal elements of *Acanthaster planci* were observed throughout the Holocene sediments stretching back for at least the last 8,000 years. Consequently, the authors argued, population outbreaks of this predator had occurred naturally in the system for multiple millennia, long before notable anthropogenic changes affected the region. Although the validity of the specific methodology used by Walbran et al. (1989a) has been questioned (e.g., Keesing et al., 1992; Fabricius and Fabricius, 1992; Brodie et al., 2005), the approach is certainly valid in general terms and should yield useful insights when applied stringently.

Taphonomy

Preservational quality of skeletal remains (their taphonomic attributes) can potentially inform us as to the health of local ecosystems and intensity of anthropogenic changes. This approach is particularly noteworthy because of its biologically non-invasive nature: collecting modest samples of shells or bones is unlikely to affect living communities, especially when comparing with approaches that require extensive sampling of live organisms.

Because taphonomic processes vary greatly depending on type of skeletal remains, nature of depositional environments, and many other parameters, specific research strategies need to be developed for each environmental setting (or depositional system) individually and their applicability may be limited to certain types of biominerals. Here, I briefly discuss two examples that illustrate direct application of taphonomy to conservation biology and historical ecology.

Brown et al. (2005) studied death assemblages of freshwater mussels in the North Fork Holston River, Virginia (an EPA Superfund Site). This river system has been affected by extensive mercury pollution downstream of the town of Saltville, where Hg was used to produce chlorine and caustic soda from 1950 to 1972. The analysis of 366 shells, collected more than 30 years after the factory had closed, showed that shell alteration varied predictably across sites with different extirpation histories, but did not relate to a local stream gradient. At Saltville, where living mussel communities have been absent for at least three decades, shells were heavily altered and fragmented. In contrast, upstream of Saltville, where reproducing mussel populations were still present, fresh-looking shells were abundant in death assemblages, despite a relatively steeper stream gradient. Downstream of Saltville, in areas where new populations had been reintroduced (Henley and Neves, 1999) before the study was conducted, shells displayed intermediate taphonomic signatures (Fig. 5). It is noteworthy that taphonomic signatures of shells not only discriminated sites with different extirpation histories, but also did not require sampling of *any* live mussels. Brown et al.'s study is a straightforward example of taphonomic research augmenting routine strategies of environmental monitoring, an approach that may prove particularly useful in areas with unknown extirpation and pollution histories. However, rivers and streams represent complex taphonomic environments, so care must be taken to consider other confounding factors that may affect shell preservation patterns (Kotzian and Simões, 2006; see also Briggs et al., 1990; Cummins, 1994).

In a study of the staghorn coral (*Acropora cervicornis*) in Discovery Bay (Jamaica), Wapnick et al. (2004) compared taphonomic characteristics of subfossil rubble extracted from cores (440–1260 yrs BP) against present-day skeletal material found on the surface at the same location. The subfossil bioskeletal material was much less affected by internal bioerosion

→

FIGURE 4.—Four separate examples of multi-centennial to multi-millennial age distributions assembled by AAR dating of individual valves of macrobenthic shellfish from various environments and regions of the world. *1*, 45 valves of bivalves (mostly from the genus *Pitar*) collected from tropical carbonate and siliciclastic environments of the San Blas Shelf off the coast of Panama, data from Kidwell et al. (2005); *2*, 103 valves of the brachiopod *Bouchardia rosea* collected from multiple sites (10 to 30m in depth) on the inner part of the southern Brazilian shelf, data combined from Carroll et al., 2003 and Krause et al., in press; *3*, 250 valves of the bivalve *Tellina casta* from two short cores (<1.5m) collected from a shallow carbonate lagoon, Great Barrier Reef, Australia, data from Kosnik et al. 2007; *4*, 165 valves of the bivalve *Chione fluctifraga* from macrotidal beach ridges of the Lower Colorado River Delta, Gulf of California, data from Kowalewski et al. (1998). Grey lines highlight the historical context of the data by using an arbitrary set of cultural/technological developments, prominent historical figures, and major events in the history of our civilization. All hand drawings were prepared, and are copyrighted to, James Schiffbauer (Virginia Tech).

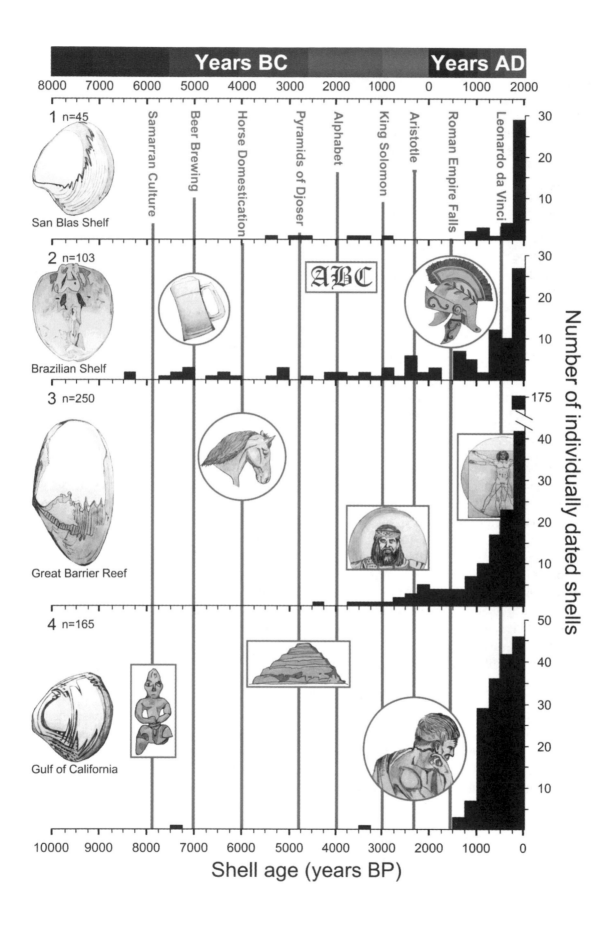

than the present-day surficial rubble. Wapnick and colleagues attributed the much less intense bioerosion in the subfossil material to a shorter exposure time prior to burial. This shorter exposure time was, in turn, tied to faster burial rates in the past due to continuous production of coral rubble by healthy, steadily growing reefs. The much more intense bioerosion observed in recent coral rubble was interpreted as reflecting more prolonged exposure due to the lack of production of new rubble since the 1980s, when human-related perturbations (especially overfishing) disrupted the growth and burial of coral populations in Discovery Bay. Wapnick et al. (2004) used bioerosion data from cores to argue that *A. cervicornis* populations grew continuously (and were buried rapidly thereof) during most centuries of the preceding millennium.

Ichnology

Some trace fossils (drill holes, repair scars, etc.) offer quantifiable records of predator-prey interactions that can be used to evaluate trophic importance of dominant faunas in a given ecosystem or assess intensity of interactions between multiple species. This approach may help to assess the impact of anthropogenic changes on the intensity and nature of ecological interactions. This strategy may also allow us to assess the trophic importance of populations that declined or underwent extirpation (i.e., went locally extinct) due to anthropogenic changes.

For example, McIntyre et al. (2005) used multiple lines of evidence to evaluate the impact of anthropogenic sedimentation on endemic snails in Lake Tanganyika, East Africa. Among others, they compared frequencies of predation between sediment-disturbed and reference localities. Scars left on snail shells by predatory crabs were significantly less frequent at disturbed sites, suggesting a shift in the intensity of biotic interactions. McIntyre and colleagues noted that assemblage-level parameters (e.g., diversity and evenness) did not reveal any major differences across sites and argued that incorporating individual-level measures can enhance the sensitivity of anthropogenic surveys by quantifying disturbance effects on important interspecific interactions.

In a somewhat different approach, Cintra-Buenostro et al. (2005) used traces of predation in subfossil hard parts to quantify the trophic importance of prey species that were abundant prior to anthropogenic changes. They focused on the bivalve *Mulinia coloradoensis* (Fig. 1.2), which was the dominant mollusk of the Colorado River Estuary prior to the construction of upstream dams and water diversions (the species is now nearly extinct). They demonstrated that drill holes made by predatory snails and marginal shell damage induced by crabs were frequent (34% of specimens) in shells sampled from accumulations that predated water diversions and upstream dams. Clearly, this clam species was a major food source for multiple predatory organisms and its decline likely triggered substantial changes higher in the trophic chain, including population declines and/or prey-switching. This study shows that it is possible to use subfossil material to assess anthropogenic effects on trophic webs from fossilized remains of prey alone, even when pre-impact historical surveys are absent and data on the fossil record of predators have not been or cannot be collected.

The predation traces may also be used to complement or re-evaluate other approaches. For example, the above discussed study of Walbran et al. (1989a), which focused on the historical importance of the predatory starfish *A. planci* around the Great Barrier Reef, was re-evaluated recently using scars left on massive corals by the starfish (Devantier and Done, 2007). In contrast to Walbran et al. (1989a), this new study found evidence for a very recent increase in predation intensity (i.e., frequency of scars increased within the last half century, when compared to earlier times).

Most recently, Dietl and Alexander (2009) pointed out that traces of predation can also be used to study environmental gradients in biotic interactions, thus potentially providing data that can be used to evaluate anthropogenically-induced changes in spatial dynamics of ecological processes.

Geochronology

Time series provided by growth rings of individual specimens (sclerochronology) or assembled by individual dating of numerous skeletal remains not only can serve as a source of geochemical data that can be placed in a rigorous temporal context (e.g., stable isotopes extracted along the growth axis of a long-lived mollusk specimen), but can themselves provide us with direct data pertaining to anthropogenic changes.

For example, Kowalewski et al. (2000) estimated that at least 2 trillion ($2*10^{12}$) shells (= 4 trillion disarticulated valves) of the above mentioned bivalve

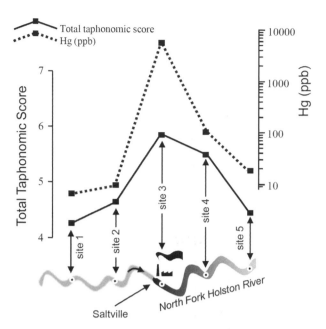

FIGURE 5.—Joint application of taphonomy (quality of shells preservation) and geochemistry (Hg concentration in shells) to valves of freshwater mussels collected from 5 sites along North Fork Holston River (Virginia), an EPA superfund site affected by extensive mercury pollution downstream of Saltville. A semi-quantitative taphonomic score is an averaged score for all valves collected at a given site (increasingly higher scores reflect poorer preservation – a combined effect of fragmentation, disarticulation, edge abrasion, and loss of external shell luster). Hg estimates represent maximum values at a given site (note the logarithmic time scale). Both methods provide a consistent spatial pattern congruent with historical data on the pollution of this river system. That is, Sites 1 and 2 (upstream from pollution source, where healthy mussel populations are still present) are characterized by presence of well preserved shells (low average taphonomic score) and none of the analyzed shells have notable quantities of mercury. Site 5, far downstream from the pollution site, and with mussel populations reintroduced, is comparable to the two upstream sites, both taphonomically and geochemically. In contrast, Sites 3 and 4 are characterized by much poorer preservation of mussel valves and include specimens with Hg concentrations exceeding background conditions by multiple orders of magnitude (all data after Brown et al., 2005).

Mulinia coloradoensis are present in the supratidal beach ridges of the lower Colorado River Delta, Mexico. Using ^{14}C-calibrated AAR rates, they dated individually 125 mollusk valves collected from those ridges. The age structure of dated specimens demonstrated that those 2 trillion shells derived primarily from clam populations lived continuously in the delta between 950

and 1950 A.D. (that is, prior to the artificial shutdown of the river). The decline in shell frequency in older age classes of the age distribution (Fig. 4) was interpreted as reflecting decay-like loss of older specimens (e.g., Flessa et al., 1993; Olszewski, 1999) or sequential erosional unroofing of the tidal flat sediments, from the youngest to the oldest. The most conservative calculations based on these and related data suggested that during the time of natural river flow, an average population density of bivalve mollusks was at least ~50/m^2. The estimate is as much as 20 times higher than the present-day abundance of shelly macrobenthos in the delta (3/m^2, 1999–2000) (see also Avila-Serrano et al., 2006). Thus, by both counting and dating the shells, it was possible to estimate quantitatively the loss of benthic bioproductivity resulting from diversion of the river's flow. Moreover, these estimates also suggested that partial restoration efforts (intermittent water releases into the delta started in 1981) have been woefully inadequate.

Sclerochronological approaches can also be used directly to study temporal changes in biological processes by quantifying changes in the rate of shell growth. Thus, for example, Schöne et al. (2003a), in a detailed analysis of growth rings in the bivalve mollusk *Chione,* demonstrated that shells secreted by individuals that pre-dated anthropogenic changes grew at much smaller pace than individuals from the same species alive today in the same area. They linked the increase in the annual rate of shell growth to anthropogenic changes (reduced freshwater influx) and pointed out that different factors controlled the shell growth rates during pre-anthropogenic times (i.e., seasonal changes in salinity) versus post-anthropogenic times (i.e., temperature and nutrients).

Geochemistry

Stable isotope and trace element signatures archived in skeletal remains of accretionarily growing organisms are commonly extracted from subfossil materials (see numerous references cited above in the section on "Major Types of Data Recoverable from Skeletal Remains"). Here, only a few examples are highlighted briefly, a meager sample of the existing literature (for other relevant examples and references for various setting and biomineral types; see Lazareth et al., 2000; Rodriguez et al., 2001; Dettman et al., 2005; Goewert et al., 2007).

Numerous studies dealt with wide arrays of trace elements that can be extracted from shells. For example, to evaluate utility of freshwater bivalves as an archive of annual metal inputs into rivers, Markich et al. (2002) used secondary ion mass spectroscopy (SIMS) to measure Cu, Mn, Zn, U, Ni, Co, Pb, and Fe/Ca ratios across the annual shell laminations of the longest-lived specimens of the bivalves *Velesunio angasi* sampled in 1996 from the Finniss River in tropical northern Australia. The specimens were collected from 10 localities that included sites both exposed and nonexposed to acid rock drainage from a copper-uranium mine with elevated metal concentrations. At uncontaminated sites relatively constant and similar metal ion yields were found for all analyzed metals in shell growth rings dating as far back as 1965. At contaminated sites elevated SIMS-acquired trace metal concentrations were evident for Cu, Mn, Zn, Ni, and Co in the specimens with shell records also extending back multiple decades. The temporal patterns of Cu, Zn, and Mn recorded in shells from the most contaminated localities paralleled trends based on annual dissolved loads measured in the surface waters. Concentrations of multiple metals retrieved from shells revealed multi-fold decreases moving away from the mine in the downstream direction, which was also consistent with spatial pattern established for water and sediment chemistry. Markich and colleagues showed conclusively that the freshwater bivalve shells found in the polluted system of the Finnis River provide a reliable mutli-decadal record of metal pollution history of surface waters.

Similarly, in the above discussed study by Brown et al. (2005), atomic adsorption spectroscopy was employed successfully to demonstrate that shells of freshwater mussels from the North Fork Holston River provided a spatial record of mercury pollution consistent with the location of the contamination site and the historical records on pollution history. In addition, Brown et al. (2005) were able to show that the geochemical and taphonomic data provided a consistent spatial signal of anthropogenic changes (Fig. 5).

Stable isotopes extracted from biomineral remains, especially $\delta^{13}C$ and $\delta^{18}O$, have been used even more widely than trace elements. They can record global processes, as in the example of a sclerosponge discussed above (Lazareth et al., 2000; Fig. 2), or local environmental change (e.g., Dettman et al., 2004), or even both effects simultaneously.

For example, Surge et al. (2003) used shells of oysters *Crassostrea virginica* to evaluate pre-disturbance conditions in channelized watersheds of Florida. ^{14}C-calibrated, AAR dated specimens of subfossil oysters as well as live-collected specimens were analyzed for $\delta^{18}O$ and $\delta^{13}C$ values. Among others, Surge and colleagues uncovered notable shifts in isotopic signatures from subfossil to modern shells. These changes may have reflected both global (increase in CO_2 to the atmosphere from fossil fuels) and local (change in the dominant carbon sources triggered by vegetation changes) environmental changes.

CLOSING REMARKS

In recent years, we have witnessed unprecedented advances in geochronological, geochemical, and microsampling techniques that allow us to analyze ever smaller aliquots of biomineral solids. And, we can do so with increasing precision and accuracy, at a faster pace and at lower costs. Concurrently, equally dramatic progress has been made in terms of our understanding of taphonomic and temporal characteristics of bioskeletal remains that abound in many terrestrial and marine environments. We realize now that shells, bones, tests and other skeletal remains are not only a rich source of diverse data (from taphonomy to geochemistry), but can also provide multi-centennial to multi-millennial archives of those diverse data.

Numerous case studies highlighted above represent a forceful statement as to the utility of such data for conservation biology and related research themes of direct societal relevance. Using various strategies, the youngest fossil record can aid us in evaluating the current state of habitats, forecasting their future, and guiding restoration efforts. After all, bioskeletal remains contain data that neither ecologists nor climate modelers can ever access. And those data are critical for establishing the natural range of ecological and environmental variations, both locally and globally. Ultimately, this knowledge can aid us directly in assessing inherent stability of Holocene ecosystems and making informed predictions about consequences of their continuing use (and abuse) by humans.

As the existing knowledge grows, both locally and globally, data for different regions can be compared in search for commonalities that may reflect imprints of global trends onto regional variations. Such interregional comparisons should offer us an independent assessment system for evaluating the importance and nature of global changes over centennial-to-millennial time scales. Is the long-term history of local ecosystems shaped mainly by unique local circumstances or does it mainly reflect global trends? The answers to these and related questions will further improve our understanding of processes controlling environmental trends on our planet.

Unfortunately, except for the most recent decades (and even then only for some regions), we have failed to keep a systematic record of the organisms and environments that surround us, or surrounded us until recently. Fortuitously, organisms themselves have left behind a continuous diary of the past. Our ability to read those natural archives, and our appreciation as to their scientific value, have both been growing exponentially in recent years.

ACKNOWLEDGMENTS

I thank Greg Dietl and Karl Flessa for invitation to participate in the Paleontological Society Short Course, and the companion special publication, on "Conservation Paleobiology." I am grateful to Karl Flessa, John Huntley, Marcello G. Simões, and Yurena Yanes for providing field photos used here to assemble Figure 1. I am also exceedingly grateful to Jim Schiffbauer for sharing so selflessly his artistic talent and drafting numerous beautiful drawings that now adorn Figure 4. Troy Dexter, Jim Schiffbauer, Peter Voice, and Shuhai Xiao (all at Virginia Tech) offered numerous useful comments on various drafts of this manuscript. The paper was further improved thanks to careful reviews by Matt Kosnik and an anonymous reviewer. This paper benefited from recent projects conducted by the author and his collaborators, including support from the National Science Foundation (EAR-0125149 and OCE-0602375), Petroleum Research Fund (35021-G and 40735-AC2), and the Virginia Water Resources Research Center.

REFERENCES

ALEXANDER, R. R., AND G. P. DIETL. 2001. Latitudinal trends in naticid predation on *Anadara ovalis* (Bruguiere, 1789) and *Divalinga quadrisulcata* (Orbigny, 1842) from New Jersey to the Florida Keys. American Malacological Bulletin, 16:179-194.

ALEXANDER, R. R., AND G. P. DIETL. 2003. The fossil record of shell-breaking predation on marine bivalves and gastropods, p. 141-176. *In* P. H. Kelley, M. Kowalewski, and T. A. Hansen (eds.), Predator-Prey Interactions in the Fossil Record. Kluwer Academic/Plenum Publishers, New York.

ALIN, S. R., AND A. S. COHEN. 2004. The live, the dead, and the very dead: taphonomic calibration of the recent record of paleoecological change in Lake Tanganyika, East Africa. Paleobiology, 30(1):44-81.

AULER, A. S., L. B. PILÓ, P. L. SMART, X. WANG, D. HOFFMANN, D. A. RICHARDS, R. L. EDWARDS, W. A. NEVES, AND H. CHENG. 2006. U-series dating and taphonomy of Quaternary vertebrates from Brazilian caves. Palaeogeography, Palaeoclimatology, Palaeoecology, 240:508-522.

AVILA-SERRANO, G. E., K. W. FLESSA, M. A. TELLEZ-DUARTE, AND C. E. CINTRA-BUENROSTRO. 2006. Distribution of the intertidal macrofauna of the Colorado River Delta, northern Gulf of California, Mexico. Ciencias Marinas, 32(4):649-661.

BARATS, A., D. AMOUROUX, L. CHAUVAUD, C. PECHEYRAN, A. LORRAIN, J. THEBAULT, T. M. CHURCH, AND O. F. X. DONARD. 2009. High frequency Barium profiles in shells of the Great Scallop *Pecten maximus*: A methodical long-term and multi-site survey in Western Europe. Biogeosciences, 6(2):157-170.

BARBOUR WOOD, S. L., R. A. KRAUSE, JR., M. KOWALEWSKI, J. F. WEHMILLER, AND M. G. SIMÕES. 2006. Aspartic acid racemization dating of Holocene brachiopods and bivalves from the Southern Brazilian Shelf, South Atlantic. Quaternary Research, 66:323-331.

BAUER, G. 1992. Variation in the life-span and size of the fresh-water pearl mussel. Journal of Animal Ecology, 61(2):425-436.

BECKER, B., B. KRONER, AND P. TRIMBORN. 1991. A stable isotope tree-ring timescale of the Late Glacial/Holocene boundary. Nature, 353:647-49.

BEHRENSMEYER, A. K., AND D. WESTERN. 2009. High fidelity bone taphonomy in the Amboseli ecosystem of southern Kenya. 9th North American Paleontological Convention Abstracts. Cincinnati Museum Center Scientific Contributions, 3:141.

BEHRENSMEYER, A. K., S. M. KIDWELL, AND R. A. GASTALDO. 2000. Taphonomy and paleobiology. *In* D. H. Erwin and S. L. Wing (eds.), Deep Time: Paleobiology's Perspective. Paleobiology, 26:103-147.

BERNSTEIN, R.E., P. R. BETZER, R. A. FEELY, R. H. BYRNE, M. F. LAMB, AND A. F. MICHAELS. 1987. Acantharian fluxes and Strontium to chlorinity ratios in the North Pacific Ocean. Science, 237:1490-1494.

BERNSTEIN, R.E., R. H. BYRNE, P. R. BETZER, AND A. M. GRECO. 1992. Morphologies and transformations of celestite in seawater: the role of acantharians in Strontium and Barium geochemistry. Geochimica et Cosmochimica Acta, 56:3273-3279.

BERRY, W. B. N., AND R. M. BARKER. 1968. Fossil bivalve shells indicate longer month and year in Cretaceous than present. Nature, 217:938-939.

BEST, M. M. R., AND S. M. KIDWELL. 2000. Bivalve taphonomy in tropical mixed siliciclastic-carbonate settings; I, Environmental variation in shell condition. Paleobiology, 26:80-102.

BRETT. C. E., AND G. C. BAIRD. 1986. Comparative taphonomy: A key to paleoenvironmental interpretation based on fossil preservation. Palaios 1(3):207-227.

BRIGGS, D. J., D. D. GILBERTSON, AND A. L. HARRIS. 1990. Molluscan taphonomy in a braided river environment and its implications for studies of Quaternary cold-stage river deposits. Journal of Biogeography, 17:623-637.

BRIFFA, K. R., T. S. BARTHOLIN, D. ECKSTEIN, P. D. JONES, W. KARLEN, F. H. SCHWEINGRUBER, AND P. ZETTERBERG. 1990. A 1,400-year tree-ring record of summer temperatures in Fennoscandia. Nature, 346:434-39.

BRIFFA, K. R., P. D. JONES, F. H. SCHWEINGRUBER, AND T. J. OSBORN. 1998. Influence of volcanic eruptions on Northern Hemisphere summer temperature over the past 600 years. Nature, 393:450-455.

BRIOT, D. 2008. Sr isotopes of the shells of the euryhaline gastropod *Potamides lamarcki* from the Oligocene of the French Massif Central and Paris Basin: A clue to its habitats. Palaeogeography, Palaeoclimatology, Palaeoecology, 268(1-2):116-122.

BRODIE, J., K. FABRICIUS, G. DE'ATH, AND K. OKAJI. 2005. Are increased nutrient inputs responsible for more outbreaks of crown-of-thorns starfish? An appraisal of the evidence. Marine Pollution Bulletin 51(1-4): 266-278.

BROMLEY, R. G. 1996. Trace Fossils: Biology, Taphonomy and Applications. 2nd Ed. Chapman and Hall, London, 361 p.

BROMLEY, R. G., N. M. HANKEN, AND U. ASGAARD. 1990. Shallow marine bioerosion: Preliminary results of an experimental study. Bulletin of the Geological Society of Denmark, 38:85-99.

BROWN, M. E., M. KOWALEWSKI, R. J. NEVES, D. S. CHERRY, AND M. E. SCHREIBER. 2005. Freshwater mussel shells as environmental chronicles: Geochemical and taphonomic signatures of mercury-related extirpations in the North Fork Holston River, Virginia. Environmental Science and Technology, 39:1455-1462.

CADEÉ, G. C., S. E. WALKER, AND K W. FLESSA. 1997. Gastropod shell repair in the intertidal of Bahia La Choya (N. Gulf of California). Palaeogeography Palaeoclimatology Palaeoecology, 136(1-4):67-78.

CARMICHAEL R. H., T. HATTENRATH, I. VALIELA, R. H., AND MICHENER. 2008. Nitrogen stable isotopes in the shell of *Mercenaria mercenaria* trace wastewater inputs from watersheds to estuarine ecosystems. Aquatic Biology, 4(2):99-111.

CARRÈ, M., I. BENTALEB, M. FONTUGNE, AND D. LAVALLEE. 2005. Strong El Niño events during the early Holocene: stable isotope evidence from Peruvian sea shells. Holocene, 15(1):42-47.

CARRIKER M. R. 1981. Shell penetration and feeding by Naticacean and Muricacean predatory gastropods: A synthesis. Malacologia, 20(2):403-422.

CARROLL, M., AND C. S. ROMANEK. 2008. Shell layer variation in trace element concentration for the freshwater bivalve *Elliptio complanata*. Geo-Marine Letters, 28:369-381.

CARROLL, M., M. KOWALEWSKI, M. G. SIMÕES, AND G. A. GOODFRIEND. 2003. Quantitative estimates of time-averaging in brachiopod shell accumulations from a modern tropical shelf. Paleobiology, 29(3):381-402.

CARROLL, M., C. S. ROMANEK, AND L. PADDOCK. 2006. The relationship between the hydrogen and oxygen isotopes of freshwater bivalve shells and their home streams. Chemical Geology, 234(3-4):211-222.

CASEY, M. M., AND D. CHATTOPADHYAY. 2008. Clumping behavior as a strategy against drilling predation: Implications for the fossil record. Journal of Experimental Marine Biology and Ecology, 367(2):174-179.

CHAUVAUD, L., A. LORRAIN, R. B. DUNBAR, Y. M. PAULET, G. THOUZEAU, F. JEAN, J-M. GUARINI, AND D. MUCCIARONE. 2005. Shell of the Great Scallop *Pecten maximus* as a high-frequency archive of paleoenvironmental changes. Geochemistry, Geophysics, Geosystems, 6:Q08001.

CINTRA-BUENROSTRO, C. E., K. W. FLESSA, AND G. AVILA-SERRANO. 2005. Who cares about a vanishing clam? Trophic importance of *Mulinia coloradoensis* inferred from predatory damage. Palaios, 20:296-302.

COHEN, A. S. 1989. Taphonomy of gastropod shell accumulations in large lakes: An example from Lake Tanganyika, Africa. Paleobiology, 15(1):26-45.

CUMMINS, R. H. 1994. Taphonomic processes in modern freshwater molluscan death assemblages: Implications for the freshwater fossil record. Palaeogeography, Palaeoclimatology, Palaeoecology, 108:55-73.

CUTLER, A. H. 1995. Taphonomic implications of shell surface textures in Bahia la Choya, northern Gulf of California. Palaeogeography, Palaeoclimatology, Palaeoecology, 114:219-240.

DAVIES, D. J., E. N. POWELL, AND J. R. J. STANTON. 1989. Taphonomic signature as a function of environmental process: Shells and shell beds in a hurricane-influenced inlet on the Texas coast. Palaeogeography, Palaeoclimatology, Palaeoecology, 72:317-356.

DETTMAN, D. L., AND K. C. LOHMANN. 1995. Microsampling carbonates for stable isotope and minor element analysis; Physical separation of samples on a 20 micrometer scale. Journal of Sedimentary Research, 65(A):566-569.

DETTMAN D. L., K. W. FLESSA, P. D. ROOPNARINE, B. R. SCHÖNE, AND D. H. GOODWIN. 2004. The use of oxygen isotope variation in shells of estuarine mollusks as a quantitative record of seasonal and annual Colorado River discharge. Geochimica et Cosmochimica Acta, 68(6):1253-1263.

DETTMAN, D. L., M. R. PALACIOS-FEST, H. H. NKOTAGU, AND A. S. COHEN. 2005. Paleolimnological investigations of anthropogenic environmental change in Lake Tanganyika: VII. Carbonate isotope geochemistry as a record of riverine runoff. Journal of Paleolimnology, 34(1):93-105.

DEVANTIER, L. M., AND T. J. DONE. 2007. Inferring past outbreaks of the crown of thorns seastar from scar patterns on coral heads, p. 85-125. In R. Aronson (ed.), Geological Approaches to Coral Reef Ecology. Springer, New York.

DE VILLIERS, S., G. T. SHEN, AND B. K. NELSON. 1994. The Sr/Ca relationship in coralline aragonite: Influence of variability in $(Sr/Ca)_{seawater}$ and skeletal growth parameters. Geochimica et Cosmochimica Acta, 58:197-208.

DE VILLIERS, S., B. K. NELSON, AND A. R. CHIVAS. 1995. Biological controls on coral Sr/Ca and $\delta^{18}O$ reconstructions of sea surface temperatures. Science, 269:1247-1249.

DIETL, G. P., AND R. R. ALEXANDER. 2009. Patterns of unsuccessful shell-crushing predation along a tidal gradient in two geographically separated salt marshes. Marine Ecology-An Evolutionary Perspective, 30:116-124.

DIETL, G. P., AND G. S. HERBERT. 2005. Influence of alternative shell-drilling behaviours on attack duration of the predatory snail, Chicoreus dilectus. Journal of Zoology, 265(2):201-206.

DIETL, G. P., P. H. KELLEY, R. BARRICK, AND W. SHOWERS. 2002. Escalation and extinction selectivity: Morphology versus isotopic reconstruction of bivalve metabolism. Evolution, 56(2):284-291.

DUNBAR, R. B., AND J. E. COLE. 1993. Coral records of ocean-atmosphere variability. NOAA Climate and Global Change Program Special Report, 10:1-38.

DUNCA, E., B. R. SCHÖNE, AND H. MUTVEI. 2005. Freshwater bivalves tell of past climates: But how clearly do shells from polluted rivers speak. Palaeogeography, Palaeoclimatology, Palaeoecology, 228(1-2):43-57.

EDINGER. E. N., G. S. BURR, J. M. PANDOLFI, AND J. C. ORTIZ. 2007. Age accuracy and resolution of quaternary corals used as proxies for sea leavel. Earth and Planetary Science Letters, 253(1-2):37-49.

ETHERIDGE, D.M., L.P. STEELE, R.L. LANGENFELDS, R.J. FRANCEY, J.-M. BARNOLA, AND V.I. MORGAN. 1998. Historical CO_2 records from the Law Dome DE08, DE08-2, and DSS ice cores: in Trends: A compendium of data on global change: Oak Ridge, Tennessee, Oak Ridge National Laboratory, Carbon Dioxide Information Analysis Center. Available from: http://cdiac.esd.oml.gov/trends/co2/lawdome.html.

FABRICIUS, K. E., AND F. H. FABRICIUS. 1992. Re-assessment of ossicle frequency patterns in sediment cores: Rate of sedimentation related to Acanthaster planci. Coral Reefs, 11:109-114.

FERGUSON, C. A. 2008. Nutrient pollution and the molluscan death record: Use of mollusc shells to diagnose environmental change. Journal of Coastal Research, 24:250-259.

FERGUSON, C. A., AND A. I. MILLER. 2007. A sea change in Smuggler's Cove? Detection of decadal-scale compositional transitions in the subfossil record. Palaeogeography, Palaeoclimatology, Palaeoecology, 254(3-4):418-429.

FIEBIG J., B. R. SCHÖNE, AND W. OSCHMANN. 2005. High-precision oxygen and carbon isotope analysis of very small (10-30 mu g) amounts of carbonates using continuous flow isotope ratio mass spectrometry. Rapid Communications in Mass Spectrometry, 19(16):2355-2358.

FLESSA, K. W. 1998. Well-traveled cockles: Shell transport during the Holocene transgression of the southern North Sea. Geology, 26:187-190.

FLESSA, K. W. 2002. Conservation paleobiology. American Paleontologist, 10:2-5

FLESSA, K. W., AND S. T. JACKSON. 2005. Forging a common agenda for ecology and paleoecology. Bioscience, 55(12):1030-1031.

FLESSA, K. W., AND M. KOWALEWSKI. 1994. Shell survival and time averaging in nearshore and shelf environments: estimates from the radiocarbon literature. Lethaia, 27:153-165.

FLESSA, K. W., A. H. CUTLER, AND K. H. MELDAHL. 1993. Time and taphonomy: Quantitative estimates of time-averaging and stratigraphic disorder in a shallow marine habitat. Paleobiology, 19:266-286.

FLESSA, K. W., S. T. JACKSON, J. D. ABER, M. A. ARTHUR, P. R. CRANE, D. H. ERWIN, R. W. GRAHAM, J. B. C. JACKSON, S. M. KIDWELL, C. G. MAPLES, C. H. PETERSON, AND O. J. REICHMAN. 2005. The Geological Record of Biosphere Dynamics. Understanding the Biotic Effects of Future Environmental Change. National Research Council, The National Academies Press, Washington, D.C., 200 p.

FOX, D. L., D. C. FISHER, S. VARTANYAN, A. N. TIKHONOV, D. MOL, AND B. BUIGUES. 2007. Paleoclimatic implications of oxygen isotopic variation in late Pleistocene and Holocene tusks of Mammuthus primigenius from northern Eurasia. Quaternary International, 169:154-165.

FRIEDLI, H., H. LÖTSCHER, H. OESCHGER, U. SIEGENTHALER, AND B. STAUFFER. 1986. Ice core record of the $^{13}C/^{12}C$ ratio of atmospheric CO_2 in the past two centuries: Nature, 324:237-238.

FRY, B. 2007. Stable Isotope Ecology. Springer, New York, 308 p.

FÜRSICH, F. T., AND K. W. FLESSA. 1987. Taphonomy of tidal flat mollusks in the northern Gulf of California: Paleoenvironmental analysis despite the perils of preservation. Palaios, 2:543-559.

GILLIKIN, D. P., F. DEHAIRS, A. LORRAIN, D. STEENMANS, W. BAEYENS, AND L. ANDRE. 2006. Barium uptake into the shells of the common mussel (*Mytilus edulis*) and the potential for estuarine paleo-chemistry reconstruction. Geochimica et Cosmochimica Acta, 70(2):395-407.

GOEWERT A., D. SURGE, S. J. CARPENTER, AND J. DOWNING. 2007. Oxygen and carbon isotope ratios of *Lampsilis cardium* (Unionidae) from two streams in agricultural watersheds of Iowa, USA. Palaeogeography, Palaeoclimatology, Palaeoecology, 252(3-4):637-648.

GOODFRIEND, G. A. 1987. Chronostratigraphic studies of sediments in the Negev Desert, using amino acid epimerization analysis of land snail shells. Quaternary Research, 28:374-392.

GOODFRIEND, G. A. 1989. Complementary use of amino acid epimerization and radiocarbon analysis for dating mixed-age fossil assemblages. Radiocarbon, 31:1041-1047.

GOODFRIEND, G. A., AND S. J. GOULD. 1996. Paleontology and chronology of two evolutionary transitions by hybridization in the Bahamian land snail *Cerion*. Science, 274(5294):1894-1897.

GOODFRIEND, G. A., K. W. FLESSA, AND P. H. HARE. 1997. Variation in amino acid epimerization rates and amino acid composition among shell layers in the bivalve *Chione* from the Gulf of California. Geochimica et Cosmochimica Acta, 61:1487-1493.

GOODFRIEND, G. A., M. J. COLLINS, M. L. FOGEL, S. A. MACKO, AND J. F. WEHMILLER. 2000. Perspectives in Amino Acid and Protein Geochemistry: Oxford University Press, Oxford, 384 p.

GOODWIN, D. H., K. W. FLESSA, B. R. SCHÖNE, AND D. L. DETTMAN. 2001. Cross-calibration of daily growth increments, stable isotope variation, and temperature in the Gulf of California bivalve mollusk *Chione cortezi*: Implications for environmental analysis. Palaios, 16:387-398.

GREENSTEIN B. J., A. H. CURRAN, AND J. M. PANDOLFI. 1998. Shifting ecological baselines and the demise of Acropora cervicornis in the western North Atlantic and Caribbean Province: A Pleistocene perspective. Coral Reefs, 17(3):249-261.

HADLY, E. A. 1999. Fidelity of terrestrial vertebrate fossils to a modern ecosystem. Palaeogeography, Palaeoclimatology, Palaeoecology, 149:389-409.

HENDERSON, W. G., L. C. ANDERSON, AND C. R. MCGIMSEY. 2002. Distinguishing natural and archaeological deposits: Stratigraphy, taxonomy, and taphonomy of Holocene shell-rich accumulations from the Louisiana chenier plain. Palaios, 17(2):192-205.

HENLEY, W. F., AND R. J. NEVES. 1999. Recovery status of freshwater mussels in the North Fork Holston River, Va. American Malacological Bulletin, 5:65-73.

HOEFS, J. 2009. Stable Isotope Geochemistry. 6th ed., Springer, New York, 288 p.

HOLMES, K. E., E. N. EDINGER,G. V. HARIYADI, E. LIMMON, AND M. J. RISK. 2000. Bioerosion of live massive corals and branching coral rubble on Indonesian coral reefs. Marine Pollution Bulletin, 40(7):606-617.

HUNTLEY, J. W., AND M. KOWALEWSKI. 2007. Strong coupling of predation intensity and diversity in the Phanerozoic fossil record. Proceedings of National Academy of Sciences of the United States of America, 104:15006-15010.

HUNTLEY, J. W., Y. YANES, M. KOWALEWSKI, C. CASTILLO, A. DELGADO-HUERTAS, M. IBÁÑEZ, M. R. ALONSO, J. E. ORTIZ, AND T. DE TORRES. 2008. Testing limiting similarity in Quaternary terrestrial gastropods. Paleobiology, 34:378-388.

IVANY, C. L., B. H. WILKINSON, AND D. S. JONES. 2003. Using stable isotopic data to resolve rate and duration of growth throughout ontogeny: An example from the surf clam, *Spisula solidissima*. Palaios, 18:126-137

JACKSON, J. B. C., M. X. KIRBY, W. H. BERGER, K. A. BJORNDAL, L. W. BOTSFORD, B. J. BOURQUE, R. H. BRADBURY, R. COOKE, J. ERLANDSON, J. A. ESTES, T. P. HUGHES, S. KIDWELL, C. B. LANGE, H. S. LENIHAN, J. M. PANDOLFI, C. H. PETERSON, S. R. STENECK, M. J. TEGNER, AND R. R. WARNER. 2001. Historical overfishing and the recent collapse of coastal ecosystems. Science, 293(5530):629-638.

JONES, D. S., M. A. ARTHUR, AND D. J. ALLARD. 1989. Sclerochronological records of temperature and growth from shells of *Mercenaria mercenaria* from Narragansett Bay, Rhode Island. Marine Biology, 102:225-234.

KAUFMAN, D. S., AND W. F. MANLEY. 1998. A new procedure for determining DL amino acid ratios in fossils using reverse phase liquid chromatography. Quaternary Science Reviews, 17:987-1000.

KEESING, J. K., R. H. BRADBURY, L. M. DEVANTIER, M. J. RIDDLE, AND G. DE'ATH. 1992. The geological evidence for recurring outbreaks of the crown-of-thorns starfish: A reassessment from an ecological perspective. Coral Reefs, 11:79-85.

KELLEY, P. H., AND T. A. HANSEN. 1993. Evolution of the naticid gastropod predator-prey system: An evaluation of the hypothesis of escalation. Palaios, 8:358-375.

KELLEY, P. H., AND T. A. HANSEN. 2003. The fossil record of drilling predation on bivalves and gastropods, p. 113-

139. *In* P. H. Kelley, M. Kowalewski, and T. A. Hansen, (eds.), Predator-Prey Interactions in the Fossil Record. Kluwer Academic/Plenum Publishers, New York.

KENNISH, M. J., AND R. K. OLSSON. 1975. Effects of thermal discharges on the microstructural growth of *Mercenaria mercenaria*. Environmental Geology, 1:41-64.

KIDWELL, S. M. 2007. Discordance between living and death assemblages as evidence for anthropogenic ecological change. Proceedings of the National Academy of Sciences of the United States of America, 104(45):17701-17706.

KIDWELL, S. M. 2008. Ecological fidelity of open marine molluscan death assemblages: effects of post-mortem transportation, shelf health, and taphonomic inertia. Lethaia, 41(3):199-217.

KIDWELL, S. M. 2009. Evaluating human modification of shallow marine ecosystems: Mismatch in composition of molluscan living and time-averaged death assemblages. *In* Dietl, G. P. and K. W, Flessa, (eds.), Conservation Paleobiology: Using the Past to Manage for the Future. The Paleontological Society Papers, 15 (this volume).

KIDWELL, S. M., T. A. ROTHFUS, AND M. M. R. BEST. 2001. Sensitivity of taphonomic signatures to sample size, sieve size, damage scoring system, and target taxa. Palaios, 16(1):26-52.

KIDWELL, S. M., M. M. R. BEST, AND D. S. KAUFMAN. 2005. Taphonomic trade-offs in tropical marine death assemblages: Differential time averaging, shell loss, and probable bias in siliciclastic vs. carbonate facies. Geology, 33:729-732.

KITCHELL, J. A. 1986. The evolution of predator-prey behavior: naticid gastropods and their molluscan prey, p. 88-110. *In* M. Nitecki and J. A. Kitchell, (eds). Evolution of Animal Behavior: Paleontological and Field Approaches. Oxford University Press, Oxford.

KITCHELL, J. A., C. H. BOGGS, J. F. KITCHELL, AND J. A. RICE. 1981. Prey selection by naticid gastropods: Experimental tests and application to the fossil record. Paleobiology, 7:533-552.

KOSNIK, M. A., AND D. S. KAUFMAN. 2008. Identifying outliers and assessing the accuracy of amino acid racemization measurements for use in geochronology: II. Data screening. Quaternary Geochronology, 3(4):328-341.

KOSNIK, M. A., D. JABLONSKI, R. LOCKWOOD, AND P. M. NOVACK-GOTTSHALL. 2006. Quantifying molluscan body size in evolutionary and ecological analyses: Maximizing the return on data-collection efforts. Palaios, 21(6):588-597.

KOSNIK, M.A., Q. HUA, G.E. JACOBSEN, D.S. KAUFMAN, AND R. A. WÜST. 2007. Sediment mixing and stratigraphic disorder revealed by the age-structure of *Tellina* shells in Great Barrier Reef sediment. Geology, 35:811-814.

KOSNIK, M. A., D. S. KAUFMAN, AND Q. HUA. 2008. Identi-

fying outliers and assessing the accuracy of amino acid racemization measurements for use in geochronology: I. Age calibration curves. Quaternary Geochronology, 3(4):308-327.

KOTZIAN, C. B., AND M. G. SIMÕES. 2006. Taphonomy of Recent freshwater molluscan death assemblages, Touro Passo Stream, Southern Brazil. Revista Brasileira de Paleontologia, 9(2):243-260.

KOWALEWSKI, M. 1996. Time-averaging, overcompleteness and the geological record. Journal of Geology, 104:317-326.

KOWALEWSKI, M. 2002. The fossil record of predation: An overview of analytical methods. *In* M. Kowalewski and P. H. Kelley, (eds.), The Fossil Record of Predation. Paleontological Society Special Papers, 8:3-42.

KOWALEWSKI, M., AND R. K. BAMBACH. 2003. The limits of paleontological resolution, p. 1-48. *In* P. J. Harries (ed.), Approaches in High-resolution Stratigraphic Paleontology. Kluwer Academic/Plenum Publishers, New York.

KOWALEWSKI, M., AND K. W. FLESSA. 2000. Seasonal predation by migratory shorebirds recorded in shells of lingulid brachiopods from Baja California, Mexico. Bulletin of Marine Science, 66:405-416.

KOWALEWSKI, M., K. W. FLESSA, AND D. P. HALLMAN. 1995. Ternary taphograms: Triangular diagrams applied to taphonomic analysis. Palaios, 10:478-483.

KOWALEWSKI, M., G. A. GOODFRIEND, AND K. W. FLESSA. 1998. High-resolution estimates of temporal mixing within shell beds: The evils and virtues of time-averaging. Paleobiology, 24(3):287-304.

KOWALEWSKI, M., G. E. AVILA SERRANO, K. W. FLESSA, AND G. A. GOODFRIEND. 2000. Dead delta's former productivity: Two trillion shells at the mouth of the Colorado River. Geology, 28:1059-1062.

KRAUSE, R. A., JR., J. STEMPIEN, M. KOWALEWSKI, AND A. I. MILLER. 2007. Body size estimates from the literature: Utility and potential for macroevolutionary studies. Palaios, 22:63-76.

KRAUSE, R. A., JR., S. L. BARBOUR WOOD, M. KOWALEWSKI, D. S. KAUFMAN, C. S. ROMANEK, M. G. SIMÕES, AND J. F. WEHMILLER. In press. Quantitative estimates and modeling of time averaging in bivalve and brachiopod shell accumulations. Paleobiology.

KURATA K., H. MINAMI, AND E. KIKUCHI. 2001. Stable isotope analysis of food sources for salt marsh snails. Marine Ecology-Progress Series, 223:167-177.

LAZARETH, C. E., P. WILLENZ, J. NAVEZ, E. KEPPENS, F. DEHAIRS, AND L. ANDRÉ. 2000. Sclerosponges as a new potential record of environmental changes: Lead in *Ceratoporella nicholsoni*. Geology, 28:515-518.

LEA, D. W., G. T. SHEN, AND E. A. BOYLE. 1989. Coralline Barium records temporal variability in equatorial Pacific upwelling. Nature, 340:373-376.

LEIGHTON, L. R. 2002. Inferring predation intensity in the marine fossil record. Paleobiology, 28(3):328-342.

LESCINSKY, H. L. 1997. Epibiont communities: Recruitment and competition on North American Carboniferous brachiopods. Journal of Paleontology, 71(1):34-53.

LESCINSKY, H. L., E. EDINGER, AND M. J. RISK. 2002. Mollusc shell encrustation and bioerosion rates in a modern epeiric sea: Taphonomy experiments in the Java Sea, Indonesia. Palaios, 17(2):171-191.

LIEBIG, P. M., T. A. TAYLOR, AND K.W. FLESSA. 2003. Bones on the beach: Marine mammal taphonomy of the Colorado Delta, Mexico. Palaios, 18:168-175.

LOCKWOOD, R., AND L. A. WORK. 2006. Quantifying taphonomic bias in molluscan death assemblages from the Upper Chesapeake Bay: Patterns of shell damage. Palaios, 21(5):442-450.

MARCHITTO, T. A., G. A. JONES, G. A. GOODFRIEND, AND C. R. WEIDMAN. 2000. Precise temporal correlation of Holocene mollusk shells using sclerochronology. Quaternary Research, 53:236-246.

MARKICH, S. J., R. A. JEFFREE, AND P. T. BURKE. 2002. Freshwater bivalve shells as archival indicators of metal pollution from a copper-uranium mine in tropical northern Australia. Environmental Science and Technology, 36(5):821-832.

MARTIN, R. E., J. F. WEHMILLER, M. S. HARRIS, AND W. D. LIDDEL. 1996. Comparative taphonomy of bivalves and foraminifera from Holocene tidal flat sediments, Bahia la Choya, Sonora, Mexico (northern Gulf of California): taphonomic grades and temporal resolution. Paleobiology, 22:80-90.

MARTINELL, J., AND R. DOMÈNECH. 2009. Commensalism in the fossil record: Eunicid polychaete bioerosion on Pliocene solitary corals. Acta Palaeontologica Polonica, 54(1):143-154.

MCCULLOCH, M. T., AND G. E. MORTIMER. 2008. Applications of the U-238-Th-230 decay series to dating of fossil and modern corals using MC-ICPMS. Australian Journal of Earth Sciences, 55(6-7):955-965

MCCULLOCH, M. T., M. K. GAGAN, G. E. MORTIMER, A. R. CHIVAS, AND P. J. ISDALE. 1994. A high-resolution Sr/Ca and δ^{18}O coral record from the Great Barrier Reef, Australia, and the 1982-1983 El Niño. Geochimica et Cosmochimica Acta, 58:2747-2754.

MCINTYRE, P. B., E. MICHEL, K. FRANCE, A. RIVERS, P. HAKIZIMANA, AND A. S. COHEN. 2005. Individual- and assemblage-level effects of anthropogenic sedimentation on snails in Lake Tanganyika. Conservation Biology, 19(1):171-181.

MELDAHL, K. H., AND K. W. FLESSA. 1990. Taphonomic pathways and comparative biofacies and taphofacies in a recent intertidal shallow shelf environment. Lethaia, 23:43-60.

MELDAHL, K. H., K. W. FLESSA, AND A. H. CUTLER. 1997. Time averaging and postmortem skeletal survival in benthic fossil assemblages: Quantitative comparisons among Holocene environments. Paleobiology, 23:207-229.

METREF, S., D. D. ROUSSEAU, I. BENTALEB, M. LABONNE, AND M. VAINEY-LIAUD. 2003. Study of the diet effect on δ^{13}C of shell carbonate of the land snail Helix aspersa in experimental. Earth and Planetary Science Letters, 211:381-393.

MUROZUMI, M., T. J. CHOW, AND C. PATTERSON. 1969. Chemical concentrations of pollutant lead, aerosols, terrestrial dusts, and sea salts in Greenland and Antarctic snow strata. Geochimica et Cosmochimica Acta, 33:1247-1294.

NEFTEL, A., E. MOOR, H. OESCHGER, AND B. STAUFFER. 1985. Evidence from polar ice cores for the increase in atmospheric CO_2 in the past two centuries: Nature, 315:45-47.

NORRIS, R. D., AND R. M. CORFIELD. 1998. Isotope Paleobiology and Paleoecology: Paleontological Society Papers, 4, Paleontological Society, 285 p.

NOVACK-GOTTSHALL, P. M. 2008. Using simple body-size metrics to estimate fossil body volume: Empirical validation using diverse Paleozoic invertebrates. Palaios, 23(3):163-173.

OLSZEWSKI, T. D. 1999. Taking advantage of time-averaging. Paleobiology, 25:226-238.

OLSZEWSKI, T. D. 2004. Modeling the influence of taphonomic destruction, reworking, and burial on time-averaging in fossil accumulations. Palaios, 19(1):39-50.

PANDOLFI, J. M., R. H. BRADBURY. E. SALA, T. P. HUGHES, K. A. BJORNDAL, R. G. COOKE, D. MCARDLE, L. MCCLENACHAN, M. J. H. NEWMAN, G. PAREDES, R. R. WARNER, AND J. B. C. JACKSON. 2003. Global trajectories of the long-term decline of coral reef ecosystems. Science, 301(5635):955-958.

POMIÈS, C., G. R. DAVIES, AND S. M. H. CONAN. 2002. Neodymium in modern foraminifera from the Indian Ocean: Implications for the use of foraminiferal Nd isotope compositions in paleo-oceanography. Earth and Planetary Science Letters, 203(3-4):1031-1045.

POWELL, E. N., AND D. J. DAVIES. 1990. When is an 'old' shell really old? Journal of Geology, 98:823-844.

POWELL, E. N., A. LOGAN, R. J. STANTON, D. J. DAVIES, AND P. E. HARE. 1989. Estimating time-since-death from the free amino acid content of the mollusc shell: A measure of time averaging in modern death assemblages? Description of the technique. Palaios, 4(1):16-31.

POWELL, E. N., W. R. CALLENDER, W. RUSSELL, G. M. STAFF, K. M. PARSONS-HUBBARD, C. E. BRETT, S. E. WALKER, A. RAYMOND, AND K. A. ASHTON-ALCOX. 2008. Molluscan shell condition after eight years on the sea floor: Taphonomy in the Gulf of Mexico and Bahamas. Journal of Shellfish Research, 27(1):191-225.

RICIPUTI L. R., B. A. PATERSON, AND R. L. RIPPERDAN. 1998. Measurement of light stable isotope ratios by SIMS: Matrix effects for oxygen, carbon, and sulfur isotopes in minerals. International Journal of Mass Spectrometry, 178:81-112.

RODLAND, D. L., M. KOWALEWSKI, D. L. DETTMAN, K. W. FLESSA, V. ATUDOREI, AND Z. SHARP. 2003. High resolution analysis of $\delta^{18}O$ in the biogenic phosphate of modern and fossil lingulid brachiopods. Journal of Geology, 111:441-454.

RODLAND, D. L., M. KOWALEWSKI, M. CARROLL, AND M. G. SIMÕES. 2004. Colonization of a "lost world": Encrustation patterns in modern subtropical brachiopod assemblages. Palaios, 19(4):381-395.

RODLAND, D. L., B. R. SCHÖNE, S. HELAMA, J. K. NIELSEN, AND S. BAIER. 2006a. Ultradian rhythms in bivalve activity revealed by digital photography. Journal of Experimental Marine Biology and Ecology, 334(2):316-323.

RODLAND, D. L., M. KOWALEWSKI, M. CARROLL, AND M. G. SIMÕES. 2006b. The temporal resolution of epibiont assemblages; are they ecological snapshots or overexposures? Journal of Geology, 114(3):313-324.

RODRIGUES, S. C. 2007. Biotic interactions recorded in shells of recent rhynchonelliform brachiopods from San Juan Island, USA. Journal of Shellfish Research, 26(1):241-252.

RODRIGUES, S. C., M. G. SIMÕES, M. KOWALEWSKI, M. A. V. PETTI, E. F. NONATO, S. MARTINEZ, AND C. J. DEL RÍO. 2008. Biotic interaction between spionid polychaetes and bouchardiid brachiopods: Paleoecological, taphonomic and evolutionary implications. Acta Palaeontologica Polonica, 53:657-668.

RODRIGUEZ C. A., K. W. FLESSA, M. A. TELLEZ-DUARTE, D. L. DETTMAN, AND G. A. AVILA-SERRANO. 2001. Macrofaunal and isotopic estimates of the former extent of the Colorado River estuary, upper Gulf of California, Mexico. Journal of Arid Environments, 49(1):183-193.

ROHLING. E. J., M. SPROVIERI, T. CANE, J. S. L. CASFORD, S. COOKE, I. BOULOUBASSI, K. C. EMEIS, R. SCHIEBEL, M. ROGERSON, A. HAYES, F. J. JORISSEN, AND D. KROON. 2004. Reconstructing past planktic foraminiferal habitats using stable isotope data: a case history for Mediterranean Sapropel S5. Marine Micropaleontology, 50(1-2):89-123.

ROWELL, K., K. W. FLESSA, D. L. DETTMAN, AND M. ROMÁN. 2005. The importance of Colorado River flow to nursery habitats of the Gulf corvina Cynoscion othonopterus. Canadian Journal of Fisheries and Aquatic Sciences, 62:2874-2885.

SATO, S. 1997. Shell microgrowth patterns of bivalves reflecting seasonal change of phytoplankton abundance.

Paleontological Research, 1:260-266.

SCARPONI, D., AND KOWALEWSKI, M. 2004. Stratigraphic paleoecology: Bathymetric signatures and sequence overprint of mollusk associations from the upper Quaternary sequences of the Po Plain, Italy. Geology, 32:989-992.

SCHIFFBAUER, J., AND S. H. XIAO, S. H. In press. Novel application of focused ion beam electron microscopy (FIB-EM) in preparation and analysis of microfossil ultrastructures: a new view of complexity in early eukaryotic organisms. Palaios.

SCHIFFBAUER, J. D., Y. YANES, C. L. TYLER, M. KOWALEWSKI, AND L. R. LEIGHTON. 2008. The microstructural record of predation: A new approach for identifying predatory drill holes. Palaios, 23:810-820.

SCHÖNE, B. R. 2003. A 'clam-ring' master-chronology constructed from a short-lived bivalve mollusc from the northern Gulf of California, USA. Holocene, 13:39-49.

SCHÖNE, B. R., AND D. SURGE. 2005. Looking back over skeletal diaries: High-resolution environmental reconstructions from accretionary hard parts of aquatic organisms. Palaeogeography, Palaeoclimatology, Palaeoecology, 228(1-2):1-192.

SCHÖNE, B. R., K. W. FLESSA, D. L. DETTMAN, AND D. H. GOODWIN. 2003a. Upstream dams and downstream clams: Growth rates of bivalve mollusks unveil impact of river management on estuarine ecosystems (Colorado River Delta, Mexico). Estuarine Coastal and Shelf Science, 58(4):715-726.

SCHÖNE, B. R., K. TANABE, D. DETTMAN, AND S. SATO. 2003b. Environmental controls on shell growth rates and delta O-18 of the shallow-marine bivalve mollusk Phacosoma japonicum in Japan. Marine Biology, 142(3):473-485.

SCHÖNE, B. R., E. DUNCA, AND H. MUTVEI. 2004. A 217-year record of summer air temperature reconstructed from freshwater pearl mussels (M. margarifitera, Sweden). Quaternary Science Reviews, 23(16-17):1803-1816.

SCHÖNE, B. R., J. FIEBIG, M. PFEIFFER, R. GLESS, J. HICKSON, A. L. A. JOHNSON, W. DREYER, AND W. OSCHMANN. 2005. Climate records from a bivalved Methuselah (Arctica islandica, Mollusca; Iceland). Palaeogeography, Palaeoclimatology, Palaeoecology, 228(1-2):130-148.

SCHOPF, T. J. M. 1976. Fossilization potential of an intertidal fauna: Friday Harbor, Washington. Paleobiology, 4(3):261-270.

SCHWEINGRUBER, F. H. 1996. Tree Rings and Environment: Dendroecology. Swiss Federal Institute for Forest, Snow and Landscape Research, and Paul Haupt Verlag, 609 p.

SCUDERI, L. A. 1993. A 2000-year tree ring record of annual temperatures in the Sierra Nevada Mountains. Science, 259:1433-1436.

SIMÕES, M. G., S. C. RODRIGUES, AND M. KOWALEWSKI. 2005. Dead-live fidelity of brachiopod assemblages from a subtropical shelf (southern Brazil): Implications for paleoecology and conservation paleobiology. 19th Annual Meeting of the Society for Conservation Biology, Brasília, Book of Abstracts, p. 198-199.

SHARP, Z. 2006. Principles of Stable Isotope Geochemistry. Prentice Hall, New York, 360 p.

SHEN, G.T., J. E. COLE, D. W. LEA, L. J. LINN, E. A. McCONNAUGHEY, AND R. G. FAIRBANKS. 1992. Surface ocean variability at Galapagos from 1936-1982; calibration of geochemical tracers in corals. Paleoceanography, 7:563-588.

SPEYER, S. E., AND C. E. BRETT. 1986. Trilobite taphonomy and Middle Devonian taphofacies. Palaios, 1(3):312-327.

STAFF, G., E. N. POWELL, R. J. STANTON JR., AND H. CUMMINS. 1985. Biomass: Is it a useful tool in paleocommunity reconstruction? Lethaia, 18:209-232.

STECHER, H.A., III, D. E. KRANTZ, C. J. LORD, G. W. LUTHER, III, AND K. W. BOCK. 1996. Profiles of strontium and barium in *Mercenaria mercenaria* and *Spisula solidissima* shells. Geochimica et Cosmochimica Acta, 60:3445-3456.

STEMPIEN, J. A. 2005. Brachyuran taphonomy in a modern tidal-flat environment: Preservation potential and anatomical bias. Palaios, 20(4):400-410.

STEMPIEN, J. A. 2007. Detecting avian predation on bivalve assemblages using indirect methods. Journal of Shellfish Research, 26(1):271-280.

STEPHENSON, A. E., J. J. DEYOREO, L. WU, K. J. WU, J. HOYER, J., AND P. M. DOVE. 2008. Peptides enhance magnesium signature in calcite: Insights into origins of vital effects. Science, 322 (5902):724-727.

SURGE, D. M., K. C. LOHMANN, AND G. A. GOODFRIEND. 2003. Reconstructing estuarine conditions: Oyster shells as recorders of environmental change, Southwest Florida. Estuarine, Coastal and Shelf Science, 57(5-6):737-756.

SWART, P. K., R. E. DODGE, AND H. J. HUDSON. 1996. A 240-year stable oxygen and carbon isotopic record in a coral from south Florida: Implications for the prediction of precipitation in southern Florida. Palaios, 11:362-75.

TAYLOR, P. D., AND M. A. WILSON. 2003. Palaeoecology and evolution of marine hard substrate communities. Earth-Science Reviews, 62(1-2):1-103.

TERRY, R. C. 2007. Holocene small mammals of the Great Basin: tracking recent richness declines through live/dead analysis of raptor-generated faunal remains. Geological Society of America Abstracts with Programs, 39(6):168.

TERRY, R. C. 2008. Modeling the effects of predation, prey cycling, and time averaging on relative abundance in raptor generated small mammal death assemblages. Palaios, 23(6):402-410.

TOMAŠOVÝCH, A., AND S. M. KIDWELL. 2009a. Fideli-

ty of variation in species composition and diversity partitioning by death assemblages: Time-averaging transfers diversity from beta to alpha levels. Paleobiology, 35(1):94-118.

TOMAŠOVÝCH, A., AND S. M. KIDWELL. 2009b. Preservation of spatial and environmental gradients by death assemblages. Paleobiology, 35(1):119-145.

TOMAŠOVÝCH, A., AND T. A. ROTHFUS. 2005. Differential taphonomy of modern brachiopods (San Juan Islands, Washington State): Effect of intrinsic factors on damage and community-level abundance. Lethaia, 38(3):271-292.

TSUNOGAI. S., AND Y. WATANABE. 1981. Calcium in the North Pacific water and the effect of organic matter on the calcium-alkalinity relation. Geochemical Journal, 15:95-107.

VANDER PUTTEN, E., W. BAEYENS, F. DEHAIRS, AND E. KEPPENS. 2000. High resolution distribution of trace elements in the calcite shell layer of modern *Mytilus edulis*: Environmental and biological controls. Geochimica et Cosmochimica Acta, 64:997-1011.

VERMEIJ, G. J. 1983. Traces and trends of predation, with special reference to bivalved animals. Palaeontology, 26:455-465.

VERMEIJ, G. J. 1987. Evolution and Escalation. Princeton University Press, Princeton, New Jersey, 544 p.

VERMEIJ, G. J., D. E. SCHINDEL, AND E. ZIPSER. 1981. Predation through geological time: Evidence from gastropod shell repair. Science, 214:1024-1026.

VOIGHT, J. R., AND J. D. SIGWART. 2007. Scarred limpets at hydrothermal vents: Evidence of predation by deep-sea whelks. Marine Biology, 152(1):129-133.

WALBRAN, P. D., R. A. HENDERSON, A. J. T. JULL, AND M. J. HEAD. 1989a. Evidence from sediments of long-term *Acanthaster planci* predation on corals of the Great Barrier Reef. Science, 245(4920):847-850.

WALBRAN, P. D., R. A. HENDERSON, J. W. FAITHFUL, H. A. POLACH, R. J. SPARKS, G. WALLACE, AND D. C. LOWE. 1989b. Crown-of-thorns starfish outbreaks on the Great Barrier Reef: A geological perspective based upon the sediment record. Coral Reefs, 8(2):67-78.

WALKER, S. E., K. PARSONS-HUBBARD, E. N. POWELL, AND C. E. BRETT. 1998. Bioerosion or bioaccumulation? Shelf-slope trends for epi-and endobionts on experimentally deployed gastropod shells. Historical Biology, 13:61-72.

WAPNICK C. M., W. F. PRECHT, AND R. B. ARONSON. 2004. Millennial-scale dynamics of staghorn coral in Discovery Bay, Jamaica. Ecology Letters, 7(4):354-361.

WEBER, J. N. 1973. Incorporation of Strontium into reef coral skeletal carbonate. Geochimica et Cosmochimica Acta, 37:2173-2190.

WEHMILLER, J. F., AND G. H. MILLER. 2000. Aminostratigraphic dating methods in Quaternary geology, p. 187-

222. *In* J. S. Noller, J. M. Sowers, and W. R. Lettis (eds.), Quaternary Geochronology, Methods and Applications. American Geophysical Union Reference Shelf.

WEHMILLER, J. F., L. L. YORK, AND M. L. BART. 1995. Amino-acid racemization geochronology of reworked Quaternary mollusks on U.S. Atlantic coast beaches: Implications for chronostratigraphy, taphonomy, and coastal sediment transport. Marine Geology, 124:303-337.

YANES, Y., M. KOWALEWSKI, J.E. ORTIZ, C. CASTILLO, T. TORRES, AND J. NUEZ. 2007. Scale and structure of time-averaging (age mixing) in terrestrial gastropod assemblages from Quaternary eolian deposits of the eastern Canary Islands. Palaeogeography, Palaeoclimatology, Palaecology, 251:283-299.

YANES, Y., A. DELGADO, C. CASTILLO, M. R. ALONSO, M. IBÁÑEZ, J. DE LA NUEZ, AND M. KOWALEWSKI. 2008a. Stable isotope (δ^{18}O, δ^{13}C, and δD) signatures of Recent terrestrial communities from a low-latitude, oceanic setting: Endemic land snails, plants, rain, and carbonate sediments from the Eastern Canary Islands. Chemical Geology, 249:377-392.

YANES, Y., A. TOMAŠOVÝCH, M. KOWALEWSKI, C. CASTILLO, J. AGUIRRE, M. R. ALONSO, AND M. IBÁÑEZ. 2008. Taphonomy and compositional fidelity of Quaternary fossil assemblages of terrestrial gastropods from carbonate-rich environments of the Canary Islands. Lethaia, 42:235-256.

ZHAO, J. X., D. T. NEIL, Y. X. FENG, K. F. YU, AND J. M. PANDOLFI. 2009. High-precision U-series dating of very young cyclone-transported coral reef blocks from Heron and Wistari reefs, southern Great Barrier Reef, Australia. Quaternary International, 195:122-127.

ZUSCHIN M., M. STACHOWITSCH, P. PERVESLER, AND H. KOLLMANN. 1999. Structural features and taphonomic pathways of a high-biomass epifauna in the northern Gulf of Trieste, Adriatic Sea. Lethaia, 32(4):299-317.

CHAPTER TWO

CONSERVATION BIOLOGY AND ENVIRONMENTAL CHANGE: A PALEOLIMNOLOGICAL PERSPECTIVE

JOHN P. SMOL

Paleoecological Environmental Assessment and Research Lab (PEARL), Department of Biology, Queen's University, Kingston, Ontario K7L 3N6 Canada

ABSTRACT.—Lake and river sediments archive a vast array of data that can be used by paleolimnologists to track past ecological and environmental conditions. Much of this long-term information has considerable application for conservation biology efforts. This chapter reviews some of the general approaches used by paleolimnologists and then provides several case studies and examples where paleolimnological data are used to assist ecologists in conservation efforts. Specific examples include the ecological effects of acidification, eutrophication, erosion, climatic change, contaminant transport, species invasions and extirpations, and the reconstruction of past wildlife population dynamics.

INTRODUCTION

ONE OF the biggest challenges faced by ecologists and other environmental managers addressing complex conservation issues is a general lack of long-term monitoring data. Instrumental and other direct observational data are typically very short in duration, with only a small percentage of monitoring studies extending over three years in duration (Cohen, 2003; Smol, 2008), and so cannot answer critical ecological questions such as: Has this ecosystem changed from a natural or pre-disturbance state? If so, what were the drivers of the change, and what was the nature of community and ecosystem responses? What are realistic recovery targets for this ecosystem? Because most environmental assessments only occur after a problem has been identified, the historical data required to answer these questions must often be inferred using indirect techniques. Paleolimnology, the multi-disciplinary science that uses the physical, chemical, and biological information preserved in lake and river sediments (Smol, 2008), provides conservation ecologists and ecosystem managers with powerful tools to reconstruct these missing data sets.

Lakes, ponds, rivers, and wetlands are dominant features of most landscapes. Each of these ecosystems potentially contains an important sedimentary record of past environmental changes. The information contained in sediments is often referred to as proxy data, which researchers can use to reconstruct past populations as well as to infer past environmental conditions. The paleolimnologist's main task is to retrieve this sedi-

mentary "history book" and the data it contains, and interpret this information in a manner that is meaningful to other scientists and the public-at-large. This chapter introduces some of the common approaches used by paleolimnologists, and then provides a few example applications that are related to conservation biology. Although a diverse array of geochemical and physical records may be analyzed from lake sediments, my focus in this chapter will be on biological approaches. Many of the subjects addressed in this short chapter are dealt with in far more detail in my recent textbook (Smol, 2008[1]), and other publications (Cohen, 2003). For further information on paleolimnological techniques, volumes in the *Developments in Paleoenvironmental Research* book series contain detailed methodological approaches in paleolimnology (e.g., Last and Smol, 2001a, b; Smol et al., 2001a, b), as well as other syntheses that are relevant to the use of paleolimnology in conservation biology (e.g., Leng, 2006).

With respect to conservation issues, paleolimnological data can typically be used in two important ways. The first application is to use information preserved in lake sediments to reconstruct past populations of specific organisms or groups of organisms that comprise important components of aquatic ecosystems. As described later, plant macrofossils may be used to track the arrival or extirpation of specific taxa of interest (e.g., Birks, 2002), or the chitinous remains of invertebrates can be used to track specific invertebrate populations (e.g., Hall and Yan, 1997). Other indirect approaches may use geochemical markers, such as stable nitrogen isotopes, in conjunction with biologi-

In *Conservation Paleobiology: Using the Past to Manage for the Future, Paleontological Society Short Course, October 17th, 2009. The Paleontological Society Papers, Volume 15*, Gregory P. Dietl and Karl W. Flessa (eds.). Copyright © 2009 The Paleontological Society.

cal microfossils, including diatoms and cladocerans, to track past changes in anadromous salmon populations (e.g., Gregory-Eaves et al., 2009). Most paleolimnological studies, however, typically use the information preserved in sediments to quantitatively reconstruct past environmental conditions, such as the use of diatom species assemblages to reconstruct past pH levels, or the use of chironomid (non-biting midge) head capsules, to reconstruct past benthic oxygen conditions. Such approaches provide important information on habitat changes that are critical to the survival of organisms. Overall, paleolimnological studies can often provide realistic targets for mitigation efforts, as well as important information on the trajectories of environmental change over time. Paleolimnology, therefore, can address questions such as "What were conditions like before human impacts?", or "What level of stress is required before important ecosystems changes occur?".

The overall paleolimnological approach

Paleolimnological approaches are fairly straightforward. For simplicity, I will focus on lake ecosystems in this chapter, but the methods are generally similar for reconstructing ecological and environmental information from river, pond, and other aquatic environments (Smol, 2008). The raw materials used by paleolimnologists, namely sediment cores, can be retrieved and subsampled using a wide range of equipment (Glew et al., 2001). For most paleolimnological studies, a sediment core is sectioned into appropriate temporal intervals, and the variables contained in the chosen sediment intervals are analyzed. The time scale can be set to varying degrees by choosing the size of sediment intervals. This analytical phase can often be very time consuming, especially if it involves extensive taxonomic identification of indicators using microscopy and subsequent enumeration and analysis. Dating control is often provided by radioisotopes, such as ^{14}C or ^{210}Pb, although other approaches using independent dating markers are available (Last and Smol, 2001a).

Virtually everything that lived in the lake since its formation has left some sort of body fossil remain that can be studied using a microscope, or a biogeochemical remnant, such as fossil pigments, DNA, etc. A synopsis of the major indicator groups are summarized in a series of chapters collated by Last and Smol (2001b), Smol et al. (2001a, b), and textbooks by Cohen (2003) and Smol (2008) provide additional details. For exam-

ple, past changes in terrestrial and, to a more limited extent, aquatic vegetation can be tracked using the pollen grains, seeds, and other macrofossils left by these organisms. Past aquatic microbiota are often well represented in the sedimentary record. For example, the siliceous, species-specific valves of diatoms, the scales and cysts of chrysophytes, as well as the structural or physical remains of other algae (e.g., *Pediastrum*) are often used in paleolimnological studies. For those algae and cyanobacteria that do not leave reliable morphological remains, past populations can often be reconstructed using biogeochemical techniques, such as fossil pigment analyses (Leavitt and Hodgson, 2001). Invertebrates are well represented by a variety of chitinous body parts, such as cladoceran body parts and chironomid head capsules, as well as by a variety of resting stages, such as *Daphnia* ephippia (the sexual resting stages of Cladocera). Vertebrate remains in lake sediments, such as fish, are less common – but the population dynamics of these organisms can sometimes be reconstructed using more indirect approaches, such as stable isotope analyses. New insights are being made through paleoenvironmental studies of DNA (e.g., Duffy et al., 2000). Of course, in addition to biological remains, sediments archive a vast library of physical and chemical proxy data that can also be used in paleolimnological investigations (Last and Smol, 2001a, b). Granted the paleolimnological record is incomplete, but it is often sufficiently comprehensive to reach defendable conclusions that can be applied to ecological and conservation issues.

In most paleolimnological applications, the presence or absence of an individual taxon is not the major focus of the investigation, but rather it is to use the biological and other information contained in lake sediments to reconstruct past ecosystem conditions, such as climate, lakewater pH, or limnetic nutrient concentrations. Because different organisms have different ecological niches, paleolimnologists can use the known environmental optima and tolerances of taxa under modern conditions to infer such variables in paleoecological reconstructions. In many cases, this is done qualitatively, as researchers have a reasonable understanding of the ecological requirements of the indicators they are using. However, more quantitative approaches are increasingly being used. One common method used extensively in applied paleolimnology, especially as it relates to issues such as acidification and eutrophication, is the surface sediment calibration set or training set approach, coupled with the construction of transfer functions (Smol, 2008, Chapter 6). For example, suppose an investigator wished to develop a transfer function to reconstruct past lakewater pH from diatoms preserved in regional lake-sediment cores. First, a calibration set would be constructed by choosing 50 or more lakes that have limnological data spanning the environmental gradient of interest, in this case, pH. For example, the calibration set might include lakes that have current pH levels of 4.5 to 7.5. This calibration set, with associated limnological data, would comprise the environmental matrix required for the development of the transfer function. From each calibration lake, typically 0.5-1.0 cm of surface sediment next would be collected and analyzed for the sedimentary indicator, in this example, diatom species composition. The resulting diatom counts, typically expressed as relative frequency data, would represent the second data matrix, the species data. Then, a variety of statistical approaches could be applied to both data matrixes and a transfer function could be developed that would quantitatively relate biological assemblage changes to the environmental variable of interest, in this case diatoms predicting lakewater pH. Downcore diatom assemblages could then be studied and the transfer function applied to infer past lakewater pH. A detailed overview of these statistical approaches can be found in Birks (1998).

Setting the time scale.—One powerful aspect of the paleolimnological approach is that the investigator can set the timescale and temporal resolution of the analysis to some degree. Applied paleolimnology began to flourish during the 1980s, especially with problems related to acid precipitation. Using coring and sectioning techniques developed at that time, sediment cores can now be collected and sectioned with a high degree of temporal resolution, and dated accurately with radioisotopic methods, such as ^{210}Pb geochronology (Appleby, 2001).

Most paleolimnological studies examine continuous records of indicators at close sediment intervals throughout a sediment core to obtain time-series data of sedimentary variables. Some regionally based paleolimnological studies, however, use a "before-and-after" type of assessment, which is often colloquially referred to as the "top - bottom approach". Using this type of assessment, typically only two sediment samples are analyzed from each core. The "top" sample refers to the most recent surface sediments, and typically

it comprises the top 0.5-1.0 cm of sediment accumulation. Indicators contained in this material represent recent environmental conditions. The "bottom" sediment sample represents material that pre-dates the period of putative human impacts, such as pre-1850 sediment if one wanted to assess limnological conditions prior to anthropogenic atmospheric acid deposition in North America. The advantage of this top-bottom paleolimnological approach is that only two samples are analyzed from each core, thus many more lakes can be studied, and one can more readily develop a regional assessment of environmental change. The disadvantage of this approach, however, is that there are no continuous data on the timing or rate of changes occurring between the two points in time represented by those samples.

SOME PALEOLIMNOLOGICAL APPLICATIONS RELATED TO CONSERVATION BIOLOGY

Acidification

Lake acidification was a highly-politicized, and to some extent, a scientifically contentious issue in the 1980s and early 1990s. One of the earliest applications of biological paleolimnology to a major environmental problem was the use of lake sediments to track pH declines in lakes, as well as the associated limnological and ecological changes. Much has been written about using paleolimnology to track lake acidification and its effects in North America and Europe (Battarbee et al., 1990; Smol, 2008).

As noted in the earlier general example, the most common indicators used to reconstruct lakewater pH levels are sub-fossil diatom assemblages (Battarbee et al., 1999). However, other indicators, such as siliceous chrysophyte algal scales (Smol, 1995), are also used. Statistically robust and ecologically sound transfer functions have been developed for these organisms from lake sediment calibration sets developed across the most acid sensitive regions in North America and Europe. For example, Figure 1 shows the very strong relationship (r=0.91) between measured and diatom-inferred lakewater pH for a surface-sediment calibration set developed from 71 lakes from Adirondack Park, New York. As explained earlier, paleolimnologists use indicators preserved in the recent sediments of well-studied lakes to define the ecological optima of taxa, from which they can then develop transfer functions to infer environmental variables. From these data, paleolimnologists can reconstruct, with a relatively high degree of certainty, past lakewater pH levels based on the assemblages preserved in dated lake sediment cores. For example, in a pioneering application of the top-bottom paleolimnological approach to study the regional effects of acidification in lakes from Adirondack Park, Cumming et al. (1992) examined diatoms and chrysophyte assemblages from the sediments of 37 randomly selected park lakes, and showed that anthropogenic acidification was a major problem. Every study lake that had current pH levels below 6.0 had acidified since the early 1800s (Fig. 2).

Because lake sediments archive a wide array of other organisms, the broader ecological effects of pH declines and related limnological changes can be tracked using the sediment record. For example, one of the most common questions posed during the acid rain debates was "Has this lake lost its fish populations as a result of acidification?" Given the lack of reliable fisheries data for many of these remote lakes, direct observational records could not be used. While past lakewater pH can be reconstructed using diatom and chrysophyte assemblages, vertebrates such as fish rarely leave reliable fossils in lake sediments, although some studies have effectively used fish remains in paleolim-

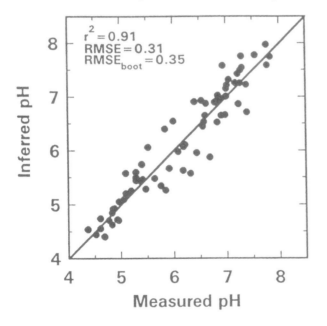

FIGURE 1.—The relationship between measured and diatom-inferred lakewater pH for a 71-lake surface sediment calibration set (see text for details) from Adirondack Park, New York. RMSE = root mean squared error of prediction; RMSE$_{boot}$ = bootstrapped root mean squared error of prediction. Modified from Dixit et al. (1993).

nological studies (e.g., Davidson et al., 2003). Certain invertebrate populations, however, can provide important clues concerning past fish presence. For example, Uutala (1990) noted that the chitinized mandibles of Chaoboridae (Insecta) larvae are well preserved in lake sediments. Of course, a variety of environmental factors influence the distribution of chaoborid larvae in lakes (Sweetman and Smol, 2006), but the aquatic larvae of one taxon, *Chaoborus americanus* Johannsen, also known as the phantom midge, from eastern North American lakes can rarely, if ever, co-exist with fish as it is highly vulnerable to fish predation. Uutala (1990) suggested that the 20th century emergence of this species in acidified lake sediments was an indication of fish extirpations, a relationship that was subsequently used in a number of other paleolimnological studies tracking fish losses in lakes (Sweetman and Smol, 2006).

Paleolimnological approaches can also be used to determine if fish populations are under specific environmental threats. For example, the Atlantic whitefish (*Coregonus huntsamni* Scott), the only endemic fish species in Nova Scotia, Canada, is an endangered species whose last remaining habitats are three lakes near the town of Bridgewater, Nova Scotia. Because the remaining lake habitat for Atlantic whitefish in Nova Scotia is situated in an area of potential human impacts (e.g., acidification, eutrophication, recent climate change), there is considerable scientific and public interest in knowing if the environmental conditions are changing in these three lakes. This prompted Ginn et al. (2008) to undertake a detailed diatom-based paleolimnological study from these lakes. They found that, contrary to initial concerns, the lakes had not yet been affected by acidic precipitation, but that other potential stressors, such as increasing temperatures and associated limnological changes, were already affecting the lakes, with possible ramifications for the Atlantic whitefish.

Another application of paleolimnological indicators of pH has been to examine the efficacy of remediation programs in reducing acidic precipitation and facilitating aquatic ecosystem recovery. Some of the most striking examples of partial ecosystem recovery following reductions in acidic deposition have been observed in increasing pH levels in Sudbury (Ontario) lakes (Dixit et al., 1992). Sudbury has been a major mining and smelting region of Canada since the 19th century, with concomitant increases in sulfur dioxide emissions from large smoke stacks. Lakes in the are suffered massive acidification and associated ecosystem damages over the years. With recent legislated changes in the application of technology, however, there have been equally impressive declines in emissions (Gunn, 1995). Diatom- and chrysophyte-based paleolimnological approaches have reliably tracked both the acidification and recovery trajectories in many of these lakes (Smol, 2008). Such dramatic recoveries in lakewater pH, however, have not occurred in all lake regions. Smol et al. (1998), using comparative paleolimnological approaches, suggested that the rate of recovery of lake pH levels across acidified systems may be related to background (i.e., pre-acidification) pH levels, which

could be linked to bedrock geology.

Paleolimnological data are also being used by conservation scientists to define restoration goals for species conservation and re-introductions. For example, there are several lakes north of the major mining and smelting installations in Sudbury, Ontario, that once supported an endangered strain of brook trout (*Salvelinus fontinalis* Mitchill), the aurora trout. Presumably due to the acidification in the region, aurora trout had disappeared entirely from its home range by the late 1960s. Conservation biologists made a final attempt to save this fish by maintaining populations in a hatchery, with the goal of eventually re-introducing

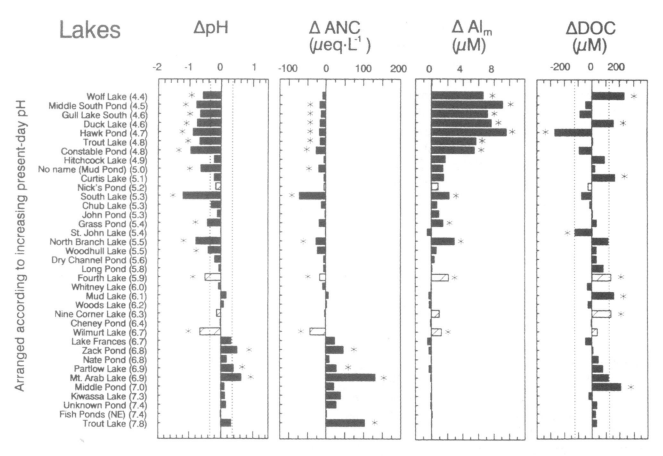

FIGURE 2.—Diatom-based estimates of historical change (pre-industrial to the present) in pH, acid neutralizing capacity (ANC), total monomeric aluminum (Al_m), and dissolved organic carbon (DOC) from 37 randomly selected Adirondack Park (New York, USA) lakes. ANC is a measure of the overall buffering capacity of water against acidification; Al_m is a form of aluminum that is especially toxic to some organisms, such as fish; DOC may represent a natural source of acidity. The estimates are presented as the differences in inferred water chemistry between the top (0-1 cm, representing recent lake conditions) and bottom (> 20 cm, usually > 25 cm, representing pre-acidic precipitation periods) sediment core intervals. The lakes are arranged according to increasing present-day measured pH. Lakes that were limed within five years before coring and/or assessment of water chemistry are indicated by hatched bars. The dotted lines represent ± bootstrap root mean squared error ($RMSE_{boot}$) of prediction for pH and DOC. Estimates of change in ANC and Al_m are presented as differences between backtransformed estimates since these models were developed on log_e data. Asterisks denote changes that are > $RMSE_{boot}$ for the various inference models. From Cumming et al. (1992); used with permission.

these trout into "rehabilitated" lakes. However, with no existing pre-impact limnological data, the only way to determine the strain's "natural habitat" was to use paleolimnological techniques. Using diatom-based transfer functions, Dixit et al. (1996) determined that acidification was the most likely root cause of the declines, but that pre-impact conditions were similar in the trout lakes and in other Sudbury lakes, so other lakes could likely support this strain as well. Reclamation attempts via liming and reduced emissions have largely been successful, and as a result of paleolimnological identification of suitable historical habitat, hatchery aurora trout have now been re-introduced in the native lakes and the fish are naturally reproducing.

As noted earlier, one of the main conservation applications for paleolimnology is to set realistic mitigation goals. Paleoenvironmental data provide critical information on past lake conditions and so can help determine whether a lake requires any mitigation, and if so, what a realistic target might be. Continuing with the acidification examples described above, lake liming was used extensively in Sweden as an attempt to mitigate the detrimental effects of lake acidification. A recent diatom-based paleolimnological study (Norberg et al., 2008), however, suggested that 5 of 12 study lakes that had been previously limed had never really acidified, and had naturally low pH levels. It would appear in these examples that liming, the supposed mitigation measure, was unnecessary and had likely shifted these lakes away from their "natural" conditions, a situation that would never have been recognized without a paleolimnological perspective.

Eutrophication

The largest water quality problem affecting freshwater ecosystems is cultural eutrophication, or the fertilization of lakes and rivers with excessive nutrients, such as from sewage, agricultural runoff, and industrial outputs. Eutrophication can result in excessive algal blooms and growth of aquatic macrophytes (plants), which can result in many water-quality problems, such as taste and odor problems, esthetic issues, and declining deepwater oxygen levels due to the decomposition of excessive organic matter. For most temperate lakes, phosphorus is the limiting nutrient, and so often the first challenge is to reconstruct how past phosphorus levels may have changed in a lake system due to natural or anthropogenic activities. Intuitively, one might think that simply undertaking a geochemi-

cal analysis of phosphorus concentrations in a dated sediment profile would provide these data, but such an approach is fraught with difficulty as sediment phosphorus can be highly mobile under certain conditions (Engstrom and Wright, 1984; Boyle, 2001). Instead, paleolimnologists can develop reliable transfer functions based on the environmental optima of different diatom species to lakewater phosphorus concentrations (Hall and Smol, 1999).

Reconstructing past lakewater phosphorus concentrations is a major focus of eutrophication research. Similar to the problems related to acidification, common questions posed by lake managers include "What were nutrient levels before European-style agriculture began in the catchment?", or "When in a lake's history did cultural activities exceed some water quality guidelines related to eutrophication?" However, many other types of questions can be posed and answered using paleolimnological approaches (Smol, 2008).

One of the most common problems related to cultural eutrophication is declining dissolved oxygen levels in the hypolimnia, or deep water layers of lakes, because of the decomposition of excess organic matter, such as algae and macrophytes. Declining oxygen levels have serious repercussions for deepwater fish populations and other benthic biota. Diatoms and other algal indicators cannot infer past oxygen levels, but deep water invertebrates, such as the larvae of Chironomidae (Insecta) possess defined optima and tolerances to deepwater oxygen levels. The heavily scleroterized head capsules of these larvae are well preserved in lake sediments and can often be identified to the family level, and sometimes even to the genus or species level based on the morphology of the mouth parts. Chironomids have been used extensively to develop transfer functions that infer past deepwater oxygen levels (Quinlan and Smol, 2002; Brodersen et al., 2004).

Erosion and sediment discharge

In addition to causing eutrophication, cultural activities in a lake's catchment can result in accelerated erosion and subsequent increased sediment transport to receiving waters (Dearing and Jones, 2003), which can have serious repercussions for aquatic communities. A variety of physical paleolimnological approaches, such as paleomagnetic studies (Maher and Thompson, 1999) and other methods (Dearing and Jones, 2003), can be used to track past sedimentation changes as a result of increased erosion. These data can then be used in conjunction with the biological approaches reviewed in this chapter to determine the effects of these disturbances on aquatic biota.

For example, Cohen et al. (2005a, b) summarized a wide spectrum of paleolimnological data that were used to study the effects of deforestation and other disturbances on different parts of Lake Tanganyika, the largest of the African rift lakes. Not surprisingly, rates of sedimentation in the lake were higher in cores taken near heavily populated sections of the watershed. Deltas that were relatively unaffected by deforestation continued to support diverse benthic invertebrate and fish communities, whereas lower diversity, and even some local extirpations, were recorded in deltas that were more heavily influenced.

Species introductions and extirpations

One of the most serious issues facing conservation biologists is the introduction by humans (often inadvertently) of non-native or so-called exotic species. These introductions can have very serious ecological (Elton, 2000) and economic (Pimentel et al., 2000) repercussions. In some cases, paleolimnological techniques can be used to track the introduction and consequences of these invasions. For example, Stoermer et al. (1985) and Edlund et al. (2000) tracked the invasion of exotic diatom species to the Great Lakes, likely through the ballast waters of ships, using lake sediments. Harper (1994) used similar techniques to suggest that the commonly occurring diatom, *Asterionella formosa* Hassall, may have been introduced to New Zealand lakes by Europeans in the 1860s. The spiny water flea, *Bythotrephes longimanus*, a large, predaceous cladoceran zooplankter, is believed to have invaded Lake Ontario in 1982. The chitinized caudal process of this taxon is well preserved in sediments, and can be used to track this taxon's distribution, both in space and time, using lake sediments (Hall and Yan, 1997; Forman and Whiteside, 2000).

Species extirpations can similarly be studied using paleolimnological methods. For example, Birks (2002) used plant macrofossils in dated sediment cores from three large lagoons in the Nile Delta (Egypt) to document the relatively recent extirpation of the floating water-fern *Azolla nilotica* Decaisne *ex*. Mettenius from Egypt. The probable cause of this extirpation was linked to changes in salinity from the control of the annual flow of the Nile River and increasing eutrophication (Birks et al., 2001). In a comprehensive paleoeco-

logical study of a lake from Easter Island, Flenley et al. (1991) tracked the destruction of the island's trees by the local population, and the eventual collapse of this society.

Salmon and other wildlife

As noted earlier, vertebrates such as fish rarely leave reliable macro-fossil records in lake sediments, although there are notable exceptions where fish fossils such as otoliths, bones and scales provide important paleoecological information (Panfili et al., 2002; Davidson et al., 2003). In some cases, however, paleolimnologists have used indirect methods to infer past populations. One such example is the reconstruction of past sockeye salmon (*Oncorhynchus nerka* Walbaum) population dynamics in west-coast nursery lakes (Gregory-Eaves et al., 2009). Pacific salmon are anadromous (i.e., spend part of their life cycle in both the ocean and in freshwaters) and most have a semelparous (i.e., spawn once and then die) life cycle. Sockeye salmon, originally hatched in freshwater nursery lakes, spend almost all of their adult life in the Pacific Ocean where they accumulate over 95% of their body mass, but then return to their nursery systems, with remarkable fidelity, to reproduce and die. In some regions, sockeye salmon return in the millions to spawn, and so the decaying carcasses may be an important source of nutrients, especially phosphorus and nitrogen, to the nursery lake. Paleolimnologists have capitalized on the sockeye salmon's lifecycle, and the profound influence that the decaying carcasses may have on nursery lake ecosystems, to reconstruct past salmon runs (Gregory-Eaves et al., 2009). First, as sockeye salmon feed at relatively high positions in Pacific Ocean food webs, they are enriched in the rarer stable isotope of nitrogen (^{15}N relative to ^{14}N), and the resultant $\delta^{15}N$ signal archived in nursery lake sediments, can be used as an indicator of changing population abundances. Second, the nutrients released by decaying carcasses can be tracked using diatoms and other nutrient-sensitive indicators to cross-validate isotopic inferences of population dynamics. Concomitant increases in eutrophic taxa with increases in $\delta^{15}N$ are likely related to the numbers of returning salmon, assuming no other disturbances had enriched the water with phosphorus and nitrogen. Finney et al. (2000) demonstrated that sedimentary $\delta^{15}N$ was not only related to escapement through time, but that similar patterns were present from a wide spectrum of Alaskan nursery lakes. Diatom and other paleolimnological data gathered for these lakes further demonstrated marked changes in nutrients that could be linked to salmon-derived nutrients. Several examples have subsequently been published on reconstructing sockeye salmon populations (Gregory-Eaves et al., 2009). Figure 3 shows one such application for Karluk Lake in Alaska. Although the proxy data clearly show a decline in sockeye salmon related to recent harvesting and other environmental changes, there are also sustained and pronounced variations in at least some salmon populations over the past ~2200 years that could not be linked to human activities, and that are likely related to climate-driven changes in oceanic conditions (Finney et al., 2002).

Selbie et al. (2007) demonstrated the utility of paleolimnological approaches to the conservation of endangered salmon in Idaho, with the Snake River sockeye salmon. The authors demonstrated catastrophic population declines over the 20th century in two historical nursery systems, and they identified the complex factors most likely responsible for the endangerment of this stock, which included commercial fishing, hydroelectric damming, and fish introductions. By examining stable nitrogen isotopes and bio-indicators representing multiple trophic levels, Selbie et al. (2007) also were able to identify changes in the freshwater food web that likely impede ongoing recovery efforts, and that result from human interventions, such as a shift towards a less energetically beneficial zooplankton community because of kokanee and non-native sockeye salmon introductions. Moreover, the paleolimnological record provided evidence of atmospheric nitrogen deposition and climate change, which might reinforce problematic rearing conditions.

Similar paleolimnological procedures have now been used to track other animal groups. For example, working with Arctic archeologists, Douglas et al. (2004) tracked the limnological effects that Thule Inuit whalers had on local pond ecosystems over several centuries. The Thule depended almost entirely on bowhead whales for food, oil, and for building material, and whale bones were used for shelter structural support in a region with no trees. Douglas et al. (2004) used $\delta^{15}N$ as a tracer for past whale inputs, an approach similar to salmon nutrient studies, and they showed that Inuit whalers markedly affected Arctic pond ecosystems as a result of the fertilizing effects of nutrients released from butchered whales and other activities. Similarly, the nutrifying effects of guano from past seabird popula-

tions can be traced using both stable isotopes, diatoms, and other indicators (Michelutti et al., 2009). Working at the opposite pole, Hodgson and Johnston (1997) and Hodgson et al. (1998) tracked past Antarctic fur seal populations in Sub-Antarctic lake sediment cores using a variety of paleolimnological indicators, including seal hairs preserved in the sedimentary profiles.

The above studies also lend themselves to research on contaminant transport, and the potential influence that biovectors have on aquatic ecosystems and biota (Blais et al., 2007). Anadromous fish and seabirds can transport both nutrients and contaminants from the ocean back to land. For example, Krümmel et al. (2003, 2005, 2009) showed how paleolimnological approaches can be used to track the delivery of marine-derived persistent organic pollutants (POPs) into Alaskan sockeye salmon nursery lakes via the decay of salmon carcasses. Blais et al. (2005), Evenset et al. (2007), and Michelutti et al. (2009) used similar approaches to demonstrate how nesting Arctic seabirds were delivering both nutrients and POPs to high Arctic ponds.

Climatic change

Paleoenvironmental approaches are being used increasingly in studies of long-term climate change, potentially the most important and pervasive stressor relevant to conservation biologists. Such studies and approaches have been reviewed in Smol and Cumming (2000) and chapters in Battarbee et al. (2004), as well as various sections in Cohen (2003). In general, these studies can be divided into two major types of applications. The first type uses paleolimnological indicators to track past changes in climate, and the second type uses what is known about past climate, such as available meteorological data and lake ice records, to determine how past climate changes have affected lake communities.

Some of the earlier paleolimnological work on climatic change focused on closed-basin, saline (athalassic) lakes. The sediments of such lakes archive an important record of past shifts in evaporation (E) to precipitation (P) ratios, and hence past droughts. In its simplest form, periods of drought (i.e., increased E:P ratios) will result in lower water levels and increased salinity concentrations. The reverse should occur under cooler and wetter conditions. Because many bio-indicators, but most notably diatoms, are reliable indicators of lakewater salinity and water levels (Fritz et al., 1999), paleolimnological methods can be used to reconstruct past E:P ratios. Of course, other variables (e.g., changes in groundwater input) can affect lakewater salinity, and so multi-indicator studies, using comparative approaches, provide the most robust interpretations. However, as changing salinity levels can have profound influences on the distribution of taxa, these studies have important implications for conservation biologists.

Non-saline lakes also contain important records of past climatic change in their sediments. For example, a recent meta-analysis of over 200 diatom profiles from North American and Western European lakes summarized some of the wide-scale changes that were most

likely attributable to recent climatic warming (Rühland et al., 2008). The authors provided a focus on the limnological effects of decreased ice cover and/or shifts in the duration and amount of thermal stratification. Such limnological changes can profoundly affect many other aspects of lake ecology with cascading effects throughout aquatic food webs. For example, with longer periods of thermal stratification, deep-water anoxia levels may become a great management problem, especially as it relates to deepwater fish populations. Complaints concerning recent increases in blue–green algal (cyanobacterial) blooms may also be linked to recent climate warming and thermally stratified lakes.

Some of the largest climate-related changes in

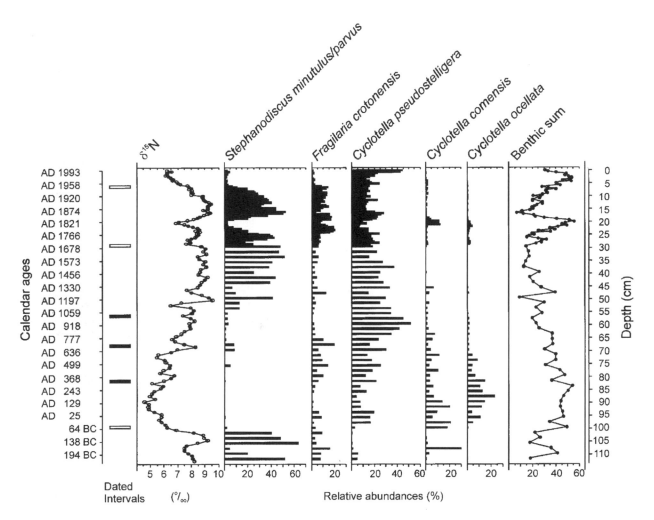

FIGURE 3.—Paleolimnological reconstruction of past sockeye salmon abundances in Karluk Lake, a salmon nursery lake in Alaska. As noted in the text, increases in $\delta^{15}N$ indicate higher sockeye salmon returns to this lake. In addition, the changes in diatom species composition indicate changing nutrient levels, linked to the decaying carcasses of spawned salmon. For example, the increases in *Stephanodiscus* species are linked to higher nutrient levels, whilst *Cyclotella* taxa tend to indicate lower nutrient levels. These profiles indicate that sockeye salmon numbers in Karluk Lake have changed markedly in the past, even before human interventions. From Finney et al. (2002); used with permission.

freshwater habitats have occurred in polar regions, such as the Canadian Arctic. However, due to the scarcity of long-term temperature readings in the Arctic, paleolimnological approaches are especially important in these regions for reconstructing past environmental change. Douglas et al. (1994) completed the first detailed diatom-based paleolimnological study that attempted to reconstruct if and how climate variability was affecting shallow High Arctic ponds from Cape Herschel on Ellesmere Island. They recorded striking successional changes, believed to have started in the 19th century, which Douglas et al. (1994) interpreted as indicative of warming-induced ecosystem changes. Initially these findings were considered highly controversial. However, as recently reviewed by Smol and Douglas (2007a), a large number of paleolimnological studies have since been completed which test alternative hypotheses for these species changes, but recent climatic warming remained the only factor examined that could reasonably explain these assemblage changes. For example, subsequent paleolimnological analyses from across the circum-polar Arctic (Smol et al., 2005) have reached similar conclusions to the original 1994 study. As predicted by Smol and Douglas in the 1990s, many of the Cape Hershel ponds now desiccate totally in summer, due to the extended periods of warmer conditions (Smol and Douglas, 2007b). Sadly, entire ecosystems are disappearing due to climatic warming.

CONCLUSIONS

Paleolimnology has enjoyed tremendous progress over the last two decades, and many of these advancements can be directly applied to questions related to conservation biology. The examples summarized here represent only a small portion of the many potential applications. While paleolimnological approaches continue to get stronger with new and more robust methods, the challenges faced by conservation scientists continue to increase in numbers and complexities. For example, the problem of multiple stressors, where several environmental stressors are occurring in unison, pose important challenges to paleolimnologists (Jeziorski et al., 2008). These challenges are not necessarily insurmountable, provided that comparative paleolimnological studies are carefully planned and executed (Smol, in press). Given the magnitude and complexity of the environmental problems humans currently face, it seems inevitable that the need for such data for ecosystem managers and policy makers will only increase in the foreseeable future.

ACKNOWLEDGMENTS

Many helpful comments on this manuscript were provided by Neal Michelutti, Kathleen Rühland, Irene Gregory-Eaves, Daniel Selbie, Joshua Thienpont, Thomas Whitmore, and an anonymous reviewer. Research in my laboratory is primarily provided by the Natural Sciences and Engineering Research Council of Canada.

NOTE

[1]All of the paleolimnology figures in the Smol (2008) textbook are available as a free download of PowerPoint slides, with notes, for use in teaching. Please contact the author for information on downloading these lecture slides

REFERENCES

APPLEBY, P. G. 2001. Chronostratigraphic techniques in recent sediments, p. 171-203. In W. M. Last, and J. P. Smol (eds.), Tracking Environmental Change Using Lake Sediments. Volume 1: Basin Analysis, Coring, and Chronological Techniques. Kluwer Academic Publishers, Dordrecht.

BATTARBEE, R. W., D. F. CHARLES, S.S. DIXIT, AND I. RENBERG. 1999. Diatoms as indicators of surface water acidity, p. 85-127. In E. F. Stoermer and J. P. Smol (eds.), The Diatoms: Applications for the Environmental and Earth Sciences. Cambridge University Press, Cambridge.

BATTARBEE, R.W., F. GASSE, AND C.E. STICKLEY. 2004. Past Climate Variability through Europe and Africa. Springer, Dordrecht, 638 p.

BATTARBEE, R. W., J. MASON, I. RENBERG, AND J. F. TALLING. 1990. Palaeolimnology and Lake Acidification. Royal Society of London, London, 219 p.

BIRKS, H. H. 2002. The recent extinction of *Azolla nilotica* in the Nile Delta, Egypt. Acta Palaeobotanica, 42:303-213.

BIRKS, H. H., S. M. PEGLAR, I. BOOMER, R. J. FLOWER, M. RAMBANI, P. G. APPLEBY, A. E. BJUNE, S. T. PATRICK, M. M. KRAÏEM, A. A.FATHI, AND H. M. A. ABDELZAHER. 2001. Palaeolimnological responses of nine North African lakes in the CASSARINA Project to recent environmental changes and human impact detected by plant macrofossil, pollen, and faunal analyses. Aquatic Ecology, 35:405-430.

BIRKS, H. J. B. 1998. Numerical tools in palaeolimnology - Progress, potentialities, and problems. Journal of Paleolimnology, 20:307-332.

BLAIS, J. M., L. E. KIMPE, D. McMAHON, B. E. KEATLEY, M. L. MALLORY, M. S. V. DOUGLAS, AND J. P. SMOL. 2005. Arctic seabirds transport marine-derived contaminants. Science, 309:445.

BLAIS, J. M., R. W. MACDONALD, D. MACKAY, E. WEBSTER, C. HARVEY, AND J. P. SMOL. 2007. Biologically mediated transport of contaminants to aquatic ecosystems. Environmental Science & Technology, 41:1075-1084.

BOYLE, J. F. 2001. Inorganic geochemical methods in paleolimnology, p. 83-41. *In* W. M. Last and J. P. Smol (eds.), Tracking Environmental Change Using Lake Sediments. Volume 2: Physical and Geochemical Methods. Kluwer Academic Publishers, Dordrecht.

BRODERSEN, K.P., O. PEDERSEN, C. LINDEGAARD, AND K. HAMBURGER. 2004. Chironomids (Diptera) and oxy-regulatory capacity: An experimental approach to paleolimnological interpretation. Limnology and Oceanography, 49: 1549-1559.

COHEN, A. S. 2003. Paleolimnology: The History and Evolution of Lake Systems. Oxford University Press, Oxford, 500 p.

COHEN, A. S., M. R. PALACIOS-FEST, J. McGILL, P. W. SWARZENSKI, D. VERSCHUREN, R. SINYINZA, T. SONGORI, B. KAKAGOZO, M. SYAMPILA, C. M. O'REILLY, AND S. R. ALIN. 2005a. Paleolimnological investigations of anthropogenic environmental change in Lake Tanganyika: I. An introduction to the project. Journal of Paleolimnology, 34:1-18.

COHEN, A.S., M. R. PALACIOS-FEST, E.S. MSAKY, S. R. ALIN, B. McKEE, C. M. O'REILLY, D. L. DETTMAN, H. NKOTAGU, AND K. E. LEZZAR. 2005b. Paleolimnological investigations of anthropogenic environmental change in Lake Tanganyika: IX. Summary of paleorecords of environmental change and catchment deforestation at Lake Tanganyika and impacts on the Lake Tanganyika ecosystem. Journal of Paleolimnology, 34:125-145.

CUMMING, B. F., J. P. SMOL, J. C. KINGSTON, D. F. CHARLES, H. J. B. BIRKS, K. E. CAMBURN, S. S. DIXIT, A. J. UUTALA, AND A. R. SELLE. 1992. How much acidification has occurred in Adirondack region (New York, USA) lakes since pre-industrial time? Canadian Journal of Fisheries and Aquatic Sciences, 49:128-141.

DAVIDSON, T. A., C. D. SAYER, M. R. PERROW, AND M. L. TOMLINSON. 2003. Representation of fish communities by scale sub-fossils in shallow lakes: Implications for inferring percid-cyprinid shifts. Journal of Paleolimnology, 30:441-449.

DEARING, J. A. AND R. T. JONES. 2003. Coupling temporal and spatial dimensions of global sediment flux through lake and marine sediment records. Global and Planetary Change, 39:147-168.

DIXIT, A. S., S. S. DIXIT, AND J. P. SMOL. 1992. Algal microfossils provide high temporal resolution of environmental trends. Water, Air, and Soil Pollution, 62:75-87.

DIXIT, A. S., S. S. DIXIT, AND J. P. SMOL. 1996. Long-term trends in limnological characteristics in the Aurora trout lakes, Sudbury, Canada. Hydrobiologia, 335:171-181.

DIXIT, S.S., B. F. CUMMING, J. C. KINGSTON, J. P. SMOL, H. J. B. BIRKS, A. J. UUTALA, D. F. CHARLES, AND K. CAMBURN. 1993. Diatom assemblages from Adirondack lakes (N.Y., USA) and the development of inference models for retrospective environmental assessment. Journal of Paleolimnology, 8:27-47.

DOUGLAS, M. S. V., J. P. SMOL, AND W. BLAKE, JR. 1994. Marked post-18th century environmental change in high Arctic ecosystems. Science, 266:416-419.

DOUGLAS, M. S. V., J. P. SMOL, J. M. SAVELLE, AND J. M. BLAIS. 2004. Prehistoric Inuit whalers affected Arctic freshwater ecosystems. Proceedings of the National Academy of Sciences of the United States of America, 101:1613-1617.

DUFFY, M. A, L. J. PERRY, C. M. KEARNS, L. J. WEIDER, AND N. G. HAIRSTON, JR. 2000. Paleogenetic evidence for a past invasion of Onondaga Lake, New York, by exotic *Daphnia curvirostris* using mtDNA from dormant eggs. Limnology and Oceanography, 45:1409-1414.

EDLUND, M. B., C. M. TAYLOR, C. L. SCHELSKE AND E. F. STOERMER. 2000. *Thalassiosira baltica* (Bacillariophyta), a new exotic species in the Great Lakes. Canadian Journal of Fisheries and Aquatic Sciences, 54:610-615.

ELTON, C. S. 2000. The Ecology of Invasions by Animals and Plants. University of Chicago Press, Chicago, Illinois, 181 p.

ENGSTROM, D., AND E. H. WRIGHT, J. R. 1984. Chemical stratigraphy of lake sediments, p. 11-67. *In* E. Haworth and J. Lund (eds.), Lake Sediments and Environmental History. University of Minnesota Press, Minneapolis, Minnesota.

EVENSET, A., G. N. CHRISTENSEN, J. CARROLL, A. ZABORSKA, U. BERGER, D. HERZKE, AND D. GREGOR. 2007. Historical trends in persistent organic pollutants and metals recorded in sediment from Lake Ellasjoen, Bjornoya, Norwegian Arctic. Environmental Pollution, 146:196-205.

FINNEY, B. P., I. GREGORY-EAVES, M. S. V. DOUGLAS, AND J. P. SMOL. 2002. Fisheries productivity in the northeastern Pacific Ocean over the past 2,200 years. Nature, 416:729-733.

FINNEY, B. P., I. GREGORY-EAVES, J. SWEETMAN, M. D. DOUGLAS, AND J. P. SMOL. 2000. Impacts of climatic change and fishing on Pacific salmon abundance over the past 300 years. Science, 290:795-799.

FLENLEY, J. R., S. M KING, J. JACKSON, C. CHEW, J. T. TELLER, AND M. E. PRENTICE. 1991. The late-Quaternary vegetational climatic history of Easter Island. Journal of Quaternary Science, 6:85-115.

FORMAN, M. R., AND M. C. WHITESIDE. 2000. Occurrence of

Bythotrephes cederstroemi in inland lakes in northeastern Minnesota as indicated from sediment records. Verhandlungen der Internationalen Vereinigung von Limnologen, 27:1552-1555.

FRITZ, S. F., B. F. CUMMING, F. GASSE, AND K. LAIRD. 1999. Diatoms as indicators of hydrologic and climatic change in saline lakes, p. 41-72. *In* E.F. Stoermer, and J.P. Smol (eds.) The Diatoms: Applications for the Environmental and Earth Sciences. Cambridge University Press, Cambridge.

GINN, B. K., L. GRACE, B. F. CUMMING, AND J. P. SMOL. 2008. Tracking anthropogenic- and climatic-related environmental changes in the remaining habitat lakes of the endangered Atlantic whitefish *(Coregonus huntsmani)* using paleolimnological techniques. Aquatic Conservation: Marine and Freshwater Ecosystems, 18:1217-1286.

GLEW, J. R., J. P. SMOL, AND W. M. LAST. 2001. Sediment core collection and extrusion, p. 73-105. *In* W. M. Last, and J. P. Smol (eds.), Tracking Environmental Change Using Lake Sediments. Volume 1: Basin Analysis, Coring, and Chronological Techniques. Kluwer Academic Publishers, Dordrecht.

GREGORY-EAVES, I., D. T. SELBIE, J. N. SWEETMAN, B. P. FINNEY, AND J. P. SMOL. 2009. Tracking sockeye salmon population dynamics from lake sediment cores: A review and synthesis, p. 379-393. *In* A. J. Haro, K. L. Smith, R. A. Rulifson, C. M. Moffitt, R. J. Klauda, M. J. Dadswell, R. A. Cunjak, J. E. Cooper, K. L. Beal, and T. S. Avery (eds.), Challenges for Diadromous Fishes in a Dynamic Global Environment. American Fisheries Society, Symposium 69, Bethesda, Maryland.

GUNN, J. M. 1995. Restoration and Recovery of an Industrial Region. Springer-Verlag, New York, 358 p.

HALL, R. I., AND J. P. SMOL. 1999. Diatoms as indicators of lake eutrophication, p.128-168. *In* E. F. Stoermer and J. P. Smol (eds.), The Diatoms: Applications for the Environmental and Earth Sciences. Cambridge University Press, Cambridge.

HALL, R. I., AND N. D. YAN. 1997. Comparing annual population growth of the exotic invader *Bythrotrephes* by using sediment and plankton records. Limnology and Oceanography, 42:112-120.

HARPER, M. A. 1994. Did Europeans introduce *Asterionella formosa* Hassall to New Zealand?, p. 479-484. *In* J.P. Kociolek (ed.), Proceedings of the 11th International Diatom Symposium, California Academy of Sciences, San Francisco, California.

HODGSON, D. A., AND N. M. JOHNSTON. 1997. Inferring seal populations from lake sediments. Nature, 387:30-31.

HODGSON, D. A., N. M. JOHNSTON, A. P. CAULKETT, AND V. J. JONES. 1998. Palaeolimnology of Antarctic fur seal *Arctocephalus gazella* populations and implications for Antarctic management. Biological Conservation,

83:145-154.

JEZIORSKI, A., N. D.YAN, A. M. PATERSON, A. M. DESELLAS, M. A. TURNER, D. S. JEFFRIES, B. KELLER, R. C. WEEBER, D. K. MCNICOL, M. E. PALM ER, K. MCIVER, K. ARSENEAU, B. K. GINN, B. F. CUMMING, AND J. P. SMOL. 2008. The widespread threat of calcium decline in fresh waters. Science, 322:1374-1377.

KRÜMMEL, E., R. MACDONALD, L.E. KIMPE, I. GREGORY-EAVES, J.P. SMOL, B. FINNEY, AND J. M. BLAIS. 2003. Delivery of pollutants by spawning salmon. Nature, 425:255-256.

KRÜMMEL, E., I. GREGORY-EAVES, R. MACDONALD, L. E. KIMPE, R. J. DEMERS, J. P. SMOL, B. FINNEY, AND J. M. BLAIS. 2005. Concentrations and fluxes of salmon derived PCBs in lake sediments. Environmental Science & Technology, 39:7020-7026.

KRÜMMEL, E. M., M. SCHEER, R. GREGORY-EAVES, R. W. MACDONALD, L. E. KIMPE, J. P. SMOL, B. F. FINNEY, AND J. M. BLAIS. 2009. Historical analysis of salmon-derived polychlorinated biphenyls (PCBs) in lake sediments. Science of the Total Environment, 407:1977-1989.

LAST, W. M., AND J. P. SMOL. 2001a. Tracking Environmental Change Using Lake Sediments. Volume 1: Basin Analysis, Coring, and Chronological Techniques. Kluwer Academic Publishers, Dordrecht, 548 p.

LAST, W. M., AND J. P. SMOL. 2001b. Tracking Environmental Change Using Lake Sediments. Volume 2: Physical and Geochemical Methods. Kluwer Academic Publishers, Dordrecht, 504 p.

LEAVITT, P. R. AND HODGSON, D. A. 2001. Sedimentary pigments, p. 295-325. *In* J. P. Smol, H. J. B. Birks, and W. M. Last, (eds.), Tracking Environmental Change Using Lake Sediments. Volume 3: Terrestrial, Algal, and Siliceous Indicators. Kluwer Academic Publishers, Dordrecht.

LENG, M. J. 2006. Isotopes in Palaeoenvironmental Research. Springer, Dordrecht, 307 p.

MAHER, B. A., AND R. THOMPSON. 1999. Quaternary Climates, Environments and Magnetism. Cambridge University Press, Cambridge, 402 p.

MICHELUTTI, N., B. E. KEATLEY, S. BRIMBLE, J. M. BLAIS, H. LIU, M. S. V. DOUGLAS, M. L. MALLORY, AND J. P. SMOL. 2009. Seabird-driven shifts in Arctic pond ecosystems. Proceedings of the Royal Society London, B, 276:591-596.

NORBERG, M., C. BIGLER, AND I. RENBERG. 2008. Monitoring compared with paleolimnology: implications for the definition of reference condition in limed lakes in Sweden. Environmental Monitoring and Assessment, 146:295-308.

PANFILI, J., H. D. E PONTUAL, H. TROADEC, AND P. J. WRIGHT. 2002. Manual of Fish Sclerochronology. Ifremer-IRD coedition, Brest, 463 p.

PIMENTEL, D. L. LACH, R. ZUNIGA, AND D. MORRISON. 2000. Environmental and economic costs of non-indigenous species in the United States. Bioscience, 50: 53-65.

QUINLAN, R., AND J. P. SMOL. 2002. Regional assessment of long-term hypolimnetic oxygen changes in Ontario (Canada) shield lakes using subfossil chironomids. Journal of Paleolimnology, 27:249-260.

RÜHLAND, K., A. M. PATERSON, AND J. P. SMOL. 2008. Hemispheric-scale patterns of climate-induced shifts in planktonic diatoms from North American and European lakes. Global Change Biology, 14:2740-2745.

SELBIE, D. T., B. A. LEWIS, J. P. SMOL, AND B. P. FINNEY. 2007. Long-term population dynamics of the endangered Snake River sockeye salmon: Evidence of past influences on stock decline and impediments to recovery. Transactions of the American Fisheries Society, 136: 800-821.

SMOL, J. P. 1995. Application of chrysophytes to problems in paleoecology, p. 303-329. *In* S. D. Sandgren, J. P. Smol, and J. Kristiansen (eds.), Chrysophyte Algae: Ecology, Phylogeny, and Development. Cambridge University Press, Cambridge.

SMOL, J. P. 2008. Pollution of Lakes and Rivers: A Paleoenvironmental Perspective, 2nd edition. Oxford University Press, New York, 383 p.

SMOL, J. P. In press. The power of the past: Using sediments to track the effects of multiple stressors on lake ecosystems. Freshwater Biology.

SMOL, J. P., AND B. F. CUMMING. 2000. Tracking long-term changes in climate using algal indicators in lake sediments. Journal of Phycology, 36:986-1011.

SMOL, J. P., AND M. S. V. DOUGLAS. 2007a. From controversy to consensus: making the case for recent climate using lake sediments. Frontiers in Ecology and the Environment, 5:466-474.

SMOL, J. P., AND M. S. V. DOUGLAS. 2007b. Crossing the final ecological threshold in high Arctic ponds. Proceedings of the National Academy of Sciences of the United States of America, 104:12395-12397.

SMOL, J. P., H. J. B. BIRKS, AND W. M. LAST. 2001a. Tracking Environmental Change Using Lake Sediments. Volume 3: Terrestrial, Algal, and Siliceous Indicators. Kluwer Academic Publishers, Dordrecht, 371 p.

SMOL, J. P., H. J. B. BIRKS, AND W. M. LAST. 2001b. Tracking Environmental Change Using Lake Sediments. Volume 4: Zoological Indicators. Kluwer Academic Publishers, Dordrecht, 217 p.

SMOL, J. P., B. F. CUMMING, A. S. DIXIT, AND S. S. DIXIT. 1998. Tracking recovery patterns in acidified lakes: A paleolimnological perspective. Restoration Ecology, 6:318-326.

SMOL, J. P., A. P. WOLFE, H. J. B. BIRKS, M. S. V. DOUGLAS, V. J. JONES, A. KORHOLA, R. PIENITZ, K. RÜHLAND, S. SORVARI, D. ANTONIADES, S. J. BROOKS, M-A. FALLU, M. HUGHES, B. E. KEATLEY, T. E. LAING, N. MICHELUTTI, L. NAZAROVA, M. NYMAN, A. M. PATERSON, B. PERREN, R. QUINLAN, M. RAUTIO, E. SAULNIER-TALBOT, S. SIITONEN, N. SOLOVIEVA, AND J. WECKSTRÖM. 2005. Climate-driven regime shifts in the biological communities of arctic lakes. Proceedings of the National Academy of Sciences of the United States of America, 102:4397-4402.

STOERMER, E.F., J. WOLIN, C.L. SCHELSKE, AND D. CONLEY. 1985. An assessment of ecological changes during the recent history of Lake Ontario based on siliceous algal microfossils preserved in the sediments. Journal of Phycology, 21:257-276.

SWEETMAN, J. N., AND J. P. SMOL. 2006. Reconstructing past shifts in fish populations using subfossil *Chaoborus* (Diptera: Chaoboridae) remains. Quaternary Science Reviews, 25:2013-2023.

UUTALA, A. J. 1990. *Chaoborus* (Diptera: Chaoboridae) mandibles: Paleolimnological indicators of the historical status of fish populations in acid-sensitive lakes. Journal of Paleolimnology, 4:139-151.

CHAPTER THREE

VERTEBRATE FOSSILS AND THE FUTURE OF CONSERVATION BIOLOGY

ELIZABETH A. HADLY[1] AND ANTHONY D. BARNOSKY[2]

[1]Department of Biology, Stanford University, Stanford, CA 94305 USA
[2]Department of Integrative Biology and Museums of Paleontology and Vertebrate Zoology, University of California, Berkeley, CA 94720 USA

ABSTRACT.—The science and practice of conservation biology face new challenges in the next few decades that will require application of the vertebrate fossil record. The new challenges are how to preserve not only individual species, but also natural ecosystem function through a time that is seeing unprecedented rates of climate change, human-population growth, and habitat fragmentation. Under these circumstances, linkages between vertebrate paleontology and conservation biology are needed to: (1) define the range of normal variation that ecosystems typically experience in their lifespan; (2) provide metrics for monitoring ecosystems that are useful for conservation biologists and benchmarks for recognizing successful ecosystem management; and (3) develop effective conservation strategies for species and ecosystems. Here we summarize some ways vertebrate paleontological work on genetics, populations, species diversity, and extinction is contributing to these needs. For example, the application of ancient DNA techniques in the context of life history strategies provides a means of determining when modern populations are in trouble. Species composition, abundance, and richness in modern ecosystems all can be compared to the paleontological record to assess when an ecosystem is exhibiting unusual changes. Past extinctions offer insights as to how to avoid future ones. New conservation efforts such as assisted migration will require information on which kinds of species substitutions maintained ecological function in the past. We anticipate that as ecosystems require more and more human manipulation to sustain biodiversity, the paleontological record will become even more important as a baseline against which to assess ecological health.

"The first step in park management is historical research, to ascertain as accurately as possible what plants and animals and biotic associations existed originally in each locality."–Leopold et al. (1963)

INTRODUCTION

IN TODAY'S world, it is difficult to find a major group of organisms that does not have species in danger of extinction (Wake and Vredenburg, 2008; Davidson et al., 2009; IUCN, 2009). The poster children for endangered species, however, are usually vertebrates—amphibians, reptiles, birds, and especially the so-called charismatic mammals, like primates, tigers, elephants, and whales. Thus, it is particularly important to define the "range of normal" fluctuations that vertebrates typically experience through evolutionary and ecological history, so that we can recognize not only when various animals and communities are truly imperiled but also when the conservation goal of maintaining "naturally-functioning" ecosystems has been achieved.

Vertebrate paleontology already has contributed to those goals in numerous ways (Hadly, 2003; Lyman and Cannon, 2004; Lyman, 2006). For example, the proportions of species among orders of North American mammals show little deviation over time-scales of millions of years (Alroy, 2000). At the hundred-thousand-year scale, vertebrate paleontology has demonstrated that while mammalian communities may change in species composition, the numbers of species in trophic and size categories remain relatively constant within a given kind of ecosystem (mountain, desert, etc.), and links between ecological niches are maintained although the species filling those niches may change (Owen et al., 2000; Hadly and Maurer, 2001; Barnosky et al., 2004a; Barnosky and Shabel, 2005; McGill et al., 2005; Hadly et al., in press). At the more resolved thousand-year time scale, it becomes apparent which kinds of species in a particular community typically respond to non-human caused perturbations, notably climate change, and which kinds of species can be expected to persist or disappear through the normal course of environmental variation (Guilday, 1971, 1984; Graham and Grimm, 1990; Wood and Barnosky, 1994; Graham et al., 1996; Hadly, 1996; Graham, 1997; Grayson, 1998, 2005, 2006; Barnosky, 2004a; Blois and Hadly, 2009).

In *Conservation Paleobiology: Using the Past to Manage for the Future, Paleontological Society Short Course, October 17th, 2009. The Paleontological Society Papers, Volume 15, Gregory P. Dietl and Karl W. Flessa (eds.). Copyright © 2009 The Paleontological Society.*

Useful as these kinds of information have been in the theoretical realm, they have been little utilized by land managers simply because it has been difficult to develop direct comparisons between present and past. Where these comparisons have been made, they have been extremely powerful for the management community; for example, studies of Holocene fauna that were instrumental in assessing elk populations and restoring wolves to Yellowstone National Park (Hadly, 1996; NRC, 2002), and in establishing a context in which to manage mountain goats in Olympic National Park (Lyman, 1998).

Accordingly, here we provide information on methods that rely on the vertebrate paleontological record and that directly link past and present in order to manage for the future. At the local scale we discuss how to infer the health and prognosis of mammalian populations using analyses of fossil and modern DNA. At regional and continental scales, we provide methods for assessing "normal" levels of biodiversity in mammalian communities, in order to establish conservation benchmarks that can be relatively easily monitored. And at the global scale, we refer to the fossil record to suggest how species might be expected to respond to the current threats of habitat fragmentation and global climate change, and suggest mitigation strategies.

NEW CHALLENGES IN CONSERVATION BIOLOGY

Conservation biology has at its core the mission to save Earth's biodiversity; from a practical standpoint, the goals have been to save individual species, to save products and services of ecosystems that people need, and to save places where people can still experience a feeling of wilderness (Meine et al., 2006). From a philosophical perspective, all three of those goals have heretofore been achievable by simply setting aside a large enough tract of land and attempting to manage it in a way that keeps it in its "normal" state. "Normal" has been implicitly defined by the landmark Leopold Report (Leopold et al., 1963), which set forth the working definition: "to preserve, or where necessary to recreate, the ecologic scene as viewed by the first European visitors." Although originally formulated for management of U.S. National Park lands, this philosophy has become the guiding light for much of land management and for the conservation ethic in general.

The first lesson of the vertebrate fossil record is that this criterion for "normal" is based on a fixed point in time, generally around a century or two ago that in North America is thought to precede significant European human impact. This view of "normal" does not take into account the range of variation ecosystems experience through their existence. Thus, an initial challenge for conservation biology is to explicitly shift the paradigm for conservation from one that emphasizes holding ecosystems at a static point defined by an arbitrary temporal benchmark, to one that focuses on maintaining key ecosystem functions within the range of variation exhibited through past environmental perturbations.

In this context we suggest a workable definition for "normal" might be: "maintaining ecosystems within the limits of variation they experience in times or regions where humans were not abundant on the landscape." Although humans colonized the continents at different times, in fact our dominance of the global ecosystem is relatively recent, coinciding with the advent of agriculture about 8,000-10,000 years ago and subsequent urbanization. Thus, the global dominance of humans is of much shorter duration than is the backdrop on which most living species evolved to interact, which is on the timescale of the glacial-interglacial climatic excursions beginning some 2.6 million years ago. The time of human dominance also is considerably shorter than the life spans of most species on Earth, which in general last about one to several million years (Alroy, 1996; Avise et al., 1998; Johns and Avise, 1998; Alroy, 2000). While we might quibble about the exact wording, the key concept we endorse is that a given ecosystem is characterized by a core set of ecological niches—for example, the "ground squirrel niche," or the "grass niche"—and although the particular species filling those niches through time may change, in the absence of significant human perturbation the niches, and the connections between them, remain through at least hundreds of thousands of years. This has been demonstrated not only for vertebrates (Barnosky, 2004a; Barnosky et al., 2004a; McGill et al., 2005; Hadly et al., in press), but for plants and invertebrates as well (Pandolfi, 2002; Hughes et al., 2003; DiMichele et al., 2004; Flessa and Jackson, 2005; Jackson and Erwin, 2006; Pandolfi, 2006).

The second big challenge facing conservation biology is that abnormally rapid global warming (IPCC, 2007) has changed the rules of the conservation game, such that even the straight-forward practice of preserving species by simply setting aside a protected tract of land no longer is viable (Barnosky, 2009). The basic problem is that as climate changes rapidly within protected areas, the species we have become used to in those places lose the climate they need to survive there (e.g., McMenamin et al., 2008). Whereas during past times of rapid climate change, like glacial-interglacial transitions, species have tracked their required climate as it shifts across the landscape. This type of response is not an option for species whose last refuge is in nature reserves, which is the case for many endangered vertebrate species. At best, legislatively protected nature reserves comprise only 12% of Earth's land surface (NGS, 2006), and reserves typically are widely separated from one another by human-altered landscapes such as farms, ranches, cities, and dams. This makes each reserve essentially a small, isolated island.

The problems species face on that kind of landscape are threefold. First, there is little, if any, suitable habitat immediately outside the boundaries of most nature reserves, so species cannot track their climate zone from one to the next, even if they could disperse fast enough. Second, the rate of warming is so much faster than even the most rapid climate changes that species have seen in their past—some 10% to 300% faster than changes experienced at the Medieval Warm Period or glacial-interglacial transitions (e.g., Barnosky et al., 2003; Barnosky, 2009; Blois and Hadly, 2009)—that it is unknown whether many species are capable of dispersing at fast enough rates, even if suitable habitat corridors existed. Third, Earth is actually entering a new climate state in respect to what extant species and ecosystems have evolved in. No matter which IPCC scenario (IPCC, 2007) plays out, by approximately 2050 the global mean temperature will be hotter than it has been since *Homo sapiens* first evolved some 160,000 years ago, and under the more carbon-intensive A2 scenario, by 2100 Earth could be hotter than it has been in 3 million years, longer than many vertebrate species, and virtually all mammal species, have been in existence (Alroy, 1996; Avise and Walker, 1998; Avise et al., 1998; Alroy, 2000).

Those realities mean that saving individual species—that is, preserving biodiversity—will become increasingly difficult without intensive human manipulation of species ranges. Saving species as their climate disappears in a given reserve may well require moving them to other reserves where suitable climate exists; such "assisted migrations" already are underway for insects and in active discussion phases for plants and other organisms (McLachlan and Hellmann, 2007; Zimmer, 2009). However, such human manipulation of species distributions could be the antithesis of maintaining naturally functioning ecosystems, if the assisted migrations result in ecological interactions that are outside the range of normal as we define it above (e.g., outside the variations a given ecosystem experiences through thousands of years in the absence of significant human intervention).

To sum up: today and in the coming decades, the successful practice of conservation biology requires (1) the recognition that ecosystems experience natural variation through time; (2) a means of distinguishing when the boundaries of that natural variation have been exceeded, and (3) understanding that the goal of saving species may soon be decoupled from the goal of maintaining ecosystems whose species interactions are not primarily orchestrated by humans. Therefore, the first essential contribution of vertebrate paleontology is to establish, in metrics that are easy to understand and possible to apply to modern systems, the range of variation that can be considered normal in the absence of significant human impact. The second essential contribution is to use what we know about past ecosystem dynamics to help structure biodiversity preserves that will become increasingly manipulated by humans, as we strive to keep endangered species alive.

DEFINING THE NATURAL RANGE OF ECOSYSTEM VARIATION THROUGH TIME

Ecosystems experience many kinds of variation through time. Over human lifetimes, for example, populations within species are known to wax or wane; new species may enter into an area because of natural dispersal or deliberate or unintentional introduction by people, or species may be lost from an area (extirpation) or globally (extinction). The relevant question for most land managers is a straightforward one: When are these kinds of observed changes a signal that something is going wrong with the ecosystem under their jurisdiction, and when do the observed changes simply reflect natural variation in a highly complex system? Here we give examples of how the vertebrate-fossil record can be used to separate signal (a change that indicates an ecological problem) from noise (natural variation in the ecosystem) using three different kinds of metrics that are relevant and commonly considered in ecosystem management and conservation biology: population genetics, species composition and diversity, and extinction risk.

Population genetics

Species are comprised of populations, and together, those populations determine the geographic distribution of the species. The environment acts on the individuals in populations through time to influence where they can persist and where they thrive. All species have evolved suites of physiological, morphological, and behavioral attributes that they use to handle changes to their abiotic and biotic environments (Blois and Hadly, 2009). Variation in these attributes usually means that species can handle a greater range of challenges, and usually is underlain by genetic variation. Hence, assessing genetic variation within and between populations is a key aspect of assessing the viability of species within ecosystems (Gilpin and Soulé, 1986; Soulé, 1987). Genetic variability also provides evidence suggestive of the abbreviated history of events that populations experience through time (Ramakrishnan and Hadly, in press). Particularly large and recent events can leave profound marks on the genetic diversity of a species. For example, a severe reduction in population size of a species can eliminate almost all its genetic diversity through the process of random drift such as occurred, for example, in the cheetah (Menotti-Raymond and O'Brien, 1993), the elephant seal (Hoelzel et al., 1993) and the Wollemi pine (Peakall et al., 2003). Thus, not only does a severe population bottleneck event leave its historic mark on the genetic diversity of a species, it also increases the susceptibility of the species to stochastic environmental catastrophes because the species has fewer tools in its genetic toolkit. However, over long periods of time (hundreds of millennia to millions of years), if the species recovers and persists, even evidence of severe bottlenecks can be erased or mitigated since mutation and recombination will eventually contribute genetic novelty to the species. Our own species provides such an example, since *Homo sapiens* experienced very recent population expansion about 50,000 years ago (Atkinson et al., 2008). Without a temporal record of genetic diversity, discerning how frequently recovery from these bottleneck events occurs is impossible, yet it is critical since many species and their populations are now imperiled (Ceballos et al., 2005).

Where the right preservation conditions exist for long periods of time, fossil and subfossil specimens of vertebrates can harbor sufficient organic material to preserve DNA and thus temporal variations in genetic diversity, gene flow, and population size can be accessed. These prehistoric and historic samples of DNA, called 'ancient DNA' can reveal not only the identity of species in the absence of characteristic morphological traits, but also much about the genetic diversity of species and their populations through time. The fossil samples provide an empirical benchmark to which similar genetic assessments of extant populations can be compared and modeled, in order to identify any current population-genetic attributes that are out of the ordinary. This ancient genetic diversity can also reveal a "moving picture" of the population through time, indicating how and if species responded to environmental events in their history.

The practical applications of data from aDNA are diverse. For example, changes in genetic diversity of populations can be used to unravel specific responses to past climatic events, which is a powerful way to use the past to predict the future of certain species in the face of the current global warming crisis. While only a few species have been examined in this manner, it is becoming apparent that a species' genetic response to climate change is tightly tied to its life history strategy, and that fossil data are essential to interpreting what genetic variation in modern populations reveals about their future viability. In Yellowstone Park, for instance (Hadly et al., 2004), species characterized by low dispersal rates, such as pocket gophers, maintain similar haplotypes and low genetic diversity through climatic fluctuations that include a ~1°C local warming during the Medieval Warm Period and subsequent cooling during the Little Ice Age, and current populations are genetically very similar to ancient ones in spite of climate fluctuations. In contrast, species characterized by high dispersal per generation, such as voles, exhibit dramatic differences in particular genotypes between current and past populations, and highlight how misleading it may be to assume high genetic diversity in itself indicates sustainable populations. As population-size of certain species of *Microtus* decreased locally in northern Yellowstone Park during the Medieval Warm Period, their apparent diversity markedly increased. Modeling these data showed that the only way to explain this anomaly was that the local populations were replaced by extralocal ones, signaling that the area had become a genetic sink rather than a source of genetic diversity. From the standpoint of conservation biology and ecosystem management, these fossil data provide critical insights that are simply impossible to gain by looking only at extant populations, namely: (1) low genetic diversity in pocket gophers (and by inference, in other kinds of low-dispersal species) is not necessarily a cause for concern in the face of global warming; and (2) high genetic diversity in voles (and by inference, other high-dispersal species) does not ensure that they are not under stress from habitat loss; in fact, it can indicate just the opposite. For high-dispersal species, landscape connectivity is clearly essential for population movement and recolonization.

In the bigger picture, studies on a variety of species hint at how changing environments, particularly climatic changes of the last 20 millennia, have governed past population movements (gene flow) in mammals, and thus what we might expect to see in extant species as environmental changes unfold in the future. Among the species for which ancient genetic information is now available are: brown bears (Barnes et al., 2002; Calvignac et al., 2008; Valdiosera et al., 2008), cave bears (Hofreiter et al., 2002; Orlando et al., 2002; Hofreiter et al., 2007), foxes (Dalen et al., 2007), gophers (Hadly et al., 1998), gray whales (Alter et al., 2007), horses (Orlando et al., 2006), humans (Lalueza-Fox et al., 2005; Weaver and Roseman, 2005; Belle et al., 2006; Lalueza-Fox et al., 2006), northern fur seals (Newsome et al., 2007), rats (Barnes et al., 2006), squirrels (van Tuinen et al., 2008), and voles (Hadly et al., 2004).

Ancient and historic data from museum specimens can also help to unravel histories of human manipulation of populations including hybridization between species and the impacts of overharvesting (e.g., Leonard, 2008). Studies detailing ancient genetic data for invertebrates and plants are much more limited, but can reveal the biotic compositions of ancient communities (e.g., Willerslev et al., 2007) and plant domestication (e.g, Gugerli et al., 2005).

Ancient DNA studies also have tracked genetic diversity associated with climatic change before and during population size change, in organisms such as bison (Shapiro et al., 2004), voles and gophers (Hadly et al., 2004), tuco-tucos (Chan et al., 2005, 2006), wolves and dogs (Leonard et al., 2002), horses (Weinstock et al., 2005; Orlando et al., 2008), and mammoths (Barnes et al., 2007). In some cases, nuclear genetic analysis has been used to ascertain phenotypic traits characteristic of particular environments such as light versus dark coat color (Rompler et al., 2006).

In North American bison, a lineage that shows dwarfing during the latest Pleistocene to Holocene (Hill et al., 2008), ancient genetic data assembled from both North America and Asia from the past 150,000 years were interpreted to show evidence of population growth until approximately 37,000 years ago, when the population suffered loss in genetic diversity consistent with population size decline (Shapiro et al., 2004). Shapiro et al. (2004) concluded that the decline was likely due to climatic change since the timing of the decline was coincident with initiation of the last glacial maximum in Beringia and preceded evidence of human migration into the Americas. A subsequent, more sensitive analysis using the same data detected another more subtle event: population size in bison was at its minimum ~10,000 years ago (Drummond et al., 2005), a time coincident with extinction of megafauna in North America (Barnosky et al., 2004b; Koch and Barnosky, 2006), the peak in the last glacial maximum, and arrival of humans to North America. The analysis by Drummond et al. (2005) revealed that humans may also have played a role in the reduction in bison populations in Beringia, albeit subsequent to the initial effects of climatic change. Bison after 10,000 years ago showed population size recovery in North America until intensive historic hunting by European settlers. Thus, unraveling the timing of population size contraction and expansion using ancient genetic data and population genetic modeling enables us to better understand which factors may be responsible for events in species population histories. In the case of bison, these analyses also revealed that the species narrowly missed the fate of extinction that other megafauna suffered, and that the genetic diversity of this species today is but a vestige of its relatively recent past.

Besides its direct practical application in comparing present with past genetic variation, the study of ancient DNA has matured to the point where it is beginning to contribute substantially to population genetic theory, which underpins assessments of population viability in conservation applications. Empirical temporal data has been critical, because genetic change within lineages is a result of population-level processes that take place over generations, with cumulative effects building up over hundreds to thousands of years (Ramakrishnan and Hadly, in press). The key processes are recombination, mutation, selection, random genetic drift and gene flow, with all but mutation strongly influenced by population size (Ramakrishnan and Hadly, in press). A persistent problem in population genetics has been sorting out which of these population-level processes dominates in explaining the genetic signature of a given population, especially when trying to assess whether environmental change has influenced (or is influencing) a genetic signature, because each process (except possibly recombination) is known to be influenced by the abiotic and biotic environments. For example, mutagenesis has been shown to vary by environment in some organisms, and further, environmental correlates such as rates of ultraviolet radiation may play an important role (Pawlowski et al., 1997). Environmental changes cause individuals to move in search of favorable habitats; as a result, individuals may disperse from one population to the next (gene flow); populations can increase or decrease in numbers of individuals (population size), and/or they may shuffle the percentage of adaptive or maladaptive traits (selection).

To untangle the role of environmental change from stochastic process, it is now possible to examine the expectations of genetic diversity using theoretical models of mutation, selection pressure, population size and gene flow that were developed to analyze ancient DNA sequence data (Ramakrishnan and Hadly, in press). When genetic change is directly observed through time, it becomes possible to disentangle which population processes were most likely responsible for the observed genetic diversity (Anderson et al., 2005; Drummond et al., 2005; Ramakrishnan et al., 2005; Chan et al., 2006). Such models, when supported by empirical temporal data, can discriminate whether populations were connected or isolated during periods of climatic change (Hadly et al., 2004; Hofreiter et al., 2004), the probable size of ancient populations (Chan et al., 2006) and whether changes in population size are concordant with what is known about the population biology of the species (Hadly et al., 2004). Indeed, the addition of ancient temporal data has been shown to increase the probability of revealing the correct evolutionary history of populations and species over modern data alone (Ramakrishnan et al., 2005), and to markedly enhance our ability to determine whether and how populations and species might adapt to climatic or other environmental changes.

Ancient DNA from species that became extinct at the Pleistocene-Holocene transition, such as mammoth or giant ground sloth (Poinar et al., 1998, 2003; Barnes et al., 2007; Hofreiter, 2008a), or in the very recent past such as dodos, moas, and other birds (Sorenson et al., 1999; Huynen et al., 2003, 2008; Bunce et al., 2005; Allentoft et al., 2009), has yielded insights about how climatic or human perturbations affected the last populations of the species, and in turn inform us about signs of impending extinction in today's animals. Such ancient samples are not yet sufficient to quantify how much reduction in genetic variation signals impending extinction, although studies of threatened animals show that the majority suffer loss in genetic diversity before they are driven to extinction (Spielman et al., 2004). Ancient genetic studies have the potential to help us understand whether the majority of species experiencing extinction did so prior to losing genetic diversity because of events affecting populations, or whether the loss of genetic diversity itself makes species more extinction-prone.

Obtaining information from aDNA is not without its challenges. It requires a dedicated lab physically separated from facilities where modern DNA is extracted and sequenced, and extraordinary procedures to guard against contamination. Reconstructing the ancient genetic sequences of individuals of course also requires adequate fossil localities that preserve fossil vertebrates (usually bones or teeth). In order to preserve DNA, specimens must be rapidly buried and protected from wide swings in temperature. Favorable conditions are most frequently found at high latitudes where bones may be preserved in permafrost, or within sheltered environments such as caves, or where the environment is exceptionally arid (Hofreiter et al., 2001). Tropical areas generally are not optimal for long-term organic preservation, and thus ancient DNA from the tropics is rare. Additionally, sequencing and amplifying ancient genetic data is difficult because the pieces of DNA that manage to be preserved for even brief periods of time after the death of an animal are usually short (~100-300 basepairs) and over time, the number of copies of this sequence will decline dramatically, meaning that accessing ancient DNA sequences has a hypothesized limit of less than 100,000 years (Handt et al., 1994; Lindahl, 2000; Willerslev and Cooper, 2005). Rarely DNA may be preserved longer than 100,000 years in very cold environments (Willerslev et al., 2007), and there are reports of much older protein sequences (Asara et al., 2007). Finally, most aDNA studies are constrained to use mitochondrial DNA because of its abundance in the mammalian genome: it is at least 3600 times more common than nuclear DNA in humans (Miller et al., 2003).

Despite these limitations, molecular paleontologists already have the ability to determine the cadence of genetic variation within populations over time in important cases, as exemplified by the numerous studies cited above. Further, as sequencing technology becomes more streamlined, it should eventually be possible to reconstruct genomic-level variation, which holds great promise for understanding both variation in neutral regions of the genome as well as those genetic areas under selection (Hofreiter, 2008b).

Species composition and diversity

Three important metrics that are useful and relatively easy to monitor in ecosystems are species composition (what species are there), species abundance (how common or rare individuals of a species are), and species richness (how many species there are). Species composition is important because species are thought to conserve their ecological niches through time (Ackerly, 2003; Martinez-Meyer et al., 2004; Wiens and Graham, 2005; Hadly et al., in press). Thus, the presence of a given species indicates the presence of a given ecological niche, and in turn, the persistence of species indicates persistence of the ecological niches that define a given ecosystem. Species abundance is related to the life history attributes of a species (body size, generation time, habitat preference, trophic group, etc.) and also to the abundance of habitats within the ecosystem that can support individuals of a given species (Blois and Hadly, 2009). Pronounced changes in abundance, particularly in the relative abundance of species as compared to each other, indicates at least changes in landscape characteristics (for example, expansion of arid microhabitats at the expense of mesic ones) and can indicate major shifts in how species interact (for example, decrease in predators that in turn lead to increased herbivore populations with attendant effects on vegetation). Species richness is critical to monitor because it is correlated with important ecosystem attributes such as productivity, disturbance regime, and habitat heterogeneity (Rosenzweig, 1995; Barnosky et al., 2001; Hadly and Maurer, 2001; Cardinale et al., 2002); dramatic changes in richness thus indicate fundamental changes in the ecological niches within an ecosystem and in the connections between niches.

In the fossil record it is usually possible to identify mammal species by dental, cranial, or other osteological features, and many amphibian, reptile, and bird species also can be identified from bones. In cases where morphological features are not diagnostic, the ancient DNA techniques discussed above can often be applied to assess species identity. For example, in a study of the influence of tectonics and climate on the long-tailed vole (*Microtus longicaudus*) in the Greater Yellowstone Ecosystem, ancient DNA was used to discern the frequency of *M. montanus* vs. *M. longicaudus,* which lack diagnostic traits on isolated teeth necessary to distinguish between the two species (Spaeth et al., 2009). From the perspective of conservation biology, identifying which species were present in the past becomes critical, because the management questions of interest typically require knowing: (1) what species "should" be present in a given ecosystem in the absence of significant human manipulation; (2) is the loss or gain of a given species unusual; and (3) are the normal number of species one would expect the ecosystem to support actually there? As with analyses of ancient DNA to assess population-level changes, the vertebrate fossil record offers a rich source of answers for these species- and community-level questions, which yield both practical and theoretical information for land managers.

The most relevant fossil information to apply to the first two questions comes from paleontological deposits that accurately sample the living communities of the past, and from time frames during which extant species, or at least species that are closely related to extant ones, had already evolved. Typically these are fossil deposits that extend back a few thousands to a few hundreds of thousands of years (although essential information also comes from older deposits; see below). Some of the best community samples come from deposits in caves and rockshelters, where woodrats (*Neotoma* spp.) den and drag bone-laden pellets of carnivores, raptors, and small pieces of decomposing large-mammal skeletons into their nests. The same caves or rockshelters are often used by large carnivores for their dens, and their bones and the bones of their prey add to the sample of skeletal materials that eventually become fossilized. Typically, abundant organic material in these deposits, including the bones themselves, provide opportunity for multiple radiocarbon dates that can be used to refine chronologies, and to assess the extent of time-averaging (that is, how many years a given community-sample actually took to accumulate). Detailed analyses of modern communities sampled by similar vectors demonstrates very high fidelity between the sample and the living animals

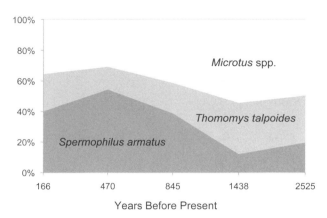

FIGURE *1.*—Relative abundance fluctuations of the three most common small mammals through the late Holocene at La-mar Cave, Yellowstone National Park, Wyoming (N=2835 *Microtus* spp. specimens; N=2210 *Spermophilus armatus* specimens; N=1552 *Thomomys talpoides* specimens). Ages are median calendar years before present of several strati-graphic levels based on a detailed radiocarbon chronology. Note that during relatively arid times such as during the Me-dieval Warm Period (~800-1200 ybp), *Spermophilus* rises in abundance at the expense of *Microtus* and *Thomomys*, and declines during cooler, wetter intervals. Nevertheless, these three taxa remain as the most common taxa in the commu-nity, which includes a total of 40 mammalian species from the cave. Data are from Hadly (1996).

on the landscape, especially for presence-absence and relative abundance of small mammals, birds, and am-phibians that live within about 5 km of the fossil site (Hadly, 1999; Porder et al., 2003; Terry, 2008). Fortu-nately, cave and rockshelters that hold such deposits are widely distributed throughout the Appalachians, Rocky Mountains, Great Basin, Pacific Northwest, and Sierra Nevada—exactly the areas that hold abundant publicly administered lands so important in conserving species and ecosystems. Where caves and rockshelters are not present, the opportunity for high-fidelity sam-ples of fossils sometimes exists in fluvial and riparian environments, as shown by recent studies that compare bone accumulations with the living community in Af-rica (Behrensmeyer et al., 2003; Western and Behrens-meyer, 2009).

When adequate fossil information is available it is relatively straightforward to determine whether mod-ern species composition has been significantly per-turbed from what existed prior to anthropogenic domi-nance. The approach of using Holocene fossil depos-its to inform land management decisions has already been used in Yellowstone National Park, where fossils from Lamar Cave demonstrated that all of the mammal species now protected in the park, including wolves, grizzly bears, and elk, species whose management has been controversial (NRC, 2002), were common in that ecosystem for at least the past 3000 years. The only mammal species that is missing is a vole (*Microtus ochrogaster*) that always was at very low abundance (Hadly, 1996, 1999). Although the birds and amphib-ians from the deposit have not been studied in as much detail, species composition in those groups likewise seems to have remained relatively stable over the past few thousand years. Examples of other areas where the Holocene fossil record is rich and useful in demonstrat-ing that the mammals on the landscape today largely reflect the species composition of the past few thou-sand years include northern California (Blois, 2009), the Great Basin (Grayson, 1998; Grayson and Madsen, 2000; Grayson, 2006; Terry, 2007), the Great Plains (Semken and Graham, 1996), and the Appalachians (Guilday, 1971, 1984).

A great value of cave and rockshelter sites that have successive layers of high-fidelity fossil depos-its—or any such stratigraphic sequence—is their utility in tracking fluctuations in abundance of various taxa through time. Usually, it is only the small-mammal component of the community for which numbers of specimens are adequate to analyze abundance changes; nevertheless, those are precisely the kinds of organisms that can also be easily monitored in modern ecosys-tems in order to compare present with past, and they have greater sensitivity to microhabitat changes than do larger mammals (Hadly, 1996). The fossil depos-its demonstrate that pronounced fluctuations in abun-dance are characteristic in a given ecosystem over the 100-year, 1000-year, 10,000-year, and 100,000-year time scales. Relative abundance fluctuations clearly are the normal, early response to climatic changes of the past. For example, in Lamar Cave in Yellowstone Park, voles of the genus *Microtus* are the most abun-dant small mammals during cool, moist periods, and ground squirrels (genus *Spermophilus*) are most abun-dant during arid times (Hadly, 1996; Hadly et al., 2004) (Fig 1.). Each abundance state lasts at least a few hun-dred years. But the same kinds of relative abundance shifts also manifest at longer time scales: for example, at Porcupine Cave, Colorado, voles and lemmings are most abundant during cool, moist glacial times, but ground squirrels increase in abundance during more arid interglacials (Barnosky, 2004b). In this case, the

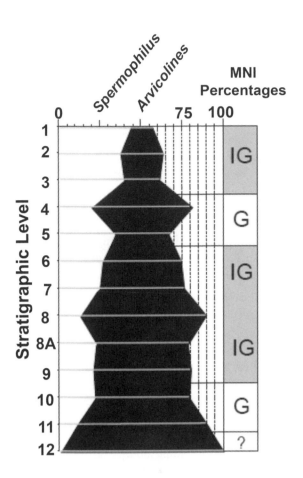

FIGURE 2.—Relative abundance of ground squirrels (*Spermophilus* spp.) in respect to voles and lemmings (arvicolines) through >100,000 years at Porcupine Cave, South Park, Colorado. Level 1 is roughly dated to 800,000 years old (but could be somewhat younger) and Level 12 to at least 900,000 years (but could be somewhat older). Climatic interpretations for each group of stratigraphic levels are interpreted from the sediments, not from the fauna, and are indicated by the abbreviations on the right of the diagram; IG=interglacial; G=glacial. The sediments indicate that the topmost interglacial (Levels 1-3) was dramatically more arid than any previous time represented by the record. Note that both ground squirrels and arvicolines are common throughout the >100,000 years represented by the record, but that during the very arid time represented by Levels 1-3, ground squirrels begin to outnumber arvicolines after tens of thousands of years of arvicolines being generally more abundant. Data are from Barnosky (2004b).

abundances appear relatively stable for thousands of years, before switching in pace with climatic changes (Fig. 2). From the land-management perspective, then, observed shifts in relative abundance of taxa provide an early warning signal that environmental changes are

affecting an ecosystem, but in themselves do not indicate an ecosystem has been perturbed from its "normal" state. Access to such paleontological data enables researchers to predict which species are likely to be affected by global change, whether they will increase or decrease in abundance, and to what extent their populations have adjusted in the past.

However, in terms of relative abundance of taxa, the paleontologic record also demonstrates that in terrestrial ecosystems where we have adequate records (Appalachians, Rocky Mountains, Great Basin, Great Plains, California), three or four genera tend to dominate in abundance throughout hundreds to thousands of years in a given place, even though rank-order abundance among the three or four most common genera fluctuates in accordance with environmental changes. For example, the top three genera in abundance throughout the entire 3000-year record for Lamar Cave are *Spermophilus* (ground squirrel), *Microtus* (vole) and *Thomomys* (pocket gopher) (Hadly, 1996) (Fig. 1); and for most of the >100,000 year record at Porcupine Cave, the dominant genera are *Mictomys* (vole), *Spermophilus* (ground squirrel), and *Neotoma* (wood rat) (Barnosky, 2004b). In fact, abundances of all small mammal genera are significantly correlated between these two sites, which are separated by 785 km and some 900,000 years (Fig. 3). These genera (listed in Fig. 3) clearly serve as important components—essentially defining a core set of taxa—in the North American montane taxon pool. Abundance changes that would indicate perturbation from a normal ecosystem state, then, are, if any of the most abundant genera dwindled to lower ranks in the abundance hierarchy, or more critically, became extinct. Actual loss of genera from an area would provide a very clear signal that an ecosystem has been perturbed into a new state, since the fossil record demonstrates constancy in genera and abundance patterns through nearly 1 million years in the studied mammalian communities (Fig. 3) (Hadly and Maurer, 2001; McGill et al., 2005).

The Holocene and Pleistocene fossil records indicate that minor changes in species composition also are normal through the course of time ecosystems typically exist. Generic similarity of communities in space and time is apparent across even isolated montane regions (Hadly and Maurer, 2001). Perhaps most indicative of this are records such as Porcupine Cave, which demonstrated species turnover within genera, but very little change in the number of species found in each

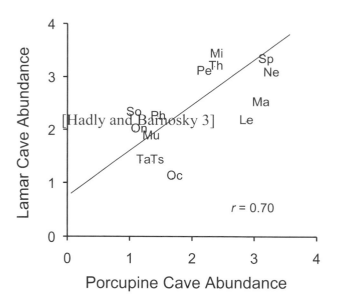

FIGURE 3.—Assessment of community composition and abundance of individuals at the taxonomic level of genus for a relatively recent paleontological site (Lamar Cave, Wyoming, <3200 years) and a relatively old site (Porcupine Cave, Colorado, >600,000-900,000 years), both in a Rocky Mountain ecosystem though widely separate both spatially and temporally. The axes are log10-scale in units of number of identified bones. Each symbol on the graph represents the number of specimens of a particular genus found at each site, summed across all levels. The Pearson r value is significant at $P = 0.001$. The line is a ranged major axis regression. Pairwise comparisons of other sites indicate a similar result, namely, that the characteristic assemblage of small mammals in mountain and intermontane ecosystems of the United States includes not only genera typically represented by abundant numbers of individuals, such as *Spermophilus*, *Microtus*, and *Thomomys*, but also "rare" genera represented by few individuals such as the western jumping mouse *Zapus*, and thus the 'species pool' of western North America has persisted with the same constituents for at least ~900,000 years (McGill et al., 2005). Abbreviations: Cl—*Clethrionomys*; Le—Leporidae; Ma—*Marmota*; Mi—*Microtus*; Mu—*Mustela*; Ne—*Neotoma*; Oc—*Ochotona*; On—*Ondatra*; Pe—*Peromyscus*; Ph—*Phenacomys*; So—*Sorex*; Sp—*Spermophilus*; Ta—*Tamias*; Ts—*Tamiasciurus*; Th—*Thomomys*; Za—*Zapus*.

trophic and body-size group through >100,000 years and at least two glacial-interglacial climatic transitions (Fig. 4) (Barnosky, 2004a, b; Barnosky et al., 2004a). This, and the abundance data summarized above, suggests that a key indicator that an ecosystem is functioning within its natural range of variation is simply maintaining adequate numbers of species within each genus, with species identity being less important. Put simply, in terms of ecosystem function, one species of pocket gopher (genus *Thomomys*) may be as good as the next, but pocket gophers are long-term dominant members of North American mammalian communities and fill a critical niche.

Such species substitutions in a given place are most evident coincident with rapid climate changes, such as the warming that accompanies the shift from glacial to interglacial climate states (Guilday, 1971, 1984; Barnosky, 2004a; Blois, 2009; Blois and Hadly, 2009). In that light, counter-intuitively, maintaining species diversity becomes particularly important, especially in the face of today's known environmental pressure of rapid climate change. This is because the only way it is possible to substitute one species for another to maintain ecosystem function as climate changes is if there is a large reservoir of species to draw on (Hadly et al., in press), and if those species can disperse between sites. Numerous paleontologic and neontologic studies have demonstrated that species divide their geographic range by climate space (Martinez-Meyer et al., 2004; Davis, 2005; Wiens and Graham, 2005; Elith et al., 2006; Hijmans and Graham, 2006; Nogues-Bravo et al., 2008; Blois, 2009; Hadly et al., in press). To continue with the pocket gopher example, while any species of *Thomomys* might be suitable for filling the pocket gopher niche in an ecosystem, only certain species seem able to thrive in specific places characterized by certain climatic parameters. Some species inhabit relatively cooler, moister parts of the entire range of the genus, whereas other congeners are restricted to relatively more arid parts of the range (Blois, 2009). For instance, in northern California, where the geographic ranges of three species of *Thomomys* are in close proximity, it is evident that the species that adapted to cool, moist microhabitats was replaced by the species adapted to drier habitats at the Pleistocene-Holocene transition (Blois, 2009). It seems likely that such "climatic plasticity" for a genus is possible only by maintaining multiple congeneric species (Hadly et al., in press).

Another important lesson of the fossil record is that the number of species in a given ecosystem—species richness—tends to vary little over at least thousands to hundreds of thousands of years, despite species turnover. For example, characteristic species richness during interglacial climates for non-volant terrestrial mammals in Rocky Mountain ecosystems seems to

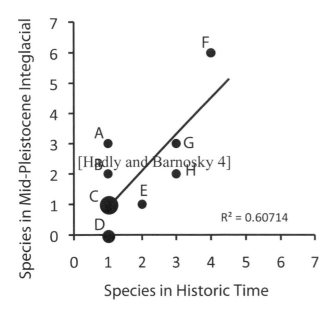

FIGURE *4.*—Comparison of numbers of species within each higher taxonomic group in the modern community around Porcupine Cave, Colorado, and the >600,000-year-old fossil community from a mid-Pleistocene interglacial (Level 1). Size of circle represents number of groups for each comparison. Abbreviations for each taxon: A—*Neotoma*; B—*Cynomys*; C—*Ochotona, Marmota, Tamiasciurus, Peromyscus, Thomomys, Ondatra, Erethizon*; D—*Sciurus, Zapus, Castor*; E—*Tamias*; F—Arvioclines (voles and lemmings); G—Leporids (rabbits); H—*Spermophilus*. While species numbers are similar for most taxonomic groups, the species in the fossil deposits are for the most part different species than those representing the corresponding genera today (Barnosky et al., 2004a).

be around 40 species as demonstrated over thousands of years in Yellowstone Park (Hadly, 1996) and 45-50 species as tracked over hundreds of thousands of years in South Park, Colorado (Barnosky et al., 2004a; Barnosky and Shabel, 2005). Thus, reduction of mammalian species much below those numbers would clearly be an ecological danger sign in a Rocky Mountain land-management unit. Similar comparison of Pleistocene and Holocene diversity with modern diversity can provide the same kind of valuable information in other regions, although we are only just now reassessing modern species richness subsequent to climatic change and the massive landscape alteration that has characterized the last few decades in North America.

Non-anthropogenic climate change seems to affect species richness primarily in the lower size and lower trophic categories. For example, the characteristic gla-cial fauna in Colorado has 29 species in those cat-egories, and the characteristic interglacial fauna has 24; little change is evident across climatic boundaries in other size and trophic groups (Barnosky et al., 2004a). In northern California, the small mammal component exhibits declines in both richness (from 13 species to 9 species) and evenness (with deer mice becoming much more common) as climate warms at the beginning of the Holocene (Blois, 2009). Decline in species richness as Holocene warming commenced has been reported for voles and shrews, voles in sites from Pennsylvania, extending south through the Appalachians into Tennessee, into Texas (Graham, 1976). Mechanistically, the loss in richness at the beginning of the Holocene seems to involve geographic range shifts of climatically sensitive species into refugia (Graham and Grimm, 1990; FAUNMAP Working Group, 1996). These observations suggest that in modern land-management units, a decline in small mammal richness and/or evenness might be an early warning sign that global warming is beginning to alter community structure, and thus, ecological relationships.

On a broader scale, with the advent of computerized databases that make it possible to map fossil mammal occurrences from just a few hundred years back through millions of years ago (Carrasco et al., 2007; MIOMAP, 2009; NEOTOMA, 2009; NOW, 2009; PBDB, 2009), it now is possible to assess the range of normal for species richness for entire biogeographic provinces and for large portions of continents. Comparisons of these regional baselines determined from the fossil record with corresponding current species richness values can reveal which landscapes are least impacted by human activities, and serve as benchmarks by which to assess the success of conservation efforts in the future.

Extinction

One of the most cogent lessons from the fossil record is that the coincidence of unusual events can elevate extinction rates to dangerous levels (Arens and West, 2008; Brook, 2008; Brook et al., 2008; Brook and Barnosky, in prep.). While each of the 'Big Five' mass extinctions (Jablonski and Chaloner, 1994) illustrates that in its own way, perhaps the most relevant example from the standpoint of conservation biology is the end-Pleistocene megafauna extinction, which deleted approximately half of the mammal species alive that were in the >44 kg body size categories. While considerable debate has raged as to the primary cause of those extinctions (Martin and Wright, 1967; Martin and Klein, 1984; MacPhee, 1999), it is now clear that most taxa succumbed only after human populations on a given continent or island reached critical mass (MacPhee and Marx, 1997; MacPhee and Fleming, 1999; Alroy, 2001; Brook and Bowman, 2002; Barnosky et al., 2004b; Wroe et al., 2004; Steadman et al., 2005; Koch and Barnosky, 2006; Brook et al., 2007; Barnosky, 2008; Brook and Barnosky, in prep.). It is also clear that end-Pleistocene global warming contributed to extinction in some places where animals couldn't migrate in response to climate change either because of natural barriers or because of intervening human populations or land alteration (Barnosky, 1986; Barnosky et al., 2004b; Koch and Barnosky, 2006). Finally, extinctions were most severe in areas where first human entry into the ecosystem coincided with end-Pleistocene warming (Barnosky et al., 2004b; Koch and Barnosky, 2006; Nogues-Bravo et al., 2008; Barnosky, 2009; Brook and Barnosky, in prep.). The general picture that emerges is that the synergistic effects of abnormally fast human population growth and abnormally fast global climate change were particularly pernicious in the past, and direct human impacts often affect animals in the highest size and trophic categories (e.g., the big animals), in contrast to climate change, where effects are most obvious in the lower size and trophic categories.

There are clear parallels with what is happening today and projections for the next few decades, as the human population rises from an already all-time high of near 7 billion to an inevitable 9 billion or so, and as global temperatures rise at unprecedented rates to take us into a global climate hotter than humans have ever seen, all by the year 2050 (IPCC, 2007). On that landscape, direct human effects will be impacting large animals dramatically and climate change will be impacting small animals and plants; ecosystems will be squeezed from both the top down and the bottom up. Non-human species are certain to have their geographic ranges fragmented even more than they already are, and will see severe diminishment of opportunities to track their needed climate envelope as Earth's climate zones shift around the landscape. Inevitably, many species can be expected to perish under this scenario without stepped-up conservation efforts that recognize both what the past has to teach us about what is "normal," and also that new strategies must be implemented to cope with the new challenges (Barnosky, 2009).

SUGGESTIONS FOR THE FUTURE

Key lessons from the vertebrate fossil record are that population genetic structure, rank-order generic abundance, and species richness tend to remain remarkably stable in ecosystems through thousands of years, with relatively minor adjustments at lower size and trophic levels triggered by climate changes. Most of the adjustments—for example, switches in which of the top-ranked genera are most abundant, the replacement of species by congeners, or loss or gain of a few small herbivore species—ultimately result from shrinking or expanding populations that build up to major alterations in geographic ranges for species as they track their habitats across a dynamically changing landscape (Blois and Hadly, 2009). We also now know that at least the mammalian component of communities is structured differently than it was for millions of years prior to anthropogenic dominance, in that the largest mammals—essentially the entire right tail of the characteristic body size distribution for mammal communities—have been lost from most ecosystems. The only remaining exceptions are in some nature preserves in Africa, which still have nearly the full complement of large-bodied species (Lyons et al., 2004).

The good news for conservation biology is that some 46% of Earth's terrestrial landscape can be classified as reasonably ecologically intact, if "ecologically intact" is defined as patches of land that are at least 10,000 km^2, inhabited by <5 people/km^2, and containing at least 70% native vegetation (Mittermeier et al., 2003). Species in some of those ecologically intact places (the 12% of Earth's surface that comprises nature reserves, as noted above) are legislatively protected. A major challenge for conservation biology in the 21st century, however, is how to keep ecosystems in such places functioning as they have for millennia, given the pressures they are facing. Long recognized threats include habitat fragmentation, invasive species, and resource consumption by ever-growing human populations, and added to these is the newest threat of abnormally rapid and severe global warming (Barnosky, 2009).

All of these pressures combine to prevent the very mechanism that has allowed ecosystems to equilibrate to environmental perturbations in the past: the expansion and contraction of populations across the landscape to rapidly adjust species ranges such that the particular ecological niches that define a given kind of ecosystem can be filled in a given locale. Species have increasingly fewer avenues open to them to relocate to novel protected geographic locations within their preferred environment. Especially problematic in this context is the rapidity of global warming and the fact that the Earth is entering a new climate state in respect to the species that now exist on it. It is unknown whether many species even have the inherent capacity to keep pace with the climate changes now in progress, even if needed dispersal routes were available.

In view of that, it will not be enough to simply try to preserve species where they now exist; for many species, it will also become necessary to predict where they will be able to exist as changing climate deletes their current habitats in the places where they now are restricted. In that context, vertebrate paleontology has been (Graham, 1984, 1997; Graham and Mead, 1987; Graham and Grimm, 1990; Wood and Barnosky, 1994; Blois, 2009; Blois and Hadly, 2009; Hadly et al., in press) and will continue to be important in developing and ground-truthing ecological niche models.

Critical to conservation will be providing mechanisms by which species can actually get to the places they need to be. On landscapes increasingly fragmented by human use, those mechanisms very likely will include "assisted migration," a conservation approach already employed for butterfly species (Zimmer, 2009), and under active discussion by the conservation community (McLachlan and Hellmann, 2007). In essence, assisted migration is an approach where people decide which species need to be moved and where they need to go, and put them in a truck or car to transport them.

The obvious problem with assisted migration is deciding which species can be inserted into a new place without disrupting the ecological structure and function that already exist there. In that context, the lessons from the vertebrate fossil record become extremely important. They show, for example, that in the face of environmental perturbations, ecosystems tend to maintain species richness by substituting congeners, particularly at lower trophic levels. Thus, as monitoring identifies certain small mammal species that may be foundering as climate change degrades their habitat, it may well be ecologically sound to introduce congeners that have demonstrated climatic suitability, but that could not disperse there naturally. On the other hand, it probably would not be ecologically sound to introduce species from size or trophic groups, or taxonomic groups, which had not been present in the ecosystem

for thousands of years, if the goal was to maintain ecosystem structure and function within the range of variation exhibited over millennia.

The vertebrate fossil record also has an important statement to make about conservation of species in Africa. Africa is the only place left on Earth that still has ecosystems that are operating much as they have for hundreds of thousands of years, in that they still have almost the full complement of animals in the largest size categories. Those ecosystems somehow escaped the ecological restructuring that humans precipitated everywhere else by contributing to extinctions of megafauna (Lyons et al., 2004; Barnosky, 2009). In that light, preservation of megafauna in Africa is doubly important, in that its loss would not only mean loss of the species themselves, but also loss of the only remaining ecosystem that gives us a hint of how Earth functioned with a fully-stocked complement of mammalian species.

Solving the problem of how to preserve those large-bodied species in Africa is particularly complex, given human needs, cultural traditions, and climate trends on the continent. One innovative suggestion originating from a group of vertebrate paleontologists and conservation biologists is a new strategy called "Pleistocene Re-wilding," which seeks to restore non-African ecosystems to their Pleistocene condition of including many species in the largest body-size categories (Donlan, 2005; Martin, 2005; Donlan et al., 2006; Caro, 2007). In brief, "Pleistocene Re-wilding" would use African species as ecological analogs to repopulate ecosystems in North America with large-bodied animals. Opponents have pointed out many serious political and ecological problems that would result (Chapron, 2005; Dinerstein and Irvin, 2005; Schlaepfer, 2005; Shay, 2005; Smith, 2005; Rubenstein et al., 2006). Not least among those is introducing new genera of animals into ecosystems that had never seen them, or even seen their ecological analogs in more than 10,000 years. There are also the theoretical difficulties of trying to reconstruct what is essentially a cool-climate ecosystem in an interglacial time that is becoming abnormally hot, and arbitrarily choosing the pre-Holocene ecosystem as the definition for "normal." Nevertheless, if "Pleistocene Re-wilding" were restricted to large game-parks set aside outside of Africa with the explicit recognition that the goal was simply to preserve endangered species in an ecological setting that would allow individuals to maintain viable population reservoirs, it could accomplish a valuable conservation goal. It would be important to acknowledge, however, that such places were in effect large zoos, not ecosystems that were functioning within the bounds of variation that had existed in those areas through the last several millennia.

CONCLUSIONS

Vertebrate paleontology has matured to the extent that it has much to offer conservation biology. Foremost is that the vertebrate fossil record provides a practical and theoretically sound way to define 'natural' for the purposes of establishing conservation targets: essentially, as maintaining the ecological structure and function of a given area within the range of variation that existed before humans dominated the landscape. Since we live in an interglacial, a practical guideline would be to maintain ecosystems within the range of variation they exhibit during interglacial times. Since we live in a rapidly warming world, another guideline would be to facilitate the adaptive response that species showed to previous times of rapid warming, exemplified by the transition from glacial to interglacial times. In practice, this will mean facilitating the dispersal of the right kinds of species into suitable refugia.

Vertebrate paleontology also offers practical guidelines for monitoring ecosystems to assess how closely they approximate the natural condition as defined above. At the population level, genetic structuring and diversity can now be sampled locally and across space, and then modern conditions compared with the temporal changes to assess whether they fall within or outside the range of variation exhibited in the past. Analyses of ancient DNA have demonstrated that interpretations of modern genetic diversity must be made in the context of life-history strategy of the species in order to be meaningfully applied towards conservation goals. At the species level, vertebrate paleontology has shown that many areas still have the complement of species that can be considered normal for an interglacial climate, and where they don't, which kinds of species are missing. It also becomes clear through the vertebrate fossil record that the rank-order abundance of genera remains fairly stable through time, and that most changes in rank-order involve primarily just two or three of the top-ranked genera whose abundance changes in step with local environmental changes. Species richness, both overall and within size and trophic category, is likewise reasonably stable through

long time periods, although species composition may change. Changes in species composition usually take place simply by replacing one species within a genus with another species of the same genus, with the replacing species being more adapted to local microhabitats. Finally, the vertebrate fossil record has shown that species richness through most of the present interglacial is depauperate with respect to glacial times, previous interglacials, and pre-Pleistocene times. In large part this is because the synergy between human population expansion and end-Pleistocene climate changes resulted in widespread megafaunal extinctions, but for small mammals, may also reflect an overall pattern of decreased diversity during warmer times, at least in the conterminous USA.

These benchmarks derived from the vertebrate fossil record provide practical ways to recognize danger signs in modern ecosystems, that is, signs that the ecosystem is changing more than it has in thousands of years. Among these danger signals are: declines in genetic diversity (taking into account life history strategy of the species of interest and the diversity baseline established through ancient DNA analyses of similar taxa); declines in the population size of species in top-ranked genera, especially when a formerly top-ranked species falls into rarer categories; loss of species without replacement by a congener; decline of species richness below normal values demonstrated by the fossil record for that ecosystem; and, especially for Africa, loss of any large mammals. Because the synergy between a fast-changing climate and increasing human populations that caused megafauna extinction in the past is ramping up again today, it is particularly critical to anticipate the elevated extinction pressures before it is too late to do anything about them.

Finally, it is clear that in the next few decades, conservation efforts will have to employ some new strategies to conserve biodiversity in an increasingly human-dominated world. In this too, information from the vertebrate fossil record will be essential to predict, among other things, the ecological effects of attempting to save species and ecosystem function and structure by assisting the migration of species. For example, past ecological adjustments in the absence of humans suggest species richness and ecosystem function might effectively be maintained by replacing a species that may be dwindling in a given ecosystem with a congener that likely would have been able to get there in the absence of human-induced habitat fragmentation.

However, there is no paleontological justification for introducing a supposed ecological analog that is taxonomically distant.

As we move into a world where extinction pressures intensify and correspondingly step up efforts to save individual species, there is a hidden danger. Increasingly, what it will take to save individual species will be the opposite of what it will take to save ecosystems that maintain some semblance of the interactions of species that operate and adjust in the absence of human domination. For example, strategies such as assisted migration and Pleistocene Re-wilding, important as they may be in keeping species alive, ultimately result in ecosystems that are manipulated by humans, which means the loss of places where ecosystem processes play out without a heavy human hand. Thus, to attain a full range of nature preservation, it will also be necessary to set aside preserves where species manipulation is off-limits, essentially Earth's control plots where we simply observe how ecosystems cope in this new world. In this vein, we suggest that vertebrate fossil record will be particularly crucial, as it will be a key yardstick by which we can ascertain how much terrestrial species, communities, and ecosystems of the future are changing, and what that means for planetary health.

ACKNOWLEDGMENTS

We thank Jessica Blois, Robert Feranec, Greg Dietl, and an anonymous reviewer for commenting on the manuscript. Thanks to Greg Dietl and Karl Flessa for inviting us to the Paleontological Society Short Course on Conservation Paleobiology. Funding that contributed to these ideas and research was provided by grants from the National Science Foundation programs in Sedimentary Geology and Paleobiology and the Ecology Cluster of Division of Environmental Biology.

REFERENCES

ACKERLY, D. D. 2003. Community assembly, niche conservatism, and adaptive evolution in changing environments. International Journal of Plant Sciences, 164(3):S165-S184.

ALLENTOFT, M. E., S. C. SCHUSTER, R. N. HOLDAWAY, M. L. HALE, E. MCLAY, C. OSKAM, M. T. P. GILBERT, P. SPENCER, E. WILLERSLEV, AND M. BUNCE. 2009. Identification of microsatellites from an extinct moa species using high-throughput (454) sequence data. Biotechniques, 46(3):195-200.

ALROY, J. 1996. Constant extinction, constrained diversification, and uncoordinated stasis in North American mammals: New perspectives on faunal stability in the fossil record. Palaeogeography, Palaeoclimatology, Palaeoecology, 127(1-4):285-311.

ALROY, J. 2000. New methods for quantifying macroevolutionary patterns and processes. Paleobiology, 26(4):707-733.

ALROY, J. 2001. A multispecies overkill simulation of the end-Pleistocene megafaunal mass extinction. Science, 292:1893-1896.

ALTER, S. E., E. RYNES, AND S. R. PALUMBI. 2007. DNA evidence for historic population size and past ecosystem impacts of gray whales. Proceedings of the National Academy of Sciences of the United States of America, 104(38):15162-15167.

ANDERSON, C. N. K., U. RAMAKRISHNAN, Y. L. CHAN, AND E. A. HADLY. 2005. Serial SimCoal: A population genetics model for data from multiple populations and points in time. Bioinformatics, 21(8):1733-1734.

ARENS, N. C., AND I. D. WEST. 2008. Press-pulse: A general theory of mass extinction? Paleobiology, 34(4):456-471.

ASARA, J. M., M. H. SCHWEITZER, L. M. FREIMARK, M. PHILLIPS, AND L. C. CANTLEY. 2007. Protein sequences from mastodon and *Tyrannosaurus rex* revealed by mass spectrometry. Science, 316(5822):280-285.

ATKINSON, Q. D., R. D. GRAY, AND A. J. DRUMMOND. 2008. mtDNA variation predicts population size in humans and reveals a major southern Asian chapter in human prehistory. Molecular Biology and Evolution, 25(2):468.

AVISE, J. C., AND D. WALKER. 1998. Pleistocene phylogeographic effects on avian populations and the speciation process. Proceedings of the Royal Society of London, B, 265:457-463.

AVISE, J. C., D. WALKER, AND G. C. JOHNS. 1998. Speciation durations and Pleistocene effects on vertebrate phylogeography. Proceedings of the Royal Society of London B, 265:1707–1712.

BARNES, I., P. MATHEUS, B. SHAPIRO, D. JENSEN, AND A. COOPER. 2002. Dynamics of Pleistocene population extinctions in Beringian brown bears. Science, 295:2267-2270.

BARNES, I., B. SHAPIRO, A. LISTER, T. KUZNETSOVA, A. SHER, D. GUTHRIE, AND M. G. THOMAS. 2007. Genetic structure and extinction of the woolly mammoth, *Mammuthus primigenius*. Current Biology, 17(12):1072-1075.

BARNES, S. S., E. MATISOO-SMITH, AND T. L. HUNT. 2006. Ancient DNA of the Pacific rat (*Rattus exulans*) from Rapa Nui (Easter Island). Journal of Archaeological Science, 33(11):1536-1540.

BARNOSKY, A. D. 1986. "Big game" extinction caused by late Pleistocene climatic change: Irish Elk (*Megaloceros giganteus*) in Ireland. Quaternary Research, 25(1):128-135.

BARNOSKY, A. D. 2004a. Climate change, biodiversity, and ecosystem health: The past as a key to the future, p. 3-5. *In* A. D. Barnosky (ed.), Biodiversity Response to Environmental Change in the Early and Middle Pleistocene: The Porcupine Cave Fauna from Colorado. University of California Press, Berkeley, California.

BARNOSKY, A. D. 2004b. Faunal dynamics of small mammals through the Pit Sequence, p. 318-326. *In* A. D. Barnosky (ed.), Biodiversity Response to Climate Change in the Early and Middle Pleistocene: The Porcupine Cave Fauna from Colorado. University of California Press, Berkeley, California.

BARNOSKY, A. D. 2008. Megafauna biomass tradeoff as a driver of Quaternary and future extinctions. Proceedings of the National Academy of Sciences of the United States of America, 105:11543-11548.

BARNOSKY, A. D. 2009. Heatstroke: Nature in an Age of Global Warming. Island Press, Washington, DC, 269 p.

BARNOSKY, A. D., AND A. B. SHABEL. 2005. Comparison of species richness and ecological structure in historic and middle Pleistocene Colorado mountain mammal communities. Proceedings of the California Academy of Sciences, Series 4, 56(Suppl. 1):50-61.

BARNOSKY, A. D., C. J. BELL, S. D. EMSLIE, H. T. GOODWIN, J. I. MEAD, C. A. REPENNING, E. SCOTT, AND A. B. SHABEL. 2004a. Exceptional record of mid-Pleistocene vertebrates helps differentiate climatic from anthropogenic ecosystem perturbations. Proceedings of the National Academy of Sciences of the United States of America, 101:9297-9302.

BARNOSKY, A. D., E. A. HADLY, AND C. J. BELL. 2003. Mammalian response to global warming on varied temporal scales. Journal of Mammalogy, 84(2):354-368.

BARNOSKY, A. D., E. A. HADLY, B. A. MAURER, AND M. I. CHRISTIE. 2001. Temperate terrestrial vertebrate faunas in North and South America: Interplay of ecology, evolution, and geography with biodiversity. Conservation Biology, 15(3):658-674.

BARNOSKY, A. D., P. L. KOCH, R. S. FERANEC, S. L. WING, AND A. B. SHABEL. 2004b. Assessing the causes of late Pleistocene extinctions on the continents. Science, 306:70-75.

BEHRENSMEYER, A. K., C. T. STAYTON, AND R. E. CHAPMAN. 2003. Taphonomy and ecology of modern avifaunal remains from Amboseli Park, Kenya. Paleobiology, 29(1):52-70.

BELLE, E. M. S., P. A. LANDRY, AND G. BARBUJANI. 2006. Origins and evolution of the Europeans' genome: Evidence from multiple microsatellite loci. Proceedings of the Royal Society, B, 273(1594):1595-1602.

BLOIS, J. L. 2009. Ecological Responses to Paleoclimatic Change: Insights from Mammalian Populations, Species, and Communities. Unpublished Ph.D. disseration, Stanford University.

BLOIS, J. L., AND E. A. HADLY. 2009. Mammalian response to Cenozoic climatic change. Annual Review of Earth and Planetary Sciences, 37:181-208.

BROOK, B. W. 2008. Synergies between climate change, extinctions and invasive vertebrates. Wildlife Research, 35:doi: 10.1071/wr07116.

BROOK, B. W., AND D. M. J. S. BOWMAN. 2002. Explaining the Pleistocene megafaunal extinctions: Models, chronologies, and assumptions. Proceedings of the National Academy of Sciences of the United States of America, 99(23):14624-14627.

BROOK, B. W., D. M. J. S. BOWMAN, D. A. BURNEY, T. F. FLANNERY, M. K. GAGAN, R. GILLESPIE, C. N. JOHNSON, A. P. KERSHAW, J. W. MAGEE, P. S. MARTIN, G. H. MILLER, B. PEISER, AND R. G. ROBERTS. 2007. Would the Australian megafauna have become extinct if humans had never colonised the continent? Quaternary Science Reviews, 26:560-564.

BROOK, B. W., N. S. SODHI, AND C. J. A. BRADSHAW. 2008. Synergies among extinction drivers under global change. Trends in Ecology and Evolution, 23:453-460.

BUNCE, M., M. SZULKIN, H. R. L. LERNER, I. BARNES, B. SHAPIRO, A. COOPER, AND R. N. HOLDAWAY. 2005. Ancient DNA provides new insights into the evolutionary history of New Zealand's extinct giant eagle. PLoS Biology, 3(1):44-46.

CALVIGNAC, S., S. HUGHES, C. TOUGARD, J. MICHAUX, M. THEVENOT, M. PHILIPPE, W. HAMDINE, AND C. HANNI. 2008. Ancient DNA evidence for the loss of a highly divergent brown bear clade during historical times. Molecular Ecology, 17(8):1962-1970.

CARDINALE, B. J., M. A. PALMER, AND S. L. COLLINS. 2002. Species diversity enhances ecosystem functioning through interspecific facilitation. Nature(6870):426-428.

CARO, T. 2007. The Pleistocene re-wilding gambit. Trends in Ecology and Evolution, 22(6):281-283.

CARRASCO, M. A., B. P. KRAATZ, E. B. DAVIS, AND A. D. BARNOSKY. 2007. The Miocene mammal mapping project (MIOMAP): An online database of Arikareean through Hemphillian fossil mammals. Bulletin of the Carnegie Museum of Natural History, 39:183-188.

CEBALLOS, G., P. R. EHRLICH, J. SOBERÓN, I. SALAZAR, AND J. P. FAY. 2005. Global mammal conservation: What must we manage? Science, 309:603-607.

CHAN, Y. L., C. N. K. ANDERSON, AND E. A. HADLY. 2006. Bayesian estimation of the timing and severity of a population bottleneck from ancient DNA. PloS Genetics, 2(4):451-460.

CHAN, Y. L., E. A. LACEY, O. P. PEARSON, AND E. A. HADLY. 2005. Ancient DNA reveals Holocene loss of genetic diversity in a South American rodent. Biology Letters,

1(4):423-426.

CHAPRON, G. 2005. Re-wilding: Other projects help carnivores stay wild. Nature, 437(7057):318-318.

DALEN, L., V. NYSTROM, C. VALDIOSERA, M. GERMONPRE, M. SABLIN, E. TURNER, A. ANGERBJORN, J. L. ARSUAGA, AND A. GOTHERSTROM. 2007. Ancient DNA reveals lack of postglacial habitat tracking in the arctic fox. Proceedings of the National Academy of Sciences of the United States of America, 104(16):6726-6729.

DAVIDSON, A. D., M. J. HAMILTON, A. G. BOYER, J. H. BROWN, AND G. CEBALLOSA. 2009. Multiple ecological pathways to extinction in mammals. Proceedings of the National Academy of Science of the United States of America, 106(26):10702–10705.

DAVIS, E. B. 2005. Comparison of climate space and phylogeny of *Marmota* (Mammalia: Rodentia) indicates a connection between evolutionary history and climate preference. Proceedings of the Royal Society, B, 272:519-526.

DIMICHELE, W. A., A. K. BEHRENSMEYER, T. D. OLSZEWSKI, C. C. LABANDEIRA, J. M. PANDOLFI, S. L. WING, AND R. BOBE. 2004. Long-term stasis in ecological assemblages: Evidence from the fossil record. Annual Review of Ecology, Evolution, and Systematics, 35:285-322.

DINERSTEIN, E., AND W. R. IRVIN. 2005. Re-wilding: No need for exotics as natives return. Nature, 437(7058):476-476.

DONLAN, C. J. 2005. Re-wilding North America. Nature, 436(7053):913-914.

DONLAN, C. J., J. BERGER, C. E. BOCK, J. H. BOCK, D. A. BURNEY, J. A. ESTES, D. FOREMAN, P. S. MARTIN, G. W. ROEMER, F. A. SMITH, M. E. SOULÉ, AND H. W. GREENE. 2006. Pleistocene rewilding: An optimistic agenda for twenty-first century conservation. American Naturalist, 168(5):660-681.

DRUMMOND, A. J., A. RAMBAUT, B. SHAPIRO, AND O. G. PYBUS. 2005. Bayesian coalescent inference of past population dynamics from molecular sequences. Molecular Biology and Evolution, 22:1185-1192.

ELITH, J., C. H. GRAHAM, R. P. ANDERSON, M. DUDIK, S. FERRIER, A. GUISAN, R. J. HIJMANS, F. HUETTMANN, J. R. LEATHWICK, A. LEHMANN, J. LI, L. G. LOHMANN, B. A. LOISELLE, G. MANION, C. MORITZ, M. NAKAMURA, Y. NAKAZAWA, J. M. OVERTON, A. T. PETERSON, S. J. PHILLIPS, K. RICHARDSON, R. SCACHETTI-PEREIRA, R. E. SCHAPIRE, J. SOBERON, S. WILLIAMS, M. S. WISZ, AND N. E. ZIMMERMANN. 2006. Novel methods improve prediction of species' distributions from occurrence data. Ecography, 29(2):129-151.

FAUNMAP WORKING GROUP. 1996. Spatial response of mammals to the late Quaternary environmental fluctuations. Science, 272(5268):1601-1606.

FLESSA, K. W., AND S. T. JACKSON. 2005. Forging a common agenda for ecology and paleoecology. Bioscience, 55(12):1030-1031.

GILPIN, M. E., AND M. E. SOULÉ. 1986. Minimum viable populations: The processes of species extinctions, p. 13-34. *In* M. E. Soulé (ed.), Conservation Biology: The Science of Scarcity and Diversity. Sinauer Associates, Sunderland, Massachusetts.

GRAHAM, R. W. 1976. Late Wisconsin mammalian faunas and environmental gradients of the eastern United States. Paleobiology, 2(4):343-350.

GRAHAM, R. W. 1984. Paleoenvironmental implications of the Quaternary distribution of the Eastern chipmunk (*Tamias striatus*) in central Texas. Quaternary Research, 21(1):111-114.

GRAHAM, R. W. 1997. The spatial response of mammals to Quaternary climate changes, p. 153-162. *In* B. Huntley, W. Cramer, A. V. Morgan, H. C. Prentice, and J. R. M. Allen (eds.), Past and Future Rapid Environmental Changes: The Spatial and Evolutionary Responses of Terrestrial Biota. Springer-Verlag, Berlin.

GRAHAM, R. W., AND E. C. GRIMM. 1990. Effects of global climate change on the patterns of terrestrial biological communities. Trends in Ecology and Evolution, 5(9):289-292.

GRAHAM, R. W., AND J. I. MEAD. 1987. Environmental fluctuations and evolution of mammalian faunas during the last deglaciation in North America, p. 371-402. *In* W. F. Ruddiman, and H. E. Wright (eds.), The Geology of North America, Volume K-3, North America and Adjacent Oceans During the Last Deglaciation. Geological Society of America, Boulder, Colorado.

GRAHAM, R. W., E. L. LUNDELIUS, M. A. GRAHAM, E. K. SCHROEDER, R. S. TOOMEY, E. ANDERSON, A. D. BARNOSKY, J. A. BURNS, C. S. CHURCHER, D. K. GRAYSON, R. D. GUTHRIE, C. R. HARINGTON, G. T. JEFFERSON, L. D. MARTIN, H. G. MCDONALD, R. E. MORLAN, H. A. SEMKEN, S. D. WEBB, L. WERDELIN, AND M. C. WILSON. 1996. Spatial response of mammals to late Quaternary environmental fluctuations. Science, 272(5268):1601-1606.

GRAYSON, D. K. 1998. Moisture history and small mammal community richness during the latest Pleistocene and Holocene, northern Bonneville Basin, Utah. Quaternary Research, 49:330-334.

GRAYSON, D. K. 2005. A brief history of Great Basin pikas. Journal of Biogeography, 32(12):2103-2111.

GRAYSON, D. K. 2006. The Late Quaternary biogeographic histories of some Great Basin mammals (western USA). Quaternary Science Reviews, 25(21-22):2964-2991.

GRAYSON, D. K., AND D. B. MADSEN. 2000. Biogeographic implications of recent low-elevation recolonization by *Neotoma cinerea* in the Great Basin. Journal of Mammalogy, 81(4):1100-1105.

GUGERLI, F., L. PARDUCCI, AND R. J. PETIT. 2005. Ancient plant DNA: Review and prospects. New Phytologist, 166(2):409-418.

GUILDAY, J. E. 1971. The Pleistocene history of the Appalachian mammal faunas, p. 233-262. *In* P. C. Holt, R. A. Paterson, and J. P. Hubbard (eds.), The Distributional History of the Biota of the Southern Appalachians. Part III: The Vertebrates. Research Division Monograph 4. Virginia Polytechnic Institute and State University, Blacksburg, Virginia.

GUILDAY, J. E. 1984. Pleistocene extinction and environmental change: Case study of the Appalachians, p. 250-258. *In* P. S. Martin and R. G. Klein (eds.), Quaternary Extinctions: A Prehistoric Revolution. University of Arizona Press, Tucson, Arizona.

HADLY, E. A. 1996. Influence of late-Holocene climate on northern Rocky Mountain mammals. Quaternary Research, 46(3):298-310.

HADLY, E. A. 1999. Fidelity of terrestrial vertebrate fossils to a modern ecosystem. Palaeogeography, Palaeoclimatology, Palaeoecology, 149(1-4):389-409.

HADLY, E. A. 2003. The interface of paleontology and mammalogy: Past, present, and future. Journal of Mammalogy, 84(2):347-353.

HADLY, E. A., AND B. A. MAURER. 2001. Spatial and temporal patterns of species diversity in montane mammal communities of western North America. Evolutionary Ecology Research, 3(4):477-486.

HADLY, E. A., M. H. KOHN, J. A. LEONARD, AND R. K. WAYNE. 1998. A genetic record of population isolation in pocket gophers during Holocene climatic change. Proceedings of the National Academy of Sciences of the United States of America, 95(12):6893-6896.

HADLY, E. A., U. RAMAKRISHNAN, Y. L. CHAN, M. VAN TUINEN, K. O'KEEFE, P. A. SPAETH, AND C. J. CONROY. 2004. Genetic response to climatic change: Insights from ancient DNA and phylochronology. PLoS Biology, 2(10):1600-1609.

HADLY, E. A., P. A. SPAETH, AND C. L. LI. In press. Niche conservatism above the species level. Proceedings of the National Academy of Sciences of the United States of America.

HANDT, O., M. HOSS, M. KRINGS, AND S. PAABO. 1994. Ancient DNA: Methodological challenges. Experientia, 50(6):524-529.

HILL, M. G., JR., M. G. HILL, AND C. C. WIDGA. 2008. Late Quaternary Bison diminution on the Great Plains of North America: Evaluating the role of human hunting versus climate change. Quaternary Science Reviews, 27:1752–1771.

HIJMANS, R. J., AND C. H. GRAHAM. 2006. The ability of climate envelope models to predict the effect of climate change on species distributions. Global Change Biology, 12:2272-2281.

HOELZEL, A. R., J. HALLEY, S. J. O'BRIEN, C. CAMPAGNA, T. ARNBOM, B. LEBOEUF, K. RALLS, AND G. A. DOVER. 1993. Elephant seal genetic-variation and the use of simulation-models to investigate historical population bottlenecks. Journal of Heredity, 84: 443-449.

HOFREITER, M. 2008a. Mammoth genomics. Nature, 456(7220):330-331.

HOFREITER, M. 2008b. Long DNA sequences and large data sets: Investigating the Quaternary via ancient DNA. Quaternary Science Reviews, 27:2586-2592.

HOFREITER, M., C. CAPELLI, M. KRINGS, L. WAITS, N. CONARD, S. MUNZEL, G. RABEDER, D. NAGEL, M. PAUNOVIC, G. JAMBRESIC, S. MEYER, G. WEISS, AND S. PAABO. 2002. Ancient DNA analyses reveal high mitochondrial DNA sequence diversity and parallel morphological evolution of late pleistocene cave bears. Molecular Biology and Evolution, 19(8):1244-1250.

HOFREITER, M., S. MUENZEL, N. J. CONARD, J. POLLACK, M. SLATKIN, G. WEISS, AND S. PAABO. 2007. Sudden replacement of cave bear mitochondrial DNA in the late Pleistocene. Current Biology, 17(4):R122-R123.

HOFREITER, M., D. SERRE, H. N. POINAR, M. KUCH, AND S. PAABO. 2001. Ancient DNA. Nature Reviews, Genetics, 2:353-359.

HOFREITER, M., D. SERRE, N. ROHLAND, G. RABEDER, D. NAGEL, N. CONARD, S. MUNZEL, AND S. PAABO. 2004. Lack of phylogeography in European mammals before the last glaciation. Proceedings of the National Academy of Sciences of the United States of America, 101(35):12963-12968.

HUGHES, T. P., A. H. BAIRD, D. R. BELLWOOD, M. CARD, S. R. CONNOLLY, C. FOLKE, R. GROSBERG, O. HOEGH-GULDBERG, J. B. C. JACKSON, J. KLEYPAS, J. M. LOUGH, P. MARSHALL, M. NYSTROM, S. R. PALUMBI, J. M. PANDOLFI, B. ROSEN, AND J. ROUGHGARDEN. 2003. Climate change, human impacts, and the resilience of coral reefs. Science, 301(5635):929-933.

HUYNEN, L., I. LISSONE, S. SAWYER, AND D. LAMBERT. 2008. Genetic identification of moa remains recovered from Tiniroto, Gisborne. Journal of the Royal Society of New Zealand, 38(4):231-235.

HUYNEN, L., C. D. MILLAR, R. P. SCOFIELD, AND D. M. LAMBERT. 2003. Nuclear DNA sequences detect species limits in ancient moa. Nature, 425(6954):175-178.

IPCC. 2007. Climate Change 2007. Cambridge University Press, Geneva, 210 p.

IUCN. 2009. IUCN Red List of Endangered Species, 2009.1. Available from: http://www.iucnredlist.org.

JABLONSKI, D., AND W. G. CHALONER. 1994. Extinctions in the fossil record. Philosophical Transactions of the Royal Society, B, 344:11-17.

JACKSON, J. B. C., AND D. ERWIN. 2006. What can we learn about ecology and evolution from the fossil record? Trends in Ecology and Evolution, 21:322-328.

JOHNS, G. C., AND J. C. AVISE. 1998. A comparative summary of genetic distances in the vertebrates from the mitochondrial cytochrome *b* gene. Molecular Biology and Evolution, 15(11):1481-1490.

KOCH, P. L., AND A. D. BARNOSKY. 2006. Late Quaternary extinctions: State of the debate. Annual Review of Ecology, Evolution, and Systematics, 37:215-250.

LALUEZA-FOX, C., J. KRAUSE, D. CARAMELLI, G. CATALANO, L. MILANI, M. L. SAMPIETRO, F. CALAFELL, C. MARTI-NEZ-MAZA, M. BASTIR, A. GARCIA-TABERNERO, M. DE LA RASILLA, J. FORTEA, S. PAABO, J. BERTRANPETIT, AND A. ROSAS. 2006. Mitochondrial DNA of an Iberian neandertal suggests a population affinity with other European neandertals. Current Biology, 16(16):R629-R630.

LALUEZA-FOX, C., M. L. SAMPIETRO, D. CARAMELLI, Y. PUDER, M. LARI, F. CALAFELL, C. MARTINEZ-MAZA, M. BASTIR, J. FORTEA, M. DE LA RASILLA, J. BERTRANPETIT, AND A. ROSAS. 2005. Neandertal evolutionary genetics: Mitochondrial DNA data from the Iberian Peninsula. Molecular Biology and Evolution, 22(4):1077-1081.

LEONARD, J. A. 2008. Ancient DNA applications for wildlife conservation. Molecular Ecology, 17:4186–4196.

LEONARD, J. A., R. K. WAYNE, J. WHEELER, R. VALADEZ, S. GUILLEN, AND C. VILA. 2002. Ancient DNA evidence for Old World origin of New World dogs. Science, 5598:1613-1616.

LEOPOLD, A. S., S. A. CAIN, C. M. COTTAM, I. N. GABRIELSON, AND T. KIMBALL. 1963. Wildlife Management in the National Parks: The Leopold Report. The National Park Service. Available from: http://www.nps.gov/history/history/online_books/leopold/leopold.htm.

LINDAHL, T. 2000. Quick guide: Fossil DNA. Current Biology, 10(17):R616-R616.

LYMAN, R. L. 1998. White Goats, White Lies: The Misuse of Science in Olympic National Park. University of Utah Press, Salt Lake City, Utah, 278 p.

LYMAN, R. L. 2006. Paleozoology in the service of conservation biology. Evolutionary Anthropology, 15:11-19.

LYMAN, R. L., AND K. P. CANNON. 2004. Zooarchaeology and Conservation Biology. University of Utah Press, Salt Lake City, Utah, 266 p.

LYONS, S. K., F. A. SMITH, AND J. H. BROWN. 2004. Of mice, mastodons, and men: Human mediated extinctions on four continents. Evolutionary Ecology Research, 6:339-358.

MACPHEE, R. D. E. 1999. Extinctions in Near Time: Causes, Contexts, and Consequences. Kluwer Academic/Plenum Publishers, New York, London, 394 p.

MACPHEE, R. D. E., AND C. FLEMING. 1999. Requiem Aeternam: The last five hundred years of mammalian species extinctions, p. 333-371. In R. D. E. MacPhee (ed.), Extinctions in Near Time: Causes, Contexts, and Consequences. Kluwer Academic/Plenum Publishers, New York.

MACPHEE, R. D. E., AND P. A. MARX. 1997. The 40,000 year plague: Humans, hyperdiseases, and first-contact extinctions, p. 169-217. In S. M. Goodman and B. R. Patterson (eds.), Natural Change and Human Impact in Madagascar. Smithsonian Institution Press, Washington, D. C.

MARTIN, P. S. 2005. Twilight of the Mammoths: Ice Age Extinctions and the Rewilding of America. University of California Press, Berkeley, California, 269 p.

MARTIN, P. S., AND R. G. KLEIN. 1984. Quaternary Extinctions: A Prehistoric Revolution. University of Arizona Press, Tucson, Arizona, 892 p.

MARTIN, P. S., AND H. E. WRIGHT, JR. 1967. Pleistocene Extinctions: The Search for a Cause. Volume 6 of the Proceedings of the VII Congress of the International Association for Quaternary Research. Yale University Press, New Haven, Connecticut, 453 p.

MARTINEZ-MEYER, E., A. TOWNSEND PETERSON, AND W. W. HARGROVE. 2004. Ecological niches as stable distributional constraints on mammal species, with implications for Pleistocene extinctions and climate change projections for biodiversity. Global Ecology and Biogeography, 13(4):305-314.

MCGILL, B. J., E. A. HADLY, AND B. A. MAURER. 2005. Community inertia of quaternary small mammal assemblages in North America. Proceedings of the National Academy of Sciences of the United States of America, 102(46):16701-16706.

MCLACHLAN, J. S., AND J. J. HELLMANN. 2007. A framework for the debate of assisted migration in an era of climate change. Conservation Biology, 21:297-302.

MCMENAMIN, S. K., E. A. HADLY, AND C. K. WRIGHT. 2008. Climatic change and wetland desiccation cause amphibian decline in Yellowstone National Park. Proceedings of the National Academy of Sciences of the United States of America, 105(44):16988-16993.

MEINE, C., M. SOULÉ, AND R. F. NOSS. 2006. "A mission-driven discipline": The growth of conservation biology. Conservation Biology, 20(3):631-651.

MENOTTI-RAYMOND, M., AND S. J. O'BRIEN. 1993. Dating the genetic bottleneck of the African cheetah. Proceedings of the National Academy of Sciences of the United States of America, 90:3172-3176.

MILLER, F. J., F. L. ROSENFELDT, C. ZHANG, A. W. LINNANE, AND P. NAGLEY. 2003. Precise determination of mitochondrial DNA copy number in human skeletal and cardiac muscle by a PCR-based assay: Lack of change of copy number with age. Nucleic Acids Research, 31:e61.

MIOMAP. 2009. MIOMAP Database. Available from: http://www.ucmp.berkeley.edu/.

MITTERMEIER, R., C. GOETTSCH-MITTERMEIER, P. R. GIL, G. FONSECA, T. BROOKS, J. PILGRIM, AND W. R. KONSTANT. 2003. Wilderness: Earth's Last Wild Places. University of Chicago Press, Chicago, Illinois, 576 p.

NEOTOMA. 2009. NEOTOMA Paleobiology Database, Plio-Pleistocene to Holocene. Available from: http://www.neotomadb.org/.

NEWSOME, S. D., M. A. ETNIER, D. GIFFORD-GONZALEZ, D. L. PHILLIPS, M. VAN TUINEN, E. A. HADLY, D. P. COSTA, D.

J. KENNETT, T. P. GUILDERSON, AND P. L. KOCH. 2007. The shifting baseline of northern fur seal ecology in the northeast Pacific Ocean. Proceedings of the National Academy of Sciences of the United States of America, 104(23):9709-9714.

NGS. 2006. World Parks. National Geographic, October. Available from: http://ngm.nationalgeographic.com/ngm/0610/feature0611/map.html.

NOGUES-BRAVO, D., J. RODIGUEZ, J. HORTAL, P. BATRA, AND M. B. ARAUJO. 2008. Climate change, humans, and the extinction of the woolly mammoth. PloS Biology, 6(4):685-692.

NOW. 2009. Neogene of the Old World Database. Available from: http://www.helsinki.fi/science/now.

NRC (NATIONAL RESEARCH COUNCIL). 2002. Ecological Dynamics on Yellowstone's Northern Range. The National Academies Press, Washington, D. C., p. 198.

ORLANDO, L., D. BONJEAN, H. BOCHERENS, A. THENOT, A. ARGANT, M. OTTE, AND C. HANNI. 2002. Ancient DNA and the population genetics of cave bears (*Ursus spelaeus*) through space and time. Molecular Biology and Evolution, 19(11):1920-1933.

ORLANDO, L., D. MALE, M. T. ALBERDI, J. L. PRADO, A. PRIETO, A. COOPER, AND C. HANNI. 2008. Ancient DNA clarifies the evolutionary history of American late Pleistocene equids. Journal of Molecular Evolution, 66(5):533-538.

ORLANDO, L., M. MASHKOUR, A. BURKE, C. J. DOUADY, V. EISENMANN, AND C. HANNI. 2006. Geographic distribution of an extinct equid (*Equus hydruntinus*: Mammalia, Equidae) revealed by morphological and genetical analyses of fossils. Molecular Ecology, 15(8):2083-2093.

OWEN, P. R., C. J. BELL, AND E. M. MEAD. 2000. Fossils, diet, and conservation of black-footed ferrets (*Mustela nigripes*). Journal of Mammalogy, 81(2):422-433.

PANDOLFI, J. M. 2002. Coral community dynamics at multiple scales. Coral Reefs, 21(1):13-23.

PANDOLFI, J. M. 2006. Ecology - Corals fail a test of neutrality. Nature, 440(7080):35-36.

PAWLOWSKI, J., I. BOLIVAR, J. F. FAHRNI, C. DEVARGAS, M. GOUY, AND L. ZANINETTI. 1997. Extreme differences in rates of molecular evolution of foraminifera revealed by comparison of ribosomal DNA sequences and the fossil record. Molecular Biology and Evolution, 14(5):498-505.

PBDB. 2009. Paleobiology Database. Available from: http://paleodb.org/cgi-bin/bridge.pl.

PEAKALL, R., D. EBERT, L. J. SCOTT, P. F. MEAGHER, AND C. A. OFFORD. 2003. Comparative genetic study confirms exceptionally low genetic variation in the ancient and endangered relictual conifer, *Wollemia nobilis* (Araucariaceae). Molecular Ecology, 12:2331-2343.

POINAR, H. N., M. HOFREITER, W. G. SPAULDING, P. S. MARTIN, B. A. STANKIEWICZ, H. BLAND, R. P. EVERSHED, G. POSSNERT, AND S. PAABO. 1998. Molecular coproscopy: Dung and diet of the extinct ground sloth *Nothrotheriops shastensis*. Science, 281(5375):402-406.

POINAR, H. N., M. KUCH, G. MCDONALD, P. MARTIN, AND S. PAABO. 2003. Nuclear gene sequences from a late Pleistocene sloth coprolite. Current Biology, 13(13):1150-1152.

PORDER, S., A. PAYTAN, AND E. A. HADLY. 2003. Mapping the origin of faunal assemblages using strontium isotopes. Paleobiology, 29(2):197-204.

RAMAKRISHNAN, U., AND E. A. HADLY. In press. Using phylochronology to reveal cryptic population histories: Review and synthesis of four ancient DNA studies. Molecular Ecology.

RAMAKRISHNAN, U., E. A. HADLY, AND J. L. MOUNTAIN. 2005. Detecting past population bottlenecks using temporal genetic data. Molecular Ecology, 14(10):2915-2922.

ROMPLER, H., N. ROHLAND, C. LALUEZA-FOX, E. WILLERSLEV, T. KUZNETSOVA, G. RABEDER, J. BERTRANPETIT, T. SCHONEBERG, AND M. HOFREITER. 2006. Nuclear gene indicates coat-color polymorphism in mammoths. Science, 313(5783):62-62.

ROSENZWEIG, M. L. 1995. Species Diversity in Space and Time. Cambridge University Press, New York, 436 p.

RUBENSTEIN, D. R., D. I. RUBENSTEIN, P. W. SHERMAN, AND T. A. GAVIN. 2006. Pleistocene park: Does re-wilding North America represent sound conservation for the 21st century? Biological Conservation, 132(2):232-238.

SCHLAEPFER, M. A. 2005. Re-wilding: A bold plan that needs native megafauna. Nature, 437(7061):951-951.

SEMKEN, H. A., JR., AND R. W. GRAHAM. 1996. Paleoecologic and taphonomic patterns derived from correspondence analysis of zooarcheological and paleontological faunal samples, a case study from the North American prairie/forest ecotone, p. 477-490. *In* A. Nadachowski and L. Werdelin (eds.), Neogene and Quaternary Mammals of the Palaearctic. Acta Zoologica Cracoviensia 39.

SHAPIRO, B., A. J. DRUMMOND, A. RAMBAUT, M. C. WILSON, P. E. MATHEUS, A. V. SHER, O. G. PYBUS, M. T. P. GILBERT, I. BARNES, J. BINLADEN, E. WILLERSLEV, A. J. HANSEN, G. F. BARYSHNIKOV, J. A. BURNS, S. DAVYDOV, J. C. DRIVER, D. G. FROESE, C. R. HARINGTON, G. KEDDIE, P. KOSINTSEV, M. L. KUNZ, L. D. MARTIN, R. O. STEPHENSON, J. STORER, R. TEDFORD, S. ZIMOV, AND A. COOPER. 2004. Rise and fall of the Beringian steppe bison. Science, 306(5701):1561-1565.

SHAY, S. 2005. Re-wilding: Don't overlook humans living on the plains. Nature, 437(7058):476-476.

SMITH, C. I. 2005. Re-wilding: Introductions could reduce biodiversity. Nature, 437(7057):318-318.

SORENSON, M. D., A. COOPER, E. E. PAXINOS, T. W. QUINN, H. F. JAMES, S. L. OLSON, AND R. C. FLEISCHER. 1999. Relationships of the extinct moa nalos, flightless Hawaiian waterfowl, based on ancient DNA. Proceedings of the Royal Society of London, B, 266(1434):2187-2193.

SOULÉ, M. E. 1987. Where do we go from here?, p. 175-183. *In* M. E. Soulé (ed.), Viable Populations for Conservation. Cambridge University Press, Cambridge.

SPAETH, P. A., M. VAN TUINEN, Y. L. CHAN, D. TERCA, AND E. A. HADLY. 2009. Phylogeography of *Microtus longicaudus* in the tectonically and glacially dynamic central Rocky Mountains. Journal of Mammalogy, 90(3):571–584.

SPIELMAN, D., B. W. BROOK, AND R. FRANKHAM. 2004. Most species are not driven to extinction before genetic factors impact them. Proceedings of the National Academy of Sciences of the United States of America, 101:15261-15264.

STEADMAN, D. W., P. S. MARTIN, R. D. E. MACPHEE, A. J. T. JULL, H. G. MCDONALD, C. A. WOODS, M. ITURRALDE-VINENT, AND G. W. L. HODGINS. 2005. Asynchronous extinction of late Quaternary sloths on continents and islands. Proceedings of the National Academy of Sciences of the United States of America, 102(33):11763-11768.

TERRY, R. C. 2007. Inferring predator identity from small-mammal remains. Evolutionary Ecology Research, 9:199-219.

TERRY, R. C. 2008. Modeling the effects of predation, prey cycling, and time averaging on relative abundance in raptor-generated small mammal death assemblages. Palaios, 23(6):402-410.

VALDIOSERA, C. E., J. L. GARCIA-GARITAGOITIA, N. GARCIA, I. DOADRIO, M. G. THOMAS, C. HANNI, J. L. ARSUAGA, I. BARNES, M. HOFREITER, L. ORLANDO, AND A. GOTHERSTORM. 2008. Surprising migration and population size dynamics in ancient Iberian brown bears (*Ursus arctos*). Proceedings of the National Academy of Sciences of the United States of America, 105(13):5123-5128.

VAN TUINEN, M., K. O'KEEFE, U. RAMAKRISHNAN, AND E. A. HADLY. 2008. Fire and ice: Genetic structure of the Uinta ground squirrel (*Spermophilus armatus*) across the Yellowstone hotspot. Molecular Ecology, 17(7):1776-1788.

WAKE, D. B., AND V. T. VREDENBURG. 2008. Are we in the midst of the sixth mass extinction? A view from the world of amphibians. Proceedings of the National

Academy of Sciences of the United States of America, 105:11466-11473.

WEAVER, T. D., AND C. C. ROSEMAN. 2005. Ancient DNA, late Neandertal survival, and modern-human-Neandertal genetic admixture. Current Anthropology, 46(4):677-683.

WEINSTOCK, J., E. WILLERSLEV, A. SHER, W. F. TONG, S. Y. W. HO, D. RUBENSTEIN, J. STORER, J. BURNS, L. MARTIN, C. BRAVI, A. PRIETO, D. FROESE, E. SCOTT, X. L. LAI, AND A. COOPER. 2005. Evolution, systematics, and phylogeography of Pleistocene horses in the New World: A molecular perspective. PLoS Biology, 3(8):1373-1379.

WESTERN, D., AND A. K. BEHRENSMEYER. 2009. Bone assemblages track animal community structure over 40 years in an African savanna ecosystem. Science, 324(5930):1061-1064.

WIENS, J. J., AND C. H. GRAHAM. 2005. Niche conservatism: Integrating evolution, ecology, and conservation biology. Annual Review of Ecology, Evolution, and Systematics, 36:519-539.

WILLERSLEV, E., E. CAPPELLINI, W. BOOMSMA, R. NIELSEN, M. B. HEBSGAARD, T. B. BRAND, M. HOFREITER, M. BUNCE, H. N. POINAR, D. DAHL-JENSEN, S. JOHNSEN, J. P. STEFFENSEN, O. BENNIKE, J.-L. SCHWENNINGER, R. NATHAN, S. ARMITAGE, C.-J. D. HOOG, V. ALFIMOV, M. CHRISTL, J. BEER, R. MUSCHELER, J. BARKER, M. SHARP, K. E. H. PENKMAN, J. HAILE, P. TABERLET, M. THOMAS, P. GILBERT, A. CASOLI, E. CAMPANI, AND M. J. COLLINS. 2007. Ancient biomolecules from deep ice cores reveal a forested southern Greenland. Science, 317:111-114.

WILLERSLEV, E., AND A. COOPER. 2005. Ancient DNA. Proceedings of the Royal Society, B, 272:3-16.

WOOD, D. L., AND A. D. BARNOSKY. 1994. Middle Pleistocene climate change in the Colorado Rocky Mountains indicated by fossil mammals from Porcupine Cave. Quaternary Research, 41(3):366-375.

WROE, S., J. FIELD, R. FULLAGAR, AND L. S. JERMIIN. 2004. Megafaunal extinction in the late Quaternary and the global overkill hypothesis. Alcheringa, 28(1):291-331.

ZIMMER, C. 2009. As climate warms, species may need to migrate or perish. Yale environment360, 20 April 2009. Available from:http://www.e360.yale.edu/content/feature.msp?id=2142.

CHAPTER FOUR

PALEOECOLOGY AND RESOURCE MANAGEMENT IN A DYNAMIC LANDSCAPE: CASE STUDIES FROM THE ROCKY MOUNTAIN HEADWATERS

STEPHEN T. JACKSON[1,4], STEPHEN T. GRAY[2,4], AND BRYAN SHUMAN[3,4]

[1]Department of Botany, University of Wyoming, 1000 E. University Ave., Laramie, WY 82072 USA
[2]Wyoming Water Resources Data System and Wyoming State Climate Office, University of Wyoming, 1000 E. University Ave., Laramie, WY 82072 USA
[3]Department of Geology & Geophysics, University of Wyoming, 1000 E. Univ. Ave., Laramie, WY 82072 USA
[4]Program in Ecology, University of Wyoming, 1000 E. University Ave., Laramie, WY 82072 USA

ABSTRACT.—Ecosystems of the Rocky Mountain region are undergoing rapid change owing to interactions among climate change, widespread fire and pathogen outbreaks, species invasions, and human land-use. Natural-resource managers in the region face great challenges in planning and implementing effective management strategies, challenges that are amplified by the uncertainties in predicting local to regional-scale climate change. Paleoecological and paleoclimatological records can provide valuable perspectives and insights for resource management by showing how local and regional ecosystems have responded to past environmental changes of various types, magnitudes, and rates. Different patterns of change emerge across a nested hierarchy of time scales (respectively, the past 15, 150, 1500, and 15,000 years), and the dynamics at these various scales can provide realistic and concrete scenarios that reveal vulnerabilities and potential responses to future environmental variability and change.

INTRODUCTION

THE ROCKY Mountain region (Fig. 1) poses critical resource-management and conservation challenges in the face of global change. The states of Wyoming, Colorado, and Montana comprise the headwaters for three major U.S. river systems (the Missouri/Mississippi, the Snake/Columbia, and the Green/Colorado) and provide water for millions of people downstream. Often perceived as the "last frontier" of the continental U.S., the region is experiencing urban expansion, suburban encroachment into forests, woodlands, and rangelands, and large-scale fossil-fuel extraction along with ensuing landscape fragmentation. Invasive species are altering ecosystems throughout the region, particularly in the grasslands, steppes and woodlands of the foothills and intermountain basins.

Beginning in the late 1990's, severe drought further increased threats to human enterprises and ecosystems within the region, and jeopardized water resources in downstream areas dependent on runoff from Rocky Mountain watersheds. Although this drought was partly attributable to natural variability in Atlantic and Pacific sea-surface temperatures (McCabe et al., 2007), it occurred against a backdrop of steadily increasing temperatures and increasingly early onset of spring snowmelt and streamflow peaks (Mote et al., 2005; Stewart et al., 2004). The latter trends are consistent with predicted impacts of global climate change (IPCC, 2007), and recent studies suggest that recent shifts in the regional hydrologic cycle are directly attributable to increasing temperatures (Barnett et al., 2008). The dramatic impacts of recent drought and warming can also be seen in rapidly changing terrestrial ecosystems and accelerated disturbance cycles, particularly fire and pest-induced mortality, across the region (e.g., Westerling et al., 2006; Kurz et al., 2008; Worrall et al., 2008; van Mantgem et al., 2009). Droughts of increasing severity—and possibly greater frequency and duration—are predicted for the region in coming decades (Hoerling and Eischeid, 2007; Christensen and Lettenmaier, 2007).

While the scope and rapidity of recent change is startling, these events are superimposed on climate change and regional ecosystem dynamics that have been underway since the last glacial maximum ca. 21,000 years ago. Paleoecological studies show that many Rocky Mountain ecosystems are relatively young, and that all ecosystems in the region have undergone significant change in the past 15,000 years. During the

In *Conservation Paleobiology: Using the Past to Manage for the Future, Paleontological Society Short Course, October 17th, 2009. The Paleontological Society Papers, Volume 15, Gregory P. Dietl and Karl W. Flessa (eds.). Copyright © 2009 The Paleontological Society.*

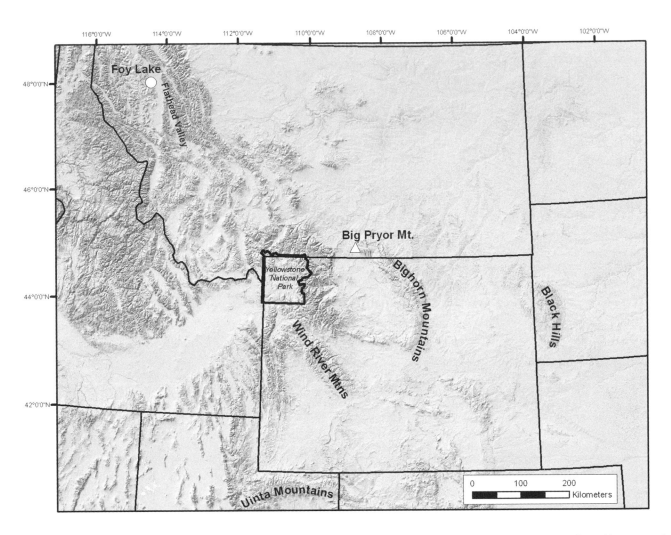

FIGURE *1.*—Northern and Central Rocky Mountain region, showing locations of sites and study areas mentioned in text and figures.

past 1500 years, ecosystems have experienced rates and magnitudes of climate variability far beyond those of the past 150 years' observational and instrumental records. Comparison of the immediate past (15 years) and the instrumental period (150 years) with the recent past (1500 years) and the post-glacial period (15,000 years) provides context and insights that bear directly on conservation and resource management in the region.

In this paper we review sources of paleoclimatic and paleoecological information that are particularly relevant to the Rocky Mountain region, focusing on three—tree-rings, packrat middens, and lake sediments—that together provide complementary information on climate change and ecological response across a wide range of time and space scales. After summarizing insights emerging from these data sources, we

discuss the controls and ecological consequences of climatic change and variability in the region over the past 15, 150, 1500, and 15,000 years. Using this nested time-series, we illustrate the contrasting climatic and ecological dynamics that emerge at each timescale, and discuss how patterns at one timescale can be better understood in the context of longer- and shorter-term dynamics. Finally, we describe a scenario-planning framework by which the long-term perspectives from paleoclimatology and paleoecology can be incorporated into formal decision-making by resource managers and policymakers.

ARCHIVES OF PAST ENVIRONMENTS AND ECOSYSTEMS

The post-glacial history of environment and ecosystems of the Rocky Mountain region is preserved in a wide variety of archives. Land-surface features such as dunefields, glacial moraines, mima mounds, playas, and frost polygons record past climate intervals (Mears, 1987; Stokes and Gaylord, 1993; Forman et al., 2001; Langford, 2003; Mayer and Mahan, 2004). Alluvial deposits provide evidence of past floods and wildfires (Pierce et al., 2002; Meyer et al., 1995; Persico and Meyer, 2009). Bone beds in caves yield information on vertebrate faunal composition (Hadly, 1996; Barnosky, 2004). Archeological sites record past cultural practices and resource utilization by humans (Frison, 1991). Documentary and photo archives reveal recent changes in climate (Mock, 1991) and vegetation (Andersen and Baker, 2006; Zier and Baker, 2006). We describe three characteristic Rocky Mountain archives that illustrate the nature of environmental and ecological change at different temporal and spatial scales: tree-rings, pack-rat middens, and lake sediments. These archives constitute some of the richest sources of management-relevant paleoecological information for the region. We present an example application for each archive, and discuss implications of these records for resource management.

Dendroclimatology and dendroecology

Description.—Trees across the Rocky Mountain region record environmental variations in their annual wood-growth, and the resulting patterns of large and small growth-rings can be used to assign precise dates to wood samples (Fritts, 1976, Cook and Kairiukstis, 1990). Tree-ring dating can in turn be used to reconstruct the history of disturbance events and demographic changes in regional forests and woodlands (Swetnam and Betancourt, 1998; Swetnam et al., 1999). Some tree species form fire scars when cambial tissues are damaged or killed by heat exposure, providing information on the frequency and timing of past surface fires. Insect defoliation, disease events, and individual recruitment dates are also recorded in tree-rings. Tree-rings can also provide precise mortality dates for well-preserved dead trees. When these techniques are applied across stands or regions, tree-rings provide information on the spatial extent and spatial patterning of disturbances, recruitment pulses and mortality events.

Tree-rings also provide annual records of past climate variations. Trees growing in open stands on thin soils or bedrock outcrops are particularly sensitive to precipitation variation, especially near lower tree-line

and on south-facing slopes. Trees can attain great ages (500 to 1000+ years) in such habitats (Fritts, 1976). Dead trees can be preserved for hundreds of years on cold and/or arid sites. Visual and statistical pattern matching allows samples from dead wood to be dated against samples of known age, extending records far beyond the age of living trees. Tree-ring records from upper tree-line may also be used to reconstruct variations in temperature.

In order to estimate past climate, ring-widths from precisely dated wood obtained from dozens of trees at multiple sites in a region are first measured and their growth records corrected for size-related biases (Fritts, 1976). Ring-widths from across each sampling site are then converted to a common scale and combined to form a master chronology. Ring-widths from the region are compared statistically to instrumental records of climate, yielding a statistical model linking ring-width and climate. Application of this model to ring-width data yields an annually resolved paleoclimate record that can span hundreds to thousands of years.

Example.—A recent study by Brown (2006) in the Black Hills region of northeastern Wyoming and adjacent South Dakota shows how tree-ring studies can reveal ecological responses to climate variability over the past 500 years (Fig. 2). The middle panel shows a time-smoothed record of precipitation variability from isolated stands of trees in the surrounding Great Plains region (Stockton and Meko, 1983) and from lower-tree-line stands in the Bighorn Basin to the west (Gray et al., 2004a). These two records show coherent patterns at decadal to centennial scales (Fig. 2). The Bighorn Basin record extends further back in time (to 1260 CE) than the Plains record owing to greater antiquity of living trees and better preservation of ancient dead trees. Key findings from these and similar studies across the region are that the 20th century was relatively wet compared to earlier centuries, and that the region is prone to droughts of greater severity and duration than those of the past 150 years.

The upper panel in Figure 2 portrays fire history in Ponderosa pine forests of the Black Hills. Each horizontal line represents an individual stand. Because a single stand can burn in any given year, fire-scar studies sample multiple stands and then seek synchrony among stands in fire events. This record reveals that widespread fires occurred during droughts (e.g., 1753, 1785, 1822-1823, 1863-1864, 1880). The lower panel in Figure 2 shows recruitment patterns of Ponderosa

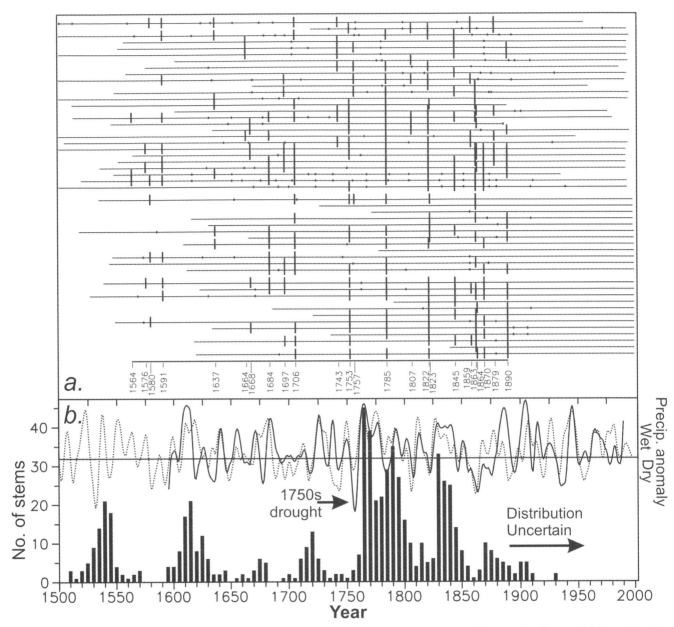

FIGURE 2.—Composite tree-ring records of climate, fire, and recruitment history in Ponderosa pine forest of the Black Hills of South Dakota and Wyoming. (a) Fire-scar chronologies from 51 sites in the region. Horizontal lines represent time-span of each record; long vertical bars denote regional fire years (identified by ≥10% of sites with fire scars from that year). Regional fire years are identified at bottom. (b) Curves represent regional precipitation records from tree-ring chronologies in the Great Plains (smooth) and Bighorn Basin (dotted), smoothed using a 10-year spline to draw out decadal-scale variation. Histograms represent regional recruitment of Ponderosa pine individuals (20th century recruits were not sampled). Reproduced with permission from the Ecological Society of America.

pines in the Black Hills, based on tree-ring sampling of multiple stands (Brown, 2006). The two most recent recruitment events occurred during unusually wet periods (1760s-1770s and 1820s-1830s) that followed closely on drought-and-wildfire periods (1753, 1822-1823) (Fig. 2). A recruitment pulse in the early 1600s also occurred during a wet period, and may have fol-lowed a major fire. Ponderosa pine recruitment has been episodic, concentrated during extended wet periods (Brown, 2006). These recruitment events may have been facilitated by local or regional tree mortality during preceding drought periods. Similar patterns of climate-pulsed recruitment and disturbance are recorded across the region (Swetnam and Betancourt, 1998;

Gray et al., 2006), and may be characteristic of other regions (Jackson et al., in press).

Packrat middens

Description.—Packrats of the genus *Neotoma* build nests and food caches in crevices and caves of rocky uplands throughout the West, using plant parts as construction materials (Betancourt et al., 1990; Betancourt, 2004). Packrat middens eventually become indurated by crystalline packrat urine and can be preserved for thousands to tens of thousands of years. Middens as old as 40,000 yr BP have been found in the central Rockies (Jackson et al., 2005).

Each midden represents a relatively brief interval of material accumulation, rarely spanning more than a few decades (Finley, 1990). Individual middens are collected and cleaned in the field, and disaggregated, sieved, and sorted in the laboratory. Plant debris is identified and quantified, and in some cases pollen is also analyzed. Individual middens are dated using ^{14}C, and a collection of dated middens from a single canyon complex can be stacked chronologically to yield a time-series of plant-macrofossil presence/absence and/or abundance, and pollen abundance (Fig. 3).

Packrats are biased in their assembly of plant materials. Comparative studies of modern midden composition and surrounding vegetation indicate that packrats are particularly fond of junipers and other conifers (Lyford et al., 2004). Most grasses, many forbs, and sagebrush are less consistently represented in middens. Although midden taphonomy is inadequately studied, studies to date indicate that middens assembled by *Neotoma cinerea*, the primary species in the central Rocky Mountains, provide reliable records of presence/absence of conifers, particularly junipers and pines (Lyford et al., 2004; M.R. Lesser and S.T. Jackson, personal observations).

Example.—Lyford et al. (2002) obtained 17 middens from south- and west-facing slopes in a series of canyons on the west flank of Big Pryor Mountain in south-central Montana. All locales were in woodlands dominated by Utah juniper (*Juniperus osteosperma*) with scattered Rocky Mountain juniper (*J. scopulorum*) and limber pine (*P. flexilis*) on north-facing slopes across the narrow canyons. The ^{14}C-dated middens yielded a time-series spanning the past 11,500 years (Fig. 3). The time-series is typical in its irregular temporal distribution of samples (Fig. 3). Some periods (e.g., 3500-1000 yr BP) have high sampling density, while significant

spans of time lack middens (e.g., 6300-3500 yr BP) or are sparsely represented (e.g., 11,500-6300 yr BP).

Common juniper (*J. communis*), currently restricted to higher elevations on Big Pryor Mountain, occurred at the study sites in the late-glacial and early Holocene but disappeared between 9000 and 7500 yr BP (Fig. 3), consistent with a transition to warmer conditions. The early Holocene occurrence of saltbush (*Atriplex*), now restricted to lower elevations, indicates conditions warmer and drier than today. Rocky Mountain juniper and limber pine have occurred throughout the past 11,500 years, although their abundance and local habitat distributions may have shifted. Finally, Utah juniper, dominant at the site today, was not established until ca. 2400 yr BP. The accompanying increase in Cupressaceae (i.e., juniper) pollen suggests that juniper density increased with the Utah-juniper invasion, consistent with the high density of Utah-juniper stands observed in the region today compared to Rocky Mountain juniper stands. Mountain-mahogany (*Cercocarpus ledifolius*) and yucca (*Yucca glauca*) also colonized after 2500 yr BP.

Lake sediments

Description.—Lake basins are largely limited to subalpine forest and alpine tundra portions of the Rockies, although a few occur in montane forests. Lakes scattered in intermountain basins and lower slopes are generally shallow and most are subject to periodic desiccation (and consequent loss of organic sediment constituents). Lakes, or portions of lakes, that have escaped desiccation in the past 10,000 or more years accumulate sediments that incorporate pollen, plant macrofossils, and charcoal derived from surrounding uplands, and diatoms, insect larvae, organic debris, and other materials produced within the lake. Sediment cores are dated using ^{14}C, and analyzed to yield time-series. Pollen sequences provide integrated records of regional vegetation, although interpretation can be hampered by windborne transport of pollen across ecotones (Jackson and Smith, 1994; Lynch, 1996). Plant macrofossils—needles, seeds, budscales, etc.—record vegetation on adjacent slopes and permit species-level discriminations not always possible with pollen (Baker, 1976; Weng and Jackson, 1999; Jackson and Booth, 2007). Charcoal particles of various sizes record wildfires at various spatial scales surrounding the lakes (Whitlock and Larsen, 2002; Gavin et al., 2007). Remains of bark beetles (Brunelle et al., 2008), may prove useful in

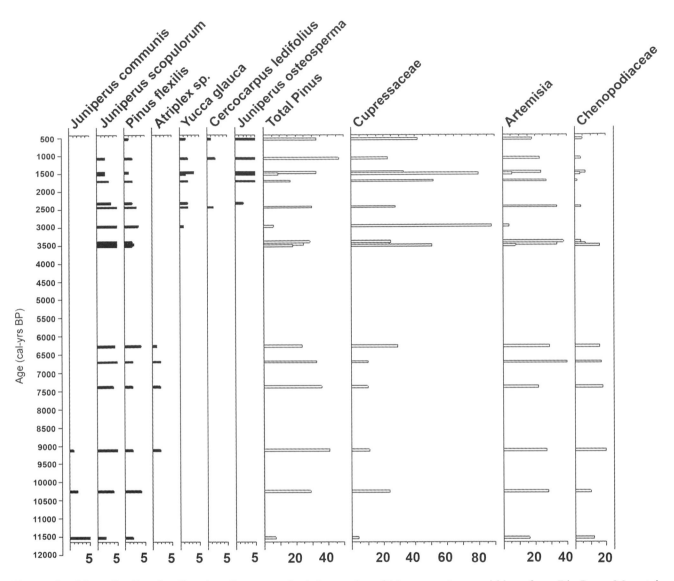

FIGURE 3.—Macrofossil and pollen data from a stacked time series of *Neotoma cinera* middens from Big Pryor Mountain, south-central Montana. Each horizon represents a single ^{14}C-dated midden. Modified from Lyford et al. (2002).

assessing prehistoric pest outbreaks.

Lake sediments can also preserve evidence of past fluctuations in lake water-levels. Geomorphic features preserve past shorelines formed during wet periods when lakes reach high stands (e.g., Langford, 2003; Pierce et al., 2002), and submerged stratigraphic features detected with geophysical surveys and sediment cores document shorelines from low water periods (e.g., Hofman et al., 2006; Pierce et al., 2002; Shuman et al., 2009).

Example.—A lake-sediment record of pollen and charcoal from Foy Lake shows vegetational, climatic, and ecosystem changes in the Flathead Valley of northwestern Montana during the past 3800 years (Fig. 4). The lake was formed more than 13,000 yr BP with ice

retreat from the Flathead Valley. Power et al. (2006) studied pollen and charcoal in annually laminated sediments deposited in the late Holocene in the deepest (39.9 m) part of the lake basin. Shuman et al. (2009) obtained a lake-level record from a transect of sediment cores along a water-depth gradient. Chronologies were based on a combination of radiocarbon dating, ^{210}Pb dating (spanning the last century), and varve counting. The lake-level record indicates a rapid rise in water level commencing ca. 2700 yr BP (Fig. 4). For example, a core from intermediate water depth (7.9 m) shows a transition from sand to fine marl at this time, together with an increase in organic carbon and an increase in sediment accumulation rate (Fig. 4).

The pollen record shows a decrease in pine (*Pinus*)

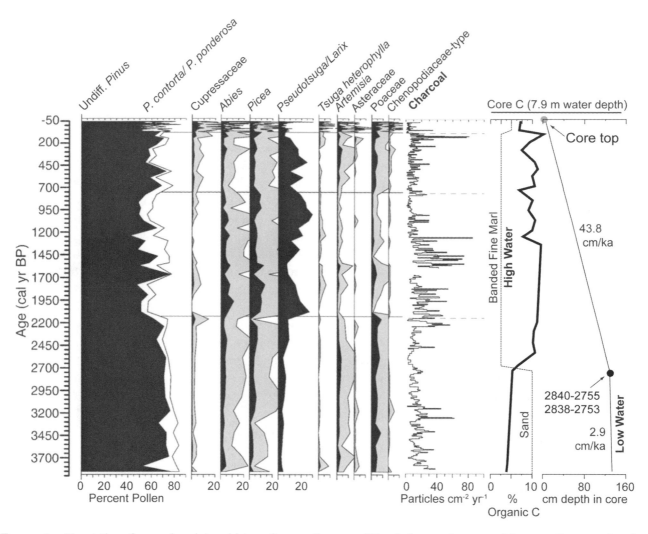

FIGURE 4.—Vegetation, fire, and moisture history from sediments of Foy Lake, northwestern Montana. Percent abundance of major pollen types found in the central Foy Lake core (collected in 39.9 m water depth) are shown in black on the left (gray indicates 5x exaggeration). Sedimentary charcoal concentration is shown from the same core (black line). Sediment characteristics of a second core (collected in 7.9 m water depth) provide evidence of lake-level change including sediment facies (thin black line), percent organic carbon (thick line), and net sediment accumulation rates (given as cm/ka). Banded marls with >5% organic carbon, and >10 cm/ka net accumulation indicate submergence of the core site and high water; the absence of fine sediment accumulation (in 7.9 m of water today) indicates low water before two calibrated radiocarbon ages (black circle) of ca. 2800 cal yr B.P. (1-SD age ranges are listed). The onset of varve accumulation in the central core after ca. 2200 cal yr BP indicates that the lake probably continued to rise until that time (when tree taxa such as *Abies, Picea, Pseudotsuga, Larix*, and *Tsuga heterophylla* reached their maximum abundances). Data derived from Power et al. (2006) and Shuman et al. (2009).

and grass (*Poaceae*) and increases in spruce (*Picea*), fir (*Abies*), western hemlock (*Tsuga heterophylla*), and either or both of palynologically indistinguishable Douglas-fir (*Pseudotsuga*) and larch (*Larix*) following the water-level increase (Fig. 4). These data indicate a transition from a mosaic of steppe and pine forest/woodland to mixed conifer forest (Power et al., 2006), consistent with the increase in effective moisture (probably increased winter precipitation) indicated by

the lake-level record. Modest increases in grass, pine, and sagebrush (*Artemisia*) at the expense of spruce and Douglas-fir/larch (Fig. 4) mark development of a steppe/parkland/forest patchwork in the valley ca. 700 yr BP (Power et al., 2006). Euro-American forest clearance of the late 19th and early 20th centuries is marked by the final increase of grass and sagebrush and decrease of pine pollen (Fig. 4).

Each of the vegetation transitions is accompanied

by a transition in charcoal deposition patterns (Fig. 4). Magnitude and frequency of charcoal peaks are highest during the forest period 2200-1200 yr BP, suggesting a transition from a surface-fire regime during the steppe/parkland period to stand-replacing crownfires during the forest period. The charcoal transition at 1200 yr BP does not correspond to a major vegetation change, but a shift in diatom composition at this time suggests a change in hydrological regime (Stevens et al., 2006). Fire suppression during the historical period is apparent in the reduced charcoal deposition at the top of the sequence.

The Foy Lake record, like the Black Hills and Big Pryor Mountain records, indicate climate change and accompanying ecological responses. Vegetation composition and fire regime at the time of Euro-American invasion, often used as the benchmark for ecological restoration, were recent phenomena, established only 600 years before. The Flathead Valley experienced several major vegetation transitions in the last 3800 years, and fire, the dominant disturbance regime, underwent transitions, some corresponding to vegetation transitions (e.g., 2200 yr BP) and others not (e.g., 1200 yr BP). Fires of the historic period, including the catastrophic 1910 fire (Pyne, 2001), were unremarkable in the context of the past 750 years, and far more destructive fires were routine as recently as 1200 years ago (Power et al., 2006).

Summary: implications for resource management

The records from Black Hills tree-rings, Big Pryor Mountain packrat-middens, and Foy Lake sediments are in different regions with contrasting vegetation, cover different time periods with different degrees of temporal precision, represent different spatial scales of sampling, and range from qualitative to quantitative in nature. Despite these differences, the records share features directly relevant to resource managers, features widely shared by similar records across the Rocky Mountain region (Woodhouse and Overpeck, 1998; Lyford et al., 2003; Gray et al. 2004a, b, 2007; Woodhouse et al., 2006; Meko et al., 2007, Whitlock et al., 2008a; Shuman et al., 2009; Watson et al., 2009) as well as the rest of North America (Swetnam and Betancourt, 1998; Swetnam et al., 1999; Jackson, 2006; Whitlock et al., 2008b; Jackson et al., in press).

Historical and instrumental baselines are inadequate for assessing variability in environment and ecosystem properties.—Risk of severe, sustained drought is greatly underestimated by the instrumental record of the past 100-150 years. Droughts of far greater magnitude and duration than those of the historic period have occurred under natural variability within the past 500 years (Fig. 2). Risk of catastrophic wildfires may be similarly underestimated; fires of magnitude far exceeding those of the historic period have occurred in the Foy Lake region (Fig. 4).

Climate events are often synchronous across the region, but not always uniform.—Decadal-scale droughts and wet periods frequently coincided between the Great Plains and Bighorn Basin regions over the past 500 years (Fig. 2). However, magnitude and direction of changes were not the same everywhere, and ecosystems show seemingly counterintuitive, contrasting responses in different regions owing to the variety of climate variables involved. The expansion of Utah juniper at Big Pryor Mountain coincided with the expansion of forest at Foy Lake. These two events are probably linked to the same general climate change, but with different manifestations in different places. The Foy Lake transition is consistent with an increase in effective moisture (Shuman et al., 2009), while the Big Pryor Mountain transition coincided with an upward shift in lower treeline, indicating effectively drier conditions (Lyford et al., 2002). However, both may represent a seasonal shift in precipitation: the vegetational and lake-level changes at Foy Lake are consistent with an increase in winter precipitation. Expansion of Utah juniper and contraction of lower treeline may also represent a shift towards greater winter precipitation accompanied by increased summer drought.

Many terrestrial ecosystems of the region are young, dynamic, and historically contingent.—The pre-European ecosystems of the Foy Lake and Big Pryor Mountain regions were developed within the past 2200 years with the expansion of the current dominant species (Figs. 3, 4), and significant changes are documented at Foy Lake as recently as 700 yr BP. These transitions set into motion long-term ecological processes (e.g., population dynamics, biogeochemical changes, biomass accretion) that may have continued into the 19th century and beyond. Wet and dry events and accompanying disturbance and recruitment pulses of the past 500 years left distinct demographic imprints and other ecological legacies (Fig. 2).

Rapid climate transitions and ecosystem transformations can occur.—Climate change and ecosystem responses are often abrupt. Tree-ring records show rapid transitions from severe drought to persistent wet periods (e.g., the transition ca. 1760 in Fig. 2), and ecological responses (e.g., fire and recruitment pulses) can rapidly transform the landscape (e.g., the increase in Black Hills Ponderosa-pine stand density following the late 18th and early 19th century recruitment events). Extrapolating from observations of ongoing expansions, the transition from steppe and open juniper woodland to dense Utah-juniper woodland at Big Pryor may have occurred in a few decades (Fig. 3). The changes in vegetation and fire regime near Foy Lake may have spanned only a few decades (Fig. 4).

Euro-American land-use has left a strong imprint on the regional landscape.—Fire suppression starting in the late 19th century is manifest in fire-scar (Fig. 2) and charcoal records (Fig. 4). Livestock grazing has interacted with fire suppression to drive persistent changes in vegetation composition and structure over much of the region. Forest density and homogeneity has increased across the Black Hills, for example, as a result of these processes (Brown, 2006).

Ecosystem restoration to historical baselines may be difficult or impossible.—Ecological restoration and conservation activities often rely on historical baselines as targets. In western North America, these baselines typically comprise ecosystem properties of the early-mid 19th century, just preceding widespread Euro-American land use. But historical ecology indicates that historical baselines are inconstant – natural climate change and variability shift the targets, and the state of an ecosystem at any given time represents not only current conditions but legacies of previous events and ecosystem states. Human activities of the past century—fire suppression, grazing, forest clearance, predator removal, introduction of non-native species—impose further constraints on restoration. Continuing or accelerating climate change will render historical targets increasingly elusive in the future (Jackson and Hobbs, 2009).

ECOSYSTEMS OF THE FUTURE IN THE CONTEXT OF THE PAST

As illustrated by the examples in Figures 2-4, environmental and ecological changes in the Rocky Mountain region have occurred over temporal scales from annual/decadal to multimillennial. These changes can be organized into a nested hierarchy of time scales, which puts the climatic and ecological dynamics of the

immediate past into broader temporal context (Fig. 5). Although many factors have driven ecosystem dynamics during the past 15,000 years, all derived ultimately from climate dynamics or interacted with prevailing climate. Fire, for example, is determined by fuel, ignition, and fire weather. Fuel is in turn influenced by hydroclimate (via net primary productivity, vegetation structure, decomposition rate, and dryness of surface and crown fuels), as are ignition (lightning, human activity) and fire weather (temperature, humidity, wind velocity and turbulence). Hydroclimate represents a climate-system response to various forcings, some external (insolation, dust, CO_2 concentration) and some internal (circulation patterns that influence transport of heat and water vapor). The linkages between climate forcing, hydroclimate response, and ecosystem response also comprise a nested hierarchy (Fig. 5).

Climate forcing

Atmospheric greenhouse gas concentrations, most prominently carbon dioxide, represent a key external driver and have undergone major changes in the past 15,000 years. Carbon dioxide concentrations were at a minimum of ~180 ppm during the last glacial maximum ca. 21,000 yr BP, increased rapidly but irregularly until 11,000 yr BP, and changed relatively little during most of the Holocene (Indermühle et al., 1999b; Monnin et al., 2001) (Fig. 5.1). Climate changes of the last deglaciation were driven by a number of factors, including orbital forcing and interactions among the atmosphere, ice sheets, and oceans, but changing carbon dioxide concentrations had an important influence (Liu et al., 2009). Holocene climate changes have been dominated by other controls (orbital variation, volcanic events, solar variability, ocean-atmosphere interactions and feedbacks) (Hansen et al., 1984; Bartlein et al., 1998; Harrison et al., 2003; Diffenbaugh et al., 2006; Shin et al., 2006). Little change in carbon dioxide occurred in the last 1500 years until onset of the Industrial Revolution 200 years ago. Pre-industrial climate change and variability during this period were forced primarily by ocean-atmosphere interactions, land-surface feedbacks, and solar and volcanic events.

Human land clearance and fossil fuel combustion dominate climate forcing during the last 150 years, contrasting with the past 11,000 years when the carbon cycle was dominated by non-human processes (Indermühle et al., 1999b; Monnin et al., 2001). At the scale of the past 15 years, forcing appears as a steady rise in carbon dioxide concentrations, with transient annual drawdown from growing-season uptake by Northern Hemisphere plants and cold-season release from decomposition. The net increase during the past 15 years is equivalent to the net increase between 12,000 and 11,000 yr BP – thus the rate of change in the past 15 years is two orders of magnitude faster than during the late Pleistocene (Monnin et al., 2001), a period of rapid

change and widespread ecosystem disruption (Shuman et al., 2005).

Hydroclimate response

Climate forcings, whether internal or external, and whether global or regional, are manifested climatically in different ways in different places as a result of interactions among controls (McCabe et al., 2008) as well

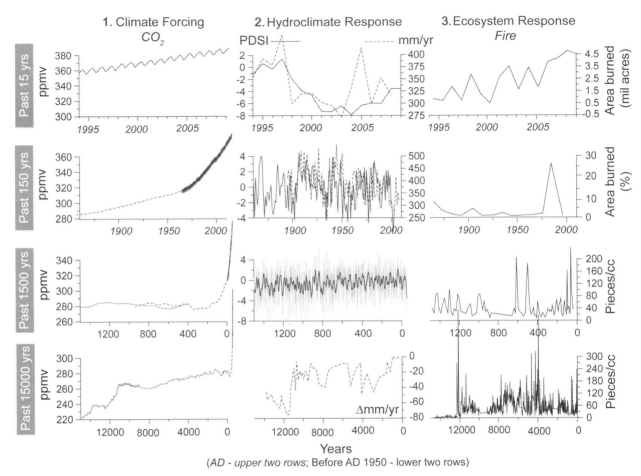

Years
(*AD - upper two rows*; Before AD 1950 - lower two rows)

FIGURE 5.—Multiple time scales of *1*, climate forcing, *2*, climate response, and *3*, ecosystem response. Atmospheric carbon dioxide concentrations; *1*, From ice cores (Monnin et al., 2004, gray line; Indermühle et al., 1999a, dotted line; Etheridge et al., 2001, dashed line) and direct measurements (Tans, 2009, solid line) represent scales of variation in one of several major external climate drivers (forcing) with seasonal and anthropogenic trends evident over the past 15 years, anthropogenic trends dominant over 150 years, limited natural variability for most of the past 1500 years, and important variations related to deglaciation after the last ice age evident over the past 15,000 years. The Palmer Drought Severity Index (PDSI, solid line) and annual precipitation (dashed line) for the region near Yellowstone National Park; *2*, Derived from measurements that show historic droughts in the past 15 and 150 years (NCDC, 1994), from dendroclimate reconstruction of PDSI that show intrinsic climate variability in the absence of strong forcing over the past 1500 years (Cook and Krusic, 2004; data for 107.5W, 40.0N), and from inference from fossil pollen data that show externally forced variations at Blacktail Pond, Wyoming (Gennett and Baker, 1986) based on the methods of Williams and Shuman (2008). Historic area burned measurements for all U.S. Fire history; *3*, Captures a representative ecosystem response to hydroclimate change: Measurements of area burned on all U.S. Federal lands in the past 15 years (NICC, 2008) and for a 129,600-ha subalpine study area in central Yellowstone (Romme and Despain,1989) show broad scale linkage with climate trends but the local heterogeneity of individual fires. Pieces of charcoal in lake sediments over the past 1500 and 15,000 years at Slough Creek Pond, Wyoming (Millspaugh, 1997), also capture the local heterogeneity but demonstrate the potential for fire regimes unlike those that exist today under other climate scenarios.

as influences of local to regional factors (e.g., geography, topography, land-cover) (Mock, 1996). Column 2 of Figure 5 focuses on the Yellowstone Plateau region of northwest Wyoming, and emphasizes hydroclimatic responses, which influence fire and other ecosystem processes in the region. These responses are derived from different sources depending on the time scale – pollen-based for the last 15,000 years, tree-rings for the last 1500 years, and instrumental records for the past 150 and 15 years.

Hydroclimate in the Yellowstone region has varied substantially during the past 15,000 years (Whitlock, 1993; Shuman et al., 2009). The rise in precipitation accompanying the late-glacial rise in carbon dioxide resulted in part from reduction in radiative cooling (i.e., an early greenhouse effect), which led to greater atmospheric capacity to retain moisture. Glacial-age low temperatures, resulting in part from reduced CO_2, maintained continental and alpine ice sheets (including the Yellowstone plateau glacier until 14,000 yr BP). The continental ice sheets and cold oceans shifted storm tracks relative to the Holocene (Bartlein et al., 1998). Winter storms that today bring large amounts of snow to the Yellowstone region were shifted south at the time, contributing to large lakes in the Great Basin (e.g., Lakes Lahontan and Bonneville). Ice retreat, together with ocean and atmospheric warming, led to a northward shift in the jet stream, leading to increased Yellowstone winter precipitation (Bartlein et al., 1998).

Precipitation changes during the last 11,000 years have largely resulted from multimillennial changes in orbital forcing (which in turn influenced ocean and atmospheric circulation), together with higher-frequency patterns in ocean circulation (e.g., El Niño-Southern Oscillation (ENSO) dynamics), with some possible imprint of volcanic-emission events (Hansen et al., 1984; Bartlein et al., 1998; Harrison et al., 2003; Diffenbaugh et al., 2006; Shin et al., 2006). These natural climate variations were capable of inducing prolonged, multi-century droughts with precipitation >40 mm/yr lower than today (Fig. 5.2).

Although orbital and radiative forcings were relatively constant during the past 1500 years (excepting greenhouse gas increase over the past ~100 years), substantial interannual to multidecadal precipitation variability occurred in the region (Fig. 5.2). This variation resulted largely from intrinsic variability in the climate system (e.g., ocean/atmosphere/land-surface interac-

tions) (Gray et al., 2004a, 2007). Finally, the past 15 years have witnessed a transition from a relatively wet period to a severe drought commencing in 1999. This transition and the ensuing drought may represent natural variability of the kind experienced in the past 1500 years, but CO_2-driven ocean warming has been implicated as a contributing factor (Barnett et al., 2008). The reduction in precipitation in the Yellowstone region during the current drought is of equivalent magnitude to the reductions during early and mid-Holocene droughts that spanned centuries. Persistent drought of this type in western North America is predicted by climate models under CO_2-doubling scenarios (Hoerling and Eischeid, 2007).

Ecosystem responses

Virtually all ecosystem properties are ultimately contingent on climate, which determines not only the dominant plant forms and structural features, but also mediates rates and pathways of material and energy transfer. In our discussion, we focus on fire patterns, which comprise an important, highly visible, and economically important property of Rocky Mountain ecosystems, and for which paleoecological records are readily available (Fig. 5.3). Fire history depends on vegetation composition, structure, and history, which are also influenced by hydroclimate. But vegetation is also influenced by fire; thus many terrestrial ecosystems are determined by interaction among climate, vegetation, and fire regime.

Lake-sediment charcoal records from the Yellowstone region show low fire occurrence during the late-glacial period, with a marked regime shift to higher charcoal influx and higher variability ca. 11,000 yr BP. This corresponds to the rapid increase in available moisture (Fig. 5.2), and to an equally rapid shift from spruce parkland to dense pine forest dominated by lodgepole pine (*Pinus contorta*) (Fig. 6). The change in fire regime resulted from increased fuel availability and connectivity, perhaps accompanied by increased summer temperature and drought. Although the Yellowstone Plateau has been continuously occupied by lodgepole pine forest throughout the Holocene (Whitlock, 1993) (Fig. 6), it has experienced a number of different fire regimes—contrasting in amplitude and frequency—driven by changes in seasonal precipitation and temperature (Whitlock and Bartlein, 2004; Brunelle et al., 2005; Marlon et al., 2006; Millspaugh et al., 2000; Power et al., 2008; Whitlock et al., 2008b).

This variation is manifest at higher resolution during the past 1500 years (Power et al., 2006; Whitlock et al., 2008a). Peaks and troughs in fire occurrence represent a combination of climate variation (e.g., the peak between 1000 and 800 yr BP corresponds to a series of regional multidecadal droughts) and legacy effects of prior disturbances (post-fire succession and fuel accumulation required to support another major fire event). A key lesson for resource managers is that similar ecosystems can be characterized by a wide variety of disturbance regimes depending on climate variability. Direct observation and experience of the past century provides only a small sample of the range of variability possible under natural conditions.

The record of the past 150 years shows the effect of fire suppression, culminating in the epic 1988 fires (Romme, 1982; Romme and Despain, 1989). 1988 was a drought year (Trenberth et al., 1988), but in many areas the severity and extent of the fires also resulted from the previous century's fuel accumulations under fire suppression (Romme and Despain, 1989). At this and finer scales, fire-climate relationships become obscured by local factors, but patterns re-emerge at broader spatial scales. A regional trend toward larger fires has been identified over the past few decades, resulting from increasingly earlier onset of snowmelt and thus to earlier onset of fire seasons (Westerling et al., 2006). Global climate change may be implicated in this recent warming (IPCC, 2007). The recent fire-size trend has not yet been identified in subalpine forests, which require centuries for fuel accumulation sufficient to support fires.

Transitions in vegetation composition and structure also occur at various timescales (Fig. 6). The past 15,000 years reveal an abrupt transition 11,000 yr BP, representing inception of lodgepole pine forest, a novel ecosystem in the sense that it did not exist in the region before. That ecosystem underwent some variation in density and patchiness over the past 11,000 years, but the dominant species and overall structure remained the same. Other portions of the Rocky Mountain region show more Holocene turnover in vegetation composition (Figs. 3, 4).

Pollen records may mask subtle but important changes in landscape structure. For example, Yellowstone pollen records of the past 1500 years show little change, but tree-ring data spanning the past few centuries reveal substantial changes in the relative amounts of recently burned forest, even-aged early-successional

forests, and mixed-age old stands. The repeated fires indicated by the sedimentary charcoal data were likely the dominant cause of these landscape dynamics, with subsequent modification by successional dynamics. Other disturbance processes (wind throw, pest and pathogen outbreaks) may have contributed. Because these processes happen with sufficient frequency to maintain a mosaic of different types of patches across the landscape, the regional forest composition (and thus the pollen deposited in lakes) was not altered by the repeated disturbances.

In the context of the previous century, the widespread 1988 fires were a surprise, rapidly altering landscape structure across a broad area within less than a month. Scientists and managers concerned with real-time records of fire history did not anticipate them. Such large and widespread fires are in no way unusual, of course, in the context of the past 1500 or 15,000 years. In fact such widespread fires might be viewed as inevitable in the long run, requiring only a concatenation of simultaneous widespread fuel availability and opportunity in the form of persistent, severe fire weather. Such "inevitable surprises" can often be anticipated only by looking at deep historical records. Similar surprises may be in store given recent climate forcing and precipitation trends; global climate changes may be expressed ecologically in rapid local changes (severe disturbances, post-disturbance expansion of novel native or exotic species).

One such surprise may be underway at the time of this writing. Mountain pine beetle outbreaks are driving forest mortality from southern Colorado to western Canada. These outbreaks were initiated by recent drought together with long-term warming of recent decades, and could have widespread and persistent ecosystem consequences. The consequences of the ongoing pine-beetle outbreak and the recent and ongoing large-scale fires for ecosystem properties remain to be seen—much will depend on subsequent successional dynamics, together with the trajectories of climate, disturbance, species invasions and migrations, and human activities (including management decisions) in the coming years and decades. However, given the magnitude of climate forcing of recent decades relative to that of the past 15,000 years (Fig. 5.1), it is possible that these recent events foreshadow major ecosystem transformation of the magnitude last seen in the region 11,000 years ago (Figs. 5, 6). Will coming decades see a return to forests similar to those of recent decades and centuries—perhaps with altered mosaic patterns and disturbance regimes? Or will they witness wholesale transitions to novel ecosystems?

USING THE PAST TO PLAN FOR THE FUTURE

Resource managers, stakeholders, and policymakers need scientific information to make informed decisions. The rapidity and complexity of global climate change poses challenges for information transfer and utilization. Academic scientists are usually inexperienced and often ineffective in engaging with the communities that require scientific information. A particular challenge arises from scientists' natural reluctance to make precise predictions for the future, and their tendency to embed findings in qualifications and contingencies. These communication challenges have been surmounted by the water resources community in the western United States, where paleohydrological studies are being incorporated into water management and policy. The approaches and methods adopted by this community can serve as models for other conservation and natural-resource applications of paleoecological and paleoclimatic studies.

The stage was set for the water-resources community when Harding et al. (1995) used tree-ring reconstructions of past Colorado River flow (Stockton and Jacoby, 1976) to demonstrate the potential for severe, sustained droughts far exceeding those in the instrumental record of river discharge. Further analyses of the tree-ring record showed that the early 20th century was unusually wet in the context of previous centuries, leading to over-allocation of river waters when policy was set in the 1930s (i.e., states were given legal rights to more water than often actually exists). These and other findings led to major policy shifts in the management of the Colorado River (NRC, 2007), the primary water source for over 20 million people. Tree-ring and sediment-based paleohydrological reconstructions are now being applied across the western United States in contexts ranging from drought planning and mitigation to design of municipal water systems and water-delivery infrastructure (Phillips et al., in press; Rice et al., in press).

Paleohydrology is now being applied to overcome limitations of climate models in hydrological forecasting. Precipitation forecasts for the western United States are highly uncertain even with the most sophisticated

downscaling approaches (IPCC, 2007). Paleohydrological records are being used to estimate the range of potential precipitation variability and sequences of wet/dry years, which are paired with temperature-change predictions (more robust than precipitation predictions) to predict stream discharge (Prairie et al., 2008; Gray and McCabe, in press).

A framework widely used in the business and finance community may provide an effective mechanism for use of scientific findings from paleoecological and paleoclimatic studies in decision-making. Scenario planning uses a combination of scientific input, expert opinion and forecast data to develop alternative scenarios for the future (Schwartz, 1991; van der Heijden, 1996), contrasting with more traditional attempts at developing precise, quantitative assessments of future conditions, which are hampered by compounded uncertainties and qualifications.

In scenario planning, alternative scenarios can be used as a starting point for exploring species or ecosystem vulnerabilities under a range of future conditions, and as a means for examining how conservation strategies might play out in the face of multiple drivers of change (Fig. 7). In our example, alternative futures are arrayed along two axes comprising integrators of potential climate-change (drought frequency) and potential changes in disturbance regimes (fire size). In concert with monitoring and modeling, paleoclimatic studies can define the range of drought frequency we might reasonably expect, and paleoecological studies of fire can place bounds on potential fire size. This exercise yields four quadrants, each comprising a distinct combination of climatic and fire-regime change (Fig. 7). These quadrants each provide a contrasting scenario or "story line" that can be used to explore potential impacts on species or ecosystems and for examining the relative costs and benefits of various mitigation and adaptation measures.

At one extreme, major climate change and altered disturbance regimes interact to drive emergence of novel ecosystems. Given our limited experience with ecosystem turnover in the region, paleoecology serves as a primary means for adding texture and substance to the scenario. It also points out the risk of finding ourselves in any one of the four quadrants (e.g., Figs. 4-6). For example, transition to "novel ecosystems" is analogous to the transition observed 11,000 yr BP at Yellowstone (Figs. 5, 6), while the transitions to "inevitable surprises" and "patches and fragments" are

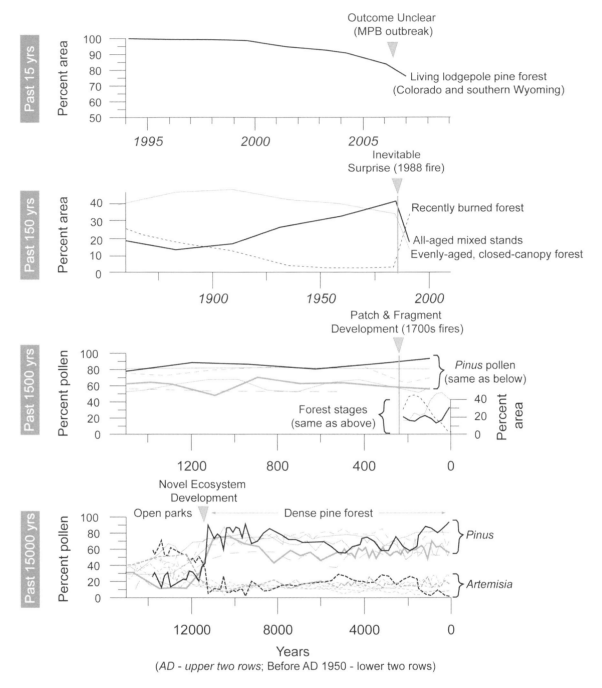

FIGURE 6.—Multiple time scales of vegetation change. At the 15 year scale, a decline in the area of living lodgepole pine forest in Colorado and Wyoming has arisen from a mountain pine beetle infestation; based on USFS data. At the 150 years scale, the percent of a 129,600-ha subalpine study area in central Yellowstone occupied by three different types of forest stages, recently burned (dashed line), even-aged (gray line), and all-age mixed stands (black line), show successional and disturbance influences on forest structure (Romme and Despain, 1989). The same data are shown at the 1500-year scale (extending back >250 yrs before AD 1950), and are compared with the percent pine (*Pinus*) pollen in pollen records from Yellowstone region, which show limited plant assemblages changes at that scale. More dramatic changes in plant assemblages are evident from the percent of pine and sagebrush (*Artemisia*) pollen over 15,000 years from the same sites. Gray arrows indicate different types of ecosystem changes. Pollen records shown here include the data used for climate reconstruction in Fig. 4 from Blacktail Pond, black line (Gennett and Baker, 1986); data from location of charcoal record in Fig. 4 from Slough Creek Pond, thick gray line (Millspaugh, 1997); data from the Romme and Despain (1989) study area from Buckbean Fen, short dashed line (Baker, 1976); Emerald Lake, long dashed line (Whitlock, 1993); and other datasets from Cygnet Lake Fen, gray line (Whitlock, 1993); Hedrick Pond, gray line (Whitlock 1993).

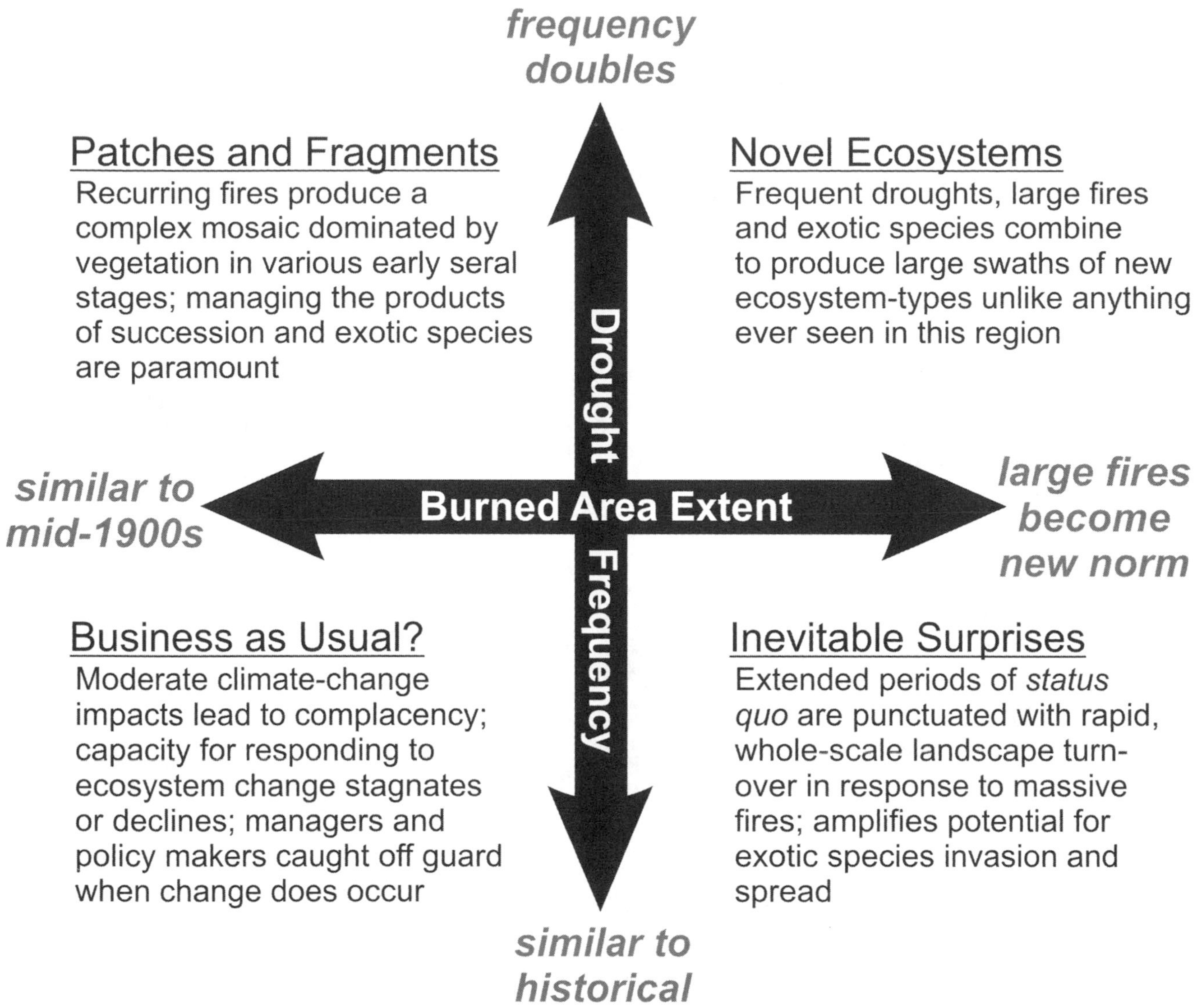

frequency doubles

Patches and Fragments

Recurring fires produce a complex mosaic dominated by vegetation in various early seral stages; managing the products of succession and exotic species are paramount

Novel Ecosystems

Frequent droughts, large fires and exotic species combine to produce large swaths of new ecosystem-types unlike anything ever seen in this region

Drought Frequency

Burned Area Extent

similar to mid-1900s

large fires become new norm

Business as Usual?

Moderate climate-change impacts lead to complacency; capacity for responding to ecosystem change stagnates or declines; managers and policy makers caught off guard when change does occur

Inevitable Surprises

Extended periods of *status quo* are punctuated with rapid, whole-scale landscape turn-over in response to massive fires; amplifies potential for exotic species invasion and spread

similar to historical

FIGURE 7.—Example of a scenario planning matrix. Each axis represents a critical driver of system change or a significant trend in the environment. In common practice, the variables chosen for analysis are likely to have the strongest influence on the system or they are associated with a high degree of uncertainty (Shoemaker, 1995). In the case presented here, the axes represent a continuum between conditions that are similar to those observed in the historical record and conditions that are significantly altered from those seen today. Combining these two drivers produces four alternative scenarios for the future conditions (e.g., frequent drought and large fires in the upper right) that can then be further developed into "storylines" that provide details about how each scenario might unfold. Depending on the application and available data, axes and the resulting storylines may be defined quantitatively, or they may be based on qualitative assessments alone.

analogous to the late Holocene transitions observed in Figures 2-4.

The greatest value in scenario planning comes from uncovering vulnerabilities and potential responses, particularly those common to multiple story lines. In our example, exotic species are likely to increase under most or all scenarios, suggesting that investing in management of invasions might be a "win-win" strategy regardless of the climate and disturbance trajectories of the future. In this case, paleoecological records enrich

the discussion by revealing the patterns and mechanisms by which species invade new territory (Lyford et al., 2003; Gray et al., 2006; Jackson et al., in press). Additional axes representing drivers (climate variability, management intervention) or stressors (pests, land-use) can be combined to produce a wider range of future scenarios, and additional matrices can be embedded within an individual storyline. In this way, integration of ecological and climatic history into scenario planning expands the range of possibilities considered in conservation efforts, and challenges conventional notions of appropriate response to environmental change. Historically based scenario planning offers a powerful decision-support tool in the face of an uncertain future.

Based on our own experience and that of others (e.g., Rice et al., in press), paleoecological and paleoclimatic perspectives are welcome additions to planning and assessment exercises. They ground future storylines within the realm of physical possibility, engaging managers, policy makers and scientists alike in the discussions. For example, stakeholders and decision-makers are more likely to consider multidecadal droughts, landscape-altering fires, and rapid species invasions as realistic possibilities for the future since they have happened in the past.

Potential applications of paleoecological perspectives in conservation and management are legion. Stream discharge, which can be reconstructed from tree-ring and other proxies, is a determinant of water temperature and chemistry, channel morphology, and other habitat variables important in management of fisheries and endangered species. Carbon sequestration and flux are linked to vegetation structure and composition, disturbance regimes, and ultimately seasonal temperature and precipitation.

All policy and management decisions are ultimately based on prevailing societal values—not only economic but also aesthetic. An historical aesthetic—the notion of the "natural" state as an ideal target—is embedded in much of conservation biology and restoration ecology. But climate change and novel stressors may render historical targets unattainable (Jackson and Hobbs, 2009). Meeting historical targets may become increasingly costly and lead to systems that are poorly suited to deal with future change (Millar et al., 2007). Paleoecology is uniquely positioned to provide long-term perspective on such issues and to deliver reality-checks in prioritization of conservation goals and development of effective management strategies.

ACKNOWLEDGMENTS

We thank Cary Mock and Tom Webb for critical comments and suggestions. The authors contributed approximately equally to the development of this paper.

REFERENCES

ANDERSEN, M. D., AND W. L. BAKER. 2006. Reconstructing Landscape-scale tree invasion using survey notes in the Medicine Bow Mountains, Wyoming, USA. Landscape Ecology, 21, 243-258.

BAKER, R. G. 1976. Late Quaternary vegetation history of the Yellowstone Lake basin, Wyoming. United States Geological Survey Professional Paper, 729-E. Washington, D.C., 48 p.

BARNETT, T. P., D. W. PIERCE, H. G. HIDALGO, C. BONFILS, B. D. SANTER, T. DAS, G. BALA, A. W. WOOD, T. NOZAWA, A. A. MIRIN, D. R. CAYAN, AND M. D. DETTINGER. 2008. Human induced changes in the hydrology of the western United States. Science, 319:1080-1083.

BARNOSKY, A. D. 2004. Biodiversity Response to Climate Change in the Middle Pleistocene: The Porcupine Cave Fauna from Colorado. University of California Press, Berkeley, California, 385 p.

BARTLEIN, P. J., K. H. ANDERSON, P. M. ANDERSON, M. E. EDWARDS, C. J. MOCK, R. S. THOMPSON, R. S. WEBB, T. WEBB III, AND C. WHITLOCK. 1998. Paleoclimate simulations for North America over the past 21,000 years: Features of the simulated climate and comparisons with paleoenvironmental data. Quaternary Science Reviews, 17:535-548.

BETANCOURT, J. L. 2004. Arid lands paleobiogeography: The rodent midden record in the Americas, p. 27-346. In M. V. Lomolino and L. R. Heaney (eds.), Frontiers of Bio-Geography: New Directions in the Geography of Nature. Sinauer, Sunderland, Massachusetts.

BETANCOURT, J. L., T. R. VAN DEVENDER, AND P. S. MARTIN. 1990. Packrat middens: the last 40,000 years of biotic change. University of Arizona Press, Tucson, Arizona, 469 p.

BROWN, P. M. 2006. Climate effects on fire regimes and tree recruitment in Black Hills Ponderosa pine forests. Ecology, 87:2500-2510.

BRUNELLE, A., G. E. REHFELDT, B. BENTZ, AND A. S. MUNSON. 2008. Holocene records of Dendroctonus bark beetles in high elevation pine forests of Idaho and Montana, USA. Forest Ecology and Management, 255:836-846.

BRUNELLE, A., C. WHITLOCK, P. BARTLEIN, AND K. KIPFMUELLER. 2005. Holocene fire and vegetation along environmental gradients in the Northern Rocky Mountains. Quaternary Science Reviews, 24:2281-2300.

CHRISTENSEN, N., AND D. P. LETTENMAIER. 2007. A multimodel ensemble approach to assessment of climate change impacts on the hydrology and water resources of the Colorado River basin. Hydrology and Earth System Science, 11:1417-1434.

COOK, E. R., AND P.J. KRUSIC. 2004. The North American Drought Atlas. Available from: http://gcmd.nasa.gov/records/GCMD_LDEO_NADA.html.

COOK, E. R., AND L. A. KAIRIUKSTIS. 1990. Methods of Dendrochronology: Applications in the Environmental Sciences. Kluwer, Dordrecht, Netherlands, 394 p.

DIFFENBAUGH, N. S., M. ASHFAQ, B. SHUMAN, J. W. WILLIAMS, AND P. J. BARTLEIN. 2006. Summer aridity in the United States: Response to mid-Holocene changes in insolation and sea surface temperature. Geophysical Research Letters, 33:L22712.

ETHERIDGE, D.M., L. P. STEELE, R. L. LANGENFELDS, AND R. J. FRANCEY. 2001. Law Dome Atmospheric CO_2 Data, IGBP PAGES/World Data Center for Paleoclimatology Data Contribution Series #2001-083. NOAA/NGDC Paleoclimatology Program, Boulder, Colorado.

FINLEY, R. B., JR. 1990. Woodrat ecology and behavior and the interpretation of paleomiddens, p. 28-42. In J. L. Betancourt, T. R. Van Devender, and P. S. Martin (eds), Packrat Middens: The Last 40,000 Years of Biotic Change. University of Arizona Press, Tucson, Arizona.

FORMAN, S. L, R. OGLESBY, AND R. S. WEBB. 2001. Temporal and spatial patterns of Holocene dune activity on the Great Plains of North America: Megadroughts and climate links. Global and Planetary Change, 29:1-29.

FRISON, G. C. 1991. Prehistoric Hunters of the High Plains. Academic Press, San Diego, California, 426 p.

FRITTS, H. C. 1976. Tree Rings and Climate. Academic Press, London, 567 p.

GAVIN, D. G., D. J. HALLETT, F. S. HU, K. P. LERTZMAN, S. J. PRICHARD, K. J. BROWN, J. A. LYNCH, P. BARTLEIN, AND D. L. PETERSON. 2007. Forest fire and climate change in western North America: insights from sediment charcoal records. Frontiers in Ecology and the Environment, 5:499-506.

GENNETT, J. A., AND R. G. BAKER. 1986. A late Quaternary pollen sequence from Blacktail Pond, Yellowstone National Park, Wyoming, U.S.A. Palynology, 10: 61-71.

GRAY, S. T., AND G. J. MCCABE. In press. Combined water balance and tree-ring approaches to understanding the potential hydrologic effects of climate change on the Yellowstone River. Water Resources Research.

GRAY, S. T., L. J. GRAUMLICH, AND J. L. BETANCOURT. 2007. Annual precipitation in the Yellowstone National Park Region since A.D. 1173. Quaternary Research, 68:18-27.

GRAY, S. T., J. L. BETANCOURT, S. T. JACKSON, AND R. G. EDDY. 2006. Role of multidecadal climatic variability in a range extension of pinyon pine. Ecology, 87:1124-1130.

GRAY, S. T., C. L. FASTIE, S. T. JACKSON, AND J. L. BETANCOURT. 2004a. Tree-ring based reconstructions of precipitation in the Bighorn Basin, Wyoming since 1260 A.D. Journal of Climate, 17:3855-3865.

GRAY, S. T., S. T. JACKSON, AND J. L. BETANCOURT. 2004b. Tree-ring based reconstructions of interannual to decadal-scale precipitation variability for northeastern Utah

since 1226 A.D. Journal of the American Water Resources Association, 40:947-960.

HADLY, E. A. 1996. Influence of late-Holocene climate on northern Rocky Mountain mammals. Quaternary Research, 46:298–310.

HANSEN, J., A. LACIE, D. RIND, G. RUSSELL, P. STONE, I. FUNG, R. RUEDY, AND J. LERNER. 1984. Climate sensitivity: Analysis of feedback mechanisms, p. 1-35. In J. E. Hansen (ed.), Climate Processes and Climate Sensitivity. American Geophysical Union, Washington, D.C.

HARDING, B. L., T. B. SANGOYOMI, AND E. A. PAYTON. 1995. Impacts of severe sustained drought in Colorado River water resources. Water Resources Bulletin, 316:815-824.

HARRISON, S. P., J. E. KUTZBACH, Z. LIU, P. J. BARTLEIN, B. OTTO-BLIESNER, D. MUHS, I. C. PRENTICE, AND R. S. THOMPSON. 2003. Mid-Holocene climates of the Americas: A dynamical response to changed seasonality. Climate Dynamics, 20:663-688.

HOERLING, M., AND J. EISCHEID. 2007. Past peak water in the Southwest. Southwest Hydrology, 6:18-19.

HOFMANN, M. H., M. S. HENDRIX, J. N. MOORE, AND M. SPERAZZA. 2006. Late Pleistocene and Holocene depositional history of sediments in Flathead Lake, Montana: Evidence from high-resolution seismic reflection interpretation. Sedimentary Geology, 184:111-131.

INDERMÜHLE, A., B. STAUFFER, T. F. STOCKER, AND M. WAHLEN. 1999a. Taylor Dome Ice Core CO_2 Holocene Data. IGBP PAGES/World Data Center for Paleoclimatology Data Contribution Series # 1999-021. NOAA/NGDC Paleoclimatology Program, Boulder, Colorado.

INDERMÜHLE, A., T. F. STOCKER, F. JOOS, H. FISCHER, H. J. SMITH, M. WAHLEN, B. DECK, D. MASTROIANNI, J. TSCHUMI, T. BLUNIER, R. MEYER, AND B. STAUFFER. 1999b. Holocene carbon-cycle dynamics based on CO_2 trapped in ice at Taylor Dome, Antarctica. Nature, 398:121.

IPCC (INTERGOVERNMENTAL PANEL ON CLIMATE CHANGE). 2007. Climate Change 2007: The Physical Science Basis. Cambridge University Press, Cambridge, New York, 996 p.

JACKSON, S. T. 2006. Vegetation, environment, and time: The origination and termination of ecosystems. Journal of Vegetation Science, 17:549-557.

JACKSON, S. T., AND R. J. HOBBS. 2009. Ecological restoration in the light of ecological history. Science, 325:567-569.

JACKSON, S. T., AND R. K. BOOTH. 2007. Validation of pollen studies, p. 2413-2422. In S.A. Elias (ed.), Encyclopedia of Quaternary Sciences. Elsevier Scientific Publishing, Inc., Amsterdam.

JACKSON, S. T., AND S. J. SMITH. 1994. Pollen dispersal and representation on an isolated, forested plateau. New Phytologist, 128:181-193.

JACKSON, S. T., J. L. BETANCOURT, R. K. BOOTH, AND S. T. GRAY.

In press. Ecology and the ratchet of events: Climate variability, niche dimensions, and species distributions. Proceeding of the National Academy of Sciences of the United States of America.

JACKSON, S. T., J. L. BETANCOURT, M. E. LYFORD, S. T. GRAY, AND K. A. RYLANDER. 2005. A 40,000-year woodrat-midden record of vegetational and biogeographic dynamics in northeastern Utah. Journal of Biogeography, 32:1085-1106.

KURZ, W. A., C. C. DYMOND, G. STINSON, G. J. RAMPLEY, E. T. NEILSON, A. L. CARROLL, T. EBATA, AND L. SAFRANYIK. 2008. Mountain pine beetle and forest carbon feedback to climate change. Nature, 452:987-990.

LANGFORD, R. P. 2003. The Holocene history of the White Sands dune field and influences on eolian deflation and playa lakes. Quaternary International, 104:31-39.

LIU, Z., B. L. OTTO-BLIESSNER, F. HE, E. C. BRADY, R. TOMAS, P. U. CLARKE, A. E. CARLSON, J. LYNCH-STIEGLITZ, W. CURRY, E. BROOK, D. ERICKSON, R. JACOB, J. KUTZBACH, AND J. CHENG. 2009. Transient simulation of last deglaciation with a new mechanism for Bolling-Allerod warming. Science, 325:310-314.

LYFORD, M. E., S. T. JACKSON, S. T. GRAY, AND R. J. EDDY. 2004. Validating the use of woodrat (Neotoma) middens for documenting natural invasions. Journal of Biogeography, 31:333-342.

LYFORD, M.E., S.T. JACKSON, J.L. BETANCOURT, AND S.T. GRAY. 2003. Influence of landscape structure and climate variability on a late Holocene plant migration. Ecological Monographs 73:567-583.

LYFORD, M. E., J. L. BETANCOURT, AND S. T. JACKSON. 2002. Holocene vegetation and climate history of the northern Bighorn Basin, southern Montana. Quaternary Research, 58:171-181.

LYNCH, E. A. 1996. The ability of pollen from small lakes and ponds to sense fine-scale vegetation patterns in the Central Rocky Mountains, USA. Review of Palaeobotany and Palynology, 94:197-210.

MARLON, J., P. J. BARTLEIN, AND C. WHITLOCK. 2006. Fire-fuel-climate linkages in the northwestern USA during the Holocene. The Holocene, 16:1059-1071.

MAYER, J. H., AND S. A. MAHAN. 2004. Late Quaternary stratigraphy and geochronology of the western Killpecker Dunes, Wyoming, USA. Quaternary Research, 61:72-84.

MCCABE, G. J., J. L. BETANCOURT, S. T. GRAY, M. A. PALECKI, AND H. G. HIDALGO. 2008. Associations of multidecadal sea-surface temperature variability with US drought. Quaternary International, 188:31-40.

MCCABE, G. J., J. L. BETANCOURT, AND H. G. HIDALGO. 2007. Associations of decadal to multidecadal sea-surface temperature variability with Upper Colorado River flow. Journal of the American Water Resources Association, 43(1):183-192.

MEARS, B., JR. 1987. Late Pleistocene Periglacial Wedge Sites in Wyoming: An Illustrated Compendium. Geological Survey of Wyoming Memoir No. 3, Laramie, Wyoming, 77 p.

MEKO, D. M., C. A. WOODHOUSE, C. A. BAISAN, T. KNIGHT, J. J. LUKAS, M. K. HUGHES, AND M. W. SALZER. 2007. Medieval drought in the Upper Colorado River Basin. Geophysical Research Letters 34:L10705.

MEYER, G. A., S. G. WELLS, AND A. J. T. JULL. 1995. Fire and alluvial chronology in Yellowstone National Park: Climatic and intrinsic controls on Holocene geomorphic processes. Geological Society of America Bulletin, 107:1211-1230.

MILLAR, C. I., N. L. STEPHENSON, AND S. L. STEPHENS. 2007. Climate change and forests of the future: Managing in the face of uncertainty. Ecological Applications, 17:2145-2151.

MILLSPAUGH, S. H. 1997. Late-glacial and Holocene variations in fire frequency in the central plateau and Yellowstone-Lamar provinces of Yellowstone National Park. Unpublished Ph.D. dissertation, University of Oregon.

MILLSPAUGH, S. H., C. WHITLOCK, AND P. J. BARTLEIN. 2000. Variations in fire frequency and climate over the past 17000 yr in central Yellowstone National Park. Geology, 28:211-214.

MOCK, C. J. 1996. Climatic controls and spatial variations of precipitation in the western United States. Journal of Climate, 9:1111-1125.

MOCK, C. J. 1991. Drought and precipitation fluctuations in the Great Plains during the Late Nineteenth Century. Great Plains Research, 1:26-57.

MONNIN, E., E. J. STEIG, U. SIEGENTHALER, K. KAWAMURA, J. SCHWANDER, B. STAUFFER, T. F. STOCKER, D. L. MORSE, J.-M. BARNOLA, B. BELLIER, D. RAYNAUD, AND H. FISCHER. 2004. EPICA Dome C ice core high resolution Holocene and transition CO2 data. IGBP PAGES/World Data Center for Paleoclimatology Data Contribution Series # 2004-055. NOAA/NGDC Paleoclimatology Program, Boulder, Colorado.

MONNIN, E., A. INDERMÜHLE, A. DALLENBACH, J. FLUCKIGER, B. STAUFFER, T. F. STOCKER, D. RAYNAUD, AND J.-M. BARNOLA. 2001. Atmospheric CO_2 concentrations over the last glacial termination. Science, 291:112-114.

MOTE, P. W., A. F. HAMLET, M. P. CLARK, AND D. P. LETTENMAIER. 2005. Declining mountain snowpack in western North America. Bulletin of the American Meteorological Society, 86(11):39-49.

NCDC (NATIONAL CLIMATE DATA CENTER). 1994. Time bias corrected divisional temperature-precipitation-drought index. Documentation for dataset TD-9640. Available from: DBMB, NCDC, NOAA, Federal Building, 37 Battery Park Ave., Asheville, NC 28801-2733.

NICC (NATIONAL INTERAGENCY COORDINATION CENTER). 2008. Total Wildland Fires and Acres (1960-2008). Boise,

Idaho.

NRC (NATIONAL RESEARCH CENTER). 2007. Colorado River Basin Water Management: Evaluating and Adjusting to Hydroclimatic Variability. The National Academies Press, Washington, D.C., 210 p.

PERSICO, L., AND G. MEYER. 2009. Holocene beaver damming, fluvial geomorphology, and climate in Yellowstone National Park, Wyoming. Quaternary Research, 71: 340-353.

PHILLIPS, D. H., Y. REININK, T. E. SKARUPA, C. E. ESTER III, AND J. A. SKINLOV. In press. Water resources planning and management at the Salt River Project, Arizona, USA. Irrigation and Drainage Systems.

PIERCE, K. L., K. P. CANNON, G. A. MEYER, M. J. TREBESCH, AND R. D. WATTS. 2002. Post-Glacial inflation-deflation cycles, tilting, and faulting in the Yellowstone caldera based on Yellowstone Lake shorelines. U.S. Geological Survey Open-File Report 02-0142.

POWER, M., J. MARLON, N. ORTIZ, P. BARTLEIN, S. HARRISON, F. MAYLE, A. BALLOUCHE, R. BRADSHAW, C. CARCAILLET, C. CORDOVA, S. MOONEY, P. MORENO, I. PRENTICE, K. THONICKE, W. TINNER, C. WHITLOCK, Y. ZHANG, Y. ZHAO, A. ALI, R. ANDERSON, R. BEER, H. B. BEHLING, G, C. BRILES, K. BROWN, A. BRUNELLE, M. BUSH, P. CAMILL, G. CHU, J. CLARK, D. COLOMBAROLI, S. CONNOR, A. L. DANIAU, M. DANIELS, J. DODSON, E. DOUGHTY, M. EDWARDS, W. FINSINGER, D. FOSTER, J. FRECHETTE, M. J. GAILLARD, D. GAVIN, E. GOBET, S. HABERLE, D. HALLETT, P. HIGUERA, G. HOPE, S. HORN, J. INOUE, P. KALTENRIEDER, L. KENNEDY, Z. KONG, C. LARSEN, C. LONG, J. LYNCH, E. LYNCH, M. McGLONE, S. MEEKS, S. MENSING, G. MEYER, T. MINCKLEY, J. MOHR, D. NELSON, J. NEW, R. NEWNHAM, R. NOTI, W. OSWALD, J. PIERCE, P. RICHARD, C. ROWE, M. SANCHEZ GOÑI, B. SHUMAN, H. TAKAHARA, J. TONEY, C. TURNEY, D. URREGO-SANCHEZ, C. UMBANHOWAR, M. VANDERGOES, B. VANNIERE, E. VESCOVI, M. WALSH, X. WANG, N. WILLIAMS, J. WILMSHURST, AND J. ZHANG. 2008. Changes in fire regimes since the Last Glacial Maximum: An assessment based on a global synthesis and analysis of charcoal data. Climate Dynamics, 30:887-907.

POWER, M. J., C. WHITLOCK, P. BARTLEIN, AND L. R. STEVENS. 2006. Fire and vegetation history during the last 3800 years in northwestern Montana. Geomorphology, 75:420-436.

PRAIRIE, J., K. NOWAK, B. RAJAGOPALAN, U. LALL, AND T. FULP. 2008. A stochastic approach for streamflow generation combining observational and paleoreconstructed data. Water Resources Research, 44:WO6423, doi:10.1029/2007WR006684.

PYNE, S. J. 2001. Year of the Fires: The Story of the Great Fires of 1910. Viking Press, New York. 322 p.

RICE J. L., C. A. WOODHOUSE, AND J. J. LUKAS. In press. Science and decision-making: Water management and tree-ring data in the western United States. Journal of the American Water Resources Association.

ROMME W. H., AND D. G. DESPAIN. 1989. Historical perspective on the Yellowstone fires of 1988. Bioscience, 39:695-699.

ROMME, W. H. 1982. Fire and landscape diversity in subalpine forests of Yellowstone National Park. Ecological Monographs, 52:199-221.

SCHWARTZ, P. 1991. The Art of the Long View. Doubleday, New York, 290 p.

SHIN, S.-I., P. D. SARDESHMUKH, R. S. WEBB, R. J. OGLESBY, AND J. J. BARSUGLI. 2006. Understanding the mid-Holocene climate. Journal of Climate, 19:2801-2817.

SHOEMAKER, P. J. H. 1995. Scenario planning: A tool for strategic thinking. Sloan Management Review, 37:25-40.

SHUMAN, B., A. HENDERSON, S. M. COLMAN, J. R. STONE, L. R. STEVENS, S. C. FRITZ, M. J. POWER, AND C. WHITLOCK. 2009. Holocene lake-level trends in the Rocky Mountains, U.S.A. Quaternary Science Reviews, 28:1861-1879.

SHUMAN, B., P. J. BARTLEIN, AND T. WEBB III. 2005. The relative magnitude of millennial- and orbital-scale climate change in eastern North America during the late-quaternary. Quaternary Science Reviews, 24:2194-2206.

STEVENS, L. R. J. R. STONE, J. CAMPBELL, AND S. C. FRITZ. 2006. A 2200-yr record of hydrologic variability from Foy Lake, Montana, USA, inferred from diatom and geochemical data. Quaternary Research, 65:264-274.

STEWART, I. T., D. R. CAYAN, AND M. D. DETTINGER. 2004. Changes in snowmelt runoff timing in western North America under a `business as usual' climate change scenario. Climatic Change, 62:217-232.

STOCKTON, C. W., AND D. M. MEKO. 1983. Drought recurrence in the great plains as reconstructed from long-term tree-ring records. Journal of Applied Meteorology, 22:17-29.

STOCKTON C. W., AND G. C. JACOBY. 1976. Long-term surface-water supply and streamflow trends in the Upper Colorado River Basin. Lake Powell Research Project Bulletin. No. 18. National Science Foundation, Arlington, Virginia.

STOKES, S., AND D. R. GAYLORD. 1993. Optical dating of holocene dune sands in the Ferris Dune Field, Wyoming. Quaternary Research, 39:274-281.

SWETNAM T. W., AND J. L. BETANCOURT. 1998. Mesoscale disturbance and ecological response to decadal climatic variability in the American Southwest. Journal of Climate, 11:3128-3147.

SWETNAM, T. W., C. D. ALLEN, AND J. L. BETANCOURT. 1999. Applied historical ecology: Using the past to manage for the future. Ecological Applications, 9:1189-1206.

TANS, P. 2009. Mauna Loa CO_2 monthly mean data. NOAA/ESRL Available from: www.esrl.noaa.gov/gmd/ccgg/trends.

TRENBERTH, K. E., G. W. BRANSTATOR, AND P. A. ARKIN. 1988.

Origins of the 1988 North American drought. Science 242:1640-1645.

VAN DE WATER, P. K., S. W. LEAVITT , AND J. L. BETANCOURT. 1994. Trends in stomatal density and $^{13}C/^{12}C$ ratios of *Pinus flexilis* needles during the last glacial-interglacial cycle. Science, 264:239-243.

VAN DER HEIJDEN, K. 1996. Scenarios: The Art of Strategic Conversation. Wiley & Sons, New York, 380 p.

VAN MANTGEM, P. J, N. L. STEPHENSON, J. C. BYRNE, L. D. DANIELS, J. F. FRANKLIN, P. Z. FULÉ, M. E. HARMON, A. J. LARSON, J. M. SMITH, A. H. TAYLOR, AND T. T.VEBLEN. 2009. Widespread increase of tree mortality rates in the western United States. Science, 323: 521-524.

WATSON, T. A. F. A. BARNETT, S. T. GRAY, AND G. A. TOOTLE. 2009. Reconstructed streamflows for the headwaters of the Wind River, Wyoming, United States. Journal of the American Water Resources Association, 45:224-236.

WENG, C., AND S. T. JACKSON. 1999. Late-glacial and Holocene vegetation and climate history of the Kaibab Plateau, northern Arizona. Palaeogeography, Palaeoclimatology, Palaeoecology, 153:179-201.

WESTERLING, A. L., H. G. HIDALGO, D. R. CAYAN, AND T. W. SWETNAM. 2006. Warming and earlier spring increase western U.S. forest wildfire activity. Science, 313:940-943.

WHITLOCK, C. 1993. Postglacial vegetation and climate of Grand Teton and southern Yellowstone National Parks. Ecological Monographs, 63:173-198.

WHITLOCK, C., AND P. J. BARTLEIN. 2004. Holocene fire activity as a record of past environmental change, p. 479-489. *In* A. Gillespie and S. C. Porter (eds), Developments in Quaternary Science Volume 1: The Quaternary Period in the United States. Elsevier, Amsterdam.

WHITLOCK, C., AND C. P. S. LARSEN. 2002. Charcoal as a fire proxy. p. 75-98 *In* J. P. Smol, H. J. B. Birks, and W. M. Last (eds.), Tracking Environmental Change Using Lake Sediments. Volume 3. Terrestrial, Algal, and Siliceous Indicators. Kluwer, Dordrecht.

WHITLOCK, C., W. DEAN, J. ROSENBAUM, L. STEVENS, S. FRITZ, B. BRACHT, AND M. POWER. 2008a. A 2650-year-long record of environmental change from northern Yellowstone National Park based on a comparison of multiple proxy data. Quaternary International, 188:126-138.

WHITLOCK, C., J. MARLON, C. BRILES, A. BRUNELLE, C. LONG, AND P. BARTLEIN. 2008b. Long-term relations among fire, fuel, and climate in the north-western US based on lake-sediment studies. International Journal of Wildland Fire, 17:72-83.

WILLIAMS, J. W., AND B. SHUMAN. 2008. Obtaining accurate and precise environmental reconstructions from the modern analog technique and North American surface pollen dataset. Quaternary Science Reviews, 27:669-687.

WOODHOUSE, C. A., AND J. T. OVERPECK. 1998. 2000 years of drought variability in the central United States. Bulletin of the American Meteorological Society, 79:2693-2714.

WOODHOUSE, C. A., S. T. GRAY, AND D. M. MEKO. 2006. Updated streamflow reconstructions for the Upper Colorado River Basin. Water Resources Research, 42:W05415.

WORRALL, J. J., L. EGELAND, T. EAGER, R. A. MASK, E. W. JOHNSON, P. A. KEMP, AND W. D. SHEPPERD. 2008. Rapid mortality of *Populus tremuloides* in southwestern Colorado, USA. Forest Ecology and Management, 255:686-696.

ZIER, J. L., AND W. L. BAKER. 2006. A century of vegetation change in the San Juan Mountains, Colorado: An analysis using repeat photography. Forest Ecology and Management, 228:251-262.

HISTORICAL ECOLOGY FOR THE PALEONTOLOGIST

JEREMY B. C. JACKSON[1,2] AND LOREN McCLENACHAN[1]

[1]Center for Marine Biodiversity and Conservation, Scripps Institution of Oceanography, University of California, San Diego, La Jolla, CA 92093-0244 USA
[2]Center for Tropical Paleoecology and Archeology, Smithsonian Tropical Research Institute, P.O. Box 0843-03092 Panama, Republic of Panama

ABSTRACT.—Human exploitation and pollution have shifted the baselines of contemporary marine ecosystems almost beyond recognition from their formerly pristine state. Structural organisms like corals, seagrasses, and kelps are greatly reduced, megafauna virtually eliminated, and food webs altered from dominance by large, apex predators to much smaller fishes and invertebrates. Geological, archeological, historical, and early scientific data are essential to reconstruct the pristine composition, structure, and function of recent marine ecosystems before human disturbance intensified. Archeological, historical, and fisheries data have the advantage of first hand human observation, but the disadvantage inherent in human choice. The fossil record is comparatively immune to such human bias but suffers potentially equivalent taphonomic effects. Nevertheless, the geochemical and fossil record provide the only direct test of the extent to which natural environmental change versus human impacts have altered marine communities. Deeper time perspective provides valuable tools for conservation and management, but only so long as the rules about the ways ecosystems function do not drastically change. Allee effects and positive feedbacks are increasingly pushing ecosystems into alternative states. The basic question is how far this degradation can go before we cannot retrieve what we have lost.

INTRODUCTION

CHARLES LYELL famously taught that the present is the key to the past. The idea served geology well for nearly a century and a half until incontrovertible evidence of sudden changes and catastrophes in deep time forced us to try to imagine worlds untold numbers of standard deviations beyond the mean of modern experience. Similar ecological uniformitarianism dominated ecology throughout most of the 20th century. Ecologists accepted the way the world is at face value, and the very foundations of modern ecological theory about distribution and abundance, food webs, and community structure were based on the assumption that what we can observe today is all that matters. Then, slowly, ecologists discovered history and the changes that fluctuations in climate and human activities had wrought on ecosystems before we began to study them. The new science of historical ecology attempts to bring such longer-term dynamics and people back into the ecological equation, where they always should have been, and to provide more meaningful baselines for basic ecological understanding as well as the conservation and restoration of altered ecosystems. Here we review the basic principles of historical ecology using examples drawn from ocean ecosystems with which we are most familiar, but the approach is equally applicable to the land.

BASIC PRINCIPLES

Historical ecology attempts to answer four questions: 1) Is it scientifically defensible to talk about an ecological "baseline" when ecosystems were supposedly pristine? 2) If so, what are the relevant historical data? 3) In what ways and how much have ecological processes and ecosystems changed over human history? 4) Why did they change?

Baselines

Ecologists and conservationists commonly talk about pristine ecosystems, which raises the question "pristine" in comparison to what? Environments are always changing, so that the longer the time series, the greater the variance in whatever it is we wish to measure (Davis, 1986). Moreover, since most species alive today originated in the Late Pliocene or Early Pleistocene, they have experienced a virtual roller coaster

In *Conservation Paleobiology: Using the Past to Manage for the Future, Paleontological Society Short Course, October 17th, 2009. The Paleontological Society Papers, Volume 15*, Gregory P. Dietl and Karl W. Flessa (eds.). Copyright © 2009 The Paleontological Society.

of sometimes rapidly changing environmental conditions (Petit et al., 1999) throughout their evolutionary histories (Jackson and Johnson, 2000; O'Dea and Jackson, in press). Nevertheless, increase in environmental variance over time does not describe a smooth, continuous curve, and there have been extended periods of comparative environmental stability for thousands of years. Such was the case throughout much of the Holocene when human civilization evolved, although numerous small-scale climatic shifts occurred that wreaked havoc for civilizations living at the limit of their natural resources (Lamb, 1995; Mayewski et al., 2004).

The range of Holocene variability provides a baseline with which to compare changes in climate and ecosystems wrought by humanity (Hansen et al., 2006; Lovelock, 2009). Atlantic cod moved south during the Little Ice Age and then back north again (Lamb, 1995; Fagan, 2001), and abundance of sardines, anchovies, and salmon fluctuated out of phase in the California Current and northeastern Pacific over the past several thousand years for oceanographic reasons that are still incompletely understood (Baumgartner et al., 1992; Finney et al., 2002). But there were a lot more cod, sardines, anchovies, and salmon for thousands of years than there are today because of overfishing. Figuring out how much was "a lot more" is a major goal of historical analysis.

Data

Historical ecological data are highly diverse, including isotopic, geochemical, and sedimentological proxies of past environmental conditions, fossils, archeological remains, accounts of explorers and colonial bureaucrats, naturalists, scientific expeditions, and exceptionally long scientific time series, among others (Jackson, 2001; Jackson et al., 2001; Lotze and Worm, 2009). But for purposes of this paper it is convenient to group these into four categories: geological, archeological, historical narrative, and modern scientific and fisheries data.

Many historical ecological data are qualitative or semi-quantitative, which is why ecologists tended to ignore them. But there is an enormous wealth of historical data if we know how to use them, and confidence in reconstructions of past conditions greatly increases when different kinds of primary data and proxies are in general agreement. Most importantly, failure to incorporate historical information into scientific assess-

ments because the data are imprecise risks ignoring the obvious. Terrestrial historical ecologists and environmental historians have understood this since at least the 1980s (Cronon, 1983; Davis, 1986). No one would deny the once fantastically great abundance of bison in North America before the arrival of the horse and the rifle (Isenberg, 2000) just because there are no good quantitative ecological survey data from the 16th century. But that is, in effect, what most marine ecologists have done until very recently for whales, monk seals, sea turtles, sea birds, Atlantic cod, salmon, and countless other species.

As Daniel Pauly (1995) famously said, the anecdotes are the data. An especially revealing example comes from detailed oral histories of fishers of different ages from the Gulf of California (Saenz-Arroyo et al., 2005). Older fishers stated that five times more species and fishing sites had been depleted and reported personal best catches five times greater than younger fishers who by and large believed the stocks were still healthy. Similar cases abound for virtually all fish stocks for which quantitative time series, if they are even available, do not begin before the 1980s. In such cases, insistence on traditional quantitative population data represents a kind of false "precisionism" that ignores reality and generality (*sensu* Levins, 1968), and we do so at our peril. The oceans used to be dominated ecologically by an enormous abundance of very large animals that are impossible to study with modern quantitative ecological methods because they are ecologically or biologically extinct (Jackson et al., 2001; Lotze and Worm, 2009).

Documenting change

We need to evaluate historical time series with the same analytical rigor as for more conventional ecological data. Great care is necessary to be sure we are comparing similar habitats and environments in both space and time since differences in depth, distance to shore, or latitude may exceed those from similar habitats over time. Otherwise the historical signal may be drowned out by environmental noise. Large numbers of observations are essential for statistical confidence, and in many cases can provide the basis for statistical assessment of qualitative observations.

For example, *Acropora palmata* and *A. cervicornis* were the most abundant shallow water reef coral species on Caribbean reefs at >80% of Late Pleistocene and Holocene localities in depths <10 m for *A. palmata* and >60% of localities from depths of 10-20 m for *A. cervicornis* (Jackson et al., 2001). However, this clear dominance fell to <50% of biological surveys before the 1980s and <20% to nil after the 1980s so that both species are now listed as critically endangered by the International Union for the Conservation of Nature (IUCN).

Similarly, detailed compilations of the numbers of nesting beaches of green turtles (*Chelonia mydas*) described by 17th and 18th century explorers who feasted on their eggs and meat (Jackson, 1997; McClenachan et al., 2006a) and the numbers of monk seal (*Monachus tropicalus*) rookeries exploited to kill seals for their oil (McClenachan and Cooper, 2008) can be turned into estimates of the animals' past abundance. The best estimates are roughly 90 million Caribbean green turtles compared to perhaps 300,000 today and about a quarter of a million Caribbean monk seals that are now extinct. Even presence/absence data based on species lists takes on statistical significance as a measure of past diversity by plotting collector's curves through time (Johnson et al., 2008; Cramer, in prep.), or for estimates of abundance when based on very large numbers of samples so that the numbers of occurrences converge on numerical relative abundance (Pandolfi and Jackson, 2001).

Ascribing causes of change

It is essential to formulate falsifiable hypotheses of cause and effect based on changes in environments or human activities that can be measured independently of the biological changes of interest (Jackson and Erwin, 2007). Sometimes this is as simple as demonstrating that the demise of species consistently followed closely upon the development of a fishery to exploit them. Such was the case for the serial depletions of species in the Bay of Fundy (Lotze and Milewski, 2004), Wadden Sea (Lotze et al., 2005), and for green and hawksbill (*Eretmochelys imbricata*) turtles from place to place throughout the Caribbean as one stock after the other collapsed (Jackson, 1997; McClenachan et al., 2006a).

But it is often more difficult to distinguish environmental change from human impact or the relative importance of different anthropogenic and natural environmental factors, as in the never-ending debate about the cause of megafaunal extinctions on the land (Barnosky et al., 2004). The same complexity plagues attempts to distinguish and measure the consequences of different kinds of human activities, such as the effects of overfishing, pollution, and climate change on the de-

mise of reef corals (Hughes, 1994; Knowlton, 2001; Pandolfi et al., 2005; Aronson and Precht, 2006; Bruno et al., 2009). In some ways the arguments are sterile because of interactive effects among the different factors, as we are finally beginning to understand for reef corals (Pandolfi et al., 2005; Kline et al., 2006; Smith et al., 2006; Knowlton and Jackson, 2008). But in other ways the debates are deeply important, because blind attention to the fashionable factor of the moment can distract managers and policy makers from the complexity of the problems they are trying to address (Knowlton and Jackson, 2008).

REEFS SINCE COLUMBUS REVISITED

Live coral covered more than half the surface of Caribbean coral reefs before the 1970s when corals began to decline precipitously (Gardner et al., 2003) due to overgrowth and smothering by seaweeds, coral disease, and coral bleaching (Hughes, 1994; Knowlton, 2001), so that living corals cover barely 10% of Caribbean reefs today. However, some places like Barbados lost most of their acroporid corals before the 20th century (Lewis, 1960, 1984). Moreover, other components of the reef ecosystem had declined to varying degrees well before the 1970s, as evidenced by the lack of large fish in most underwater photographs since the 1950s and the near extirpation of most megafauna even earlier. For example, one of us (Jackson) did not see a single green turtle in the more than fifty Caribbean turtlegrass meadows that he studied for his thesis during an intensive year of fieldwork in the late 1960s.

Caribbean coastal ecosystems were obviously degraded long before ecologists began to study them (Jackson, 1997; Fitzpatrick et al., 2008), and the challenge has been to reconstruct the historical trajectory and magnitude of degradation to better understand how pristine ecosystems functioned in the past (Pandolfi et al., 2003) and for management (Pandolfi et al., 2005). Five somewhat overlapping phases of historical degradation are emerging: (1) increasingly intensive pre-Columbian exploitation of tetrapods, fish, and mobile invertebrates based on archeological data and early explorers' accounts; (2) a poorly documented respite from exploitation due to the virtual annihilation of the millions of Native American inhabitants by Europeans; (3) decimation of megafauna including sea turtles, crocodiles, and marine mammals in the 17th to mid 19th centuries based primarily on historical narratives; (4) the inexorable removal of anything edible or marketable from the mid to late 19th century to the present based on fisheries statistics, scientific surveys, and newspaper accounts and photographs; and (5) the mass mortality of reef corals since the 1970s based on a combination of recent ecological surveys and paleontological data as a measure of pristine abundance.

Geological and paleontological analysis has played an important role in deciphering the timing and causes of the decline of reef corals, but we have barely scratched the surface of what could be done for other well-preserved organisms, most importantly molluscs for impacts of exploitation and foraminifera and other microfossils for changes in water quality and environments. Parallel trajectories for different taxa, and the similarities and differences among them, are necessary for better understanding of the causes of coral decline and outbreaks of epidemic disease that have been increasingly frequent since the early 20th century.

Structural organisms

Based on comparisons with Pleistocene and Holocene fossil reefs (Jackson, 1992; Pandolfi and Jackson, 2001, 2006; Pandolfi, 2002), communities of reef corals appeared to be remarkably intact around many islands in the Caribbean (Goreau, 1959; Bak and Luckhurst, 1980; Rogers et al, 1984; Liddell and Ohlhorst, 1987, 1988) until the 1970s when regional degradation began (Gardner et al., 2003). This impression received substantial support from studies of coral occurrences in push cores that extended back hundreds to a few thousand years (Aronson et al., 2002, 2004, 2005). For example, the cores showed that reefs at Belize were dominated by *A. cervicornis* for more than a thousand years until the corals died *en masse* from disease in 1986. However, historical analysis of the declining dominance of *Acropora* in the earlier 20th century suggests that considerable changes had already begun elsewhere (Jackson et al., 2001). This is especially true for Barbados where *Acropora* all but vanished at least a century ago so that coral communities in the 1950s (Lewis, 1960) to 1980s (Stearn et al., 1977) bore little resemblance to anything in the previous 250,000 years (Pandolfi and Jackson, 2006).

The Barbados story is key because the disappearance of *Acropora* there predates the advent of rising temperatures, global warming, and extensive coral bleaching that some have singled out as the only important drivers of coral decline (Aronson et al., 2005; Aronson and Precht, 2006; Bruno et al., 2009), as opposed to the hypothesis that other factors such as overfishing are also important drivers of degradation that had altered the baseline long before (Hughes, 1994; Jackson et al., 2001; Pandolfi et al., 2003; Knowlton and Jackson, 2008). In fact, the only data that support the hypothesis of little change in coral community composition before the 1980s come from the push cores that are too narrow to sample anything but small fragments of mostly branching corals, thereby ignoring most of the coral diversity on reefs. In contrast, a new preliminary study based on the excavation of 0.25 m^2 quadrats within the reef framework at Bocas del Toro, Panama (Cramer, in prep.) has revealed a 33-42% decline in coral diversity and changes in community composition over the past 60 years, including sites previously sampled by push cores (Aronson et al., 2004). The pits were only excavated to 25 cm, but new excavations up to 80 cm below the reef surface show even more profound and earlier changes in coral composition. Thus, it appears that Caribbean coral communities had already begun to change substantially a century before the tragic more recent devastation due to disease and bleaching.

Most reef sponges are not preserved as fossils or in archeological sites but there is excellent historical evidence for the virtual extirpation of commercially valuable sponges since the 19th century and for earlier changes in the relative abundance of toxic versus nontoxic sponges. The best data are for harvests of commercial sponges in Florida that show the classic pattern of rapidly increasing "catch" that peaked in 1930, followed by precipitous decline due to overfishing and outbreak of epidemic disease in 1939 (McClenachan, 2008). Before the crash, fishers had moved progressively farther out to sea as nearshore stocks were exhausted, and there is reason to believe that the loss of sponges may have played an important role in the decline of water quality due to very large reductions in suspension feeding.

Another tantalizing indication of earlier anthropogenic changes in sponge communities comes from historical accounts of the toxicity of hawksbill turtles that were hunted to near extinction for their beautiful shells since the 17th century (Jackson, 1997; Bjorndal and Jackson, 2003; McClenachan et al., 2006a). Hawksbills feed exclusively on sponges. They prefer less toxic species when available but can subsist on more toxic species when they are all that is available (León and Bjorndal, 2002). People who ate hawksbills before the 19th century invariably became sick but not afterwards. This suggests that when hawksbills were abundant nontoxic sponges were rare. But when the turtles were nearly exterminated by hunting, nontoxic sponges became so abundant that the few surviving hawksbills could afford to be picky eaters. Thus, it is likely that the great abundance and species composition of Caribbean reef sponge communities today is an artifact of overfishing hawksbill turtles.

Turtlegrass (*Thalassia testudinum*) and other seagrasses stabilize the seafloor and provide habitat for a great diversity of associated species, many of which occur rarely if ever in other habitats (Phillips, 1960; Ogden, 1980; Thayer et al., 1982; Zieman, 1982; Ogden et al., 1983; Thayer et al., 1984). In 1987, turtlegrass suffered massive die-offs in Florida Bay due to a fungal wasting disease (Zieman et al., 1988; Hall et al., 1999) similar to the disease that decimated eelgrass meadows in temperate estuaries and coastal seas (Lotze et al., 2006). Early reports by Dampier (1729) and Audubon (1832) describe turtlegrass as growing only10-15 cm high as opposed to 25-75 cm since at least the 1950s to today (Phillips, 1960; Randall, 1965; Zieman, 1982). The obvious explanation for much longer blades today is the decimation of the some 90 million green turtles that consumed roughly half of total annual turtlegrass productivity as opposed to less than 0.1% of total productivity today, with concomitantly large changes in sediment geochemistry and the deposition of organic matter (Jackson, 1997; Jackson et al., 2001; McClenachan et al., 2006a). As they grow, the older, more distal tips of the turtlegrass blades become increasingly fouled by all kinds of encrusting organisms. Fungal infection begins at these older, more heavily fouled tips of the blades, which strongly implies that outbreaks of turtlegrass wasting disease are the result of overharvesting of green turtles (Jackson et al., 2001) rather than a "natural cause" (Short et al., 2001) or the long list of anthropogenic changes in oceanographic, geochemical, and sedimentological factors that are the usual suspects (Zieman et al., 1988).

The question arises, as for *A. cervicornis*, whether comparably massive die-offs of turtlegrass occurred in the past or how turtlegrass communities may have previously changed due to natural or anthropogenic causes. Turtlegrass is rarely preserved in the fossil record but fortunately its associated species of molluscs, ostracodes, Foraminifera, and other microfossil taxa leave an excellent record and isotopic analyses of their shells record changes in oceanographic conditions, as shown for Florida Bay (Brewster-Wingard and Ishman, 1999; Fourqurean and Robblee, 1999; Halley and Roulier, 1999; Cronin et al., 2002). Circulation in the bay was restricted and salinities increased in the early 19th century, after a century of stability in the 1800s, due to railroad building that reduced circulation and early drainage efforts in south Florida that decreased freshwater inflow. Most paleoecological investigations have focused on reconstructing changes in environmental conditions rather than shifts in community composition that might better inform ecological cause and effect. It would be particularly interesting to establish when organisms restricted to the aging tips of turtlegrass blades first appear abundantly in the sediment record to help unravel the relative timing of oceanographic changes and eutrophication due to increased nutrients versus those due to changes in food webs associated with the loss of green turtles and other grazers. By analogy with reef corals, increased nutrients *per se* are unlikely to be the cause of disease rather than increases in organic matter that support microbial populations including disease pathogens (Nugues et al., 2004; Kline et al., 2006; Smith et al., 2006; Marhaver et al., 2008).

Tetrapods
The early, precipitous decline of sea turtles, marine mammals, and sharks is well documented archeologically and historically because of their ease of catch, high commercial value, and visibility that is reflected by detailed accounts in innumerable historical and commercial documents. The most extensive records are for green turtles, vividly described from the voyages of Columbus and by most subsequent explorers, and the hawksbill turtle (Jackson, 1997; McClenachan et al., 2006a). But Native Americans heavily exploited green turtles long before Columbus and had already hunted them to near local extinction in places like Jamaica and the Bahamas (Carlson, 1999; Carlson and Keegan, 2004; Frazier, 2002; Fitzpatrick et al., 2008). Sea turtles, crocodiles (*Crocodylus acutus*), manatee (*Trichechus manatus*), and several taxa of sharks comprised up to 90% of biomass of kitchen middens in Jamaica from the time of first settlement about 600 to 900 A.D., but were comparatively rare thereafter when the inhabitants progressively shifted to small terrestrial mammals and offshore fish like tuna (Hardt, 2008). The size of fish based on the size of vertebrae also declined in Jamaican middens from an average of 1,200 to 300

grams with a concomitant shift from larger predators to parrotfish. Very similar depletion occurred on islands in the Bahamas and elsewhere (Carlson, 1999; Frazier, 2002 ; Carlson and Keegan, 2004).

Prehistoric depletion of sea turtles pales in comparison to European slaughter that systematically exploited green turtles for their meat and hawksbill turtles for their shells. Calculations of past abundances are based on historic harvest data coupled with observations of populations on nesting beaches from different sites around the Caribbean (McClenachan et al., 2006a). Both species were reduced to less than 1% of their abundance in the 16th century. Twenty percent of nesting beaches were eliminated with nesting populations at half of the remaining beaches perilously close to extinction. An alarming implication for management is that all the examples of local nesting beach recovery are from baselines < 40 years duration whereas all longer time series exhibit dramatic decline (McClenachan et al., 2006a).

The extinct Caribbean monk seal was extensively distributed throughout the tropical Western Atlantic in the 17th century but was gone from the southeastern Caribbean before 1700 (McClenachan and Cooper, 2008). There were some 13 breeding colonies that were hunted out first for meat and subsequently for seal oil to be used as a lubricant for sugar factories. Estimates of abundance based on historical distribution and hunting data range from 233,000 to 338,000 seals, populations vastly greater than could be sustained by the remnant populations of reef fish and crustaceans around Caribbean reefs today. The last seals were exterminated from the western Caribbean by 1952. Historical ecological data for other marine mammals are sparse. Manatee may have been largely extirpated from Florida and the heavily populated Greater Antillean islands before European contact based on comparison with their great abundance in Central America (Dampier, 1729; McKillop, 1985) and their abundance in pre-Columbian archeological sites (Hardt, 2008). Whales and dolphins were also extensively hunted.

Fishes and sharks

Remains of reef fishes are rare in sediments but evidence of their activities can be derived from their impact on more durable organisms like corals and mollusks. For example, highly territorial three-spot damselfish (*Stegastes planifrons*) are extremely abundant on Caribbean reefs today and the question arises whether these tiny fishes were so abundant before the overfishing of their predators. The damselfish tend gardens of algae that they defend against marauding herbivores (Kaufman, 1977). Before the demise of staghorn corals in the 1980s, the damselfish preferentially tended gardens on living staghorn corals that they killed by repeated biting to provide space for the algae. The corals fought back by forming defensive skeletal barriers or "chimneys" that are preserved after the corals have died. The chimneys are abundantly preserved on fossil staghorn from Late Pleistocene deposits long before humans arrived (Kaufman, 1981), which strongly suggests that the damselfish were always abundant on Jamaican reefs.

As we have seen, archeological studies indicate extensive pre-Columbian overfishing of nearshore fishes and sharks. Populations of these animals very likely rebounded following the genocide of Native Americans throughout the region (Hardt, 2008) as evidenced by the glowing 16th to 17th century accounts of Dampier (1729) and Oviedo y Valdez (1526), among many others. Nevertheless, fisheries in Jamaica were considerably depleted again by the mid 19th century following the emancipation of slaves who were left to fend for themselves, and stocks had collapsed by the 1940s due to the use of fish traps constructed of chicken wire when there were no large fish left to catch (Thompson, 1945; Hardt, 2008).

The best quantitative historical data for reef fisheries in the greater Caribbean region come from the 130-year fishery time series for the Florida Keys beginning with the United States Fish Commission report for the 1880s (Goode, 1887a, b). One fifth of the historically commonest 20 fishery species were rare by the 1950s and one half are rare today, including four that are now listed as Critically Endangered or Endangered by the IUCN (McClenachan et al., 2006b). Fishing since WWII has mostly affected long-lived top predators and large bodied fish, most of which are rare today. This decline, as well as a near total shift from dominance by large groupers (Serranidae, Epinephelinae) to small snappers (Lutjanidae), is vividly evident in photographs of trophy fish from the 1950s to today (McClenachan, 2009a, b). Fishers also had to travel increasingly farther from Key West and more offshore to obtain their catch, and catch per unit effort (CPUE) declined to less than one third of peak values in the early 1930s (McClenachan and Jackson, 2009).

Large sharks including the great white shark

(*Carcharodon carcharias*), smooth hammerhead (*Sphyrna zygaena*), nurse shark (*Ginglymostoma cirratum*), and bull shark (*Carcharhinus leucas*), among others, were very abundant in the Florida Key. Sharks swarmed about the wharves at Key West in the 1880s, warning signs were posted around piers and beaches, and fishers complained of sharks devouring their catch before they could get the fish in their boats (McClenachan and Jackson, 2009). A shark fishery was established for leather, oil, and fins. Commercial fishers caught a hundred sharks per day and catches peaked at 3,000,000 pounds in 1934, which is the equivalent of 12,000 250-pound sharks, which was their average size in the 1930s. In contrast, catches in 2007 were just 265,000 pounds. Recreational catches also declined from 20 sharks a day from small boats in the 1930s to one an hour from modern fishing boats today with all their modern gear and paraphernalia. Smalltooth sawfish (*Pristis pectinata*) were particularly abundant in the 19th century when they were commonly observed swimming in schools. They were fished extensively by commercial and sports fishers whose individual catches exceeded 300 sawfish a year. Sawfish populations are estimated to have declined by at least 95% since European settlement (McClenachan and Jackson, in prep.).

Mobile invertebrates

Spiny lobsters (*Panularis argus*) and queen conch (*Strombus gigas*) were phenomenally abundant until the early 20th century, although earlier declines probably occurred in areas of dense human population. British surveyors in the Florida Keys reported that 5-pound *intertidal* lobsters abounded on mangrove islands and that a boat could be loaded with lobster in a few hours (McClenachan and Jackson, in prep.). Lobsters remained abundant until the 1920s when heavy fishing caused perceptible declines. Initially lobsters were so abundant that it was easier to catch them with nets and spears with catches up to 1,000 lobsters a night. Landings in the Keys peaked in 1918 at 284,000 pounds and had fallen to 53,000 pounds by 1938 when seeding operations had already begun. CPUE from traps declined 90% between 1923 and 1988 and average size also greatly declined. Today there are virtually no intertidal lobsters in the Keys and a 5-pound lobster would be considered a monster.

Queen conch were fished in fantastic numbers by Native Americans throughout the Greater Caribbean as witnessed by the enormous mountains of discarded shells from Los Roques, Venezuela to the Bahamas (Goggin, 1944; Stoner and Ray, 1996; Schapira et al., 2009)—except in the Florida Keys where most of the shells from middens were shipped abroad for manufacture of buttons. Pre-Columbian harvests numbered in the several billions (McClenachan and Jackson, in prep.). Fisheries statistics are incomplete because small conch remained a staple of people throughout the Bahamas and Florida Keys well into the 20th century and most of the harvest was for personal subsistence. Recorded landings peaked at about 20,000 pounds of meat in the 1940s but subsequently crashed because of overfishing and disease. Commercial harvest was banned in 1975 and recreational harvest in 1986, but stocks have failed to recover.

The story of the long-spined black sea urchin *Diadema antillarum* resembles that for acroporid corals in terms of extraordinarily great abundance on Caribbean reefs until 1983-1984 when populations were reduced by >95% throughout the entire tropical western Atlantic because of epidemic disease (Lessios et al., 1984; Lessios, 1988; Chiappone et al., 2002; Newman et al., 2006). Most populations have failed to recover except at isolated sites in the Antilles (Carpenter and Edmunds, 2006). *Diadema* abundance averaged about 5/m² on shallow reefs throughout the Caribbean (Lessios et al., 1984) and >10/m² in Jamaica (Jackson, 1991). Their disappearance was followed closely by population explosions of macroalgae that began to overgrow and smother corals within one year (Hughes et al., 1985; Carpenter, 1990; Hughes, 1994). The great abundance of *Diadema* before the die-off was attributed to overfishing (Levitan, 1992; Hughes, 1994) based on comparisons of known increases in sea urchin abundance on heavily fished reefs in Kenya (McClanahan et al., 1996) and temperate kelp forests following the removal of their predators (Estes and Palmisano, 1974; Steneck et al., 2002). However, historical descriptions clearly demonstrate that *Diadema* were strikingly abundant to all who ran across them as far back as 1688. *Diadema* were also the most abundant sea urchins in Late Pleistocene reef deposits, 100,000 years before people had come to the Americas (Jackson, 1997).

The paleoecological studies of seagrass decline in Florida Bay demonstrate the power of retrospective analysis of invertebrates with preservable hard parts (Brewster-Wingard and Ishman, 1999), but there has been virtually no investigation of changes in invertebrate communities from cores or larger excavations elsewhere in the Caribbean except for reef corals (Aronson et al., 2002, 2004; Cramer, in prep.). However, Edgar and Samson (2004) have elegantly demonstrated the power of this approach using 10-cm-diameter pushcores to measure changes in mollusk communities since 1890 in eastern Tasmania due to the scallop fishery. Collector's curves show that numbers of species declined by half with clear shifts in species composition demonstrated by ordination. There is enormous potential for more of this kind of work in coastal ecosystems around the world (Kidwell, 2007), especially wherever sedimentation rates are sufficiently high to preserve relatively unmixed assemblages down core.

ESTUARIES AND COASTAL SEAS

Like Florida Bay, the historical ecology of several large estuaries and coastal seas is exceptionally well documented based on the full gamut of geological, archeological, historical, and modern ecological data (Jackson et al., 2001; Rabalais et al., 2002; Lotze and Milewski, 2004; Lotze et al., 2005, 2006). Geochemical and paleontological data from cores have been more successfully exploited than for coral reefs, largely because high rates of sedimentation preserve environmental events in estuaries, such as the advent of eutrophication or hypoxia (Rabalais et al., 2007; Raymond et al., 2008), much more clearly than for more extensively bioturbated reef sediments (Best and Kidwell, 2000).

The sequence and patterns of degradation for 13 different sites worldwide (Lotze et al., 2006) closely resembles the global pattern for coral reefs (Pandolfi et al., 2003) with the removal of large bodied tetrapods and fishes before smaller fishes and invertebrates, except for oysters that were consistently among the first organisms to be intensively overfished (Jackson et al., 2001; Kirby, 2004). However, exploitation did not result in dramatic declines in fisheries species or eutrophication in estuaries until the development of market based economies two to three centuries ago—a difference that most likely reflects the vastly greater productivity of estuaries and coastal seas compared with coral reefs. Eutrophication developed at several sites up to half a century before the application of large quantities of industrial fertilizer due to increased deforestation, sediment runoff, and extirpation of oysters and planktivorous fish like menhaden that were the most important biological filters in estuarine ecosystems (Jackson et al., 2001; Lotze et al., 2006).

ECOSYSTEM FUNCTION IN PRISTINE COASTAL ECOSYSTEMS

Important historical generalities are emerging about the ecological structure and function of pristine versus degraded ecosystems, including changes in the structure of food webs, the roles of keystone species and ecosystem engineers, and long-term community stability.

Historical analysis and surveys of remote, relatively pristine localities have conclusively shown that the iconic trophic pyramid with the preponderance of biomass at the lowest trophic levels is an artifact of overfishing in the oceans (Jackson et al., 2001; Friedlander and DeMartini, 2002; Jackson, 2006, 2008; Sandin et al., 2008). This is energetically possible because lower trophic levels turn over much more rapidly because of much shorter generation times than their predators, a long neglected fact clearly pointed out in Eugene Odum's (1953) pioneering ecology textbook. But the shifts in biomass among trophic levels are just the tip of the iceberg because of the phenomenon of trophic cascades, whereby removal of predators at trophic level X results in increases in abundance at level X-1, which in turn reduces abundance of prey at level X-2, and so on (Estes and Palmisano, 1974; Paine, 1980; Carpenter and Kitchell, 1988). Most emphasis on trophic cascades in the ocean has been on apex predators, fishes, mobile invertebrates, and macroalgae. But it is increasingly apparent that trophic cascades don't stop with organisms visible to the naked eye and instead extend all the way down to the microbial loop and viruses (Parsons and Lalli, 2002; Jackson, 2006; Kline et al., 2006). This has the profound potential to shift the flux of primary production from the grazing food chain to the microbial loop, with catastrophic consequences for fisheries (Diaz, 2001; Parsons and Lalli, 2002; Jackson, 2006; Diaz and Rosenberg, 2008). The most dramatic examples of this are the explosive dominance of jellyfish along overfished upwelling coasts (Lynam et al., 2008) and, still very controversially, the explosion in outbreaks of epidemic disease of corals, sea urchins, sponges, conchs, seagrasses, sea turtles, and many more (Kline et al., 2006; Smith et al., 2006), as opposed to simple changes in environmental conditions or pollution (Harvell et al., 1999, 2002). Historical ecology has much to offer in these studies by revealing the structure and inferred function of marine food webs before extreme overfishing began and even longer before adequate modern fisheries statistics began to be compiled.

A second major contribution of historical analysis is the renewed attention to formerly important keystone species and ecosystem engineers whose ecological roles were never suspected or long forgotten. Caribbean seagrass and reef sponge communities are drastically different today because of the extirpation of green and hawksbill turtles (Jackson, 1997; McClenachan et al., 2006a), and we can only guess at the role of monk seals that must have consumed fantastic quantities of all those lobsters in the Florida Keys as well as reef fishes and whatever else (McClenachan and Cooper, 2008). The same goes for sharks that were greatly feared throughout the tropical western Atlantic before the mid 20th century when they were rapidly overfished (Baum and Myers, 2004; McClenachan and Jackson, 2009). Sharks ate native pearl divers in Venezuela in the early 1500s (Morón, 1954) and it seems unlikely that reef ecologists could have safely done underwater surveys by SCUBA in 1492 without being attacked by sharks. Seagrass meadows and 3-dimensional acroporid reefs were essential hiding places and nurseries for fish and invertebrates as increasingly seen when the corals and seagrasses die (Fourqurean and Robblee, 1999). And so on.

Finally, marine communities were more stable than they are today, with species composition and relative abundance virtually unchanged for hundreds to tens of thousands of years (Jackson, 1992; Pandolfi, 1996, 1999, 2002; Aronson et al., 2002, 2004). Such stability says little about the long debate about coevolved, so-called equilibrium communities, but rather reflects the fundamental niches and life-history characteristics of component species in a physically stable environment (Jackson and Erwin, 2007). There *was* a baseline that we will almost certainly never see again.

CODA

Historical ecology provides the only baseline for the way pristine marine ecosystems used to be and, as such, provides a target for ecological restoration and conservation. It is folly, of course, to imagine we can turn back the clock in the Caribbean to the time of Columbus or in the Chesapeake Bay to 1608 when Captain John Smith first explored the Bay. But knowing what these ecosystems were like then tells us what was possible in the past, what might be possible in the future,

and provides the basic data to help us to decide where along the gradient from pristine to severely degraded we want to be. Many of the most dramatic changes throughout the world oceans happened within just the last 50 to 75 years. Yet baselines for most fisheries, such as the goliath grouper (*Epinephelus itajara*) in the Florida Keys, begin only in the 1980s (McClenachan, 2009b), and the same is true for most fisheries species and plans for "ecosystem restoration" worldwide.

However, what happened in the Holocene before the intensification of human disturbance is relevant to what is possible in the future only so long as the fundamental rules about the ways ecosystems work do not drastically change. The basic question is how far we can degrade ecosystems and still be able to put them back together again if we want to. There are two basic problems (Scheffer et al., 2001; Knowlton, 2004). First, populations may fail to recover if they have fallen below some critical density so that rates of fertilization, reproduction, or recruitment are too low for populations to become reestablished because of Allee effects. Possible examples include the failure of *Diadema*, *Acropora*, or *Strombus* to recover after catastrophic mortality from disease or overfishing (Knowlton, 1992; Stoner and Ray-Culp, 2000). Second, positive feedback mechanisms may stabilize the transition to an alternative state, as seems to be the case for the overgrowth of corals by seaweeds due to overfishing (Knowlton, 1992; Hughes, 1994) because, as we have now learned, the seaweeds leak organic matter that helps to fuel coral disease that inhibits recovery of corals (Kline et al., 2006; Smith et al., 2006; Knowlton and Jackson, 2008). Likewise, eutrophication fuels populations of pathogenic microbes that cause oyster disease that inhibit their recovery from overfishing and helps to perpetuate eutrophication (Jackson et al., 2001; Lotze et al., 2006). In all such cases, the time required for recovery, if indeed recovery is possible, greatly exceeds the time for degradation to occur in the first place.

The problem of alternative states is particularly acute for the consequences of global warming due to the burning of fossil fuels—not only for shifts in ecosystem states, but more ominously the permanent global shift into a hothouse world (Lovelock, 2009). One of the two most striking positive feedback processes associated with global warming is the increasingly rapid melting of Arctic summer ice that decreases the albedo, and therefore further increases the rate of global warming. The second is the intensification of vertical stratification of the oceans (Barnett et al., 2001; Sarmiento et al., 2004) that decreases upwelling of cooler, nutrient-rich deep water that further accelerates surface warming and decreases surface productivity (Roemmich and McGowan, 1995; Polovina et al., 2008); as well as the downwelling of oxygen that may result in anoxia of the global ocean and mass extinction (Wilson and Norris, 2001). Melting of Arctic ice, vertical stratification, and sea level rise are all proceeding more rapidly than expected from the ensemble mean predictions of the Intergovernmental Panel on Climate Change (Rahmstorf et al., 2007; Lovelock, 2009). If and when we enter a hothouse world, Holocene experience will become irrelevant, although paleontology and earth history will continue to provide the only albeit imperfect roadmaps for the future. But in all respects, the prospects for ecosystems are appalling as what is left of nature and humanity vies for space and resources (Smith et al., 2009).

REFERENCES

ARONSON, R. B., AND W. F. PRECHT. 2001. White-band disease and the changing face of Caribbean coral reefs. Hydrobiologia, 460:25-38.

ARONSON, R. B., AND W. F. PRECHT. 2006. Conservation, precaution, and Caribbean reefs. Coral Reefs, 25:441-450.

ARONSON, R. B., I. G. MACINTYRE, W. F. PRECHT, T. J. T. MURDOCH, AND C. M. WAPNICK. 2002. The expanding scale of species turnover events on coral reefs in Belize. Ecological Monographs, 72:150-166.

ARONSON, R. B., I. G. MACINTYRE, C. M. WAPNICK, AND M. W. O'NEILL. 2004. Phase shifts, alternative states, and the unprecedented convergence of two reef systems. Ecology, 85:1876-1891.

ARONSON, R. B., I. G. MACINTYRE, AND W. F. PRECHT. 2005. Event preservation in lagoonal reef systems. Geology, 33:717-720.

AUDUBON, J. J. 1832. Three Floridian episodes, 1945 reprint. Tequesta, 5:52-68.

BAK, R. P. M., AND B. E. LUCKHURST. 1980. Constancy and change in coral reef habitats along depth gradients at Curaçao. Oecologia, 47:145-155.

BARNOSKY, A. D., P. L. KOCH, R. S. FERANEC, S. L. WING, AND A. B. SHABEL. 2004. Assessing the causes of Late Pleistocene extinctions on the continents. Science, 306:70-75.

BARNETT, T. P., D. W. PIERCE, AND R. SCHNUR. 2001. Detection of anthropogenic climate change in the world's oceans. Science, 292:270-274.

BAUM, J. K., AND R. A. MYERS. 2004. Shifting baselines and the decline of pelagic sharks in the Gulf of Mexico. Ecology Letters, 7:135-145.

BAUMGARTNER, T. R., A. SOUTAR, AND V. FRERREIRA-BARTRINA. 1992. Reconstruction of the history of Pacific sardine and northern anchovy populations over the past two millennia from sediments of the Santa Barbara Basin, California. California Cooperative Fisheries Institute Reports, 33:24-40.

BEST, M. M. R., AND S. M. KIDWELL. 2000. Bivalve taphonomy in tropical mixed siliciclastic-carbonate settings: I. Environmental variation in shell condition. Paleobiology, 26:80-102.

BJORNDAL, K. A., AND J. B. C. JACKSON. 2003. Roles of sea turtles in marine ecosystems: Reconstructing the past, p. 259-273. In P. L. Lutz, J. A. Musick, and J. Wyneken (eds.), The Biology of Sea Turtles, II. CRC Press, Boca Raton, Florida.

BREWSTER-WINGARD, G. L., AND S. E. ISHMAN. 1999. Historical trends in salinity and substrate in central Florida Bay: A paleoecological reconstruction using modern analogue data. Estuaries, 22:369-383.

BRUNO, J. F., H. SWEATMAN, W. F. PRECHT, E. R. SELIG, AND V. G. W. SCHUTTE. 2009. Assessing evidence of phase shifts from coral to macroalgal dominance on coral reefs. Ecology, 90:1478-1484.

CARLSON, L. A. 1999. Aftermath of a feast: Human colonization of the southern Bahamian archipelago and its effects on the indigenous fauna. Unpublished Ph.D. dissertation, University of Florida.

CARLSON, L. A., AND W. F. KEEGAN. 2004. Resource depletion in the prehistoric northern West Indies, p. 85-107. In S. M. Fitzpatrick (ed.), Voyages of Discovery: Archeology of Islands. Praeger Publishers, Westport, Connecticut.

CARPENTER, R. C. 1990. Mass mortality of Diadema antillarum. I. Long-term effects on sea urchin population-dynamics and coral reef algal communities. Marine Biology, 104:67-77.

CARPENTER, R. C., AND P. J. EDMUNDS. 2006. Local and regional scale recovery of Diadema promotes recruitment of scleractinian corals. Ecology Letters, 9:271-280.

CARPENTER, S. R., AND J. K. KITCHELL. 1988. Consumer control of lake productivity. Bioscience, 38:764-769.

CHIAPPONE, M. D., D. W. SWANSON, AND S. L. MILLER. 2002. Density, spatial distribution and size structure of sea urchins in Florida Keys coral reef and hard-bottom habitats. Marine Ecology-Progress Series, 235:117-126.

CRONIN, T. M., G. S. DWYER, S. B. SCHWEDE, C. D. VANN, AND H. DOWSETT. 2002. Climate variability from the Florida Bay sedimentary record: Possible teleconnections to ENSO, PNA and CNP. Climate Research, 19:233-245.

CRONON, W. 1983. Changes in the Land. Hill and Wang, New York, 241 p.

DAMPIER, W. 1729. A New Voyage Round the World (1968 reprint). Dover Publications, New York, 376 p.

DAVIS, M. B. 1986. Climatic instability, time lags, and community disequilibrium, p. 269-284. In J. M. Diamond and T. J. Case (eds.), Community Ecology. Harper and

Row, New York.

DIAZ, R. J. 2001. Overview of hypoxia around the world. Journal of Environmental Quality, 30:275-281.

DIAZ, R. J., AND R. ROSENBERG. 2008. Spreading dead zones and consequences for marine ecosystems. Science, 321:926-929.

EDGAR, G. J., AND C. R. SAMSON. 2004. Catastrophic decline in mollusc diversity in eastern Tasmania and its concurrence with shellfish fisheries. Conservation Biology, 18:1579-1588.

ESTES, J. A., AND J. F. PALMISANO. 1974. Sea otters: Their role in structuring nearshore communities. Science, 185:1058-1060.

FAGAN, B. M. 2001. The Little Ice Age: How Climate Made History, 1300-1850. Basic Books, New York, 273 p.

FINNEY, B. P., I. GREGORY-EAVES, M. S. V. DOUGLAS, AND J. P. SMOL. 2002. Fisheries productivity in the northeastern Pacific Ocean over the past 2,200 years. Nature, 416:729-733.

FITZPATRICK, S. M., W. F. KEEGAN, AND K. S. SEALEY. 2008. Human impacts on marine environments in the West Indies during the middle to late Holocene, p. 147-164. *In* T. C. Rick and J. M. Erlandson (eds.), Human Impacts on Ancient Marine Ecosystems. University of California Press, Berkeley, California.

FOURQUREAN, J. W., AND M. B. ROBBLEE. 1999. Florida Bay: A history of recent ecological changes. Estuaries, 22:345-357.

FRAZIER, J. 2002. Marine turtles of the past: A vision for the future?, p. 103-116. *In* R. C. G. M. Lauwerier and I. Plug (eds.), The Future from the Past: Proceedings of the 9th Conference of the International Council of Archaeozoology, Durham, August 2002. Oxbow Books Ltd., Oxford.

FRIEDLANDER, A. M., AND E. E. DEMARTINI. 2002. Contrasts in density, size, and biomass of reef fishes between the northwestern and the main Hawaiian Islands: The effects of fishing down apex predators. Marine Ecology-Progress Series, 230:253-264.

GARDNER, T. A., I. M. CÔTE, J. A. GILL, A. GRANT, AND A. R. WATKINSON. 2003. Long-term region-wide declines in Caribbean corals. Science, 301:958-960.

GOGGIN, J. M. 1944. Archeological investigations of the upper Florida Keys. Tequesta, 4:13-35.

GOODE, G. B. 1887a. A geographical review of the fisheries industries and fishing communities for the year 1880. The Fisheries and Fishery Industries of the United States, Government Printing Office, Washington, D. C.

GOODE, G. B. 1887b. The fishing grounds of North America. The Fisheries and Fishery Industries of the United States, Government Printing Office, Washington, D. C.

GOREAU, T. F. 1959. The ecology of Jamaican coral reefs. I. Species composition and zonation. Ecology, 40:67-90.

HALL, M. O., M. J. DURAKO, J. W. FOUQUREAN, AND J. C. ZIEMAN. 1999. Decadal changes in seagrass distribution and abundance in Florida Bay. Estuaries, 22:445-459.

HALLEY, R. B., AND L. M. ROULIER. 1999. Reconstructing the history of eastern and central Florida Bay using mollusk-shell isotope records. Estuaries, 22:358-368.

HANSEN, J., M. SATO, R. RUEDY, K. LO, D. W. LEA, AND M. MEDINA-ELIZADE. 2006. Global temperature change. Proceedings of the National Academy of Sciences of the United States of America, 103:14288-14293.

HARDT, M. J. 2008. Lessons from the past: The collapse of Jamaican coral reefs. Fish and Fisheries, 10:1-16.

HARVELL, C. D., K. KIM, J. M. BURKHOLDER, R.R. COLWELL, P. R. EPSTEIN, D. J. GRIMES, E. E. HOFMANN, E. K. LIPP, A. D. M. E. OSTERHAUS, R. M. OVERSTREET, J. W. PORTER, G. W. SMITH, AND G. R. VASTA. 1999. Emerging marine diseases: Climate links and anthropogenic factors. Science, 285:1505-1510.

HARVELL, C. D., C. E. MITCHELL, J. R. WARD, S. ALTIZER, A. P. DOBSON, R. S. OSTFELD, AND M. D. SAMUEL. 2002. Climate warming and disease risks for terrestrial and marine biota. Science, 296:2158-2162.

HUGHES, T. P. 1994. Catastrophes, phase shifts, and large-scale degradation of a Caribbean coral reef. Science, 265:1547-1551.

HUGHES, T.P., B. D. KELLER, J. B. C. JACKSON, AND M. J. BOYLE. 1985. Mass mortality of the echinoid *Diadema antillarum* Philippi in Jamaica. Bulletin of Marine Science, 36:377-384.

ISENBERG, A. C. 2000. The Destruction of the Bison: An Environmental History, 1750-1920. Cambridge University Press, Cambridge, 218 p.

JACKSON, J. B. C. 1991. Adaptation and diversity of coral reefs. Bioscience, 41:475-482.

JACKSON, J. B. C. 1992. Pleistocene perspectives on coral reef community structure. American Zoologist, 32:719-731.

JACKSON, J. B. C. 1997. Reefs since Columbus. Coral Reefs 16, (supplement):S23-S32.

JACKSON, J. B. C. 2001. What was natural in the coastal ocean? Proceedings of the National Academy of Sciences of the United States of America, 98:5411-5418.

JACKSON, J. B. C. 2006. When ecological pyramids were upside down, p. 27-37. *In* J. A. Estes (ed.), Whales, Whaling, and Ocean Ecosystems. University of California Press, Berkeley, California.

JACKSON, J. B. C. 2008. Evolution and extinction in the brave new ocean. Proceedings of the National Academy of the United States of America, 105 (supplement 1):11458-11465.

JACKSON, J. B. C., AND D. H. ERWIN. 2007. What can we learn about ecology and evolution from the fossil record? Trends in Ecology and Evolution, 21:322-328.

JACKSON, J. B. C., AND K. G. JOHNSON. 2000. Life in the last few million years. Paleobiology, 26 (supplement):221-235.

JACKSON, J. B. C., M. X. KIRBY, W. H. BERGER, K. A. BJORNDAL, L. W. BOTSFORD, B. J. BOURQUE, R. H. BRADBURY, R. COOKE, J. ERLANDSON, J. A. ESTES, T. P. HUGHES, S.

KIDWELL, C. B. LANGE, H. S. LENIHAN, J. M. PANDOLFI, C. H. PETERSON, R. S. STENECK, M. J. EGNER, AND R. R. WARNER. 2001. Historical overfishing and the recent collapse of coastal ecosystems. Science, 293:629-638.

JOHNSON, K. G., J. B. C. JACKSON, AND A. F. BUDD. 2008. Caribbean reef development was independent of coral diversity over 28 million years. Science, 319:1521-1523.

KAUFMAN, L. S. 1977. The threespot damselfish: Effects on benthic biota of Caribbean coral reefs. Proceedings 3rd International Coral Reef Symposium. I. Biology:559-564.

KAUFMAN, L. S. 1981. There was biological disturbance on Pleistocene coral reefs. Paleobiology, 7:527-532.

KIDWELL, S. M. 2007. Discordance between living and death assemblages as evidence for anthropogenic ecological change. Proceedings of the National Academy of Sciences of the United States of America, 104:17701-17706.

KIRBY, M. X. 2004. Fishing down the coast: Historical expansion and collapse of oyster fisheries along continental margins. Proceedings of the National Academy of Sciences of the United States of America, 101:13096-13099.

KLINE, D. I., N. M. KUNTZ, M. BREITBART, N. KNOWLTON, AND F. ROHWER. 2006. Role of elevated organic carbon levels and microbial activity in coral mortality. Marine Ecology-Progress Series, 314:119-125.

KNOWLTON, N. 1992. Thresholds and multiple stable states in coral reef community dynamics. American Zoologist, 32:674-682.

KNOWLTON, N. 2001. The future of coral reefs. Proceedings of the National Academy of Sciences of the United States of America, 98:5419-5425.

KNOWLTON, N. 2004. Multiple "stable" states and the conservation of marine ecosystems. Progress in Oceanography, 60:387-396.

KNOWLTON, N., AND J. B. C. JACKSON. 2008. Shifting baselines, local impacts, and global change on coral reefs. PloS Biology, 6:0215-0220.

LAMB, H. H. 1995. Climate, History and the Modern World, 2nd edition. Routledge, New York, 410 p.

LEON, Y. M., AND K. A. BJORNDAL. 2002. Selective feeding in the hawksbill turtle, an important predator in coral reef ecosystems. Marine Ecology-Progress Series, 245:249-258.

LESSIOS, H. A. 1988. Mass mortality of Diadema antillarum in the Caribbean: What have we learned? Annual Reviews of Ecology and Systematics, 19:371-393.

LESSIOS, H. A., D. R. ROBERTSON, AND J. D. CUBIT. 1984. Spread of Diadema mass mortality through the Caribbean. Science, 226:335-337.

LEVINS, R. 1968. Evolution in Changing Environments. Princeton University Press, Princeton, New Jersey, 132 p.

LEVITAN, D. R. 1992. Community structure in times past: Influence of human fishing pressure on algal-urchin inter-

actions. Ecology, 73:1597-1605.

LEWIS, J. B. 1960. The coral reefs and coral communities of Barbados, W. I. Canadian Journal of Zoology, 38:1133-1145.

LEWIS, J. B. 1984. The Acropora inheritance: a reinterpretation of the development of fringing reefs in Barbados, West Indies. Coral Reefs, 3:117-122.

LIDDELL, W. D., AND S. L. OHLHORST. 1987. Patterns of reef community structure, north Jamaica. Bulletin of Marine Science, 40:311-329.

LIDDELL, W. D., AND S. L. OHLHORST. 1988. Comparison of western Atlantic coral reef communities. Proceedings 6th International Coral Reef Symposium, 3:281-286.

LOTZE, H. K., K. REISE, B. WORM, M. VAN BEUSEKOM, M. BUSCH, A. AHLERS, D. HEINRICH, R. C. HOFFMAN, P. HOLM, C. JENSEN, O. S. KNOTTNERUS, N. LANGHANKI, W. PRUMMEL, M. VOLLMER, AND W. J. WOLFF. 2005. Human transformations of the Wadden Sea ecosystem through time: A synthesis. Helgoland Marine Research, 59:84-95.

LOTZE, H. K., AND I. MILEWSKI. 2004. Two centuries of multiple human impacts and successive changes in a north Atlantic food web. Ecological Applications, 14:1428-1447.

LOTZE, H. K., AND B. WORM. 2009. Historical baselines for large marine animals. Trends in Ecology and Evolution, 24:254-262.

LOTZE, H. K., H. S. LENIHAN, B. J. BOURQUE, R. H. BRADBURY, R. G. COOKE, M. C. KAY, S. M. KIDWELL, M. X. KIRBY, C. H. PETERSON, AND J. B. C. JACKSON. 2006. Depletion, degradation, and recovery potential of estuaries and coastal seas. Science, 312:1806-1809.

LOVELOCK, J. 2009. The Vanishing Face of Gaia. Basic Books, New York, 288 p.

LYNAM, C. P., M. J. GIBBONS, B. E. AXELSEN, C. A. J. SPARKS, J. COETZEE, B. G. HEYWOOD, AND A. S. BRIERLEY. 2008. Jellyfish overtake fish in a heavily fished ecosystem. Current Biology, 16:R492-R493.

MARHAVER, K. L., R. A. EDWARDS, AND F. ROHWER. 2008. Viral communities associated with healthy and bleaching corals. Environmental Microbiology, 10:2277-2286.

MAYEWSKI, P. A., E. E. ROHLING, J. C. STAGER, W. KARLEN, K. A. MAASCH, L. D. MEEKER, E. A. MEYERSON, F. GASSE, S. VAN KREVELD, K. HOLMGREN, J. LEE-THORP, G. ROSQVIST, F. RACK, M. STAUBWASSER, R. R. SCHNEIDER, AND E. J. STEIG. 2004. Holocene climate variability. Quaternary Research, 62:243-255.

McCLANAHAN, T. R., AND N. A. MUTHIGA. 1988. Changes in Kenyan coral reef community structure and function due to exploitation. Hydrobiologia, 166:269-276.

McCLANAHAN, T. R., A. T. KAMUKURU, M. A. MUTHIGA, Y. M. GILAGABHER, AND D. OBURA. 1996. Effect of sea urchin reductions on algae, coral, and fish populations. Conservation Biology, 10:136-154.

McCLENACHAN, L. 2008. Social conflict, overfishing and disease in the Florida sponge fishery, 1849-1939, p. 25-46. In D. Starkey, P. Holm, and M. Barnard (eds.),

Oceans Past: Management Insights from the History of Marine Animal Populations. Earthscan Publications Limited, London.

McCLENACHAN, L. 2009a. Documenting loss of large trophy fish from the Florida Keys with historical photographs. Conservation Biology, 23:636-643.

McCLENACHAN, L. 2009b. Historical declines in goliath grouper in south Florida. Endangered Species Research, 7:175-181.

McCLENACHAN, L., AND A. B. COOPER. 2008. Extinction rate, historical population structure and ecological role of the Caribbean monk seal. Proceedings of the Royal Society, B, 275:1351-1358.

McCLENACHAN, L., J. B. C. JACKSON, AND M. J. H. NEWMAN. 2006a. Conservation implications of historic sea turtle nesting beach loss. Frontiers in Ecology and the Environment, 4:290-296.

McCLENACHAN, L., M. J. H. NEWMAN, AND G. PAREDES. 2006b. Florida Keys coral reef fish communities, then and now. Gulf and Caribbean Fisheries Institute Proceedings, 59:1-7.

McKILLOP, H. 1985. Prehistoric exploitation of the manatee in the Maya and circum-Caribbean areas. World Archeology, 16:337-353.

MORÓN, G. 1954. Los orígenes históricos de Venezuela. I. Introducción al Siglo XVI. Consejo Superior de Investigaciones Científicas. Madrid, Spain, 385 p.

NEWMAN, M. J. H., G. A. PAREDES, E. SALA, AND J. B. C. JACKSON. 2006. Structure of Caribbean coral reef communities across a large gradient of fish biomass. Ecology Letters, 9:1216-1227.

NUGUES, M. M., G. W. SMITH, R. J. HOOIDONK, M. I. SEABRA, AND R. P. M. BAK. 2004. Algal contact as a trigger for coral disease. Ecology Letters, 7:919-923.

O'DEA, A., AND J. B. C. JACKSON. In press. Environmental change drove macroevolution in cupuladriid bryozoans. Proceedings of the Royal Society, B.

ODUM, E. P. 1953. Fundamentals of Ecology. W. B. Saunders Co., Philadelphia, Pennsylvania, 546 p.

OGDEN, J. C. 1980. Faunal relationships in Caribbean seagrass beds, p. 173-198. In R. C. Phillips and C. P. McRoy (eds.), Handbook of Seagrass Biology: An Ecosystem Perspective. Garland STP Press, New York.

OGDEN, J. C., L. ROBINSON, H. WHITLOCK, H. DAGANHART, AND R. CEBULA. 1983. Diel foraging patterns in juvenile green turtles (Chelonia mydas) in St. Croix, United States, Virgin Islands. Journal of Experimental Marine Biology and Ecology, 66:199-205.

OVIEDO Y VALDEZ, G. F. DE. 1526. Natural History of the West Indies (1959 translation by S. A. Stoudemire), University of North Carolina Press, Chapel Hill, North Carolina, 140 p.

PAINE, R. T. 1980. Food webs: Linkage, interaction strengths, and community infrastructure. Journal of Animal Ecology, 49:667-685.

PANDOLFI, J. M. 1996. Limited membership in Pleistocene reef coral assemblages from the Huon Peninsula, Papua New Guinea: Constancy during global change. Paleobiology, 22:152-176.

PANDOLFI, J. M. 1999. Response of Pleistocene coral reefs to environmental change over long temporal scales. American Zoologist, 39:113-130.

PANDOLFI, J. M. 2002. Coral community dynamics at multiple scales. Coral Reefs, 21:13-23.

PANDOLFI, J. M. AND J. B. C. JACKSON. 2001. Community structure of Pleistocene coral reefs of Curaçao, Netherlands Antilles. Ecological Monographs, 71:49-67.

PANDOLFI, J. M., AND J. B. C. JACKSON. 2006. Ecological persistence interrupted in Caribbean coral reefs. Ecology Letters, 9:818-826.

PANDOLFI, J. M., R. H. BRADBURY, E. SALA, T. P. HUGHES, K. A. BJORNDAL, R. G. COOKE, D. MCARDLE, L. MCCLENACHAN, M. J. H. NEWMAN, G. PAREDES, R. R. WARNER, AND J. B. C. JACKSON. 2003. Global trajectories of the long-term decline of coral reef ecosystems. Science, 301:955-958.

PANDOLFI, J. M., J. B. C. JACKSON, N. BARON, R. H. BRADBURY, H. M. GUZMAN, T. P. HUGHES, C. V. KAPPEL, F. MICHELI, J. C. OGDEN, H. P. POSSINGHAM, AND E. SALA. 2005. Are U. S. coral reefs on the slippery slope to slime? Science, 307:1725-1726.

PARSONS, T. R., AND C. M. LALLI. 2002. Jellyfish population explosions: Revisiting a hypothesis of possible causes. La Mer, 40:111-121.

PAULY, D. 1995. Anecdotes and the shifting baselines syndrome in fisheries. Trends in Ecology and Evolution, 10:430.

PETIT, J. R., J. JOUZEL, D. RAYNAUD, N. I. BARKOV, J. -M. BARNOLA, I. BASILE, M. BENDER, J. CHAPPALLAZ, M. DAVIS, G. DELAYGUE, M. DELMOTTE, V. M. KOTLYAKOV, M. LEGRAND, V. Y. LIPENKOV, G. LORIUS, L. PEPIN, C. RITZ, E. SALTZMAN, AND M. STIEVENARD. 1999. Climate and atmospheric history of the past 420,000 years from the Vostok ice core, Antarctica. Nature, 399:429-436.

PHILLIPS, R. C. 1960. Observations on the ecology and distributions of Florida sea grasses. Professional Papers Series of the Florida Board of Conservation, 2:1-72.

POLOVINA, J. J., E. A. HOWELL, AND M. ABECASSIS. 2008. Ocean's least productive waters are expanding. Geophysical Research Letters, 35:L03618.

RABALAIS, N. N., R. E. TURNER, AND W. J. WISEMAN, JR. 2002. Gulf of Mexico hypoxia, a. k. a. "The Dead Zone." Annual Reviews of Ecology and Systematics, 33:235-263.

RABALAIS, N. N., R. E. TURNER, B. K. SEN GUPTA, E. PLATON, AND M. L. PARSONS. 2007. Sediments tell the history of eutrophication and hypoxia in the northern Gulf of Mexico. Ecological Applications, 17 (supplement):S129-S143.

RAHMSTORF, S., A. CAZNAVE, J. A. CHURCH, J. E. HANSEN, R. F. KEELING, D. E. PARKER, AND R. C. J. SOMERVILLE. 2007. Recent climate observations compared to projections. Science, 316:709.

RANDALL, J. E. 1965. Grazing effect on seagrasses by herbivorous reef fish in the West Indies. Ecology, 46:255-260.

RAYMOND, P. A., N.-H. OH, R. E. TURNER, AND W. BROUSSARD. 2008. Anthropologically enhanced fluxes of water and carbon from the Mississippi River. Nature, 451:449-452.

ROEMMICH, D. AND, J. MCGOWAN. 1995. Climatic warming and the decline of zooplankton in the California Current. Science, 267:1324-1326.

ROGERS, C. S., H. C. FITZ III, M. GILNACK, J. BEETS, AND J. HARDIN. 1984. Scleractinian coral recruitment patterns at Salt River submarine canyon, St. Croix, U. S. Virgin Islands. Coral Reefs, 3:69-76.

SAENZ-ARROYO, A., C. M. ROBERTS, J. TORRE, M. CARINO-OLVERA, AND R. R. ENRIQUEZ-ANDRADE. 2005. Rapidly shifting environmental baselines among fishers of the Gulf of California. Proceedings of the Royal Society, B, 272:1957-1962.

SANDIN, S. A., J. E. SMITH, E. E. DEMARTINI, E. A. DINSDALE, S. D. DONNER, A. M. FRIEDLANDER, T. KONOTCHICK, M. MALAY, J. E. MARAGOS, D. OBURA, O. PANTOS, G. PAULAY, M. RICHIE, F. ROHWER, R. E. SCHROEDER, S. WALSH, J. B. C. JACKSON, N. KNOWLTON, AND E. SALA. 2008. Baselines and degradation of coral reefs in the northern Line Islands. PloS ONE, 3:e1548.

SARMIENTO, J. L., R. D. SLATER, R. T. BARBER, L. BOPP, S. C. DONEY, A. C. HIRST, J. KLEYPAS, R. MATEAR, U. MIKOLAJEWICZ, P. MONFRAY, V. SOLDATOV, S. A. SPALL, AND R. J. STOUFFER. 2004. Response of ocean ecosystems to climate warming. Global Biogeochemical Cycles, 18:GB3003.

SCHAPIRA, D., I. A. MONTANO, A. ANTCZAK, AND J. POSADA. 2009. Using shell middens to assess effects of fishing on queen conch (*Strombus gigas*) populations in Los Roques Archipelago National Park, Venezuela. Marine Biology, 156:787-795.

SCHEFFER, M., S. CARPENTER, J. A. FOLEY, C. FOLKE, AND B. WALKER. 2001. Catastrophic shifts in ecosystems. Nature, 413:591-596.

SHORT, F. T., R. G. COLES, AND C. PERGENT-MARTINI. 2001. Global seagrass distribution, p. 5-30. *In* F. T. Short and R. G. Coles (eds.), Global Seagrass Research Methods. Elsevier Science B. V., Amsterdam.

SMITH, J. A., M. SHAW, R. A. EDWARDS, D. OBURA, O. PANTOS, E. SALA, S. A. SANDIN, S. SMIRGA, M. HAYATA, AND F. L. ROHWEHR. 2006. Indirect effects of algae on coral: Algae-mediated, microbe-induced coral mortality. Ecology Letters, 9:835-845.

SMITH, J. B., S. H. SCHNEIDER, M. OPPENHEIMER, G. W. YOHE, W. HARE, M. D. MASTRANDREA, A. PATWARDHAN, I. BURTON, J. CORFEE-MORLOT, C. H. D. MAGADZA, H.-M. FÜSSEL, A. B. PITTOCK, A. RAHMAN, A. SUAREZ, AND J.-P. YPERSELE. 2009. Assessing dangerous climate change through an update of the Intergovernmental Panel on Climate Change (IPCC) "reasons for concern." Proceedings of the National Academy of Sciences of the United States of America, 106:4133-4137.

STEARN, C. W., T. P. SCOFFIN, AND W. MARTINDALE. 1977. Calcium carbonate budget of a fringing reef on the west coast of Barbados. I. Zonation and productivity. Bulletin of Marine Science, 27:479-510.

STENECK, R. S., M. H. GRAHAM, B. J. BOURQUE, D. CORBETT, J. M. ERLANDSON, J. A. ESTES, AND M. J. TEGNER. 2002. Kelp forest ecosystems: Biodiversity, stability, resilience, and future. Environmental Conservation, 29:436-459.

STONER, A. W., AND RAY-CULP. 2000. Evidence for Allee effects in an over-harvested marine gastropod: density-dependent mating and egg production. Marine Ecology-Progress Series, 202:297-302.

THAYER, G. W., D. W. ENGEL, AND K. A. BJORNDAL. 1982. Evidence for short-circuiting of the detritus cycle of seagrass beds by the green turtle *Chelonia mydas*. Journal of Experimental Marine Biology and Ecology, 62:173-183.

THAYER, G. W., K. A. BJORNDAL, J. C. OGDEN, S. L. WILLIAMS, AND J. C. ZIEMAN. 1984. Role of large herbivores in seagrass communities. Estuaries, 7:351-376.

THOMPSON, E.F. 1945. The Fisheries of Jamaica. Development and Welfare in the West Indies, Bulletin 18. Bridgetown, Barbados, 102 p.

WILSON, P. A., AND R. D. NORRIS. 2001. Warm tropical ocean surface and global anoxia during the mid-Cretaceous period. Nature, 412:425-429.

ZIEMAN, J. C. 1982. The Ecology of the Seagrasses of South Florida: A Community Profile. U. S. Fish and Wildlife Service, Office of Biological Serv., Washington, D.C.

ZIEMAN, J. C., J. W. FOURQUREAN, M. B. ROBBLEE, M. DURAKO, P. CARLSON, L. YARBRO, AND G. POWELL. 1988. A catastrophic die-off of seagrasses in Florida Bay and Everglades National Park: Extent, effects and potential causes. Eos, 69:1111.

ZIEMAN, J. C., J. W. FOURQUREAN, AND R. L. IVERSON. 1989. Distribution, abundance and productivity of seagrasses and macroalgae in Florida Bay. Bulletin of Marine Science, 44:292-311.

THE ISOTOPIC ECOLOGY OF FOSSIL VERTEBRATES AND CONSERVATION PALEOBIOLOGY

PAUL L. KOCH[1], KENA FOX-DOBBS[2], AND SETH D. NEWSOME[3]

[1]Department of Earth and Planetary Sciences, University of California, Santa Cruz, CA 95064 USA
[2]Department of Geology, University of Puget Sound, Tacoma, WA 98416 USA
[3]Geophysical Laboratory, Carnegie Institution of Washington, 5251 Broad Branch Road NW, Washington, D.C. 20015 USA

ABSTRACT.—Over the past 50,000 years, global ecosystems have experienced substantial shifts in composition and function as a result of climate change and the direct and/or indirect impact of humans. Studies of how species and ecosystems responded to these changes and characterization of the ecology of past ecosystems provide unique perspectives for conservation biology and restoration ecology. Stable isotope analysis is a powerful tool for such studies, as it can be used to trace energy flow, the strength of species interactions, animal physiology, and movement patterns. Here, we provide a brief summary of the isotopic systems used to study ecology and physiology, past and present. We highlight four examples in which isotopic data have played a central role in characterizing ecological shifts in marine and terrestrial faunas over the past 50,000 years, and end by discussing the role isotopic paleoecology could play in characterizing interactions among extinct Pleistocene megafauna. Such data would be integral to assessing the viability of Pleistocene 'rewilding' of North America.

INTRODUCTION

OVER ROUGHLY the past one million years, ecosystems have experienced dramatic shifts between glacial and interglacial climates as well as a host of disturbances set in motion by members of the genus *Homo*. The past 40,000 years have been especially eventful. Earth entered then exited one of the most profound glacial periods of the Plio-Pleistocene icehouse, and a wave of extinctions engulfed terrestrial communities as modern humans spread around the planet. These extinctions, which preferentially eliminated large, slow-breeding species on continents, and all types of species on islands, were set in motion by human impacts (especially hunting and introduced species), though regional climate change likely affected the timing and pace of extinction for particular populations or species (Barnosky et al., 2004; Burney and Flannery, 2005; Koch and Barnosky, 2006). Paleobiological exploration of how extinct and extant species responded to these changes in their physical and biological environment can produce unique theory and data of use in conservation biology and restoration ecology.

Biogeography has received the greatest attention, including issues such as how fast geographic ranges can shift, or how, mechanistically, a species or biome shifts its range. Paleobiological research has roles to play on four additional fronts. First, it provides examples of actual ecosystems, some of which persisted for millions of years or existed under different climatic regimes. If we can understand the functional links in such ecosystems, they offer targets for restoration ecology unlike those adopted from the historical record, and they can be modeled to determine if they have attributes that make them more resilient to change than our current defaunated ecosystems (Berlow et al., 2008). Second, it offers a sound ecological baseline against which to judge ongoing ecological shifts. Third, paleobiology can reveal behaviors of species and species interactions that are not occurring today, either because of very recent disturbance or some event earlier in the Holocene or Pleistocene. Finally, it reveals how aspects of species' ecology besides geographic range (e.g., trophic interactions, reproductive strategies, growth dynamics) respond to the typical perturbations of the glacial-interglacial world, as well as the unusual perturbation of Quaternary extinctions.

Here, we'll explore the purchase on such issues that can be gained through study of the isotopic biogeochemistry of fossils and sub-fossils. Our focus will be on the isotopic ecology and conservation biology of vertebrates. We'll begin by briefly describing an isotopic tool kit—the types of tissues that can be used for isotopic analysis and the aspects of animal ecology that can be deduced from isotopic data. Then we'll turn to case studies that use isotopic data to answer different

In *Conservation Paleobiology: Using the Past to Manage for the Future, Paleontological Society Short Course, October 17th, 2009. The Paleontological Society Papers, Volume 15*, Gregory P. Dietl and Karl W. Flessa (eds.). Copyright © 2009 The Paleontological Society.

types of questions in conservation biology.

There are many reasons why isotopic methods are useful for studies at the intersection between paleobiology and conservation biology. Similar approaches can be applied to newly collected specimens, materials in zoological collections, and fossils, forming a seamless temporal bridge from the present to the past. Isotopic composition often serves as a natural label of mass flow, yielding quantitative data on interaction strengths in trophic studies. Most importantly, isotopic methods can provide information on the behavior and physiology of individuals. Consider, for example, what we might mean when we say that a species is a dietary 'generalist', which might make it more resilient in the face of environmental change. If we imagine that the species can consume three types of prey, being a generalist might mean that all individuals in a species consume roughly equal amounts of the three foods, or that different populations specialize on different mixes, or that different individuals within the same population have different diets, or even that an individual consumes different foods at different times (seasonally, ontogenetically). All these situations can (potentially) be discriminated through isotopic analysis. More generally, many questions central to conservation

biology require an understanding of how a species' niche does (or does not) change in space and time. Isotopic data are amenable to conversion into niche axes, including both bionomic (defined by resources an animal uses) and scenopoetic (defined by bio-climatic limits) dimensions (Newsome et al., 2007a).

Before proceeding, we want to emphasize three points. While we are concerned with whole ecosystems and species interactions, many of our case studies are autecological. In some cases, we are looking at top carnivores (e.g., seals, wolves), so changes in their ecology may indicate shifting conditions lower in the trophic chain, or they have important top-down impacts; in other instances, we are looking at species of great value economically (e.g., salmon) or culturally (e.g., condors). Second, we would never advocate for an exclusive reliance on isotopic data in studies of conservation paleobiology. Several of the studies we discuss are vastly more powerful because they couple data from isotopes, morphology, ancient DNA, etc. Finally, perhaps more so than for other types of paleobiological research, the starting point in conservation paleobiology should be an extremely careful analysis of modern systems. This conservative approach is warranted

TABLE *1.*—Summary information on materials used as substrates for isotopic analysis of vertebrates.

Tissue	Molecule	Temporal window*	Isotope systems in material	Preservation window (years)
Hair	Keratin	Accretion	H, C, N, O, S	10^4
Feather	Keratin	Accretion	H, C, N, O, S	10^4
Bone	Bioapatite	Years	CO_3 - C, O; PO_4 - O; Sr, Nd, Pb	10^3 - 10^8? 10^3
Bone	Collagen	Years	H, C, N, O, S	10^5-10^6 (10^8?)
Bone	Lipid	Weeks-months	H, C	
Enamel	Bioapatite	Accretion	CO_3 - C, O; PO_4 - O; Sr, Nd, Pb	10^8 10^7
Dentin	Bioapatite	Accretion	Same as bone	Same as bone
Dentin	Collagen	Accretion	H, C, N, O, S	Same as bone
Egg shell	Carbonate	Days-weeks	C, O	10^7-10^8
Egg shell	Protein	Days-weeks	C, N	10^4
Otolith	Carbonate	Accretion	C, O	10^7

Compound	Signal window		Isotope systems	Preservation window
Amino acid	Depends on tissue type		H, C, N	10^6?
Cholesterol	Weeks-months		H, C	10^7?
Fatty acid	Weeks-months		H, C	10^4?

*Materials that grow by accretion have the potential for offering either short or long temporal windows, depending on how they are sampled.

when the information being gathered may be used to set policy or develop conservation strategies.

AN ISOTOPIC TOOL KIT FOR CONSERVATION PALEOBIOLOGY

Conventions

The isotopic ratio of an element (X) is expressed as the ratio of the heavy (H) to light (L) isotope using the delta (δ) notation through comparison of samples to international standards: $\delta^H X = ((^H X/^L X_{sample} \div ^H X/^L X_{standard})-1) \times 1000$ (Sharp, 2007). We will consider the following systems: hydrogen ($^2H/^1H$), carbon ($^{13}C/^{12}C$), nitrogen ($^{15}N/^{14}N$), oxygen ($^{18}O/^{16}O$), sulfur ($^{34}S/^{32}S$), and strontium ($^{87}Sr/^{86}Sr$). Units are parts per thousand (per mil, ‰). The fractionation of isotopes between two substances, A and B (e.g., diet and animal, or tissue A and tissue B within an animal), is expressed in many ways. Most common are 1) the fractionation factor: $\alpha_{A-B} = (1000 + \delta^H X_A)/(1000 + \delta^H X_B)$; 2) difference or discrimination: $\Delta_{A-B} = \delta^H X_A - \delta^H X_B$; and 3) enrichment or epsilon: $\varepsilon_{A-B} = 1000(\alpha_{A-B} - 1)$. Note that the sign (for Δ or ε) or magnitude (for α) of these terms depends on the order of the subscripts, and that $\Delta_{A-B} \approx \varepsilon_{A-B}$.

Isotopic substrates.—Vertebrate bodies offer many substrates for analysis (unless otherwise noted, this section is based on Kohn and Cerling, 2002; Koch, 2007). Tissues differ in their macromolecular and elemental compositions, styles of growth, and preservation potential (Table 1). Soft tissues such as hair and feathers contain protein and lipids and may persist for 10^3 to 10^4 years in unusual settings (e.g., mummification, permafrost). Mineralized tissues such as bone, tooth enamel and dentin, eggshell, and otoliths last much longer. Bone, enamel, and dentin are composites of mineral, protein, and lipid. The mineral is a form of hydroxyapatite ($Ca_{10}[PO_4]_6[OH]_2$) we refer to as bioapatite. Bioapatite has a few weight percent (wt %) carbonate and various cations (e.g., Sr) substituting for calcium. Bone is composed of tiny bioapatite crystals intergrown with an organic matrix (chiefly the protein collagen) that makes up ~30% of its dry weight. Enamel is much less porous, has <5 wt % organic matter, and has much larger crystals. The crystal size and organic content of dentin resemble bone, but its porosity is intermediate between enamel and bone. Bird eggshells are composed of tiny crystals (chiefly calcite) secreted around a honeycomb of fibrous organic sheets (~3 wt %, chiefly protein). Otoliths, the mineralized structures

in the vertebrate inner ear, vary in mineralogy—aragonite occurs in jawed fish and amphibians, and calcite occurs in amniotes. Various isotopic analyses can be performed on both the organic and mineral components of vertebrate tissues (Table 1). Regarding the former, a great deal of information can be obtained if isotopic analysis can be conducted on individual organic molecules (amino acids, sterols, fatty acids), rather than bulk tissue (reviewed by Evershed et al., 2007).

The time interval integrated within an isotopic sample (Table 1) depends on the mode of growth and turnover rate of each tissue (or the pools from which tissues are synthesized). Experiments and models focusing on soft tissues (blood, muscle) suggest that the rates of isotopic incorporation (i.e., the time it takes for a tissue to equilibrate after a change in input) depend on rates of tissue growth and turnover, as well as on body mass (Martínez del Rio et al., 2009). Bone growth is complex, involving cartilage ossification, accretion, and remodeling, so a bulk sample of bone mineral or collagen has material that is time-averaged over months to years. Dentin grows by accretion with little remodeling and can be microsampled to generate time series. Enamel also has incremental features, but time lags in its maturation lead to substantial time averaging. In all mammals except those with ever-growing teeth, enamel mineralization takes place early in life, over a period of months to years. All other vertebrates replace their teeth throughout life, so teeth collected from a dead individual formed over the last year or two of life, depending on replacement rate. Eggshells form rapidly, so their carbonate and protein represent short time intervals (days to weeks). Growth of feathers and hair may be continuous or episodic, and depending on sampling strategy can represent weeks to about a year. Otoliths have incremental laminae that can be microsampled for time series or bulk sampled for a lifetime average.

Isotopic studies of animal ecology require an understanding of the controls on fractionation between diet and tissue or drinking water and tissue. This complex topic has been reviewed by Koch (2007) and Martínez del Rio et al. (2009). Briefly, C, N and S in tissues are supplied by diet, O and H in tissues and body water (from which biominerals form) are supplied by diet, water, and O_2, and cations (Sr, Pb, etc.) are ingested as food and water. Animal soft tissues show only slight enrichments in ^{13}C and ^{34}S isotopes relative to diet, whereas ^{15}N enrichment is stronger and more consistent. Enrichment magnitude depends on tissue type (especially for C),

Table 2.—A summary of common isotope systems and expected patterns in δ-values used to examine scenopoetic and bionomic dimensions of ecological niche space (from Newsome et al., 2007a).

Gradient	Isotopes	High δ-values	Low δ-values	Sce.	Bio.
Trophic level	C, N	High levels	Low levels		✓
C3 vs. C4 plants	C, O, H	C4	C3	✓	✓
C3 vs. C4 animals	C	C4	C3		✓
C3 functional types	C	Thick cuticle (e.g., evergreen trees)	Thin cuticle (e.g., deciduous forbs)		✓
Canopy position	C, O	High canopy	Forest floor	✓	✓
Plant fungal symbioses	N	Non-mycorrhizal	Mycorrhizal		✓
N_2-fixation - terrestrial	N	Typically indicates little N fixation, but so do values much less than 0‰	Strong fixation when values near 0‰		✓
Marine plants/ consumers	C	Seagrass, kelp	Planktonic algae	✓	✓
Marine plants	N	Most marine 1° producers	N-fixers	✓	
Benthic-pelagic	C, S	Benthic	Pelagic	✓	✓
Marine-terrestrial	H, C, O, N	Marine	Terrestrial	✓	✓
Marine-terrestrial	S, Sr	Variable	Variable	✓	✓
Latitude - terrestrial	H, O	Low latitude	High latitude	✓	
Latitude - marine	C	Low latitude	High latitude	✓	
Latitude - marine	N	Low latitude (except areas with N_2 fixers)	High latitude	✓	
Altitude	H, O	Low altitude	High altitude	✓	
Altitude (C3 ecosystems)	C	High altitude	Low altitude	✓	
Aridity	H, O, C, N	Xeric	Mesic	✓	
Nearshore-Offshore	C	Nearshore, Upwelling	Offshore	✓	
Temperature - terrestrial	H, O	High temperature	Low temperature	✓	
Geologic substrate	Sr	Old rocks, granite	Young rocks, basalt	✓	
Methanogenesis	C	Photosynthetic carbon	Methanogenic carbon	✓	✓
Denitrification	N	Strong denitrification	Little/no denitrification	✓	
N_2-fixation - marine	N	Little fixation	Strong fixation	✓	

digestive physiology, diet quality, and physiological state (anabolic vs. catabolic). Carbon in bioapatite is derived from body water bicarbonate. It reflects bulk dietary C but shows strong ^{13}C-enrichment associated with carbonate equilibria, as well as variation related to microbial fermentation of food. Oxygen in biominerals is ^{18}O-enriched relative to body water by the amount expected for systems near isotopic equilibrium. Fractionation of O and H isotopes between diet, water, and soft tissues is a current topic of study; it is too early to summarize. It is usually asserted that Sr, Pb and other heavy isotopes are incorporated from diet with no fractionation.

Diet

Table 2 summarizes the vertebrate niche dimensions that can be discriminated using different isotopic systems (discussion based on Koch, 2007; Newsome et al., 2007a). Carbon in land plants is supplied by the well-mixed atmospheric CO_2 reservoir, so there is a degree of uniformity in $\delta^{13}C$ values across ecosystems. $\delta^{13}C$ values are commonly used to distinguish consumption of C3 versus C4 plants (or animals that feed on C3 versus C4 plants). C3 photosynthesis occurs in all trees, most shrubs and herbs, and grasses in regions with a cool growing season. C4 photosynthesis occurs in grasses where the growing season is warm, and in some sedges and dicots. C3 plants have lower $\delta^{13}C$ values than C4 plants. Crassulacean acid metabolism (CAM) occurs in succulent plants, producing $\delta^{13}C$ values that range between those for C3 and C4 plants, but are often similar to C4 plants. There are consistent $\delta^{13}C$ differences among C3 plants related to light level, moisture, position in the canopy, and other factors that affect the rate of photosynthesis and stomatal resistance to CO_2 diffusion into the plant. Within C3 ecosystems, $\delta^{13}C$ values differ among plant functional types in relation to stomatal resistance, which correlates with cuticle thickness. For example, in a boreal system Brooks et al. (1997) found that $\delta^{13}C$ values in evergreen trees > shrubs/deciduous trees > moss > deciduous forbs.

The substrates for H, O, N and S in the plants and waters at the base of terrestrial food webs vary spatially, so care must be taken when comparing consumer isotope values among ecosystems. Within ecosystems, leaf water and dry matter δ^2H and $\delta^{18}O$ values vary with water stress, canopy height, and plant functional type, chiefly because of differences in evaporation, which tend to leave the plant enriched in the heavy isotope. For example, for leaf water and tissue δ^2H values, C3 < C4 < CAM, and for leaf water $\delta^{18}O$ values, C3 dicots < C3 grass < C4 grass. Because of strong ^{15}N-enrichment moving up food webs, $\delta^{15}N$ values are used to study trophic level. Within ecosystems, plants from wetter habitats have lower $\delta^{15}N$ values than those from dry habitats, and they show $\delta^{15}N$ differences among functional types related to rooting depth, symbioses with nitrogen-fixing bacteria (which shift plant values to 0‰), and mycorrhizal associations. In boreal ecosystems, non-mycorrhizal plants (graminoids, forbs) have higher $\delta^{15}N$ values than mycorrhizal plants (trees and shrubs), mosses and lichens.

In marine ecosystems, $\delta^{13}C$ values vary among plant functional types (and their consumers) such that phytoplankton < macroalgae < sea grass, which translates into a difference between pelagic vs. benthic production. These differences are ultimately related to diffusional limitations on CO_2 supply to plants, and are affected by factors such as algae size, shape, and, most importantly, photosynthetic rate. In some case, CO_2 limitations are severe enough that plants take up HCO_3^{2-}, rather than CO_2, leading to higher tissue, $\delta^{13}C$ values. Benthic-pelagic differences are also present in $\delta^{34}S$ values, but otherwise marine $\delta^{34}S$ values are relatively invariant compared to terrestrial and freshwater values. Marine plants do not differ greatly in $\delta^{15}N$ value by functional type. In most systems, they have values ≥5‰ (the signature of deep water nitrate), though N_2-fixing cyanobacteria have values close to 0‰. Because marine food webs are often long, however, $\delta^{15}N$ values are a powerful tool for studying trophic level. Freshwater submerged aquatic plants have variable $\delta^{13}C$ values, depending on the mix of CO_2 from the atmosphere, plant decomposition, and oxidation of bacterially-generated methane (a source of ^{13}C-depleted nutrients). Finally, because most isotope systems differ between marine and terrestrial/freshwater ecosystems, they are useful for discriminating between the consumption of such resources at the interface of these systems (estuaries, coastal sites).

Habitat preferences and migratory patterns

Isotopic systems that differ within and among biomes provide information on habitat preferences and migration (discussion based on Koch, 2007; Newsome et al., 2007a; Hobson, 2008). Dietary isotope differences sometimes translate cleanly into habitat differences. For example, C4 feeding is a sign that an animal spends time on grasslands. In dense forests, $\delta^{13}C$ and $\delta^{18}O$ values of plant biomass increase from the forest floor to the top of the canopy, and as such can reveal the height at which an herbivore spends time feeding. Foraging location is also revealed by isotopic systems that discriminate consumption of marine versus terrestrial resources. In other cases, isotopic gradients among habitats are controlled by climatic or oceanographic factors (wholly scenopoetic niche dimensions; Table 2). δ^2H and $\delta^{18}O$ of precipitation and hence body water are lower in colder and higher altitude terrestrial habitats. Evaporation causes higher δ^2H and $\delta^{18}O$ values in plants, and water stress increases stomatal resistance, raising $\delta^{13}C$ values. Soil and plant $\delta^{15}N$ values also increase with decreasing rainfall. Thus for all these systems, increasing aridity leads to higher isotopic values in plants that cascade up food webs to consumers. In marine systems, higher photosynthetic rates in nearshore plants lead to higher $\delta^{13}C$ values relative to those offshore. Low $\delta^{15}N$ values occur where N-fixation is important (e.g., nutrient-poor tropical gyres), whereas high values occur where denitrification is active (e.g., highly productive regions with well developed oxygen minimum zones). In areas where primary production is limited by other factors (light, micronutrients, etc.), marine $\delta^{15}N$ and $\delta^{13}C$ values are low because of incomplete nitrate utilization and low photosynthetic rates, respectively, which may explain low $\delta^{13}C$ and $\delta^{15}N$ values in marine plants at high latitude relative to those at middle latitudes.

All these isotopic systems that vary by habitat, biome, or climate can be used to assess animal migration. The greatest amount of work has been done on bird migration, tracking latitudinal treks against δ^2H and $\delta^{18}O$ landscapes (isoscapes), supplemented with $\delta^{13}C$ and $\delta^{15}N$ values (Bowen and West, 2008). Other studies have examined time series of samples collected from teeth, otoliths, and other accreted materials (e.g., Koch et al., 1995; Kennedy et al., 2000). Elements such as Sr or Pb, which vary in relation to geologic factors, offer isotopic gradients uncorrelated to climatic or biome differences and are, therefore, extremely useful in migration studies in the terrestrial realm, or between marine and terrestrial systems.

In marine systems, movements of mammals, birds, teleosts, and sharks have all been tracked isotopically. In general, differences in isotopic composition at the base of the food web due to oceanographic differences map out very large regions that can be used to discriminate foraging zones among marine consumers. For example, there is a 2 to 3‰ decrease in food web $\delta^{13}C$ and $\delta^{15}N$ values from temperate (30 to 35°N) to high-latitude (>50°N) northeast Pacific pelagic ecosystems (discussion based on Graham et al., in press). Higher temperatures and upwelling lead to higher phytoplankton growth rates (and higher $\delta^{13}C$ values of phytoplankton and higher level consumers) in the California Current relative to the Gulf of Alaska. Higher productivity in coastal systems along the entire eastern Pacific Ocean and southern Bering Sea lead to higher ecosystem $\delta^{13}C$ values when compared to offshore systems, and zooplankton $\delta^{13}C$ and $\delta^{15}N$ values decrease from east to west across the shelf-slope break in the southeastern Bering Sea. These regional gradients have been used

to characterize migration patterns for modern and fossil marine mammals and birds in the region. Note that for expected gradients in marine $\delta^{13}C$ values, we have focused on the organic carbon fixed by primary producers. Expected patterns in the $\delta^{13}C$ value of dissolved inorganic carbon are complex, related to the balance between atmospheric and oceanic mixing, remineralization, and local productivity. These variations are not very relevant to studies of vertebrates, because with the exception of fish otoliths, most C in vertebrates is supplied by diet (organic carbon), not inorganic carbon.

Reproductive biology

Three aspects of reproductive biology have been studied using isotopic methods: nutrient allocation to reproduction, nursing/weaning, and pregnancy. Inger and Bearhop (2008) reviewed avian studies using isotopes to quantify the roles of endogenous reserves (capital) versus recently ingested nutrients (income) in forming eggs and chicks. This approach works best if there is a sharp isotopic contrast between female foraging zones immediately before nesting and those surrounding the nesting sites. Koch (2007) reviewed isotopic studies of nursing, lactation, and pregnancy in mammals. If lactating mothers catabolize their own tissues to make milk, $\delta^{15}N$ and $\delta^{13}C$ values from nursing offspring should indicate they are feeding at a higher trophic level than their mothers. As expected, nursing young are ^{15}N-enriched by 3 to 5‰ relative to their mothers in many extant terrestrial and marine mammals. The prediction is complicated for C by the fact that milk is rich in lipids, and lipids are ^{13}C-depleted relative to proteins. Weaning age can be reconstructed by assessing age at the transition from nursing to weaned isotope values by sampling tissues that turn over rapidly or by microsampling tissues that grow by accretion, or by sampling bone from growth series (though the latter turns over more slowly). The isotopic consequences of lactation and pregnancy are not as well understood. In a study of pregnant humans, Fuller et al. (2004) found no significant effects of pregnancy on $\delta^{13}C$ values, but that $\delta^{15}N$ values dropped from conception to birth, and that the magnitude of the drop correlated positively to the baby's birth weight as well as the amount of weight gained by the mother. If the patterns associated with pregnancy and lactation are common among mammals, they offer the potential to study inter-birth interval, neonatal survival rate, and other critical aspects of reproductive biology.

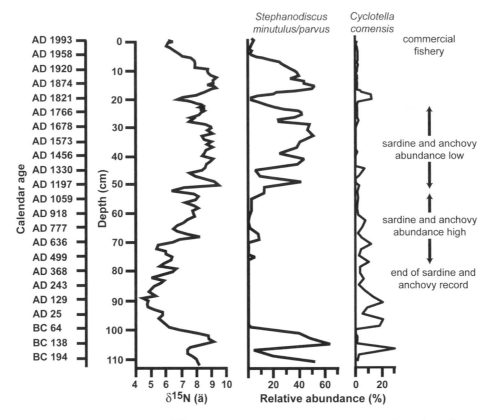

FIGURE 1.—Temporal variations in sedimentary $\delta^{15}N$ values and the relative abundances of *Stephanodiscus minutulus/parvus* (a mesotrophic to eutrophic diatom) and *Cyclotella comensis* (an oligotrophic diatom) from Karluk Lake, Alaska (modified from Finney et al., 2002).

Nutritional state

It has long been argued that during fasting or nutritional 'stress', $\delta^{15}N$ values in consumers might rise (Hobson et al., 1993; reviewed in Koch 2007; Martínez del Rio et al., 2009). Trophic fractionation of N is thought to relate to excretion of nitrogenous wastes that are ^{14}N-enriched relative to the body N pool by enzymatic effects associated with transamination and deamination. Mass balance models imply that, at steady state, the greater the fraction of N lost as ^{14}N-enriched waste, the higher the body ^{15}N-content (i.e., $\Delta^{15}N_{body-diet}$ increases). The models are supported by experiments showing that $\Delta^{15}N_{body-diet}$ increases with increasing protein content in the diet of individual species, and by the fact that $\Delta^{15}N_{body-diet}$ is slightly higher for carnivores than herbivores. When animals are out of N balance, $\Delta^{15}N_{body-diet}$ values may decrease for animals in anabolic states (growth) and increase for animals in catabolic states (fasting, starvation). If we assume that the $\delta^{15}N$ value of their diets didn't change following conception, then Fuller et al.'s (2004) observation that pregnant women have lower $\delta^{15}N$ values supports this view; lower values signal a proportionally reduced loss of ^{14}N-enriched urine as pregnant females achieve positive N balance. Fuller et al. (2005) studied pregnant women with morning sickness who lost weight early in their pregnancies and found strong ^{15}N-enrichment (consistent with a rise in $\Delta^{15}N_{body-diet}$ value), as expected given these models. Differences in $\delta^{15}N$ values based on diet protein quantity or growth vs. starvation have yet to be extensively exploited in ecological or paleoecological research.

CASE STUDIES IN CONSERVATION PALEOBIOLOGY

We will explore four case studies that use isotopic data to illuminate the autecology of extant species. In the first, isotopic data are used to reconstruct Holocene fluctuations in the abundance of an economically and environmentally critical species, sockeye salmon, in relation to climatic change. In the second, we again explore Holocene abundance variations, but this time in

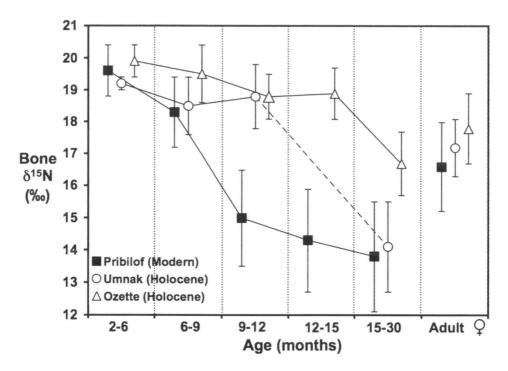

FIGURE 2.—Nitrogen isotope evidence from bone collagen for a large change in weaning age in northern fur seals. $\delta^{15}N$ values (mean ± one standard deviation) are presented for animals in different age classes from a modern rookery on the Pribilof Islands, and Holocene populations from the Olympic Peninsula (Ozette), and the Aleutians (Umnak) (modified from Newsome et al., 2007b).

a pinniped, the northern fur seal. Isotopic data are used to understand the foraging, migratory, and reproductive ecology of the species; they point to recent changes in behavior that have ramifications for the conservation biology of the species. The third study focuses squarely on feeding ecology, using isotopic data from fossils of the highly endangered California condor to understand how it survived the late Pleistocene extinction, and to consider what this understanding might imply for the ongoing restoration of the species. The fourth study also examines feeding ecology, this time for of one of the few large terrestrial mammals to survive the late Pleistocene extinction in North America, the grey wolf. Paleogenetic data reveal unsuspected dynamics for the wolves of Alaska, and isotopic and morphological data show how the species evolved behaviorally and physically to respond to its shifting prey base. The final example is a thought exercise. We consider how these different types of information might help scientists identify ecological analogs for extinct megafauna. Overall, the case studies illuminate how isotopic data, when coupled with other sources of information, can reveal the ecological roles and responses of species, as well as the existence and magnitude of important species interactions that occurred in past ecosystems.

Shifting baselines in animal abundance: Alaskan sockeye

Paleobiology can reveal whether abundance shifts documented by wildlife biologists are typical or unusual. Among large vertebrates, some of the most dramatic changes have been at higher latitudes, such as the decline of marine mammal species in the north Pacific and Bering Sea (Springer et al., 2003), and shifts in penguin abundance around Antarctica (Ainley et al., 2005). The roles of climate change (anthropogenic or natural) vs. top-down forcing by industrial whaling is debated in both cases (Ainley et al., 2007; Springer et al., 2008). The context supplied by paleobiological records, both for estimates of the magnitude of past oscillations and about the relationship between abundance shifts and environmental factors, is needed to disentangle this nexus of causes. In Antarctica, ancient DNA analysis, geochronological studies, and sedimentary isotope records are being used to reconstruct abundance changes in pinnipeds and sea birds (Baroni and Orombelli, 1994; Liu et al., 2004; Hall et al., 2006; Emslie et al., 2007).

Work on sockeye salmon abundance in Alaska over the past several millennia is an elegant example of the way that isotopic paleoecology contributes to such questions (Finney et al., 2000, 2002). Salmon are crucial components of northern Pacific ecosystems, and are important economical and cultural resources. After hatching, juvenile sockeye (*Oncorhyncus nerka*) forage for one to three years in their natal lakes or streams. They then migrate to the ocean, where they forage for one to four years and put on 99% of their adult mass before returning to their natal lakes and streams to spawn and die. Nitrogen isotope data have shown that marine-derived nutrients from salmon carcasses are very significant in some freshwater systems. A drop in salmon-derived nutrients due to overfishing or environmental factors can set in motion a negative feedback, leading to lower freshwater productivity and further reductions in salmon recruitment.

In nursery lakes on Kodiak Island and the Alaska Peninsula, Finney et al. (2000) found a positive linear relationship between salmon abundance and the $\delta^{15}N$ values of zooplankton, salmon smolts, and sediments. Given this modern calibration, Finney et al. (2000, 2002) reconstructed long-term variations in sockeye abundance by analyzing sedimentary $\delta^{15}N$ values and compared their results to historical records of catch and abundance, to shifts in benthic diatom assemblages, and to basin-scale shifts in fish abundance in relation to climate regime changes (Fig. 1). Over the last 300 years, they discovered regionally coherent shifts in sockeye abundance on decadal time scales that point to large-scale forcing from the ocean-atmosphere system. In Karluk Lake, where salmon contribute a large fraction of the nutrients, a drop in sockeye abundance in the early 1800s was followed by a swift recovery at a time when the natural feedback was operating, whereas the dramatic loss of salmon-derived nutrients associated with commercial fishing in the late 20th century seems to have had a strongly negative impact on lake productivity that may be preventing a similar recovery today.

On longer time scales, Finney et al. (2002) found that eutrophic diatom species dominated when $\delta^{15}N$ values were high (i.e., when salmon-derived nutrients were abundant), and that cycles in salmon abundance were greater than in the observational record in the period before the establishment of commercial fish-

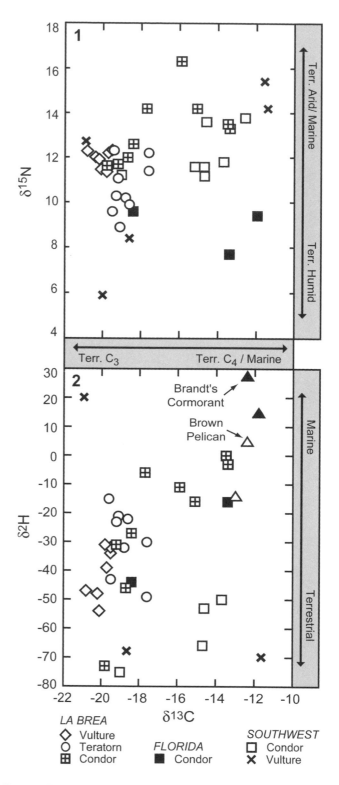

FIGURE 3.—Isotopic variations in Pleistocene condors and associated avifauna in North America; *1*, Collagen $\delta^{13}C$ versus $\delta^{15}N$ values; *2*, Collagen $\delta^{13}C$ versus δ^2H values (including modern marine birds). Environmental axes corresponding to isotopic axes are labeled (modified from Fox-Dobbs et al., 2006).

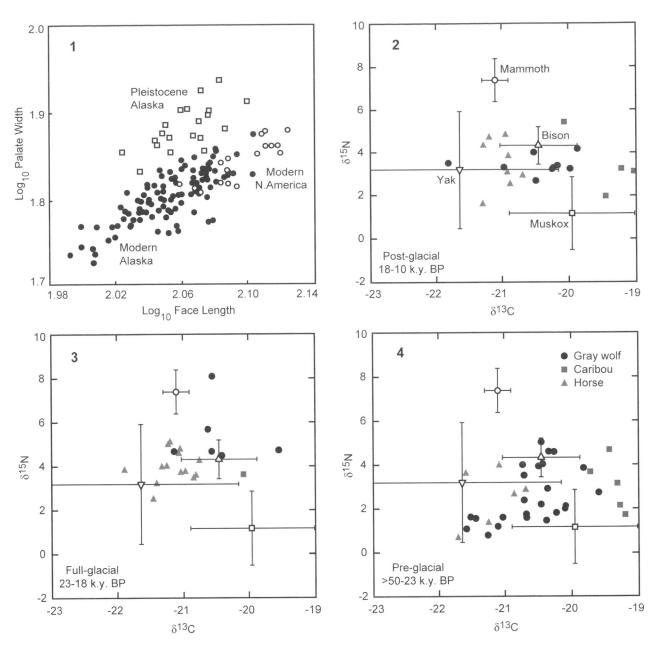

FIGURE *4.*—Morphology and isotopic data for Beringian and modern wolves and their prey; *1*, Palate width versus face length for late Pleistocene Alaskan (Beringian) and modern grey wolves. Beringian wolves had shorter and broader palates than modern wolves, indicative of a more robust skull and stronger jaws (modified from Leonard et al., 2007); 2 to 4, Collagen δ^{13}C versus δ^{15}N values for Beringian wolves and megafaunal prey. Wolf δ^{13}C and δ^{15}N values have been corrected for trophic isotope differences (-1.3 and -4.6, respectively). Each panel presents individual data for wolves (filled circle), caribou (filled square), and horses (filled triangle) for three time intervals. Bison, yak, muskox and mammoth data are not interval-specific and are presented as means (± one standard deviation) in each panel (modified from Leonard et al., 2007; Fox-Dobbs et al., 2008).

eries. Historical records over the last century suggest that climate forces salmon production and abundance, but the isotopic abundance record points to contrasting patterns at different time scales. In the twentieth century, there were antagonistic oscillations in marine productivity in the eastern Pacific Ocean to the north and south of the North Pacific Current (~45°N). On a decadal time scale, Alaskan salmon abundance covaries with that of sardines from off the California coast, whereas both are out of phase with anchovy abundance from California. These shifts have been attributed to oceanic regime shifts that affect the entire north Pacific Basin. Yet on the millennial time-scale, isotope data suggest much larger shifts in abundance of all taxa, and salmon abundance is negatively correlated with the abundance of both sardines and anchovies. These shifts are coherent with regional paleoclimatic data, pointing to basin-scale shifts in the ocean-atmosphere system as the driver. The nature of these links is complex, since the millennial abundance patterns don't match those on decadal or centennial time-scales. Given these data, it would clearly be difficult to extrapolate from the response of species to the small-scale fluctuations in decades-long historical records to predict their response to large-scale climatic or anthropogenic change. Longer time-scale records are essential to ground-truth models of the dynamics of these fisheries.

Shifting abundance and changing behavior: northern fur seals

Archaeological remains yield data at lower temporal resolution than the sedimentary cores used to study fish abundance, and the data have been through an anthropogenic filter. Yet such remains are proving a rich trove for all manner of information of relevance to conservation biology (Lyman, 2006). An example is recent work on northern fur seals (*Callorhinus ursinus*) from the northeastern Pacific Ocean (Burton et al., 2001; Gifford-Gonzalez et al., 2005; Moss et al., 2006; Newsome et al., 2007b). This species is common in archaeological middens greater than 500 years old from southern California to the Aleutian Islands. Yet today it breeds almost exclusively on high-latitude offshore islands, with the largest rookeries on the Pribilof Islands (Bering Sea). Animals occupy rookeries for four months, and mothers give birth and wean their pups in this interval. Pribilof breeders are highly migratory for their eight months at sea, foraging pelagically as far south as Baja. These behaviors would have made the species relatively inaccessible to native human hunters. Paleobiological study has revealed why they were more available in the past.

Prehistoric adult female northern fur seals have lower $\delta^{13}C$ values than nearshore-foraging harbor seals, suggesting that they were feeding offshore over their entire range, as they do today. These observations falsify the hypothesis that fur seals were available to prehistoric humans because they foraged closer to shore in the past. Furthermore, prehistoric female fur seals

cluster into three groups isotopically: a southern group (California), a northern group (Pacific Northwest to eastern Aleutians), and a western Aleutian group. Isotopic distinctions among seals from different regions suggest that these prehistoric females were less migratory than modern Pribilof females and confirm that fur seals from California were year-round residents. Archaeometric evidence for very young pups at sites in California, the Pacific Northwest, and the eastern Aleutians confirm the presence of temperate-latitude breeding colonies. In bone growth series from modern Pribilof seals, there is a rapid drop in $\delta^{15}N$ values between the 2-6 and 6-9 month age classes, consistent with rapid weaning when animals switch from feeding on maternally-produced tissue to foraging independently at a lower trophic level (Fig. 2). In contrast, $\delta^{15}N$ values of bone growth series from ancient temperate-latitude rookeries indicate that pups were weaned at 12 months or more, not in 4 months as today. Ancient DNA analysis confirms that slower weaning seals were not a separate species. Northern fur seals were more available to ancient human hunters because they had rookeries all along the eastern Pacific margin and because females and pups spent a much greater amount of time on land.

The relative roles of human hunting versus climatic change in explaining these ecological and behavioral shifts are unclear and are the focus of ongoing research. Whatever the cause, by 500 years ago northern fur seals were largely restricted to breeding in the Arctic, where severe winter conditions favored rapid weaning. Paleobiological analysis reveals that in the absence of this strong selection, later weaning was adaptive at rookeries outside the Bering Sea. Most other eared seals wean at a later age, as inferred for ancient northern fur seals. Later weaning is thought to buffer against inter-annual resource fluctuations (like those due to El Niño events) that negatively affect productivity in temperate and tropical marine ecosystems.

Northern fur seals provide a clear example that extant populations with relictual distributions may have experienced major ecological and behavioral changes. Over the past several decades, fur seal migrants from high-latitude rookeries have reestablished colonies at temperate latitudes, despite declines in the source population in the Bering Sea. Establishment of temperate-latitude rookeries may buffer the global population through diversification of population and genetic structure and by diffusing potential threats to viability over a larger geographical area. Yet the restriction of the species to high-latitude rookeries over the past 500 years may have eliminated the behavioral plasticity needed to respond to temperate-latitude conditions. As they re-establish rookeries across their ancient range, northern fur seals may need to 're-evolve' their ancient reproductive strategy. This study highlights the importance of understanding pre-exploitation biogeography and behavior of species whose current ecology may be shaped by recent exploitation and/or environmental change.

Dietary flexibility and post-Pleistocene survival: California condors

The California condor (*Gymnogyps californianus*) is a species of exceptional conservation concern. It is the largest survivor of the once diverse Pleistocene avian megafauna of North America, and it exists in the wild in only a few reintroduced populations. Their current existence is in stark contrast to the continent-wide distribution of condors just 10,000 years ago (Grayson, 1977; Emslie, 1987). Stable isotope data provide insight into the role of marine versus terrestrial resources in the extinction patterns of avian scavengers at the Pleistocene-Holocene boundary, and the types of resources that might sustain condors in the future.

The large-bodied avian scavengers that co-existed in North America for tens of thousands of years suffered extinction and extirpation at the Pleistocene-Holocene boundary coincident with the mass extinction of terrestrial mammalian megafauna. The $\delta^{13}C$ and $\delta^{15}N$ values in bone collagen from two extinct scavengers in southern California, the teratorn (*Teratornis merriami*) and western black vulture (*Coragyps occidentalis*), indicate that both species were dependent upon terrestrial carrion (Fig. 3.1) (Fox-Dobbs et al., 2006). In contrast, isotope values for fossil condors from the region point to a diet that contained terrestrial (C3) and marine resources. Marine megafauna (whales and seals) survived the late Pleistocene extinction and are abundant in the productive eastern boundary current of the northeastern Pacific Ocean; they apparently provided a source of carrion for Pacific coast condor populations that carried them through the loss of terrestrial megafauna. Isotope mixing models indicate that many condors had diets that were >50% marine (Chamberlain et al., 2005), and ^{14}C chronologies reveal no temporal patterns; Pacific coast condors fed across the terrestrial-marine interface throughout the late Pleistocene (Fox-Dobbs et al., 2006).

The loss of condors elsewhere in North America versus their survival on the Pacific coast can be explained by dietary differences between the two groups (Fox-Dobbs et al., 2006). Extirpated condors from the southwestern U.S. and Florida are found in association with mammalian megafauna, and these animals have $\delta^{13}C$ values that fall between those expected for scavengers living in pure-C_3 and pure-C_4 ecosystems. The $\delta^{15}N$ values are more ambiguous; some individuals with high $\delta^{13}C$ values also have $\delta^{15}N$ values extending into the range interpreted as indicative of marine feeding on the Pacific coast. Yet high $\delta^{13}C$ and $\delta^{15}N$ values are common in arid, C_4-plant dominated ecosystems. Thus with only C and N isotope data it is possible that the specimens with high values in coastal California were immigrants from arid regions, not marine-feeders, or conversely, that the animals with high values in the southwest were immigrants from coastal regions. This conundrum was resolved through H isotope analysis, which provides an independent proxy for consumption of terrestrial (typically low δ^2H values) versus marine resources (high δ^2H values). δ^2H data confirm that high $\delta^{15}N$ values for Pacific coast condors were associated with marine feeding, whereas animals with high values in the southwest were completely terrestrial (Fig. 3.2).

Accounts from the 1800's report condors scavenging on beached marine mammals. Condors seem to have thrived during the period of Spanish and subsequent Mexican control of the west coast of the United States, as cattle ranching for leather created an abundance of terrestrial carrion, effectively recreating their lost Pleistocene resource base, which included both marine and terrestrial megafauna. Condors ate terrestrial food during their period of decline from the 1850s to the 1970s, perhaps because marine mammal abundances were low at this time as well because of sealing, whaling, and persecution from fisherman (Chamberlain et al., 2005). With the rise in marine mammal populations on the California coast following passage of the Marine Mammal Protection Act in 1972, condors reintroduced to coastal California, Baja California, or the Channel Islands have the possibility of returning to their 'ancient' coastal foraging strategy. There are anecdotal reports of condors again feeding on seals and whales, so they retain the behavioral plasticity to expand their diet to marine mammal carrion. Unfortunately reintroduced condors are accumulating toxic lead loads, which have been traced via lead isotopes to ammunition (Church et al., 2006). Condors may be ingesting lead shot in

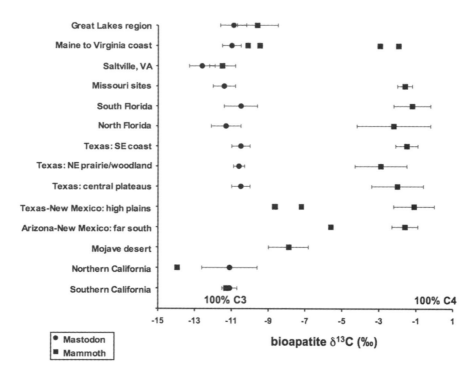

FIGURE 5.—Bioapatite δ¹³C values for Pleistocene mastodons (filled circles) and mammoths (filled squares) from different regions across North America (means ± one standard deviation or data for individual specimens if no error bars are presented). Samples span the entire late Pleistocene. Collagen δ¹³C values were converted to bioapatite values by adding +9.2‰. Expected δ¹³C values for animals on 100% C3 and 100% C4 diets are presented (data for compilation available from PLK).

gut piles left by hunters, or in the carcasses of fatally wounded animals that eluded their hunters. Whether or not consumption of marine carrion would reduce condor lead loads or induce other problems (e.g., organic contaminants) is an important open question. In any case, given the likely role of marine carrion in sustaining condor populations after the collapse of terrestrial megafauna, ecological niche modeling to prioritize areas for reintroduction (e.g., Martínez-Meyer et al., 2006) should factor in marine resource use.

Cryptic extinction and change in morphology and behavior: Alaskan grey wolves

First-order patterns of species presence and absence in the fossil record or current phylogeography do not always accurately reflect past population dynamics or diversity. Alaskan grey wolves (*Canis lupus*) are an excellent example. Grey wolves were widely distributed across Pleistocene and historic North America. Intuitively, this geographic pattern might be thought to reflect the survival of wolves across their ancient range. Within the North American mammalian carnivore guild, two factors generally correlate nega-

tively with survival across the Pleistocene-Holocene boundary: body size and degree of hypercarnivory (reliance upon a pure-meat diet) (Koch and Barnosky, 2006). Based solely on these criteria, grey wolves are unexpected survivors, but if Pleistocene wolves were prey-generalists or scavengers, they might have been able to accommodate a shifting prey-base (e.g., Romanuk et al., 2006).

However, combined stable isotope, ¹⁴C, genetic, and morphologic data from Pleistocene and modern grey wolves reveal a cryptic extinction of grey wolves in Beringia (Alaska and Yukon) (Leonard et al., 2007). Ancient DNA analyses show that modern Alaskan grey wolves are not descended from Pleistocene residents, but rather from Eurasian wolves that migrated into Alaska during the Holocene. To understand the defining features of the Beringian wolf ecomorph, and the ecological clues isotopic analyses can provide to explain its extinction, it is important to put the animal into paleoenvironmental context. For most of the glacial period from 60,000 to 10,000 years ago (including the very cold Last Glacial Maximum), the Beringian landscape was an open mosaic of steppe grassland and shrub tundra vegetation. A range of megafaunal prey was present at various times in the late Pleistocence (horse, yak, bison, caribou, muskox, and mammoth). The rapid reorganization of Beringian ecosystems at the end of the Pleistocene, which resulted in the establishment of modern tundra and boreal forest habitats, was temporally associated with the loss of most megafaunal species. The Beringian large carnivore guild was diverse and included grey wolf, American lion (*Panthera atrox*), scimitar-tooth sabercat (*Homotherium serum*), brown bear (*Ursus arctos*), and short-faced bear (*Arctodus simus*). Of these five, only gray wolves were present continuously through the entire late Pleistocene in Beringia (Fox-Dobbs et al., 2008).

Craniodental morphologic analysis suggests that Beringian wolves were either generalist large animal predators or scavengers (Fig. 4.1) (Leonard et al., 2007). They had a more robust craniodental morphology and exhibited greater tooth wear and fracture than modern wolves, suggesting that they ate more bone and perhaps took down larger prey. All morphologic indices point to heavy carcass utilization, indicating intense competition within the carnivore guild and perhaps within species. Dietary reconstructions from bone collagen $\delta^{13}C$ and $\delta^{15}N$ values are consistent with the interpretation that Beringian wolves were not specialized predators (Figs. 4.2-4.4). Wolves ate a wider range of prey than other large carnivores in each time period, matched only by lions in the late-glacial period (Fox-Dobbs et al., 2008).

Why didn't their generalized diets, which presumably contributed to their persistence through prior climatic and ecologic shifts, allow Beringian wolves to survive into the Holocene? Isotopic data from Beringian wolves and other carnivores provide an important clue. While wolves were the most generalized predators, none of the large-bodied carnivores except short-faced bear were single-prey specialists during the late Pleistocene. There was a high level of trophic overlap and interspecific competition for prey within the megafaunal carnivore guild. Brown bear, the sole survivor within the guild (Barnes et al., 2002), is the only species in which some individuals engage in substantial omnivory. Omnivory may have been the only adequate buffer when megafaunal herbivore populations crashed. Modern grey wolves are less robust than extinct Beringian conspecifics, and they appear to exhibit greater specialization on particular prey, especially cervids, which may have experienced irruptions following the

extinction of other, large herbivores. Overall, the study points to the vulnerability of highly connected, carnivore communities to food web perturbations, something that should be explored via food web modeling, and it indicates that the route to survival for modern wolves involved both morphological and behavioral evolution, as well as recolonization of range from which they had been extirpated.

Defining ecological analogs for rewilding

Janzen and Martin (1982) offered a powerful idea that is only now fully reverberating in ecology and conservation biology. They argued that the loss of Pleistocene megafauna disrupted species interactions, such as the dispersal of seeds, leaving behind organisms that are anachronisms, shaped by co-evolution for interactions that no longer occur (Barlow, 2001). We've discussed one such anachronism, the California condor, whose success at sites away from the Pacific coast was dependent on the carrion of extinct or extirpated terrestrial megafauna. As a deeper understanding has grown about the roles of megaherbivores as environmental engineers (Owen-Smith, 1988), and the roles of top carnivores in regulating terrestrial ecosystems (Terborgh et al., 1999), ecologists are beginning to speculate on the impacts of missing species on ecosystem diversity and function.

Donlan et al. (2006) followed this line of reasoning to a provocative conclusion. They suggested that missing ecological function and evolutionary potential can be restored to North America through the introduction of extant conspecifics or taxa related to extinct megafauna (Pleistocene rewilding). They envision this as a carefully managed ecosystem manipulation in which costs and benefits are measured on a case-by-case and locality-by-locality basis. The proposal has, unsurprisingly, raised a debate in conservation biology about the feasibility and wisdom of such manipulations. Would they restore lost function and interactions, or merely create potentially problematic new ones, as often results from non-native species introductions and invasions (Rubenstein et al., 2006)? We side-step this debate, and focus on a pressing matter should such experiments be attempted. What interactions were occurring in Pleistocene communities, and which modern taxa make the best analogs for missing Pleistocene species?

As an example, Levin et al. (2002) examined the indirect effects of horse grazing on salt marsh faunas to explore the types of impacts that late Pleistocene grazers might have had on coastal communities. They pointed to isotopic evidence that Pleistocene equids at coastal sites ate C4 plants, such as *Spartina* (cord-grass) in salt marshes, as do modern horses in these settings. Inadvertent re-introduction of equids to North America may have re-established species interactions and processes that had been present in North America for millions of years, but absent for the past 13,000 years. More information will be needed to confirm the hypothesis that Pleistocene horses actually ate *Spartina*, rather than upland C4 plants. Such information could come from phytoliths or plant fragments in coprolites or in dental calculus, or from $\delta^{15}N$ and $\delta^{34}S$ analysis (Peterson and Howarth, 1987).

The situation is more complex when the animal to be introduced is not conspecific or even congeneric with the extinct taxon. From the standpoint of safeguarding global species diversity and recreating important ecosystem functions, introduction of proboscideans to North America would be a top priority. In the latest Pleistocene, North America had at least three proboscidean species: *Mammuthus columbi*, the Columbian mammoth; *Mammuthus primigenius*, the woolly mammoth; *Mammut americanum*, the American mastodon. *Mammuthus columbi* and *M. americanum* were common in Pleistocene faunas south of the Laurentide ice sheet, with extensive range overlap across North America, though mammoths were more rare in eastern woodlands, and mastodon were absent from some regions on the central plains and prairies. Tooth morphology suggests that these proboscideans had different diets. Mastodon molars had blunt cusps arrayed in transverse ridges across the tooth, whereas mammoth teeth were high-crowned with flat occlusal surfaces and numerous enamel ridges. Based on these characters and rare gut content and coprolite analysis, mammoths have been reconstructed as grazers (grass eaters), whereas mastodons were considered browsers (animals that eat leaves, twigs, fruit, etc. of shrubs and trees). A growing $\delta^{13}C$ database (Koch, unpublished) reveals that the diets of these animals were, in fact, highly focused (Fig. 5). Wherever they occurred, mastodons ate a C3 plant diet, even in sites where C4 grass was available and being consumed by other taxa (e.g., Texas, Florida, Missouri). Likewise, in areas where both C4 grass and C3 browse were available, mammoths were strongly focused on grazing. In other regions, where C4 plants are absent (e.g., northern regions, California), mammoths ate C3 plants, but $\delta^{13}C$ values cannot easily distinguish between C3 grass and C3 browse.

Would surviving elephant species be appropriate analogs for these extinct taxa? Observational and isotopic data reveal that African elephants (*Loxodonta africana*, savanna elephants; *Loxodonta cyclotis*, forest elephants) have catholic diets (Koch et al., 1995; Cerling et al., 1999). While the fraction of grass versus browse in the diet varies in relation to the resources available, both species have diets focused on browse (Cerling et al., 1999). Asian elephants (*Elephas maximus*) have a dentition more similar to mammoths, and have diets richer in grass, though still with a substantial amount of browse (Cerling et al., 1999). Perhaps *L. cyclotis* would be a reasonable ecological analog for mastodons, but *L. africana* and *E. maximus* are not as focused on grazing as mammoths were. Of course, introduced elephants might evolve to a more grass-rich diet if placed on the North American prairie, as isotopic data suggest that the browse-rich diets of modern elephants are relatively recent acquisitions. Prior to one million years ago, most proboscideans in African and Asia ate C4 grass (Cerling et al., 1999). Isotopic data would be one part of any evaluation of the match between extinct species and their potential introduced replacements. Other taxa that should be examined closely are the extinct camelids (*Camelops, Hemiauchenia, Paleolama*) and the carnivores (*Smilodon, Homotherium*, and *Canis dirus*).

CONCLUSION

Isotopic data have an important role to play in conservation paleobiology because they can provide information on individuals and populations, as well as entire species. They offer proxies for abundance shifts that reveal the response of taxa to processes not operating at present, as seen in the studies on sockeye salmon. They illuminate the autecology of extinct species or extirpated populations (diet, habitat preferences, home range, reproductive ecology, etc.), documenting the full ecological potential of species that are currently in relictual distributions. Sometimes they confirm long-standing hypotheses, as was the case for the discovery of the seafood 'buffer' for coastal condors. Sometimes they reveal major surprises, such as the dramatic change in reproductive ecology in northern fur seals, or the fact that grey wolves became extinct in eastern Beringia and evolved a new diet and morphotype as they reoccupied the area. Finally, when coupled with other types of information (tooth wear, taphonomic, genetic, demographic, palynological, etc.) they can reveal the existence and magnitude of important species interactions that occurred in past ecosystems. Such information can also be used to parameterize models to test the function and stability of ecosystems following reintroductions of extirpated species or their ecological analogs. These are all essential services that isotopic paleoecology provides for conservation biology and restoration ecology, and they reinforce the conclusion that a deep temporal perspective is essential to effective and long-term conservation.

ACKNOWLEDGMENTS

We thank Greg Dietl and Karl Flessa for the invitation to participate in this short course, and Linda Ivany for a helpful review of this manuscript.

REFERENCES

AINLEY, D. G., E. D. CLARKE, K. ARRIGO, W. R. FRASER, A. KATO, K. J. BARTON, AND P. R. WILSON. 2005. Decadal-scale changes in the climate and biota of the Pacific sector of the Southern Ocean, 1950s to the 1990s. Antarctic Science, 17:171-182.

AINLEY, D., G. BALLARD, S. ACKLEY, L. K. BLIGHT, J. T. EASTMAN, S. D. EMSLIE, A. LESCROEL, S. OLMASTRON, S. E. TOWNSEND, C. T. TYNAN, P. WILSON, AND E. WOEHLER. 2007. Paradigm lost, or is top-down forcing no longer significant in the Antarctic marine ecosystem? Antarctic Science, 19:283-290.

BARLOW, C. C. 2001. The Ghosts of Evolution: Nonsensical Fruit, Missing Partners, and Other Ecological Anachronisms. Basic, New York, 291 p.

BARNES, I., P. MATHEUS, B. SHAPIRO, D. JENSEN, AND A. COOPER. 2002. Dynamics of Pleistocene population extinctions in Beringian brown bears. Science, 295:2267-2270.

BARNOSKY, A.D., P. L. KOCH, R. S. FERANEC, S. L. WING, AND A. B. SHABEL. 2004. Assessing the causes of Late Pleistocene Extinctions on the continents. Science, 306:70-75.

BARONI, C., AND G. OROMBELLI. 1994. Abandoned penguin rookeries as Holocene paleoclimatic indicators in Antarctica. Geology, 22:3-26.

BERLOW, E. L., U. BROSE, AND N. D. MARTINEZ. 2008. The "Goldilocks factor" in food webs. Proceedings of the National Academy of Science of the United States of America, 105:4079-4080.

BOWEN, G. J., AND J. B. WEST. 2008. Isotope landscapes for terrestrial migration studies, p. 79-106 . In K. A. Hobson and L. I. Wassenaar (eds.), Tracking Animal Migration with Stable Isotopes. Elsevier/Academic Press, London.

BROOKS, J. R., L. B. FLANAGAN, N. BUCHMANN, AND J. R. EHLERINGER. 1997. Carbon isotope composition of boreal plants: Functional grouping of life forms. Oecologia, 110:301-311.

BURNEY, D. A., AND T. F. FLANNERY. 2005. Fifty millennia of catastrophic extinctions after human contact. Trends in Ecology and Evolution, 20:395–401.

BURTON, R. K., J. J. SNODGRASS, D. GIFFORD-GONZALEZ, T. P. GUILDERSON, T. BROWN, AND P. L. KOCH. 2001. Holocene changes in the ecology of northern fur seals: Insights from stable isotopes and archaeofauna. Oecologia, 128:107-115.

CERLING, T.E., J. M. HARRIS, AND M. G. LEAKEY. 1999. Browsing and grazing in elephants: The isotope record of modern and fossil proboscideans. Oecologia, 120:364-374.

CHAMBERLAIN, C. P., J. R. WALDBAUER, K. FOX-DOBBS, S. D. NEWSOME, P. L. KOCH, D. R. SMITH, M. E. CHURCH, S. D. CHAMBERLAIN, K. J. SORENSON, AND R. RISEBROUGH. 2005. Pleistocene to recent dietary shifts in California condors. Proceedings of the National Academy of Sciences of the United States of America, 102:16707–16711.

CHURCH, M. E., R. GWIAZDA, R. W. RISEBROUGH, K. SORENSON, C. P. CHAMBERLAIN, S. FARRY, W. HEINRICH, B. A. RIDEOUT, AND D. R. SMITH. 2006. Ammunition is the principal source of lead accumulated by California condors re-introduced to the wild. Environmental Science and Technology, 40:6143-6150.

DONLAN, C. J., J. BERGER, C. E. BROCK, J. H. BROCK, D. A. BURNEY, J. A. ESTES, D. FOREMAN, P. S. MARTIN, G. W. ROEMER, F. A. SMITH, M. E. SOULÉ, AND H. W. GREENE. 2006. Pleistocene rewilding: An optimistic vision for 21st Century conservation. American Naturalist, 168:660-681.

EMSLIE, S. D. 1987. Age and diet of fossil California condors in Grand Canyon, Arizona. Science, 237:768–770.

EMSLIE, S. D., L. COATS, AND K. LICHT. 2007. A 45,000 yr record of Adelie penguins and climate change in the Ross Sea, Antarctica. Geology, 35:61-64.

EVERSHED, R. P., I. D. BULL, L. T. CORR, Z. M. CROSSMAN, B. E. VAN DONGEN, C. EVANS, S. JIM, H. MOTTRAM, A. J. MUKHERJEE, AND R. D. PANCOST. 2007. Compound-specific stable isotope analysis in ecological research, p. 480–540. In R. Michener and K. Lajtha (eds.), Stable Isotopes in Ecology and Environmental Science, 2nd Edition. Blackwell Publishing, Boston.

FINNEY, B. P., I. GREGORY-EAVES, M. S. V. DOUGLAS, AND J. P. SMOL. 2002. Fisheries productivity in the northeastern Pacific Ocean over the past 2,200 years. Nature, 416:729-733.

FINNEY, B. P., I. GREGORY-EAVES, J. SWEETMAN, M. S. V. DOUG-

LAS, AND J. P. SMOL. 2000. Impacts of climatic change and fishing on Pacific salmon abundance over the past 300 years. Science, 290:795-799.

FOX-DOBBS, K., J. A. LEONARD, AND P. L. KOCH. 2008. Pleistocene megafauna from eastern Beringia: Paleoecological and paleoenvironmental interpretations of stable carbon and nitrogen isotope and radiocarbon records. Palaeogeography, Palaeoclimatolatology, Palaeoecology, 261:30-46.

FOX-DOBBS, K., T. A. STIDHAM, G. J. BOWEN, S. D. EMSLIE, AND P. L. KOCH. 2006. Dietary controls on extinction versus survival among avian megafauna in the late Pleistocene. Geology, 34:685-688.

FULLER, B. T., J. L. FULLER, N. E. SAGE, D. A. HARRIS, T. C. O'CONNELL, AND R. E. M. HEDGES. 2004. Nitrogen balance and $\delta^{15}N$: Why you're not what you eat during pregnancy. Rapid Communications in Mass Spectrometry, 18:2889-2896.

FULLER, B. T., J. L. FULLER, N. E. SAGE, D. A. HARRIS, T. C. O'CONNELL, AND R. E. M. HEDGES. 2005. Nitrogen balance and $\delta^{15}N$: Why you're not what you eat during nutritional stress. Rapid Communications in Mass Spectrometry, 19:2497-2506.

GIFFORD-GONZALEZ, D., S. D. NEWSOME, P. L. KOCH, T. P. GUILDERSON, J. J. SNODGRASS, AND R. K. BURTON. 2005. Archaeofaunal insights on pinniped-human interactions in the northeastern Pacific, p. 19-38. In G. Monks (ed.), The Exploitation and Cultural Importance of Sea Mammals. Proceedings of the 9th Conference of the International Council of Archaeozoology, Durham, 2002. Oxbow Books, Oxford.

GRAHAM, B. S., P. L. KOCH, S. D. NEWSOME, K. W. MCMAHON, AND DAVID AURIOLES. In press. Using isoscapes to trace the movements and foraging behavior of top predators in oceanic ecosystems. In J. B. West, G. J. Bowen, T. E. Dawson, and K. P. Tu (eds.), Isoscapes: Understanding Movement, Pattern, and Process on Earth through Isotope Mapping. Springer, Berlin.

GRAYSON, D. K. 1977. Pleistocene avifaunas and the overkill hypothesis. Science, 195:691-693.

HALL, B. L., A. R. HOELZEL, C. BARONI, G. H. DENTON, B. J. LE BOEUF, B. OVERTURF, AND A. L. TÖPF. 2006. Holocene elephant seal distribution implies warmer-than-present climate in the Ross Sea. Proceedings of the National Academy of Sciences of the United States of America, 103:10213-10217.

HOBSON, K. A. 2008. Applying isotopic methods to tracking animal movements, p. 45-78. In K. A. Hobson and L. I. Wassenaar (eds.), Tracking Animal Migration with Stable Isotopes. Elsevier/Academic Press, London.

HOBSON, K. A., R. T. ALISAUSKAS, AND R. G. CLARK. 1993. Stable nitrogen isotope enrichment in avian tissues due to fasting and nutritional stress: Implications for isotopic analysis of diet. Condor, 95:388–394.

INGER, R., AND S. BEARHOP. 2008. Applications of stable isotope analyses to avian ecology. Ibis, 150:447-461.

JANZEN, D. H., AND P. S. MARTIN. 1982. Neotropical anachronisms: The fruits the gomphotheres ate. Science, 215:19-27.

KENNEDY, B. P., J. D. BLUM, C. L. FOLT, AND K. H. NISLOW. 2000. Using natural strontium isotopic signatures as fish markers: Methodology and application. Canadian Journal of Fisheries and Aquatic Sciences, 57:2280-2292.

KOCH, P. L. 2007. Isotopic study of the biology of modern and fossil vertebrates, p. 99-154. In R. Michener and K. Lajtha (eds.), Stable Isotopes in Ecology and Environmental Science, 2nd Edition. Blackwell Publishing, Boston, Massachusetts.

KOCH, P. L., AND A. D. BARNOSKY. 2006. Late Quaternary extinctions: State of the debate. Annual Review of Ecology, Evolution, and Systematics, 37:215-250.

KOCH, P. L., J. HEISINGER, C. MOSS, R. W. CARLSON, M. L. FOGEL, AND A. K. BEHRENSMEYER. 1995. Isotopic tracking of change in diet and habitat use in African elephants. Science, 267:1340-1343.

KOHN, M. J., AND T. E. CERLING. 2002. Stable isotope compositions of biological apatite, p. 455-488. In M. J. Kohn, J. Rakovan, and J. M. Hughes (eds.). Reviews in Mineralogy and Geochemistry, 48, Mineralogical Society of America.

LEONARD, J. A., C. VILA, K. FOX-DOBBS, P. L. KOCH, R. K. WAYNE, AND B. VAN VALKENBURGH. 2007. Cryptic extinction of Pleistocene Alaskan wolves: Evidence from genetics, isotopes, and morphology. Current Biology, 17:1146-1150.

LEVIN, P. S., J. ELLIS, R. PETRIK, AND M. E. HAY. 2002. Indirect effects of feral horses on estuarine communities. Conservation Biology, 16:1364–1371.

LIU, X., L. SUN, X. YIN, AND R. ZHU. 2004. Paleoecological implications of the nitrogen isotope signatures in the sediments amended by Antarctic seal excrements. Progress in Natural Science, 14:786-792.

LYMAN, R. L. 2006. Paleozoology in the service of conservation biology. Evolutionary Anthropology, 15:11-19.

MARTÍNEZ DEL RIO, C., N. WOLF, S. A. CARLETON, AND L. Z. GANNES. 2009. Isotopic ecology ten years after a call for more laboratory experiments. Biological Reviews, 84:91-111.

MARTÍNEZ-MEYER, E., A. T. PETERSON, J. I. SERVÍN, AND L. F. KIFF. 2006. Ecological niche modelling and prioritizing areas for species reintroductions. Oryx, 40:411–418.

MOSS, M. L., D. Y. YANG, S. D. NEWSOME, C. F. SPELLER, I. MCKECHNIE, A. D. MCMILLAN, R. J. LOSEY, AND P. L. KOCH. 2006. Historical ecology and biogeography of north Pacific pinnipeds: Isotopes and ancient DNA from three archaeological assemblages. Journal of Island and Coastal Archaeology, 1:165-190.

NEWSOME, S. D., C. MARTÍNEZ DEL RIO, D. L. PHILLIPS, AND S.

BEARHOP. 2007a. A niche for isotopic ecology. Frontiers in Ecology and the Environment, 5:429–436.

NEWSOME, S. D., M. A. ETNIER, D. GIFFORD-GONZALEZ, D. P. COSTA, M. VAN TUINEN, E. A. HADLY, D. J. KENNETT, T. P. GUILDERSON, AND P. L. KOCH. 2007b. The shifting baseline of northern fur seal ecology in the northeast Pacific Ocean. Proceedings of the National Academy of Sciences of the United States of America, 104:9709-9714

OWEN-SMITH, R. N. 1988. Megaherbivores: The Influence of Very Large Body Size on Ecology. Cambridge University Press, Cambridge, 369 p.

PETERSON, B. J., AND R. W. HOWARTH. 1987. Sulfur, carbon, and nitrogen isotopes used to trace organic matter flow in the salt-marsh estuaries of Sapelo Island, Georgia. Limnology and Oceanography, 32:1195-1213.

ROMANUK, T. N., B. BEISNER, N. D. MARTINEZ, AND J. KOLASA. 2006. Non-omnivorous generality promotes population stability. Biology Letters, 2:374-377.

RUBENSTEIN, D. R., D. I. RUBENSTEIN, P. W. SHERMAN, AND T. A. GAVIN. 2006. Pleistocene Park: Does re-wilding North America represent sound conservation for the 21st century? Biological Conservation, 132:232-238.

SHARP, Z. 2007. Principles of Stable Isotope Geochemistry. Prentice Hall, Upper Saddle River, New Jersey, 344 p.

SPRINGER, A.M., J. A. ESTES, G. B. VAN VLIET, T. M. WILLIAMS, D. F. DOAK, E. M. DANNER, K. A. FORNEY, AND B. PFISTER. 2003. Sequential megafauna collapse in the North Pacific Ocean: An ongoing legacy of industrial whaling? Proceedings of the National Academy of Sciences of the United States of America, U.S.A., 100:12223-12228.

SPRINGER, A.M., J. A. ESTES, G. B. VAN VLIET, T. M. WILLIAMS, D. F. DOAK, E. M. DANNER, AND B. PFISTER. 2008. Mammal-eating killer whales, industrial whaling, and the sequential megafaunal collapse in the North Pacific Ocean: A reply to critics of Springer et al. 2003. Marine Mammal Science, 24(2): 414–442.

TERBORGH, J., J. A. ESTES, P. PAQUET, K. RALLS, D. BOYD-HEGER, B. J. MILLER, AND R. F. NOSS. 1999. The role of top carnivores in regulating terrestrial ecosystems, p. 39-64. In M. E. Soulé and J. Terborgh (eds.), Continental Conservation: Scientific Foundations of Regional Reserve Networks. Island Press, Washington, D.C.

CHAPTER SEVEN

EVALUATING HUMAN MODIFICATION OF SHALLOW MARINE ECOSYSTEMS: MISMATCH IN COMPOSITION OF MOLLUSCAN LIVING AND TIME-AVERAGED DEATH ASSEMBLAGES

SUSAN M. KIDWELL

Department of Geophysical Sciences, University of Chicago, Chicago, IL 60637 USA

ABSTRACT.—Time-averaged death assemblages of molluscan shells sieved from seafloors during conventional biological surveys represent an important under-exploited resource for retrospective evaluation of the magnitude and nature of recent community change. Benthic biologists usually discard these death assemblages (DAs) because of concerns for postmortem bias and time averaging of multiple generations. However, quantitative synthesis of ~100 "live-dead" comparisons establish that the time-averaged nature of DAs can be used to advantage in conservation biology. The strongest correlate of poor live-dead (LD) agreement is the magnitude of anthropogenic eutrophication of the water body (AE), because the DA retains a memory of the former abundance of species that are now rare or absent in the living assemblage. In areas with a known history of AE, seagrass-dwelling species tend to be significantly more abundant dead than alive, and species that prefer organic-rich sediments or tolerate low-oxygen episodes tend to be more abundant alive than dead. Bottom-trawling (BT) of seafloors for fin- and shell-fish is also associated with LD mismatch but only in gravelly habitats where species are generally not naturally adapted to frequent physical disturbance. The overall degree of LD mismatch and the identities of the species most responsible for that mismatch can thus be used to recognize AE and BT in areas where human activities are unknown or unregulated. Further, LD agreement is remarkably high in areas unaffected by human activities, so LD analysis can also be used to identify regions that can serve as restoration baselines. The high LD agreement found in pristine settings is also the current best estimate of how confidently we can reconstruct, in presented degraded areas, what the local living community was originally like, using DAs extracted from historic layers of sediment cores. LD comparison fails to detect ecological change in almost half of all habitats where human stresses definitely exist, and thus this is a conservative tool that will tend to under-estimate human impacts. However, given the scarcity of information on present-day communities and especially on their recent past, historical ecological information is at a premium and LD comparisons will be valuable.

INTRODUCTION

ECOLOGICAL ASSESSMENT and restoration at the habitat to regional scales where conservation efforts are focused face many basic scientific challenges. These include such rudimentary questions as:

1. *What is it like now?* Unfortunately, both on land and under the sea, even "today" is usually undersampled, with knowledge rarely reflecting more than a single season of collection and taking the form of a regional checklist.

2. *Is it stable?* Past trajectories that have brought the local system up to today and thus might carry it into the future are usually unknown.

3. *What was natural?* The composition and structure of unaltered systems—both those still existing and those past—are usually unknown. Ideally, baselines for restoration efforts should be based on genuinely natural conditions and not simply the state of the system at the time it was first surveyed, which in most settings is well after the onset of human modification.

An historical perspective on the system is important for answering all three questions. How long (deep) that that time perspective needs to be depends upon the system and the question at hand. The more inherently variable the environment, the deeper the suspected roots of human impact(s), and/or the longer lived the individuals of species, then the longer the window of observation needed to evaluate the status of the biota for questions 2 and 3 (Fig. 1). Because some species are short-lived or seasonal in occurrence, even a picture of "now" (question 1) should be based on more than a single season of sampling and arguably on return visits over several years given inter-annual variability in seasons. Many regions including tropical ones also undergo decadal scale variability in response to natural

In *Conservation Paleobiology: Using the Past to Manage for the Future, Paleontological Society Short Course, October 17th, 2009. The Paleontological Society Papers, Volume 15*, Gregory P. Dietl and Karl W. Flessa (eds.). Copyright © 2009 The Paleontological Society.

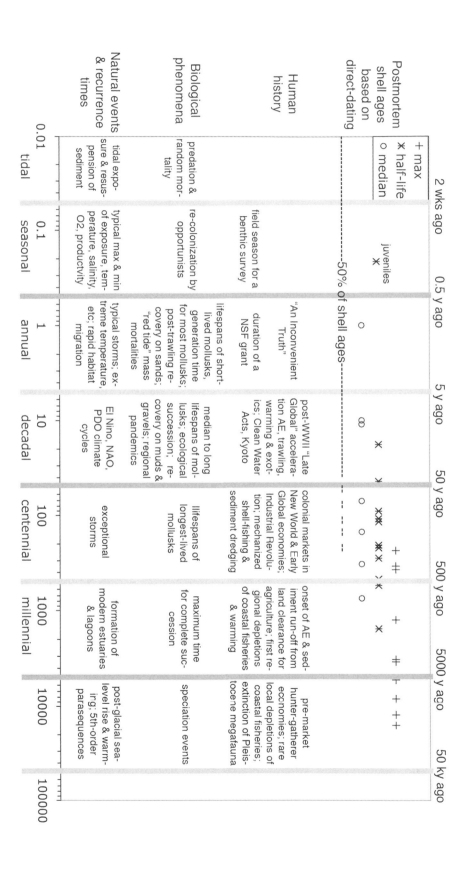

Figure 1.— Time frames of time-averaging, biological and other natural phenomena and chronology of human history relevant to shallow marine systems. Following a convention of sequence stratigraphy, time frames on a log scale (bottom) are *characterized* by their integer order of magnitude (e.g., decadal = 10^1 years) but are operationally *defined* (bounded) by mid-point values (e.g., $10^{0.5}$ and 10^5 years). Chronology (top scale) is pinned at today or, for direct-dated shells, the date of publication. Economic phases of Lötze et al. (2006); lifespans from compilation of Kidwell and Rothfus (in prep.); recovery times from bottom-trawling are generally 1-5x the generation time of the species in the community (e.g., Collie et al., 2000; NRC, 2002); maximum, half-life and median shell ages compiled from Meldahl et al., 1997; Kowalewski et al., 1998; Kidwell et al., 2005; Kosnik et al., 2007; Kidwell, 2005; Kowalewski, 2009; Kosnik et al., in press; Krause et al., in press; juvenile half-lives from Cummins et al., 1986b.

TABLE *1.*— Abbreviations and measures of LD agreement.

Abbreviations
AE = Anthropogenic eutrophication
BT = Bottom trawling
DA = Death assemblage (mollusks collected as empty shells)
LA = Living assemblage (mollusks collected alive)
LD = Live-dead

Measures of LD agreement
JC = Jaccard-Chao index of taxonomic similarity
Rho = Correlation coefficient of a Spearman rank-order test
Delta-richness = Ratio of the DA richness to LA richness, after sample-size standardization
PIE = Hurlbert's evenness metric, Probability of Interspecies Encounter
Delta-PIE = Difference between DA PIE and LA PIE

climate cycles such as El Niño/La Niña and the NAO (North Atlantic Oscillation). Habitats where few species are sampled alive at any given instant in time (i.e., have few abundant year-round occupants) may thus constitute genuine diversity hotspots when species occupation over time is considered (e.g., White et al., 2006).

Information on the composition and structure of a target community should thus minimally reflect knowledge of the system on a decadal scale even in fully natural systems, and preferably on a centennial to millennial scale given the early onset of human pressures on ecosystems and their intensification over time. For example, even in the marine realm, humans have been capable of depleting targeted food species locally for more than 5000 years based on archeological evidence and depletions of favored food and luxury species became regional in scope well before the mid-19th century Industrial Revolution (e.g., Jackson et al., 2001; Lötze et al., 2006; Rick and Erlandson, 2009; Fig. 1). Marine environmental deterioration—from excess nutrients, solid sediment runoff, and habitat destruction—as well as species depletions, pandemics, and introductions of alien species have all intensified significantly and become truly global in the last ~50 years, leaving no oceans untouched (Halpern et al., 2008; Baum and Worm, 2009).

The rhythms of biological phenomena such as in-

dividual lifespans and recolonization rates need to be compared to natural environmental rhythms and the monotonic increase in pressure on exploited species and habitats (Fig. 1). The recovery time of a community to physical disturbance, for example, is typically 1 to 5x the generation time of constituent species (e.g., Collie et al., 2000; NRC, 2002). Populations of species that grow rapidly and reproduce early will thus recover within a few months or years after a disturbance (e.g., wipe-out by an exceptional hurricane, bottom-dredging, or dynamiting). On the other hand, a minimum of years to decades are required for the community to reacquire fully functioning populations of species that do not reproduce until individuals are several years old (e.g., most mollusks, with a median lifespan of ~8 years) or that will not recolonize an area until the seafloor has reached some condition that depends on elapsed time (e.g., degree of sediment shelliness, firmness, or oxygenation; presence of grassbeds or other bio-constructed habitat). Coral reefs, which are the stereotypic image of a seafloor community, entail centuries to millennia to construct *de novo* and require at minimum decades or centuries to recover fully, assuming no further abuse.

Unfortunately, few areas have a history of scientific surveys much less the high-frequency biomonitoring that is needed to characterize natural temporal variability within systems and, with the exception of some targeted food species, quantitative data rarely extend back more than a few decades even in the most intensely studied areas. Alternative sources of quantitative historical information are thus needed to acquire basic information on the status of present-day systems, on their genuinely natural baselines, and on the biological consequences of human pressures on wild populations (NRC, 2005).

Naturally occurring death assemblages (DAs) of skeleton-bearing animals—the shells of dead invertebrates and the bones and teeth of vertebrates present in the uppermost part of the sedimentary record of a modern-day habitat—are a promising "subfossil" source of such historic information (Table 1). For example, on average, a ~10-liter sample of seafloor mud or sand will contain ~ten-times more dead molluscan individuals than living molluscan individuals (Fig. 2), representing a tantalizingly large sample size of the local fauna. The shells in molluscan DAs are clearly time-averaged over multiple generations, in contrast to the temporally acute "snapshot" of standing populations provided by a conventional biological survey. Dead individuals outnumber living individuals even on seafloors where the introduction of shells by post-mortem transportation from other habitats is extremely unlikely, range in physical condition from excellent to poor suggesting a spectrum of postmortem ages, and can include species that are almost certainly relics from past decades when different environmental conditions existed (e.g., see early "live-dead" studies such as Johnson, 1965, early quantitative assessments of benthic stocks such as Davis, 1923, and early actualistic biofacies studies such as Ladd et al., 1957).

Radiocarbon and amino-acid racemization dating of DAs sampled from the surface of modern seafloors indicate that mollusk shells can range in age up to 10-20 ky old on open continental shelves that have experienced slow net rates of sedimentation during Holocene sealevel rise (e.g., Fischer, 1961; Flessa and Kowalewski, 1994; Flessa, 1998) and can range up to several thousand years even in estuaries and lagoons that are natural sinks of sediment accumulation and have only existed for a few thousand years (see sources for Fig. 1, and see Kowalewski, 2009; Kosnik et al., in press; Krause et al., in press). Notwithstanding these maximum shell ages, surface-sampled DAs in subtidal settings tend to be *dominated* by shells from the youngest part of the age-spectrum and typically come from the

most recent decades to centuries (see median ages and half-lives in Fig.1). Although such temporal resolution seems coarse at first glance, centennial and particularly decadal resolution of community composition would provide insight into a wide range of issues important to conservation (Fig. 1). Moreover, in the absence of other historical data, even a blurry view of the past is valuable.

Here, I summarize some key findings on using molluscan death assemblages—and particularly mismatches in the species compositions of death and living assemblages—as a source of otherwise unattainable historical information on modern benthic communities. The analytic results reflect a global meta-analysis (quantitative synthesis) of the raw data from ~100 habitat-level "live-dead" studies (summarized from Kidwell, 2001, 2002a, 2002b, 2007, 2008, plus new work first reported here; Olszewski and Kidwell, 2007; see Kidwell, 2007, for complete list of raw data sources). The focus is on soft-sedimentary seafloors—various admixtures of mud, sand, and gravel, with and without seagrass—that constitute the majority of shallow marine and estuarine seafloors. The surface and upper few cm to tens of cm (exceptionally the upper meter or so) of these substrata are occupied by diverse and abundant macroscopic bottom-dwelling invertebrate life ("macrobenthos" >1 mm; Fig. 3). Polychaetes almost always constitute the majority of live-collected individuals and species in quantitative bottom samples, with mollusks ranking second or third (after arthropods). However, mollusks typically constitute the majority of living macrobenthic *biomass* owing to their relatively large body size (e.g., see summary info in Gray and Elliott, 2009) and constitute the majority of species if the DA is included (regional checklists of species almost always include species known only from empty shells).

Mollusks also play many ecological roles both during life (e.g., as suspension feeders, deposit feeders, predators, grazers, reef-builders, bioturbators, bioeroders, and prey items) and after death (e.g., dead shells as islands of hard substrata and as creators of coarser and physically more complex habitat; Kidwell and Jablonski, 1983; Palacios et al., 2000; Hewitt et al., 2005). Being able to reconstruct historical changes in the composition and structure of molluscan communities would thus give insights into a major component of shallow marine ecosystems, as well as a possible surrogate for the entire system.

WHAT IS A DEATH ASSEMBLAGE?

Death assemblages (DAs) constitute the whole and broken invertebrate shells and other organic debris (e.g., seagrass fragments, worm casings, fish otoliths) that exist in the uppermost few cm of seafloor that are sampled for living benthos. The *molluscan* death assemblage is the portion of this "grunge" or "sieve residue" that is molluscan in origin and identifiable to the species level. Benthic biologists regularly handle—and dispose of—this material when picking samples for living specimens. Dead molluscan individuals are operationally defined as whole shells or large fragments that signify a unique individual (e.g., fragments that include >50% of the bivalve hingeline, or that include the apex of a gastropod shell). Individual dead specimens can range in physical condition within a single DA from "like new" (the shell interior lacks flesh but retains its original luster) to slightly modified (dull surface, loss or flaking of periostracum, discoloration) or strongly modified (notable postmortem polishing, bioerosion, corrosion, bioencrustation). Empty shells that retain some adhering flesh are usually categorized as living individuals, both by biologists and taphonomists.

Because benthic sediment samples contain on average ~10x more dead mollusk individuals than living individuals (Fig. 2.1), molluscan DAs typically contain more species than the associated living assemblage (LA)—on average, about 3x more species (Fig. 2.2; raw values in Table 3). However, this "surplus richness" of DAs does not owe entirely to their greater sample size. When a DA is subsampled down to match the number of individuals that are present in the LA, the DA usually still contains a larger number of species (sample-size standardized values in Table 3). On average, DAs are ~1.2x richer if resampled with replacement (binomial distribution, Kidwell 2002a) and ~1.5x richer if resampled without replacement (hypergeometric distribution, Olszewski and Kidwell, 2007; Tomasovych and Kidwell, in prep.). DAs are thus clearly not simply one-to-one substitutes for conventional biological surveys of living molluscan communities.

Death assemblages sampled from the uppermost mixed layer of the sedimentary record—the 5 to 20 cm-deep "bite" of a Van Veen grab, small box-core, naturalist's dredge, or hand or air-suction excavation (Fig. 3)—are subject to a wide array of postmortem processes that are fundamentally destructive. These include *physical and biological reworking* (vertical advection that mixes generations and promotes disarticulation, fragmentation, and lateral transportation of shells), *bioerosion and bioencrustation* (especially during intervals when shells are exposed at the sediment-water interface, promoting obliteration of morphology critical to species-level identification), *microbial maceration* of microstructure (promoting shell disintegration, even when buried), and *modification by overlying and pore waters*, which may be under-saturated with respect to carbonate minerals (promoting shell dissolution) or oversaturated (promoting precipitation of minerals that tend to preserve morphology). These processes vary in strength among environments ("taphofacies") and species differ in their inherent susceptibility to these processes (variation in "intrinsic durability"; the majority of the taphonomic literature concerns these issues, see Behrensmeyer et al., 2000, for a general review). This large number of factors makes it difficult to predict, *a priori*, the degree to which a given DA might differ in composition from the local LA.

On the positive side, DAs within this uppermost mixed layer of the sedimentary record are still receiving new, freshly dead individuals from the local LA, which should continuously refresh the composition of the DA to make it to more closely match that of the local LA (Fig. 3). Infaunal (burrowing) mollusks live within this same uppermost increment of the substrata and epifaunal mollusks such as scallops, oysters, and grazing snails live on the sediment-water interface itself. The steady input of newly dead shells from the living community should promote agreement between DAs and co-sampled LAs, counteracting the biasing effects of destructive postmortem processes.

With progressive burial below the sediment-water interface, the DA encounters new pore water regimes that may be more or less favorable to shell preservation than the regime(s) within the uppermost mixed layer, but the intensity of all other postmortem processes decreases. DAs become progressively less subject to vertical mixing, and thus rates of shell loss should decrease along with rates of new-shell supply (for a probabilistic treatment, see Sadler, 1993; Fig. 3).

WHAT IS A "LIVE-DEAD" STUDY?

Live-dead (LD) studies compare the numbers of individuals occurring alive and dead for species in

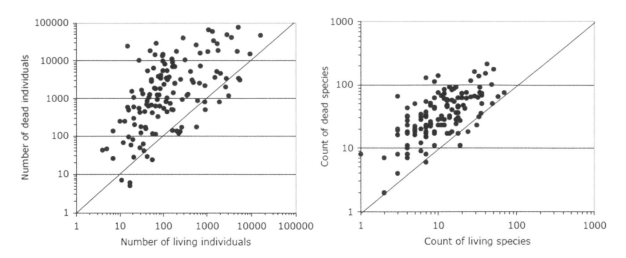

FIGURE 2.—*1*, Numbers of living and dead individuals and *2*, counts of species occurring alive and dead in subtidal soft-sedimentary seafloors, based on 115 molluscan "live-dead" datasets compiled from the global literature. Each point indicates the raw results for living and dead mollusks from the uppermost part of the seafloor, based on pooling specimens from samples at two or more stations within a single habitat. Dead individuals are on average ~10x more abundant than living individuals and represent on average ~3x more species than are sampled alive in the same sedimentary volume. Although abundant and rich, dead skeletal remains are usually ignored—in fact, discarded—in whole-fauna benthic surveys and in biomonitoring efforts. If we can be confident of their fidelity to the living assemblage, then we gain a large increase in sample size and temporal perspective on the molluscan community. Analyses here focus on the 109 datasets having >20 living and >20 dead individuals.

modern environments and are the most common means of evaluating the ability of time-averaged death assemblages (DAs) to capture the composition and structure of local living assemblages (LAs; see Kidwell and Flessa, 1995, for review). Molluscan LD datasets are generated in soft-bottoms settings by taking sediment samples of standardized volume, sieving them to remove sediment (volume and mesh size varies among studies), picking living and dead specimens from the sieve residue, and generating a list of all species encountered and the numbers of living and dead individuals for each species.

Using these raw data, LD agreement can be assessed for any number of ecological attributes. Virtually any metric developed to describe a living community can be applied to a death assemblage, and thus that attribute of the DA can be compared to that attribute of the co-sampled LA. Examples covered here include LD agreement in *species richness* (is the count of species occurring dead the same as, greater than, or less than the count of species occurring alive?), *taxonomic similarity* (do the living and dead species lists share many or few species in common?), and species *relative abundance* (e.g., in a rank-order test, are the species that are most abundant in the LA also numerically dominant in the DA, or are they instead randomly distributed through an abundance-ranked list or perhaps exclusively rare?). We can also assess LD agreement in how species composition varies along an environmental gradient, LD agreement in the magnitudes of alpha versus beta diversity, and the comparability of spatial or temporal variability (dispersion) among replicate samples of the LA or DA (e.g., Tomasovych and Kidwell, 2009a, b, c, in prep.). LD agreement can also be assessed for single species, for example the degree of similarity in the size- or age-frequency distributions of a particular species in the DA and that in the LA (e.g., Hallam, 1967; Cummins et al., 1986a; Tomasovych, 2004).

Finally, LD agreement can be assessed at any spatial scale (spatial resolution). For example, what is LD agreement within a *single sample* (e.g., Staff et al., 1986), within a single *habitat* (living and dead individuals are tallied after multiple samples from a single seafloor type in a region have been pooled; e.g., Peterson, 1976; meta-analytic focus here and in my earlier papers), and within a *multi-habitat region* (similarity of all pooled DA samples to all pooled LA samples,

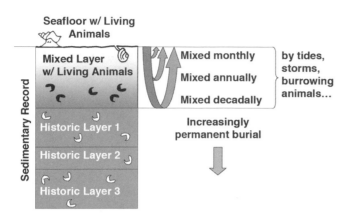

FIGURE 3.—Schematic cross-section of a seafloor. The surface and upper few cm to tens of cm (exceptionally the upper meter or so) of soft-sedimentary seafloors are occupied by diverse and abundant macroscopic bottom-dwelling invertebrate life ("macrobenthos" >1 mm), whose skeletal remains constitute the local macrobenthic death assemblage. The uppermost part of the sedimentary column is mixed with highest frequency because so many kinds of animals and physical processes intersect with it, and the deepest parts of the column are mixed least frequently (e.g., only by the burrows of deeply-penetrating infauna, during once-a-century storms, or during sealevel lowstands). Conventional biological surveys sample the upper 5 to 10 cm of the seafloor, which is generally well within the mixed layer.

e.g., Johnson, 1965; Bouchet et al., 2002, or similarity of a single DA sample to the known regional species pool, e.g., Warwick and Light, 2002; Smith, 2008)? A large number of molluscan LD studies including such classics as Warme (1971), Ekdale (1977), and Miller (1988) test the ability of DAs to reflect a facies mosaic or gradient evident in LAs within a small region (resolution of *spatial grain*).

Conservation and restoration efforts focus almost exclusively at the scale of habitats and small multi-habitat regions, and thus the ability of death assemblages to capture biological information at these scales will be stressed here.

The findings summarized here reflect analysis of the raw data from 109 habitat-level LD datasets generated by diverse authors from subtidal seafloors in ~50 different regions globally (for a complete list of sources and description of areas, see Kidwell, 2007). Each dataset reflects pooling of at least two samples of the given habitat, with a total of at least 21 living and 21 dead molluscan individuals. Habitats are defined as areas that are relatively homogeneous in sediment grain size, subaquatic vegetation, and water depth—for example,

clean sand from the margin of a bay versus broad areas ("facies") of muddy sand, sandy mud, or mud from the same bay. A single study area—e.g., Tomales Bay—thus typically yields several habitat-level datasets and one regional-level dataset. Most datasets come from siliciclastic settings. Datasets are otherwise diverse: mesh sizes used to collect living and dead individuals range from 0.3 to 5 mm, seafloors range from pure muds to gravels and grassbeds, and study areas range in latitude from 2° to 55°, i.e., from tropical to cold temperate conditions.

CONCEPT OF LIVE-DEAD MISMATCH AS A SIGNAL OF RECENT ECOLOGICAL CHANGE

Mismatch (disagreement, discordance) between the richness or species composition of a DA and the local LA is typically attributed to natural post-mortem processes and particularly to preservational bias. However, biologists and actualists alike have long been tantalized by the evidence that death assemblages with unexpected compositions might provide about recent changes in population sizes (e.g., Davis, 1923) and environmental conditions (e.g., Ladd et al., 1957), including changes wrought by human activities. For example, Bourcier (1980) noted the abundance of exclusively dead specimens of grass-dwelling mollusks in Mediterranean bays and argued that these probably signified recent and otherwise unappreciated environmental deterioration from untreated wastewater and a weakening of winter storms that might have cleansed the system. Several taphonomists have also suggested that LD mismatch within their study area indicated a recent shift in community composition driven by human activities—for example, the possible impact of shrimp-trawling on Texas shelf faunas (Staff and Powell, 1999), of point-source nutrients on Florida Bay mollusks (Ferguson, 2008), and of unknown pressures on Caribbean coral reefs (Greenstein et al., 1998).

As a tool for conservation biology, LD mismatch would take advantage of the intrinsic durability of molluscan shells, and thus the potential for the composition of a time-averaged death assemblage to lag behind any strong directional change in the composition of the living assemblage. The stronger the shift in LA composition, the lower the rate of production of dead shells from the new LA relative to the original LA, the greater the relative durability of dead shells from the original LA, or the slower the rate of permanent burial that moves the DA beyond the influence of the LA (with the last two factors determining the window for time-averaging), then the more likely that the composition of the DA will lag behind a compositional shift of the LA. Such *taphonomic inertia* of the DA to changing ecological conditions is particularly likely when the community has been modified by human activities because they tend to be more rapid, monotonic, intense, and/or multi-factorial than natural forces (e.g., Jackson, 1997; Jackson et al., 2001; Pandolfi et al., 2003; Lötze et al., 2006; Orth et al., 2006; Airoldi and Beck, 2007).

However, even in fully natural settings and in the absence of any postmortem bias, we would still expect some degree of LD mismatch to exist in seafloor samples. LAs could differ from local DAs in composition, richness, etc owing to (1) random events of mortality and colonization, which can temporarily push the LA to an unusual composition, albeit still within normal range, (2) incomplete sampling of the LA (LA samples are often quite small; Fig. 2), and (3) non-random variation in LA composition linked to seasonal and decadal cycles or other natural environmental changes (Fig. 1), all occurring within the window of molluscan time averaging. The practical question is thus whether, in general, DAs in areas of known strong human activities show higher levels of mismatch with the local living community than do DAs in areas where human activities are mild or non-existent. If so, then the effects of human activities on LD agreement (and on molluscan LAs) are stronger than the effects of natural ecological and taphonomic factors on LD agreement (and on molluscan LAs).

TURNING LD MISMATCH INTO A CONSERVATION TOOL

Although we may suspect that LD mismatch in a given area may be due to human activities rather than due to natural causes, a global meta-analysis of all available LD datasets permits the strength of the link to be assessed quantitatively, and thus indicates the confidence with which LD mismatch can be used as a practical means of recognizing strong shifts in ecological baseline in areas where historical biological data and/or knowledge of human stresses are lacking (Kidwell, 2007, 2008).

To test this correlation, I scored the intensity of human activities relevant to benthos stress for all LD study regions in the global database, using independent scientific reports and the expert knowledge of original

TABLE 2.— Semi-quantitative scale used to score anthropogenic stress in a region at the time of LD sampling. This approach permits diverse kinds of information to be incorporated in the score, e.g. government reports on trends in water quality or quantities of fish caught, academic research reports on pollution, date when local or state governments imposed regulations on pollution or fishing, economic and cultural history of the region, historical insights from sedimentary cores, and the expert but unpublished knowledge of scientists who have worked in the area. Different kinds of data were available in different regions.

Score	General meaning	Anthropogenic eutrophication	Bottom trawling
0	Stress absent or negligible	AE0 = no human settlements in area nor active land clearance	BT0 = no exploitation or only artisanal harvesting with minimal habitat destruction
0.5	Stress possibly present	AE0.5	BT0.5
1	Stress definitely present	AE1 = coastal development and/or major development in the watershed	BT1 = commercial harvesting, gear can dislodge benthos other than target species
1.5	Stress definite & possibly intense	AE1.5	BT1.5
2	Stress definitely present and intense	AE2 = near to large point source, usually in addition to diffuse sources	BT2 = especially intense commercial trawling, e.g. more than once a year

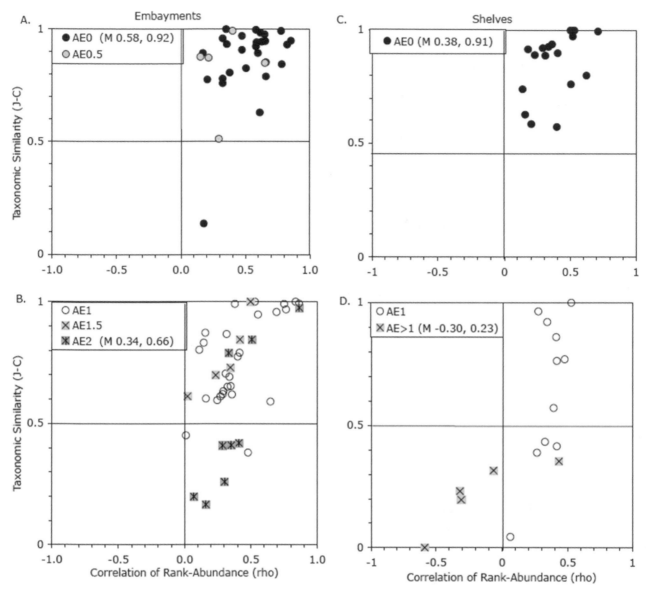

FIGURE 4.—Scatterplots of live-dead (LD) agreement in the taxonomic composition (y-axes; Jaccard-Chao index of similarity) and rank-abundance of species (x-axes; Spearman rho) of 73 datasets from coastal embayments (*1, 3*) and 34 datasets from open continental shelves (*2, 4*) under natural (AE0 or AE0.5; *1, 2*) and definitely altered nutrient conditions (AE1 to AE2; *3, 4*). In fully natural settings, LD agreement is very good (datasets fall well within the upper right quadrant of cross-plots; M = median value for AE0). LD agreement in AE≥1 datasets can be as good but ranges to significantly lower values where fewer than 50% of species are shared between LA and DA species lists (low JC similarity; bottom right quadrant) and formerly dominant species have become rare or vice versa (negative rho; bottom left quadrants of plots; M = Median value for AE>1). (Adapted from Kidwell, 2007.)

authors about human activities in the watershed at the time of their LD sampling effort (see Kidwell, 2007, for dataset-level characterizations and sources of information). Seven categories of human stresses were considered: *anthropogenic eutrophication* (AE; input of extra nutrients from cities, agriculture, coastal industries including fish-farms), *bottom-trawling* (BT) to harvest fin- and shellfish, *solid sediment runoff* from land clearance, *dredging and spoil-dumping*, *salinity modification* related to channelization or impoundment, *chemical pollution*, and *thermal pollution*.

In order to combine diverse kinds of information and compare areas, I developed a single semi-quantitative scale to apply to all stressors (Table 2). My assessments of local stresses based on the literature either matched or under-estimated those recognized by original LD authors in the instances when those authors could be interviewed. Anthropogenic eutrophication (AE) and bottom-trawling (BT) were most pervasive and other human impacts tend to increase with them, and so analyses focus on these two factors. I consider change in response to stress using two measures of molluscan *community composition* (the identity and relative abundances of species) and using two measures of *community structure* (richness and evenness):

1. Change in the identity of species that are present

Environmental degradation may lead to the local extinction of some species (these become "dead only") and also to the immigration of new species that tolerate or perhaps prefer the new conditions (the first generation will exist "alive only"). *LD agreement in species presence/absence* is measured here using the Jaccard-Chao (JC) index of taxonomic similarity, which adjusts for LD differences in sample sizes (Chao et al., 2005). When JC = 0 the LA and DA share no species in common, and when JC = 1 all species occur both alive and dead in the habitat. A JC = 0.5 indicates that ~half of all species in the habitat occur both alive and dead and the others occur either alive-only or dead-only (1 and 0 are the upper and lower bounds).

2. Change in the relative abundances of species

The first signal of environmental degradation is usually a decline in the population sizes of some species, which may decline from being numerically dominant (highly ranked in abundance within the community) to being relatively rare (low ranked; represented by only a few individuals). At the same time, other species that were formerly rare (or absent) in the community may increase in population size and even become dominant (shift from low to high ranked abundance). *LD agreement in species relative abundances* is measured here using a Spearman rank-order test, which is especially sensitive to relatively large shifts in rank (e.g., from the top to the middle or bottom of the ranked list, rather than simply shifting ranks with a few similarly ranked species). Spearman correlation coefficients (rho) are positive when each species in the DA has a similar rank to what it has in the LA (the species that dominate the DA also dominate the LA, and species that are rare in the DA are rare or absent in the LA). Rho is negative when species' ranks in the DA are flipped relative to their ranks in the LA (those that dominate the DA are rare in the LA) and rho is ~zero when ranks in one assemblage are ~random relative ranks in the other (1 and −1 are the upper and lower bounds).

3. Change in the number of species that are present (richness)

When conditions change strongly, the new community (LA) may contain a completely different set of species than were found in the original community (represented by the DA), i.e., the taxonomic similarity is nil (JC =0). However, if the new community contains the same number of species as the original, then there is no net change in species richness despite the complete turnover in species composition. *LD agreement in assemblage richness* is measured here as the ratio of dead richness to living richness after the larger sample (usually the DA, see Fig. 2) is sub-sampled to equal the size of the smaller sample (without replacement; expressed in logged units, this is "delta-S" of Olszewski and Kidwell, 2007, and the sample-size standardized "inflation" factor of Tomasovych and Kidwell, in prep.). Delta-S is 1 when the DA and LA have identical richness, is >1 when DA richness is > than LA richness, and is <1 when DA is < LA richness (no upper or lower bounds).

4. Change in the evenness (dominance) of species abundances

As environmental stress increases, fewer species find conditions tolerable. The community may thus change from containing a relatively large number of similarly abundant species (a fairly even abundance-structure) to being strongly dominated by a single species (low evenness in the distribution of individuals among species; = high dominance). *LD agreement in assemblage evenness* is measured here as "delta-PIE", which is DA evenness minus LA evenness using Hurlbert's evenness metric PIE, which stresses the proportional abundance of the most abundant species (as in Olszewski and Kidwell, 2007). Delta-PIE is positive when the DA is more even than the LA (individuals are distributed among species more evenly), negative when the DA is less even than the LA (individuals are more concentrated in a single or few species), and zero when no difference exists between the DA and LA (1 and −1 are the upper and lower bounds).

LD MISMATCH IN ASSEMBLAGE COMPOSITION INCREASES SIGNIFICANTLY WITH ANTHRO-POGENIC EUTROPHICATION (AE)

In aquatic systems, nutrient enrichment—whether natural or anthropogenic—increases phytoplankton production beyond zooplankton grazing capacity, leading to increased water turbidity, decreased light levels, and morbidity and mortality of seagrass and other non-algal subaquatic vegetation, all leading to higher input of organic matter to the benthic boundary layer and seafloor. Because of the biological oxygen demand of the microbial decomposers of these organics, hypoxia and related die-offs of benthic animals become more likely. "Dead zones" characterized by low-oxygen bottom waters and high mortality have increased in many estuaries since World War II when the use of commercial (manufactured) fertilizers accelerated—what was once an occasional summer occurrence has become more frequent, more intense, and more extensive in coastal embayments worldwide (Rabalais and Turner, 2001). In the last several decades, "dead zones" began to appear even on open continental shelves and are now known from all world oceans (Diaz and Rosenberg, 2008). Climate change is implicated in the appearance or intensity of some of these dead zones (e.g., Chan et al., 2008; Brewer and Peltzer, 2009), but most are linked to nutrient run-off from agriculture, cities, and industries in the watershed.

The results of eutrophication in coastal waters are—from the molluscan perspective—the decline and eventual loss of species that require or prefer seagrass meadows as habitat or food ("grass-dwellers" including micro-grazers and macro-herbivores), a shift from largely suspension-feeders to mixed or largely deposit-feeders, loss of attached epifauna, and the general replacement of hypoxia-intolerant species by species that are either hypoxia-tolerant or are short-lived, with rapid generation times and high (re)colonization potential (Diaz and Rosenberg, 1995; Orth et al., 2006). One thus expects change in the relative abundances of species and, if conditions become sufficiently severe, change in species' presence/absence as some species disappear entirely and new species arrive.

Analytic findings

Cross-plots of LD taxonomic similarity (JC index,

y-axis) against LD agreement in species rank-abundances (Spearman's rho, x-axis) are a useful means of displaying how LD agreement varies as a function of AE (Fig. 4). Each data point signifies LD agreement in different habitat-level dataset. LD agreement is quite high in relatively pristine (AE0) settings, both in coastal embayments (lagoons, bays, estuaries) and on open continental shelves: all but one dataset falls well within the upper right quadrant of a JC/rho cross-plot (Fig. 4.1, 4.2). These AE0 settings include areas that are naturally nutrient-poor (e.g., carbonate shelf of Yucatan) and nutrient-rich (Patagonia and Amazon shelves, both with seasonal upwelling). The scatter in AE0 data points indicates the range in LD agreement that is created in completely natural circumstances by incomplete sampling, time averaging of natural biological variability, and postmortem bias (differential preservation, out-of-habitat transportation; Table 3).

LD agreement in areas of known human impact (AE scores ≥1) can be as high as that observed in pristine areas but ranges to significantly poorer levels (Fig. 4.3, 4.4). The decline in taxonomic similarity (JC) with increasing AE is significant in both coastal embayments and open shelves despite the overlap of values with AE0 settings. The decline in rank-abundance agreement (rho) is not significant in estuaries unless datasets are partitioned by mesh size, with coarse-mesh datasets registering significant declines in rank-abundance agreement with increasing AE. In shelf settings, rank-order correlations decline to *negative* rho values in AE2 areas (study areas adjacent to pulp mills and contaminated harbors; Fig. 4.4) and is significant without partitioning datasets by treatment (Kidwell, 2007, 2008).

Variation in LD agreement along a spectrum of AE globally implies temporal changes that are consistent with taphonomic inertia. The community first shifts in the proportional abundances of species already present, eroding rank-order agreement, followed by the total loss of some species (which may become "dead-only") and, in some instances, the entry of new "alive-only" immigrants, progressively reducing taxonomic similarity. The composition of the DA equilibrates to stochastic variation within the new living community only as the remains of the previous community are destroyed and/or buried below the mixing zone, and are diluted by the input of dead shells from the new community. This same pattern is seen in biological analyses of

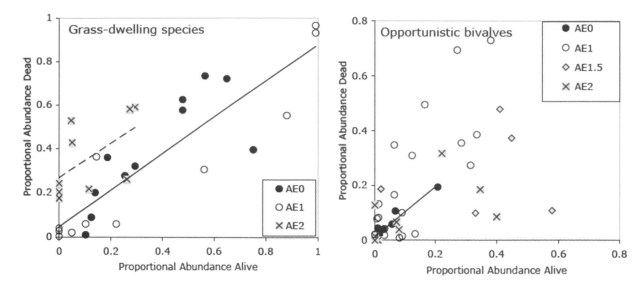

FIGURE 5.—Signals of anthropogenic eutrophication (AE) from LD mismatches in the proportional abundances of key species or sets of species. *1*, In lagoons where AE is absent or only moderate, mollusks that prefer or are found almost exclusively in seagrass beds are about as abundant in the DA as in the LA, whereas they tend to be disproportionately abundant in DAs where AE is severe (AE2), signifying the former presence of seagrass habitat. Based on 31 datasets from tropical Gulf of Mexico; *2,* The opportunistic bivalve *Mulinia lateralis* is relatively rare both alive and dead in natural lagoons (has proportional abundance <10% in all but one AE0 dataset), showing good LD agreement. In AE≥1 lagoons *Mulinia* ranges up to ~60% alive and 75% dead and is commonly among the most abundant or is the most abundant species in one assemblage or the other or both, consistent with greater stress such as from episodic hypoxia associated with AE. DAs in AE≥1 settings are not consistently enriched in *Mulinia* compared to LAs in part because of the abundance of grass-dwellers and other relictual species in many of these DAs. Based on 47 datasets from temperate and tropical Gulf of Mexico. (Both figures adapted from Kidwell, 2007.)

exclusively living assemblages. With progressive AE over time (or with increasing spatial proximity to the nutrient point source), the first changes are in species' relative abundances, followed by the complete loss of sensitive species and, in some instances, the appearance or rise to dominance of pollution-tolerant species (e.g., for review, see Gray and Elliott, 2009).

Implications for conservation biology

The strength of the association between LD agreement and AE is remarkable, given the array of natural processes, other human impacts, and methodological issues that can influence the match between a single census of the living community and the local DA. It thus implies that LD mismatch in the taxonomic composition and relative abundances of species, readily measured by taking advantage of dead shell residues in standard benthic samples, could be used to recognize significant modification of living communities by AE. This would be a significant boon for areas where historical survey data are sparse or lacking entirely, or

where human stresses are unknown or contentious.

LD comparison fails to detect AE in many instances. LD mismatch in taxonomic similarity and rank-abundance in some areas of known AE is not significantly greater than that in spatially analogous AE0 settings. About 40% of datasets from estuarine and shelf settings of certain AE fall within two standard deviations of the mean of AE0 datasets in these plots (Kidwell, 2007). That 40% can be taken as the failure rate of this method if one assumes that the LAs in these AE habitats did in fact undergo significant changes in response to AE and thus that significant LD mismatch should exist in all 64 AE≥1 datasets. LD mismatch is thus a highly conservative test of AE, yielding false negatives (fails to detect AE when AE conditions exist) rather than false positives.

Biologically, the significantly higher average mismatch encountered in AE settings implies that the magnitude of impact of human activities on benthic communities exceeds the ecological consequences of any natural change in environmental conditions occurring within the window of DA time averaging (Fig. 1). The strong impact of human activities on natural systems is well-known, but the LD results show how the magnitude of effect can be recognized in a temporal framework (the estimated duration of time-averaging of dead shells within the sampled mixed layer) without the use of sedimentary cores. The significantly higher average mismatch in AE settings implies considerable persistence of skeletal material within the mixed layer for DA composition (taphonomically interesting; corroborates independent direct-dating efforts, Fig. 1) and/or lower rates of molluscan shell production in the new community (biologically interesting).

Under AE, it clearly takes appreciable time for the "new" community to dilute the dead residua of the "old" pre-impact community—that is, to bring the time-averaged DA into equilibrium with the new LA composition. This lag contrasts with the relatively rapid temporal tracking by DAs of changes in LA composition that is implicit in the high LD agreement that characterizes fully natural conditions (i.e., the tight clustering of AE0 datasets in the upper right quadrants of Fig. 4.1, 4.2). Rapid tracking is documented explicitly in a time-lapse test of temporal stability in molluscan DA composition by Ferguson and Miller (2007), who found that the DAs of a tropical back-reef lagoon changed significantly over 20 years but in concert with comparable, apparently natural changes in LA composition over that same period. In their classic analysis of two Texas bays, Staff et al. (1986) also found rapid changes in species' relative abundances among molluscan LAs sampled at successive 6-week intervals over 18 months, mostly followed by corresponding changes in those species in DAs that then faded quite rapidly, probably because most individuals were juveniles subject to rapid postmortem destruction.

The *lack* of LD mismatch in some datasets from areas of known AE (e.g., the AE1 datasets that fall within the range of AE0 datasets in the crossplots of Fig. 4; in estuaries, even some AE1.5 and AE2 datasets fall within the pristine range) might signify several things. For example, biologically, AE might have had relatively little actual effect on LA composition locally. Species that prefer organic-rich sediments or tolerate episodic dysoxia might already have been present or even dominant in the pre-impact community, so that AE caused no significant change in the LA and thus created no unusual degree of LD mismatch. Biologists

in fact commonly find no significant impact of anthropogenic nutrient loading on benthic communities in areas that are naturally eutrophic. An alternative or additional factor may be that, by the time of sampling, the DA might have already equilibrated to the new LA, resulting in little LD mismatch. Such equilibration is especially likely in areas of long-standing AE (long-standing relative to the scale of DA time-averaging) and/or high sedimentation rates, so that the DA record of the transition of the LA by AE is already buried below the mixed layer. Sediment coring into historic layers would be required to access the DAs that contain pre-impact species. Conditions of long-standing AE and/or high sedimentation rates are especially common in coastal embayments, which often have long multi-centennial histories as harbors and industrial centers and are natural sinks of sediment, promoting relatively rapid permanent burial of shells.

LD MISMATCHES IN SPECIES IDENTITIES ARE CONSISTENT WITH AE: SPECIFIC CAUSE CAN BE INFERRED

If the increase in LD mismatch associated with AE is truly caused by AE, then LD mismatches in species identities and relative abundances should be consistent with eutrophication or other human stressors, rather than, say, with postmortem transportation, differential preservation, or natural environmental change. AE should not simply change LA composition: it should bring to dominance and drive to local extinction *particular* species, namely those sensitive to the presence of seagrass, to sediment organic content, and to hypoxia (see background information on AE above). By examining the identities and autecology of species found dead only or anomalously abundant (or scarce) in the DA relative to the LA, we should get a sense of the kind of environmental change that the area has undergone, which narrows the search for cause.

Analytic findings

A collection of 31 LD datasets from 11 tropical Mexican lagoons is a particularly powerful means of testing for this effect because all datasets reflect the same sampling procedure (1.5 mm mesh) and a single regional fauna but a range of AE conditions. In datasets from AE2 lagoons, many species that occur "dead only" or at very low abundance in the DA prefer grassy

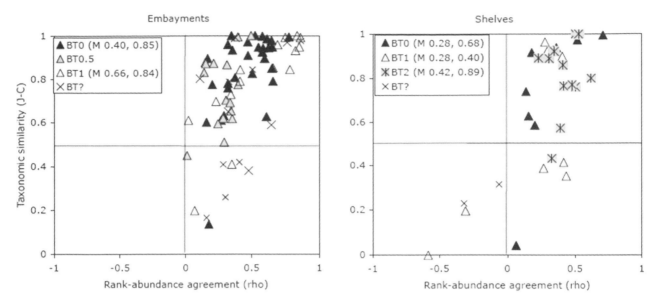

FIGURE 6.—Scatterplots of LD agreement in the taxonomic composition and rank-abundance of species as in Fig. 4 but with datasets now scored for the intensity of bottom trawling (BT). In general, areas subject to BT are also subject to AE so that these plots strongly resemble those in Fig. 3, but in itself BT has no significant effect on overall LD agreement as measured by JC and rho, either in embayments (1) or on open shelves (2). M = median rho and JC.

substrata (see Kidwell, 2007, for a complete listing of "grass-dwelling" bivalve and gastropod species). With two exceptions out of nine datasets, DAs in the five AE2 lagoons are enriched in grass-dwellers relative to living communities by 20% or more (dashed trendline in Fig. 5.1). In contrast, among datasets from pristine to mildly AE lagoons (AE0 to 1) the proportional abundance of grass-dwellers in the DA matches that in the local living community more closely. Loss of seagrass is a well-known consequence of AE (e.g., Twilley et al., 1985). The tendency for AE1 datasets to plot among pristine datasets makes sense in this regard, because fertilization initially stimulates grass growth.

The relative abundance of the opportunist bivalve *Mulinia lateralis* is also consistent with the episodic hypoxia, stunted animal growth, and pulsed mortality that can accompany eutrophication (e.g., Powers et al., 2005). Drawing on 47 datasets from both Mexican and Texas lagoons bordering the Gulf of Mexico, *Mulinia* is sparse both living and dead in the pristine AE0 lagoons (Fig. 5.2; with one exception, *Mulinia* constitutes <10% individuals). In contrast, it tends to be one of the most abundant species in AE1 and AE2 lagoons, where opportunistic colonization and strong mortality events would alternate with seasonal or other variation in water stratification. *Mulinia* is scarce both alive and dead

in some AE1 and AE2 datasets—a series of these datasets cluster among AE0 datasets in the plot of Figure 5.2. But with one exception, every dataset that contains abundant *Mulinia* either alive or dead or both alive and dead is from a lagoon where there is independent evidence of AE (wide scatter of points in Fig. 5.2).

Finally, LD mismatches in species identity in open shelf datasets also suggest eutrophication. LAs in AE1.5-2 areas (shelves immediately offshore of pulpmills or notoriously contaminated harbors) tend to have abundant deposit-feeding and chemosymbiotic bivalves that are rare in the DAs, where dead-only larger-bodied suspension feeding species are important (Kidwell, 2008). Although these mismatches might be magnified by (1) lower preservation potential among the relatively small-bodied species of the "new" community (solemyids, lucinids, tellinids, prosobranchs, thyasirids) and (2) more prolonged time averaging of death assemblages on shelves than in coastal embayments, the LD mismatches in species identities are incompatible with postmortem introduction of allochthonous shells and are not observed consistently in datasets from naturally nutrient-rich shelves. On the latter (e.g., Amazon and Patagonia shelves), the proportional abundances of organic-loving species in local LAs and DAs are more comparable than on shelves where nutrients are anthropogenic.

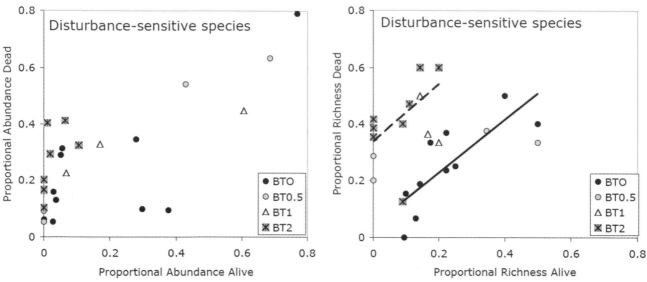

FIGURE 7.—Signals of bottom trawling (BT) from LD mismatches in (*1*) the proportional abundance and (*2*) proportional richness of bivalve species likely to be sensitive to physical disturbance, based on 25 datasets from gravels under all AE conditions. "Disturbance-sensitive species" includes all bivalves known to be facultatively or obligately byssate as adults (e.g., both semi-infaunal and epifaunal arcoids, pinnids, pteriids, isognonmonids, mytiloids, and infaunal corbulids) and all epifauna (including scallops that may swim as adults, spondylids that attach as adults, and continuously attached oysters, anomiids, plictaulids, chamids, and most limoids).

Implications for conservation biology

LD mismatches in species identities and relative abundances are consistent with eutrophication in areas where AE is judged likely, indicating that LD mismatches in key species and functional groups can be used to recognize the particular ways in which benthic communities have changed within the window of DA time averaging. Species that are dead-only or are far more abundant in the DA than in the LA probably are in fact relics of past local populations, rather than biologically meaningless artifacts of taphonomic processes.

As a caveat, the evidence from meta-analysis is statistically strong but circumstantial (inferential). It assumes that variation in LD agreement among a *spatial* spectrum of study areas reflects the ability of DAs to capture *temporal* changes locally. Long-term biomonitoring data for living communities are not available for any of the LD study areas—such data would permit useful "hard" tests of the ability of DAs to pick up on and retain a signature of AE. Moreover, information on the cultural history of LD study areas was gathered from disparate sources, including regional rather than local assessments (Kidwell, 2007, 2008). Local tests with better-constrained environmental and/or biological data are thus needed to move beyond meta-analytic findings. As a powerful example, Ferguson (2008)

took advantage of known spatial variation in nutrient input within Florida Bay that had existed for only ~20 years—a series of posts that became roosting sites for seabirds—and found that molluscan DAs captured the signature of local eutrophication and retained its fine spatial grain (sites 50 m distant from the posts in unaltered seagrass beds were compositionally distinct, non-eutrophied baselines). The modern world is full of opportunities for this kind of taphonomic field experiment, taking advantage post hoc of known human perturbations of natural systems.

LD MISMATCH IN SPECIES THAT ARE DISTURBANCE-SENSITIVE AS EVIDENCE OF BOTTOM-TRAWLING (BT)

Bottom-trawling (BT) here includes all methods of harvesting demersal fin-fish and shell-fish (*sensu lato*, i.e., mollusks, crustaceans, sea cucumbers and urchins) that disturb the sediment-water interface. Thus dredging and raking as well as bottom-trawling s.s. are included, whereas hook-and-line, pot- or trap-catching, and various mid-column netting efforts are excluded. BT tends to produce significant benthic bycatch (species other than the target species) and can do considerable damage to the physical habitat. Trawling heavy gear across a soft-sedimentary seafloor is analogous to clear-cutting or plowing land surfaces. It causes strong declines particularly in attached epifauna but also in free-living epifauna and in long-lived and/or relatively sessile infauna (e.g., low capacity to re-establish themselves in burrows), and favors the establishment and survival of short-lived, mostly soft-bodied species (e.g., Thrush et al., 1998; Gray et al., 2006; NRC, 2002). These observations provide specific predictions for LD mismatch in areas subject to bottom-trawling.

BT has a long history but was largely limited to shallow coastal waters until engines replaced sails. Mechanized harvesting methods were developed for oysters by the late 1800's, causing massive destruction of reef habitats in estuaries (e.g., Rothschild et al., 1994; Kirby, 2004). Exploitation using heavy gear expanded to increasingly deeper (shelf) waters in the early 20th century (e.g., Cranfield et al., 1999; Edgar and Sampson, 2004). The global extent and magnitude of effect of "industrial" extraction of wild fish stocks is increasingly appreciated as are the deep historical roots of "fishing down" marine foodwebs and the resulting trophic cascades and impacts on ecosystem services (e.g., Pauly, 1995; Jackson et al., 2001; Myers and Worm, 2003; Lötze et al., 2006; Lötze and Worm, 2009; Worm et al., 2006; Baum and Worm, 2009), even when bottom-trawling is not involved.

Analytic findings

Analysis includes both open shelves (covered in Kidwell, 2008) and coastal embayments (work first reported here). No LD study has included oyster-reefs s.s. although several have sampled shell-rich sediments from the flanks of reefs, which are included here. Because of uncertainty in commercial efforts in some embayments, I've added a BT0.5 category for Tomales, Galveston, and West Bays (significant oyster fisheries in the area, but not relevant to the habitats sampled for LD data). I have also shifted the Patagonian shelf to this category, out of BT0. BT scores for shelves are otherwise as used in Kidwell (2008). For a complete listing of study areas and meta-data on trawling, see Kidwell (2007).

As already reported (Kidwell, 2008), shelf datasets are notable for *not* showing a significant correlation between BT and LD agreement in species composition (JC, rho). In crossplots of taxonomic similarity (JC)

and rank-abundance agreement (rho), LD agreement datasets from BT2 areas show the same high values as those from BT0 areas, and thus no significant correlation exists even though some BT1 datasets exhibit low LD agreement (Fig. 6.2). I find the same lack of association among datasets from coastal embayments (Fig. 6.1). A significant correlation of LD agreement with BT fails to emerge even if analysis of BT is restricted to areas that are otherwise pristine (AE0 areas; not displayed).

Virtually all trawled areas in the database were subject to commercial exploitation for decades before being sampled for LD data, which might explain the lack of a trawling signal in these general (JC, rho) measures of LD agreement. Another factor—probably equal or more important than the long history of BT—is that many datasets are from sand and mud bottoms, which tend to have few epifaunal or byssate species even when undisturbed and thus exhibit only modest changes in response to trawling. Based on before-after-control studies of living communities and comparisons of trawled and untrawled seafloors (e.g., Thrush et al., 1998; Tillin et al., 2006; Hiddink et al., 2007; Hinz et al., 2009), benthic communities that are dominated naturally by short-lived and/or free-living infaunal species (mobile sands, soft muds) undergo less change and recover faster than do communities that include species that live freely on the seafloor (mobile epifauna), live attached to the seafloor (cementing and byssate epifauna), and/or live partly or fully burrowed into the sediment but stabilized using byssal threads (endo-byssate infauna and semi-infauna).

When analysis is focused exclusively on the 25 LD datasets from gravel or gravelly habitats whose communities are most likely to be sensitive to trawling, BT still leaves no distinct signal in terms of JC and rho (Fig. 6). However, significant LD mismatches exist in the proportional *abundance* and proportional *richness* of the subset of bivalve species that should be most sensitive to bottom disturbance (Fig. 7). This subset includes all epifaunal species regardless of attachment and all species that are either facultatively or obligately byssate as adults, regardless of life position (see list in caption of Fig. 7). Spearman rank correlation tests find that the difference in proportional abundance of these species in the DA versus the LA is significantly positively correlated with the BT score (rho = 0.564, p

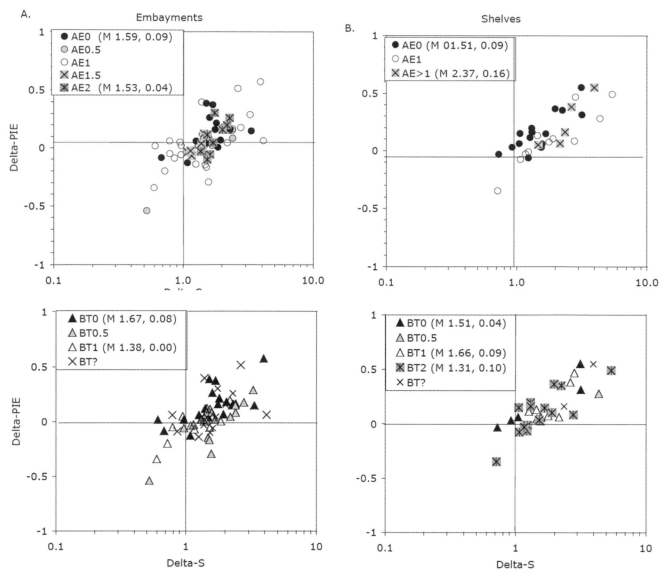

FIGURE 8.—Scatterplots of LD agreement in species richness (sample-size standardized delta-S; x-axes) and evenness (delta-PIE, y axes) in either embayments (*1, 3*) and open shelves (*2, 4*) for datasets scored by AE (*1, 2*; M = median value for AE0 and AE>1) or scored for BT (*3, 4*). DAs from areas impacted by either AE or BT are not significantly more enriched in species than are DAs from natural areas, nor do they exhibit more even abundance distributions, with the exception of shelves under strong AE (AE>1).

= 0.002), and that the difference in proportional richness of these species in the DA versus the LA is also significantly positively correlated with the BT (rho = 0.714, p << 0.001; 25 gravel datasets from all environments and states of AE).

The difference in proportional richness is particularly striking: with one exception, disturbance-sensitive species are ~30 percentage-points more abundant in BT2 DAs than in counterpart LAs, which contrasts quite cleanly with BT0 datasets where DAs tend to

match their counterpart LAs quite closely (compare trend lines in Fig. 7.2).

The same patterns result if the analysis focuses exclusively on gravel datasets from areas that are otherwise pristine. Among gravels in AE0 areas, LD mismatches in the proportional *abundance* of sensitive species are still positively correlated with BT albeit no longer significant (rho = 0.41, p = 0.10). The effect of BT on proportional *richness* remains significantly positive (rho = 0.728, p <0.001; 17 datasets).

Implications for conservation biology

There is a pressing need for methods that (1) permit recognition of BT in areas with little regulation and/or enforcement and (2) identify species that need recovery. There is also a need for methods to retrospectively (3) estimate pre-BT community composition in regions where bottom disturbance from even legal fishing is so extensive that it is difficult or impossible to identify "baseline" areas to set recovery targets (e.g., de Juan et al., 2009).

The positive results reported here indicate that LD mismatch, focusing on the subset of species whose life habits make them most sensitive to physical disturbance, would be a useful new tool for all three needs. Habitats subject to BT, especially intense BT, yield DAs that are significantly richer in disturbance-sensitive species than co-sampled LAs, indicating historical declines in the populations of these species within the scale of time averaging. Species occurring dead-only and especially those having more than trace abundance in the DA are the species that have suffered the strongest declines, and thus provide targets for judging recovery. LD analytic results are for gravelly habitats only, but biologists have already discovered that these communities undergo the strongest change and have the slowest recovery times of all soft-sedimentary seafloors.

The scatterplots provide additional insights into the biology of these settings and guidelines for recognizing human disturbance. (1) Disturbance-sensitive species can constitute anywhere up to ~80% of *individuals* in the LAs of BT0 gravels, but with one exception do not constitute more than 20% of *living* individuals in trawled gravels and in all eight intensely trawled (BT2) gravels constitute <10% of living individuals. Epifaunal and byssate infaunal mollusks thus definitely do "see" trawling and even small LA samples such as found in LD datasets can recognize the scarcity of these species (Fig. 7.1). Mollusks thus provide a reasonable surrogate for the response of the entire macrobenthos. (2) Disturbance-sensitive species constitute at least 10% of total species richness in the LAs of BTO gravels but commonly are entirely absent in the LAs of trawled gravels. In five of the 12 trawled gravels (BT0.5 to BT2) sensitive species occurred exclusively as empty shells (Fig. 7.2). Thus epifaunal and byssate taxa do not consistently dominate gravel bivalve communities

taxonomically even under completely natural conditions, but they are consistently present alive.

AE AND BT HAVE LITTLE EFFECT ON LD AGREEMENT IN COMMUNITY STRUCTURE (RICHNESS AND EVENNESS)

The richness (count of species) and evenness (degree of dominance by a few species) of benthic communities might respond in a variety of ways to human impacts. For example, richness might *decrease* owing to extermination of species by overexploitation or habitat destruction, might *increase* owing to non-native species being introduced either intentionally (e.g., transplantation for exploitation) or unintentionally (ballast water, cryptic species hitchhiking on exploited species, escapees from aquaria), or might be *unchanged* (a net balance of local extinction and introduction). Evenness is generally expected to *decrease* because human intervention commonly leads to *fewer competitors* (loss of redundancy within trophic levels or functional groups as some species are driven to functional extinction), *more stress* (narrower range of species capable of tolerating conditions of pollution, nutrient over-loading, or physical disturbance), and *more static conditions* (held at a particular level by human pressure, thereby suppressing stochastic variability and small-scale natural fluctuations in the community, allowing single species to dominate).

Analytic findings

Datasets analyzed here include all but five of the *subtidal* habitats analyzed by Olszewski and Kidwell (2007). Datasets from Mannin Bay are excluded because the richness of the death assemblage was originally under-counted in a way that cannot be corrected retrospectively, affecting delta-S but not delta-PIE.

Regardless of the type or intensity of human impacts in the study area, the majority of LD datasets have DAs that are both richer and more even than counterpart LAs—that is, most fall in the upper-right quadrant of a cross-plot of delta-PIE against delta-richness (Fig. 8, Table 3). In estuaries, AE has no significant effect on community structure. Datasets from areas with AE≥1 are centered on the cloud of AE0 datasets but show greater dispersion in both delta-richness and delta-PIE (especially datasets from AE1 areas; Fig. 8.1). On shelves, DAs from areas with AE≥1 tend to be more inflated in richness on average than are DAs from AE0

TABLE 3.—Live-Dead (LD) agreement in molluscan composition and structure, based on 109 habitat-level datasets globally (left columns) and on the subset of 40 datasets from areas with no anthropogenic eutrophication (AE0; right columns). Interquartile ranges in boldface. Mean with 95% confidence intervals calculated on the standard error.

Measure of LD agreement	All areas		Pristine (AE0) areas only	
	Median Min/ Q1/ Q3/ Max	Mean	Median Min/ Q1/ Q3/ Max	Mean
Taxonomic similarity (JC)	0.80 0/ **0.61/ 0.93/** 1	0.73 ± 0.05	0.90 0.14/**0.79/0.95**/1	0.85 ± 0.04
Rank-abundance correlation (rho)	0.36 -0.58/ **0.25/ 0.51/** 0.86	0.37 ± 0.05	0.47 0.14/ **0.32/ 0.60**/ 0.85	0.46 ± 0.06
Raw ratio of DA richness to LA richness	3.23 0.58/ **2.14/ 4.92**/ 18.4	3.94 ±0.53	2.94 1.02/ **2.17/ 5.21**/ 10.75	3.91 ± 0.86
Delta-richness (standardized)	1.53 0.52/ **1.26/ 2.07**/ 5.46	1.76 ± 0.17	1.52 0.68/ **1.30/ 1.81**/ 3.31	1.65 ± 0.21
Delta-PIE	0.06 -0.54/ **-0.03/ 0.17/** 0.57	0.08 ± 0.04	0.06 -0.35/ **0.00/ 0.16**/ 0.55	0.10 ± 0.05

areas (Fig. 8.2; see median values in key). There is no significant change in average delta-PIE among shelf datasets as a function of increasing AE. Arguably, the environmental distinctions in delta-richness arise because shelves have more richness to lose when environmental conditions deteriorate under AE, and are less likely than estuaries and lagoons to include species that are tolerant of hypoxic intervals and other stresses related to temporary nutrient pulses (see Kidwell, 2007, 2008, for discussion of mesh-size effects).

When the same datasets are evaluated for variation with trawling intensity, no significant effect emerges for either delta-richness or delta-PIE in either environment (Fig. 8.3, 8.4). Datasets from trawled (BT1) areas—even datasets from intensely trawled areas (BT2)—plot basically on top of datasets from untrawled areas (BT0). Datasets with less certain (BT0.5) or unknown trawling intensity (BT?) also occupy the same portion of the crossplots.

Implications for conservation biology

The finding here that DA richness is generally not significantly more inflated in impacted than in pristine settings suggests at first glance that there is a basic trade-off in molluscan species gained and lost with AE and BT. However, this result should *not* be interpreted

as evidence of community saturation. It arises instead from comparing time-averaged DAs with non-averaged, snapshot samples of LAs. DAs are usually richer than local LAs even in pristine settings and even when sample-size standardized (Figs. 2, 8; Table 3) because the coarser temporal resolution of the DA leads to a flatter abundance distribution (the qualitative expectation of Fürsich and Aberhan, 1990) and also allows a long tail of rare species to develop (Tomasovych and Kidwell, in prep).

The results indicate that, notwithstanding changes in the identities and relative abundances of species that occur with AE and BT (earlier sections), the magnitude of effect on LD agreement in richness and evenness is not substantively different than what is observed in pristine habitats, somewhat contrary to intuition. Within fully natural settings, natural volatility of (mostly small) populations continuously shifts species back and forth between the alternative states of being dead-only and being present both alive and dead (and see Tomasovych and Kidwell, 2009c, and Kidwell and Tomasovych, 2009, for comparisons of LD agreement with "live-live" agreement among spatially and temporally replicate samples of LAs). In addition, environmental conditions can vary considerably within the window of time averaging (see Fig. 1). These changing conditions bring additional species into the local LA (and thus also into the local DA) from the regional species pool and force some species out of the local LA (but not out of the local DA, whose richness simply continues to ratchet upward). In any single random sampling of a natural system, the richness of the LA may be seasonally low (including zero) but the DA will remain consistently higher from the inclusion of relicts from earlier natural but quite different conditions. Delta-richness can thus be naturally high. Neither AE nor BT pushes systems outside this natural range, which is quite large. Even under human stresses, the local LA and thus the local time-averaged DA continue to draw on the same regional species pool.

Recognizing AE and BT using LD mismatch thus requires an analysis of species composition—the identities and relative abundances of species, especially of functionally sensitive species, as discussed in earlier sections—and not a simple comparison of species counts or abundance structures.

DETERMINING THE COMPOSITION OF NATURAL "BASELINE" COMMUNITIES AS TARGETS FOR RESTORATION: THE NEXT FRONTIER

The high degree of LD agreement observed in pristine settings (Table 3; Fig. 4) provides the best current estimate for the reliability of DAs as time-averaged complements to the data generated by conventional biological surveys. It also represents the best current estimate, albeit an optimistic one, for the reliability of DAs from buried historic layers, from which paleoecologists try to acquire deeper-time perspectives on modern communities, both natural and human-modified. How reasonable are these two applications?

In habitats from pristine areas, LD agreement is quite high. LD agreement is not perfect—both similarity in species presence/absence (JC) and rank-abundance (rho) are less than 1 (Table 3; Fig. 4). However, focused empirical analysis and modeling of these pristine datasets now underway indicate that, although DAs and LAs differ in part because of significant postmortem bias in some instances, most of the mismatch in composition owes to spatial and temporal variability in the living community (Tomasovych and Kidwell, 2009c; Kidwell and Tomasovych, 2009). High within-habitat variability in LA composition—among sites at a

single moment and from season to season over multiple years—creates a large range of species compositions in natural settings, which makes a large (easier) target for DA composition to hit. For example, the multivariate centroid of sampled DAs within a habitat does not always fall precisely on the centroid of LA samples from that habitat, but is almost always inside the multivariate dispersion of LA samples (Tomasovych and Kidwell, in prep.).

The time-averaged nature of DAs is thus a good means of estimating average LA composition in natural settings, as anticipated by Peterson (1977) in his early treatment of the issue. Our new findings using spatial and temporal replicates of LAs, as well as the high average LD agreement summarized here from meta-analysis at the habitat-scale (Table 3; Fig. 4), indicate that LD agreement in mixed-layer sediments can be used to identify which regions might best serve as restoration baselines. Such spatial baselines—habitat similar to the impacted one but in a nearby region that has escaped or been resilient to human activities—can be difficult to identify in regions with long histories of heavy trawling or coastal development (e.g., Caddy et al., 1995; de Juan et al., 2009). Ideally, LD agreement would not be the only method used to judge the status of a given habitat or multi-habitat region, but this new approach deserves a rigorous trial via case studies.

An alternative and usually preferable baseline for setting targets for restoration is information on what the local community was like before human modification—that is, a historical perspective on the impacted habitat at hand. In the absence of a strong history of local biomonitoring, historical ecologists use past scientific surveys, interviews with elderly fishermen and other residents, archeological middens, and sedimentary cores. The mixed-layer DAs in the impacted area may show poor LD agreement, but at some depth one should encounter DAs from historic layers of sufficient geologic age that they time-averaged exclusively natural LAs.

To what extent can we trust that DAs from *historic* layers will be as faithful in composition to LAs from that past time as modern mixed-layer DAs are to modern-day LAs? This topic is a current research frontier. Part of the reason that *mixed-layer* DAs agree so well with LAs—in fact, perhaps a very large part—is that mixed-layer DAs are still open systems that receive input of dead individuals from the local LA (Fig. 3). If rates of postmortem destruction are indeed highest in the initial postmortem interval—and experiments suggest this, particularly for newly settled juvenile mollusks and other meiofauna (e.g., Cummins et al., 1986a, b; Green et al., 1998, 2004; Best et al., 2008; Best, 2009)—then the tendency of mixed-layer DAs to be dominated by recently dead shells may well be transformed into normal and even flat shell-age frequency distributions with burial down-core (e.g., as seen in shelly beach ridges positioned increasingly far from the active shoreline, Kowalewski et al., 1998). Such a change should shift the composition of the DA toward species having the highest preservation rates. To the extent that these rates correlate with species' original abundances in the LA and are not counter-balanced by variation in population turnover rates, the relative abundances of species will be either preserved or shifted (erroneously biased) toward rare species.

To test the reliability of historic-layer DAs explicitly we need, first, down-core tests of the shape of the shell age frequency-distribution. These tests must penetrate *below* the uppermost age-homogeneous mixed layer (as Terry, 2008, has done for small mammals in the stratigraphic records of rock-shelters). This can be expensive and logistically challenging given the thickness of some mixed layers, but is conceptually straightforward (e.g., Flessa et al., 1993; Kosnik et al., 2007). Second, we need down-core tests of how well DAs retain species composition and community structure. Down-core tests of fidelity present a considerable logistical challenge because we need to know, from independent evidence, the history of the living community (strongest test, based on long-term monitoring of living communities) or at the least need to know the regional environmental and cultural history quite well. We would otherwise have to assume that past LAs were *not* substantively different from those sampled locally today, which would in turn force us to (perhaps erroneously) conclude that all down-core changes in DA composition are taphonomic in origin.

One very encouraging implicit test of the compositional fidelity of historical molluscan DAs is that of Edgar and Samson (2004). Based on 13 replicate Pb-210-dated cores along a ~100 km stretch of the eastern Tasmanian shelf, they documented a ~50% decline in sample-size standardized richness over the last 60 years, coincident with an historically documented collapse of the scallop trawling industry, including a time-lag in

the onset of decline in the portion of the shelf where commercial operations began later. Because only living scallops had been directly monitored, the overall decline in regional diversity had been unappreciated.

MOVING FROM PROMISE TO APPLICATION

The findings of global meta-analysis suggest some protocols for incorporating LD analysis of seafloor sediments as a standard approach to recognizing whether and how local benthic communities have changed in response to human stressors.

1. Evidence of strong recent ecological change

Analysis of a global database of ~100 molluscan live-dead (LD) datasets indicates that LD mismatch in species composition can indeed be used to retrospectively recognize strong, recent shifts in ecosystems.

In applying this approach to conservation biology, two provisos are important. First, this method is not a silver bullet. About 40% of datasets from areas that were definitely subject to anthropogenic eutrophication at the time of LD sampling yield LD mismatches in species composition that are statistically indistinguishable from those observed in pristine areas (see overlap of scatterplots in Figs. 4 and 6). LD comparison is thus a conservative means of recognizing recent ecological changes, prone to false negatives rather than false positives. The approach is still valuable, however, especially given the dearth of even a decadal-scale temporal perspective on benthic communities in most regions. Moreover, it is possible that the LD analysis is correct that no significant change occurred within the scale of time averaging represented by the mixed-layer death assemblage in some AE areas, and/or that the DA record of pre-impact LAs is now buried (resides in historic layers, accessible only from sediment cores). In areas that are naturally eutrophic, biologists regularly find "no change" in response to significant input of new nutrients. DAs from embayments show the highest failure rate (e.g., AE2 datasets with high LD agreement), and these are in fact the settings most prone to having naturally high nutrient inputs, long histories of AE, and high rates of sedimentation, all of which would promote *minimal* LD mismatch in mixed-layer DAs even in the face of significant present-day AE.

Second, "recent ecological change" means change

that has occurred within the scale of time averaging—"recent" does not have a universal absolute value when working with time-averaged DAs. In coastal embayments, direct-dating of molluscan DAs indicates that most shells reflect input over the most recent few decades, with some shells dating to the past few centuries (and maximum shell ages of several millennia in some cases; Fig. 1; and see Kowalewski, 2009). Fewer open-shelf DAs have been analyzed as rigorously, but taphonomists generally suspect that scales of time averaging tend to be longer, with most shells reflecting input over the last few decades to centuries (and maximum shell ages of a few tens of thousands of years). Change in the species composition of the LA over these time scales, implied by LD mismatch, can still easily be anthropogenic (Fig. 1). However, in applying this LD approach to the ecological history of a given study area, direct dating of species that are anomalously abundant in the DA should be conducted to quantify how deeply in the past these former populations existed and thus to determine the absolute time-frame of significant ecological change.

2. Evidence of particular drivers of change and of how composition has changed

Although regions vary greatly in the human stressor(s) that are most important locally—for some lagoons it will be changes in salinity arising from modified inlets, and in others it might be toxic runoff or dredge-spoil dumping—eutrophication from urban, agricultural, and industrial nutrient loading leaves the strongest signal in the global molluscan database of live-dead studies (Kidwell, 2007, 2008). Meta-analysis also indicates that the impact of AE on molluscan communities in shallow coastal embayments is via the decline of seagrass as a habitat and an increased abundance of opportunistic species in all bottom types, whereas on open shelves the signature is a decline in the relative abundance of suspension-feeders and rise in chemosymbiotic, hypoxia-tolerant, and deposit-feeding species, at least where conditions are most severe. Commercial harvesting of benthos using heavy gear leaves a distinct signature only in gravelly habitats, which are more likely to have included disturbance-sensitive epifaunal and byssate infaunal species when in a natural state (Fig. 7). Declines in the proportional abundance and richness of these species are significant and evident even when the habitat has also been subject to

anthropogenic eutrophication. Using LD mismatch to assess possible change and identify unknown stresses is thus reasonable but will require surveying multiple habitats within a region because habitats differ in the sensitivity of their LAs to different kinds of stress, as is well known from biological studies. The basic protocol should be, for example, a bay-wide or cross-shelf sampling effort of LAs and DAs to intersect with a variety of seafloor types.

3. Seeing through multiple stressors

The finding here that bottom-trawling is secondary in importance to AE in creating LD mismatch does not contradict global meta-analyses using other kinds of data, where harvesting of animal populations is the anthropogenic factor that usually *initiates* significant marine ecosystem decline and is most important in explaining the depletion of commercial populations (e.g., Jackson et al., 2001; Myers and Worm, 2003; Lötze et al., 2006; the productivity of many fisheries is unaffected or improves with low-level AE, e.g., Caddy et al., 1995; Breitburg et al., 2009). As stressed by most workers (*ibid.*, plus Halpern et al., 2008; Orth et al., 2006; and many others), coastal ecosystems are in duress from multiple significant stressors, usually acting in rapid succession or simultaneously, and commonly having additive or even synergistic effects. By all accounts, AE has been hugely important in ecosystem deterioration, especially since World War II, even though many species had been depleted or driven to functional extinction by then in many regions (e.g., Lötze et al., 2006). Climate change has emerged more strongly in the last few decades, becoming recognizable as a force in an increasing number of regions, along with disease and introduced species, which are linked also with the rapid global transport of goods.

The emergence of AE as the most important factor in compositional changes in *molluscan* communities probably reflects several factors. First, most LD study areas are from embayments or nearshore shelf habitats, where AE is most likely to be strong in absolute terms and even point-sourced. The signature of LD mismatch created by AE is thus likely to overwhelm the signatures of other stressors, particularly if their effect was more deeply rooted historically and largely past (like now-collapsed fisheries) or indirect (via habitat disruption or trophic cascades associated with harvesting other species; e.g., Myers et al., 2007, on cascading effects of over-harvesting of sharks on the population increases in rays and consequent declines of their molluscan prey). Second, the vast majority of molluscan species in LD datasets are bycatch rather than targets of commercial fisheries. For example, demersal fish and shrimp are the primary targets of bottom-trawling in shelf LD study areas here, and shrimp and single bivalve species are the primary targets in most of the embayments (oysters, marsh clams; see study area descriptions in Kidwell, 2007). The strength of the signal that does emerge from bottom trawling in LD mismatch, even in areas that are also subject to AE, is thus impressive. Trawling only fails to leave a signature of LD compositional mismatch in soft mud seafloors and mobile sands where the impact of BT on the community can be difficult for biologists to recognize even using the entire macrobenthos.

4. LD mismatches in richness and evenness can be large even in natural settings

Despite their impact on the relative abundances and identities of species, neither AE nor BT are effective in themselves in explaining variation in LD agreement in richness and evenness. This lack of relationship has several practical implications. First, "taxon-free" characterizations of species diversity—simple counts of species, dominance metrics, and descriptions of the shape of the species abundance-distribution—will fail to reliably detect baseline shifts. This absence of an effect on richness and evenness does not belittle the importance of the ecological change, but rather stresses that even communities modified by human activities draw on the same regional species pool as do unimpacted communities at that site. Also, in both situations—natural and impacted—temporally acute "snapshot" samples of LAs are being compared to a time-averaged DA. These LAs will thus generally tend to be less rich than local DAs regardless of environmental conditions at the time of sampling (ongoing modeling and analytic effort with A. Tomasovych). DAs are typically 50% richer than LAs even under fully natural conditions and sample-size standardization (Table 3; 3x richer using raw data), with wide variation among individual datasets, providing a large target for impacted datasets to fall inside.

Second, the insensitivity of LD mismatch in richness and evenness to AE and BT underscores the importance of bringing taxonomic and autecologic expertise to LD analysis in the service of conservation biology. It

is not the loss (or gain) of some critical number of molluscan species or the degree to which any single species dominates locally that signifies AE or BT. Instead, it is significant change(s) in *particular* species that reveals human impacts via LD mismatch.

5. Determining the composition of natural "baseline" communities as targets for restoration

Molluscan DAs show remarkably high average agreement in composition and structure to LAs in natural settings considering the range of postmortem processes that act on DAs and the natural temporal variability in LAs encompassed by time averaging over decades to centuries (Table 3; Fig. 1). This agreement is encouraging for using mixed-layer DAs to acquire a time-averaged perspective on habitat-level communities in areas where conventional live-collected samples are sparse or unavailable. The strong spatial and temporal variability that exists *among replicate samples of LAs* under natural conditions is in itself a sobering take-home for conventional benthic surveys, underscoring the need for the averaged information captured by DAs. LD agreement in mixed-layer samples from natural settings also constitutes the best current estimate—albeit a liberal, best-case one—for the ecological fidelity of DAs from *historic* layers to the time-averaged LA of those past times in that same habitat. Moving beyond the recognition of recent ecological change in an environment (points 1 thru 4 above) to being able to recognize the composition of pre-impact baseline communities, retrospectively, from Holocene sedimentary records is a current research frontier in taphonomy applied to conservation biology.

It is thus time to add LD analysis of seafloor samples to the basic protocol of biological environmental assessment. Benthic biologists already collect mixed-layer DAs as a byproduct of sampling soft-sedimentary seafloors for LAs—DAs are simply the residual debris on sieves after living animals have been picked out. Adding LD analysis would thus entail no additional sampling, only the time to pick, identify, and count the DA already on hand. Given the scarcity of relevant and especially quantitative historical data on benthic communities globally and the pressing need to assess and rank habitats and regions for conservation or restoration, time-averaged death assemblages represent an important under-exploited resource for rapid retrospective identification of the magnitude and nature of recent community change.

ACKNOWLEDGMENTS

Many thanks to the organizers for the invitation to participate, Chicago's "Death at Noon" group and especially collaborator Adam Tomasovych, and Pete Peterson and Jeremy Jackson for inspiration. Research supported by grants NSF-EAR-0345897 and the University of Southern California Sea Grant Program, National Oceanic and Atmospheric Administration (NA07OAR4170008).

REFERENCES

AIROLDI, L., AND M.W. BECK. 2007. Loss, status and trends for coastal marine habitats of Europe. Oceanography and Marine Biology: An Annual Review, 45:345-405.

BAUM, J. K., AND B. WORM. 2009. Cascading top-down effects of changing oceanic predator abundances. Journal of Animal Ecology, 78:699-714.

BEHRENSMEYER, A. K., S. M. KIDWELL, AND R. GASTALDO. 2000. Taphonomy and paleobiology, p. 103-147. In D. H. Erwin and S. L. Wing (eds.). Deep Time: Paleobiology's Perspective. Paleobiology, Supplement to Vol. 26(4).

BEST, M. M. R. 2009. Shell taphonomy experiments across latitudes: Contrasts in net preservation, processes, and pathways. 9th North American Paleontological Convention Abstracts, Cincinnati Science Museum Scientific Contributions No. 3:220.

BEST, M. M. R., A. MLOSZEWSKA, Z. HUANG, K. BIBEAU, M. ROUGH, AND T. BROWN. 2008. Latitudinal contrasts in the net preservation of skeletal carbonate. Geological Society of America Abstracts with Programs, 40(7).

BOUCHET, P., P. LOZOUET, P. MAESTRATI, AND V. HEROS. 2002. Assessing the magnitude of species richness in tropical marine environments: Exceptionally high numbers of molluscs at a New Caledonia site. Biological Journal of the Linnean Society, 75:421–436.

BOURCIER, M. 1980. Evolution récente des peuplements macrobenthiques entre la Ciotat et les îles des Embiez (Côtes de Provence). Processus de contamination du benthos entre basins côtiers voisins. Tethys, 9:197-206.

BREITBURG, D. L., D. W. HONDORP, L. A. DAVIAS, AND R. J. DIAZ. 2009. Hypoxia, nitrogen, and fisheries: Integrating effects across local and global landscapes. Annual Review of Marine Science, 1: 329-349.

BREWER, P. G., AND E. T. PELTZER. 2009. Limits to marine life. Science, 324(5925):347-348.

CADDY, J.F., R. REFK, AND T. DO-CHI. 1995. Productivity estimates for the Mediterranean: Evidence of accelerating ecological change. Ocean and Coastal Managememt, 26:1-18.

CHAN, F., J. A. BARTH, J. LUBCHENCO, A. KIRINCICH, H. WEEKS, W.T. PETERSON, AND B. A. MENGE. 2008. Emergence of anoxia in the California Current Large Marine Ecosystem. Science, 319(5865): 920.

CHAO, A., R. L. CHAZDON, R. K. COLWELL, AND T.-J. SHEN. 2005. A new statistical approach for assessing compositional similarity based on incidence and abundance data. Ecology Letters 8:148–159.

COLLIE, J. S., S. J. HALL, M. J. KAISER, AND I. R. POINER. 2000. A quantitative analysis of fishing impacts on shelf-sea benthos. Journal of Animal Ecology, 69:785-798.

CRANFIELD, H. J., K. P. MICHAEL, AND I. J. DOONAN. 1999. Changes in the distribution of epifaunal reefs and oysters during 130 years of dredging for oysters in Foveaux Strait, southern New Zealand. Aquatic Conservation: Marine and Freshwater Ecosystems, 9:461-483.

CUMMINS, H., E. N. POWELL, R. J. STANTON, AND G. M. STAFF. 1986a. The size frequency distribution in palaeoecology: Effects of taphonomic processes during formation of molluscan death assemblages in Texas Bays. Palaeontology, 29:495-518.

CUMMINS, H., E. N. POWELL, R. J. STANTON, AND G. M. STAFF. 1986b. The rate of taphonomic loss in modern benthic habitats: How much of the potentially preservable community is preserved? Palaeogeography, Palaeoclimatology, Palaeoecology, 52: 291–320.

DAVIS, F. M. 1923. Quantitative studies on the fauna of the sea bottom. No. 1. Preliminary investigation of the Dogger Bank. Ministry of Agriculture and Fisheries, Fishery Investigations. Series II, Vol VI(2), 54 p.

DE JUAN, S., M. DEMESTRE, AND S. THRUSH. 2009. Defining ecological indicators of trawling disturbance when everywhere that can be fished is fished: A Mediterranean case study. Marine Policy, 33(3):472-478.

DIAZ, R. J., AND R. ROSENBERG. 1995. Marine benthic hypoxia: A review of its ecological effects and the behavioural responses of benthic macrofauna. Oceanography and Marine Biology: An Annual Review, 33:245-303.

DIAZ, R. J., AND R. ROSENBERG. 2008. Spreading dead zones and consequences for marine ecosystems. Science, 321(5891): 926-929.

EDGAR, G.J., AND C.R. SAMSON. 2004. Catastrophic decline in mollusk diversity in eastern Tasmania and its concurrence with shellfish fisheries. Conservation Biology, 18: 1579-1588.

EKDALE, A. A. 1977. Quantitative paleoecological aspects of modern marine mollusk distribution, northeast Yucatàn coast, Mexico, p. 195-207. In S. H. Frost, M. P. Weiss, and J. B. Sauders (eds.). Reefs and Related Carbonates: Ecology and Sedimentology. American Association of

Petroleum Geologists, Studies in Geology, 4.

FERGUSON, C. A. 2008. Nutrient pollution and the molluscan death record: Use of mollusc shells to diagnose environmental change. Journal of Coastal Research, 24:250-259.

FERGUSON, C. A., AND A. I. MILLER. 2007. A sea change in Smuggler's Cove? Detection of decadal-scale compositional transitions in the subfossil record. Palaeogeography, Palaeoclimatology, Palaeoecology, 254(3-4): 418-429.

FISCHER, A. G. 1961. Stratigraphic record of transgressing seas in light of sedimentation on Atlantic coast of New Jersey. American Association of Petroleum Geologists Bulletin, 45:1656-1666.

FLESSA, K. W. 1998. Well-traveled cockles: Shell transport during the Holocene transgression of the southern North Sea. Geology, 26:187-190.

FLESSA, K. W., AND M. KOWALEWSKI. 1994. Shell survival and time-averaging in nearshore and shelf environments: Estimates from the radiocarbon literature. Lethaia, 27:153-165.

FLESSA, K. W., A. H. CUTLER, AND K. H. MELDAHL. 1993. Time and taphonomy: Quantitative estimates of time-averaging and stratigraphic disorder in a shallow marine habitat. Paleobiology, 19: 266-286.

FÜRSICH, F. T., AND M. ABERHAN. 1990. Significance of time-averaging for paleocommunity analysis. Lethaia, 23:143-152.

GRAY, J. S., AND M. ELLIOTT. 2009. Ecology of Marine Sediments: From Science to Management (2nd edition). Oxford University Press, Oxford, 225 p.

GRAY, J. S., P. DAYTON, S. THRUSH, AND M. J. KAISER. 2006. On effects of trawling, benthos and sampling design. Marine Pollution Bulletin, 52: 840-843.

GREEN, M. A., R. C. ALLER, AND J. Y. ALLER. 1998. Influence of carbonate dissolution on survival of shell-bearing meiobenthos in nearshore sediments. Limnology and Oceanography, 43: 18-28.

GREEN, M. A., M. E. JONES, C. L. BOUDREAU, R. L. MOORE, AND B. A. WESTMAN. 2004. Dissolution mortality of juvenile bivalves in coastal marine deposits. Limnology and Oceanography, 49(3): 727-734.

GREENSTEIN, B. J., A. H. CURRAN, AND J. M. PANDOLFI. 1998. Shifting ecological baselines and the demise of *Acropora cervicornis* in the western North Atlantic and Caribbean Province: A Pleistocene perspective. Coral Reefs, 17(3):249-261.

HALLAM, A. 1967. The interpretation of size-frequency distributions in molluscan death assemblages. Palaeontology, 10:25-42.

HALPERN, B. S., S. WALBRIDGE, K. A. SELKOE, C. V. KAPPEL, F. MICHELI, C. D'AGROSA, J. F. BRUNO, K. S. CASEY, C. EBERT, H. E. FOX, R. FUJITA, D. HEINEMANN, H. S. LENI-

HAN, E. M. P. MADIN, M. T. PERRY, E. R. SELIG, M. SPALDING, R. STENECK, AND R. WATSON. 2008. A global map of human impact on marine ecosystems. Science, 319(5865):948-952.

HEWITT, J.E., S. F. THRUSH, J. HALLIDAY, AND L. DUFFY. 2005. The importance of small-scale habitat structure for maintaining beta diversity. Ecology, 86(6):1619-1626.

HIDDINK, J. G., S. JENNINGS, AND M.J. KAISER. 2007. Assessing and predicting the relative ecological impacts of disturbance on habitats with different sensitivities. Journal of Applied Ecology, 44: 405-413.

HINZ, H., V. PRIETO, AND M. J. KAISER. 2009. Trawl disturbance on benthic communities: Chronic effects and experimental predictions. Ecological Applications, 19: 761-773.

JACKSON, J. B. C. 1997. Reefs since Columbus. Coral Reefs, Supplement to Vol. 16:23–32.

JACKSON, J. B. C., M. X. KIRBY, W. H. BERGER, K. A. BJORNDAL, L. W. BOTSFORD, B. J. BOURQUE, R. H. BRADBURY, R. COOKE, J. ERLANDSON, J. A. ESTES, T. P. HUGHES, S. KIDWELL, C. B. LANGE, H. S. LENIHAN, J. M. PANDOLFI, C. H. PETERSON, R. S. STENECK, M. J. TEGNER, AND R. R. WARNER. 2001. Historical overfishing and the recent collapse of coastal ecosystems. Science, 293: 629–637.

JOHNSON, R. G. 1965. Pelecypod death assemblages in Tomales Bay, California. Journal of Paleontology, 39: 80-85.

KIDWELL, S. M. 2001. Preservation of species abundance in marine death assemblages. Science, 294:1091-1094.

KIDWELL, S. M. 2002a. Time-averaged molluscan death assemblages: Palimpsests of richness, snapshots of abundance. Geology, 30:803-806.

KIDWELL, S. M. 2002b. Mesh-size effects on the ecological fidelity of death assemblages: A meta-analysis of molluscan live-dead studies. Geobios Mémoire Special, 24:107-119.

KIDWELL, S. M. 2005. Brachiopod versus bivalve radiocarbon ages: Implications for Phanerozoic trends in time-averaging and productivity. Geological Society of America Abstracts with Programs, 37(7):117.

KIDWELL, S. M. 2007. Discordance between living and death assemblages as evidence for anthropogenic ecological change. Proceedings of the National Academy of Sciences of the United States of America, 104:17701-17706.

KIDWELL, S. M. 2008. Ecological fidelity of open marine molluscan death assemblages: Effects of post-mortem transportation, shelf health, and taphonomic inertia. Lethaia, 41(3):199-217.

KIDWELL, S. M., AND D. W. J. BOSENCE. 1991. Taphonomy and time-averaging of marine shelly faunas, p. 115-209. *In* P. A. Allison and D. E. G. Briggs (eds.). Taphonomy, Releasing the Data Locked in the Fossil Record. Plenum Press, New York.

KIDWELL, S. M., AND D. JABLONSKI. 1983. Taphonomic feedback: Ecological consequences of shell accumulation, p. 195-248. *In* M. J. S. Tevesz and P. L. McCall (eds.), Biotic Interactions in Recent and Fossil Benthic Communities. Plenum Press, New York.

KIDWELL, S. M., AND K. W. FLESSA. 1995. The quality of the fossil record: Populations, species, and communities. Annual Review of Ecology and Systematics, 26:269-299.

KIDWELL, S.M., M.M.R. BEST, AND D. KAUFMAN. 2005. Taphonomic tradeoffs in tropical marine death assemblages: Differential time-averaging, shell loss, and probable bias in siliciclastic versus carbonate facies. Geology, 33:729-732.

KIDWELL, S. M., AND A. TOMASOVYCH. 2009. Nobody's perfect: Assessing modern death assemblages as historical recorders using "live-live" comparisons. 9th North American Paleontological Convention Abstracts, Cincinnati Science Museum Scientific Contributions, 3:379-380.

KIRBY, M. X. 2004. Fishing down the coast: Historical expansion and collapse of oyster fisheries along continental margins. Proceedings of the National Academy of Sciences of the United States of America, 101:13096-13099.

KOSNIK, M., Q. HUA, G. JACOBSEN, D. S. KAUFMAN, AND R. A. WÜST. 2007. Sediment mixing and stratigraphic disorder revealed by the age-structure of *Tellina* shells in Great Barrier Reef sediment. Geology, 35: 811-814.

KOSNIK, M. A., Q. HUA, D. S. KAUFMAN, AND R. A. WÜST. In press. Taphonomic bias and time-averaging in tropical molluscan death assemblages: Differential shell half-lives in Great Barrier Reef sediment. Paleobiology.

KOWALEWSKI, M. 2009. The youngest fossil record and conservation biology: Holocene shells as eco-environmental recorders. *In* G. P. Dietl, and K. W. Flessa (eds.), Conservation Paleobiology: Using the Past to Manage for the Future. The Paleontological Society Papers, 15 (this volume).

KOWALEWSKI, M., G. A. GOODFRIEND, AND K. W. FLESSA. 1998. High-resolution estimates of temporal mixing within shell beds: The evils and virtues of time-averaging. Paleobiology, 24:287-304.

KRAUSE, R. A., JR., S. L. BARBOUR WOOD, M. KOWALEWSKI, D. S. KAUFMAN, C. S. ROMANEK, M. G. SIMÕES, AND J. F.WEHMILLER. In press. Quantitative estimates and modeling of time averaging in bivalve and brachiopod shell accumulations. Paleobiology.

LADD, H. S., J. W HEDGPETH, AND R. POST. 1957. Environments and facies of existing bays on the central Texas coast. Geological Society of America Memoir, 67:599-640.

LÖTZE, H. K., AND B. WORM. 2009. Historical baselines for large marine animals. Trends in Ecology and Evolution, 24:254-262.

LÖTZE, H. K., H. S. LENIHAN, B. J. BOURQUE, R. BRADBURY, R. G. COOKE, M. KAY, S. M. KIDWELL, M. X. KIRBY, C. H. PETERSON, AND J. B. C. JACKSON. 2006. Depletion, degradation, and recovery potential of estuaries and coastal seas worldwide. Science, 312:1806-1809.

MELDAHL, K.E., K. W. FLESSA, AND A. H. CUTLER. 1997. Time-averaging and postmortem skeletal survival in benthic fossil assemblages: Quantitative comparisons among Holocene environments. Paleobiology, 23:207-229.

MILLER, A. I. 1988. Spatial resolution in subfossil molluscan remains: Implications for paleobiological analyses. Paleobiology, 14:91-103.

MYERS, R.A., AND B. WORM. 2003. Rapid worldwide depletion of predatory fish communities. Nature, 423:280-283.

MYERS, R. A., J. K. BAUM, T. D. SHEPHERD, S. P. POWERS, AND C. H. PETERSON. 2007. Cascading effects of the loss of apex predatory sharks from a coastal ocean. Science, 315:1846-1850.

NRC (NATIONAL RESEARCH COUNCIL). 2002. Effects of Trawling and Dredging on Seafloor Habitats. National Academies Press, Washington, D.C., 136 p.

NRC (NATIONAL RESEARCH COUNCIL). 2005. The Geological Record of Ecological Dynamics. National Academies Press, Washington, D.C., 200 p.

OLSZEWSKI, T.A., AND S.M. KIDWELL. 2007. The preservational fidelity of evenness in molluscan death assemblages. Paleobiology, 33:1-23.

ORTH, R. J., T. J. B. CARRUTHERS, W. C. DENNISON, C. M. DUARTE, J. W. FOURQUREAN, K. L. HECK, A. R. HUGHES, G. A. KENDRICK, W. J. KENWORTHY, S. OLYARNIK, F. T. SHORT, M. WAYCOTT, AND S. L. WILLIAMS. 2006. A global crisis for seagrass ecosystems. Bioscience, 56:987-996.

PALACIOS, R., D. A. ARMSTRONG, AND J. ORENSANZ. 2000. Fate and legacy of an invasion: Extinct and extant populations of the soft-shell clam (*Mya arenaria*) in Grays Harbor (Washington). Aquatic Conservation: Marine and Freshwater Ecosystems, 10:279-303.

PANDOLFI, J. M., R. H. BRADBURY, E. SALA, T. P. HUGHES, K. A. BJORNDAL, R. G. COOKE, D. MACARDLE, L. McCLENAHAN, M. J. H. NEWMAN, G. PAREDES, R. R. WARNER, AND J. B. C. JACKSON. 2003. Global trajectories of the long-term decline of coral reef ecosystems. Science, 301: 955-958.

PAULY, D. 1995. Anecdotes and the shifting baseline syndrome of fisheries. Trends in Ecology and Evolution, 10:430.

PETERSON, C. H. 1976. Relative abundance of living and dead molluscs in two California lagoons. Lethaia, 9:137-148.

PETERSON, C. H. 1977. The paleoecological significance of undetected, short-term temporal variability. Journal of

Paleontology, 51:976-981.

POWERS, S. P., C. H. PETERSON, R. R. CHRISTIAN, E. SULLIVAN, M. J. POWERS, M. J. BISHOP, AND C. P. BUZZELLI. 2005. Effects of eutrophication on bottom habitat and prey resources of demersal fishes. Marine Ecology Progress Series, 302:233-243.

RABALAIS, N. N., AND R. E. TURNER. 2001. Coastal Hypoxia: Consequences for Living Resources and Ecosystems. Coastal and Estuarine Sciences, American Geophysical Union, 463 p.

RICK, T. C., AND J. M. ERLANDSON. 2009. Coastal exploitation. Science, 325:952-953.

ROTHSCHILD, B. J., J. S. AULT, P. GOULLETQUER, AND M. HÉRAL. 1994. Decline of the Chesapeake Bay oyster population: A century of habitat destruction and overfishing. Marine Ecology Progress Series, 111:29-39.

SADLER, P. M. 1993. Models of time-averaging as a maturation process: How soon do sedimentary sections escape reworking?, p. 169-187. In S. M. Kidwell and A. K. Behrensmeyer (eds.), Taphonomic Approaches to Time Resolution in Fossil Assemblages, Short Course in Paleontology, 6.

SMITH, S. D. A. 2008. Interpreting molluscan death assemblages on rocky shores: Are they representative of the regional fauna? Journal of Experimental Marine Biology and Ecology, 366:151-159.

STAFF, G. M., AND E. N. POWELL. 1999. Onshore-offshore trends in community structural attributes: Death assemblages from the shallow continental shelf of Texas. Continental Shelf Research, 19:717-756.

STAFF, G. M., R. J. STANTON, JR., E. N. POWELL, AND H. CUMMINS. 1986. Time averaging, taphonomy and their impact on paleocommunity reconstruction: Death assemblages in Texas bays. Geological Society of America Bulletin, 97:428-443.

TERRY, R. C. 2008. The scale and dynamics of time-averaging quantified through AMS ^{14}C dating of kangaroo rat bones. Geological Society of America, Abstracts with Programs, 40(7).

THRUSH, S. F., J. E. HEWITT, V. J. CUMMINGS, P. K. DAYTON, M. CRYER, S. J. TURNER, G. A. FUNNELL, R. G. BUDD, C. J. MILBURN, AND M. R. WILKINSON. 1998. Disturbance of the marine benthic habitat by commercial fishing: Impacts at the scale of the fishery. Ecological Applications, 8:866-879.

TILLIN, H. M., J. G. HIDDINK, S. JENNINGS, AND M. J. KAISER. 2006. Chronic bottom trawling alters the functional composition of benthic invertebrate communities on a sea-basin scale. Marine Ecology-Progress Series, 318:31-45.

TOMASOVYCH, A. 2004. Postmortem durability and population dynamics affecting the fidelity of brachiopod size-frequency distributions. Palaios, 19:477-496.

TOMASOVYCH, A., AND S. M. KIDWELL. 2009a. Preservation of spatial and environmental gradients by death assemblages. Paleobiology, 35:119-145.

TOMASOVYCH, A., AND S. M. KIDWELL. 2009b. Fidelity of variation in species composition and diversity partitioning by death assemblages: Time-averaging transfers diversity from beta to alpha levels. Paleobiology, 35:97-121.

TOMASOVYCH, A., AND S. M. KIDWELL. 2009c. Discriminating effects of natural and sampling variation from effects of postmortem variation on compositional fidelity of death assemblages. 9th North American Paleontological Convention Abstracts, Cincinnati Science Museum Scientific Contributions, 3:155-156.

TWILLEY, R. R., W. M. KEMP, K. W. STAVER, J. C. STEVENSON, AND W. R. BOYNTON. 1985. Nutrient enrichment of estuarine submersed vascular plant communities. 1. Algal growth and effects on production of plants and associated communities. Marine Ecology-Progress Series, 23:179-191.

WARME, J. E. 1971. Paleoecological aspects of a modern coastal lagoon. University of California Publications in Geological Sciences, 87:1-110.

WARWICK, R.M., AND J. LIGHT. 2002. Death assemblages of molluscs on St. Martin's Flats, Isles of Scilly: A surrogate for regional biodiversity? Biodiversity and Conservation, 11: 99–112.

WHITE, E. P., P. B. ADLER, W. K. LAUENROTH, R. A. GILL, D. GREENBERG, D. M. KAUFMAN, A. RASSWEILER, J. A. RUSAK, M. D. SMITH, J. R. STEINBECK, R. B. WAIDE, AND J. YAO. 2006. A comparison of the species-time relationship across ecosystems and taxonomic groups. Oikos, 112:185-195.

WORM, B., E. B. BARBIER, N. BEAUMONT, J. E. DUFFY, C. FOLKE, B. S. HALPERN, J. B. C. JACKSON, H. K. LOTZE, F. MICHELI, S. R. PALUMBI, E. SALA, K. A. SELKOE, J. J. STACHOWICZ, AND R. WATSON. 2006. Impacts of biodiversity loss on ocean ecosystem services. Science, 314:787-790.

CHAPTER EIGHT

USING A MACROECOLOGICAL APPROACH TO THE FOSSIL RECORD TO HELP INFORM CONSERVATION BIOLOGY

S. KATHLEEN LYONS AND PETER J. WAGNER

Department of Paleobiology, National Museum of Natural History, Smithsonian Institution [NHB, MRC 121],
P.O. Box 37012, Washington, D.C. 20013-7012 USA

ABSTRACT.—With or without realizing it, macroecology, paleobiology and conservation biology have been addressing similar issues using similar methods and analogous data sets. Much of what we call "paleobiology" overlaps heavily with macroecology, and their shared interest in losses in biodiversity over space and time clearly is of interest to conservation biology. Here we examine how some "classic" macroecological and paleobiological studies and techniques apply to issues that currently are of interest to conservation biology. Our examples are far from exhaustive, but include examining temporal (or possible temporal) shifts in: 1) geographic range sizes; 2) body size distributions; 3) relative abundance distributions; and, 4) morphological diversity. Reframing these issues in terms of how loss of biodiversity and richness affects particular slices of time *including* (but not limited to) the present should do much to communicate the value of macroecological and paleobiological methods and theory to conservation research.

INTRODUCTION

MACROECOLOGY IS a relatively recent subdiscipline within ecology that focuses upon larger scale statistical patterns and explores the area of overlap between multiple disciplines including paleobiology, biogeography, ecology and evolution (Brown, 1995). Although ecological theory successfully explains many different types of ecological interactions, traditional theory often fails to elucidate the interactions between different scales of biological organization (e.g., populations, communities, ecosystems and whole biotas) and has difficulty providing universal causal mechanisms for statistical patterns that are similar across large spatial, temporal or taxonomic scales (Brown and Maurer, 1989; Brown, 1995; Maurer, 1999; Gaston and Blackburn, 2000). Macroecology is increasingly filling this gap.

Macroecological theory offers expected relationships among species-level traits (e.g., body-size, geographic ranges size, abundance distributions) that vary predictably under different scenarios. Paleobiology offers a way to evaluate these expected relationships over greater time scales and under climatic regimes and other scenarios not available to modern macroecologists. Moreover, studying macroecological patterns in the fossil record can help differentiate between patterns that are a result of the anthropological stresses unique to modern ecosystems or are repeatable patterns that

are the result of long-term ecological and evolutionary processes. Importantly for conservation theory, paleobiology can give us a baseline with which to evaluate current ecological systems (e.g., Willis and Birks, 2006).

Paleobiology often appears to concern macroevolutionary theory rather than macroecological theory. However, macroevolutionary theory looks to macroecology for possible processes behind large–scale diversification and extinction patterns. Thus, macroecological theory is highly relevant to hypotheses concerning Phanerozoic–level patterns of diversity gains and loses. Macroevolutionary theory and methods pertaining to different levels of extinction might also provide insights for conservation biological theory, as many of the questions asked of modern patterns can be asked of past extinctions. Together, macroecology and paleobiology are crucial for solving the complex problems caused by anthropogenically caused environmental changes such as habitat loss, global climate change and loss of biodiversity (Smith et al., 2008).

In this paper, we review different approaches used by ecologists (both modern and paleontological) for addressing macroecological issues as well as explore how some methods hitherto used largely by paleobiologists might be useful for modern ecological studies. First we will discuss macroecological approaches typically used for modern data and discuss how they can be applied to the fossil record in the emerging field of conservation

In *Conservation Paleobiology: Using the Past to Manage for the Future, Paleontological Society Short Course, October 17th, 2009. The Paleontological Society Papers, Volume 15, Gregory P. Dietl and Karl W. Flessa (eds.). Copyright © 2009 The Paleontological Society.*

paleobiology. Then we will take the opposite tact and discuss traditional macroevolutionary methods and discuss how they can be applied to modern conservation biology to give a fuller picture of a system than would otherwise be available.

MODERN MACROECOLOGICAL METHODS AND THEIR APPLICATION TO CONSERVATION PALEOBIOLOGY

Geographic range size

The study of ecology is fundamentally about understanding the abundance and distribution of species (Begon et al., 1990). As such, a species geographic range is a basic unit of study in ecology. Using a macroecological approach can provide insight into the variation in a species geographic range by calculating correlations between a species' geographic range and other traits such as body size, life history, or abundance. Moreover, such studies of larger scale statistical patterns can provide insight into longer-term processes not typically available to reductionist ecology. For example, macroecological approaches can be used to evaluate the relationship between various species traits including geographic range size and a species extinction risk using a phylogeny and the IUCN redlist of threatened and endangered species for modern groups (e.g., Jones et al., 2003) or the fossil record for older groups (Jablonski, 1986; Jablonski and Raup, 1995), or ideally both for groups that have a substantial fossil record, but extend into the present (e.g., mammals).

Geographic range size can be characterized using a variety of methods, each with its own strengths and weaknesses (Fig. 1). The method used depends upon the type of underlying data available. When incorporating extant species, range maps are sometimes available (Fig. 1.2). Modern range maps are often based upon a few known localities that are interpolated by the creator of the map using biome or habitat information to predict where the species will occur (e.g., Hall, 1981). In many cases, the resulting range map appears to greatly exceed the reach of the locality data available. For some groups, this type of information may be the best that is available. However, it should be used with caution. If the range maps are equal-area projection maps, then geographic range size can be estimated using a planimeter (Willig and Selcer, 1989; Willig and Sandlin, 1991; Willig and Gannon, 1997). These types of range maps can also be digitized and geographic area

can be calculated using GIS programs (e.g., Patterson et al., 2003).

In some cases, the area of a geographic range is not necessary to the study and latitudinal extent can be used instead (Roy et al., 1995; Lyons and Willig, 1997; Willig and Lyons, 1998; Koleff and Gaston, 2001; Madin and Lyons, 2005; Harcourt, 2006; Ruggiero and Werenkraut, 2007; Krug et al., 2008). Latitudinal range size is usually calculated as the number of degrees of latitude between the northern and southern most extents of a species geographic range (Fig. 1.3). Latitudinal extent is positively correlated with geographic range size (Gaston, 2003; Lyons, 1994) and can be used when data allowing a reasonable estimate of geographic range size are not available. Latitudinal extent can also be reliably used when the group in question has geographic ranges that are essentially linear, e.g., marine mollusks (Roy et al., 1994, 1995; Krug et al., 2008). Finally, there are cases where latitudinal range is the unit of interest. For example, mid-domain models argue that if species' latitudinal ranges are randomly placed within a bounded domain (i.e., a land mass or closed ocean basin; Fig. 1.4), the resulting distribution and overlap of species ranges would produce a gradient in diversity with a peak in the middle of the domain (Colwell and Hurtt, 1994; Willig and Lyons, 1998). Obviously, if the mid-domain effect is the subject of a study, then latitudinal ranges are necessary.

When the underlying data are collection localities, more options are available for creating geographic ranges and estimating their area. This is particularly applicable to conservation paleobiology since much of the fossil data available are of this sort. One method involves taking the collection localities and first projecting them into an appropriate equal area projection (Fig. 1.1). Next, calculate the minimum convex hull that encloses the collection localities by identifying the limits of the collection localities and calculating the area within them (e.g., Lyons, 2003, 2005). This is akin to drawing a line around the outermost localities and calculating the area within the resulting shape. There are two potential problems with this method. First, we are unlikely to have a record of a species everywhere it occurred or even in every habitat it occupied. Therefore, this method almost certainly provides an underestimate of a species true range size. The second problem is that sampling and preservation will be uneven across species. Better-sampled species are likely to have larger estimated ranges because the minimum convex polygons

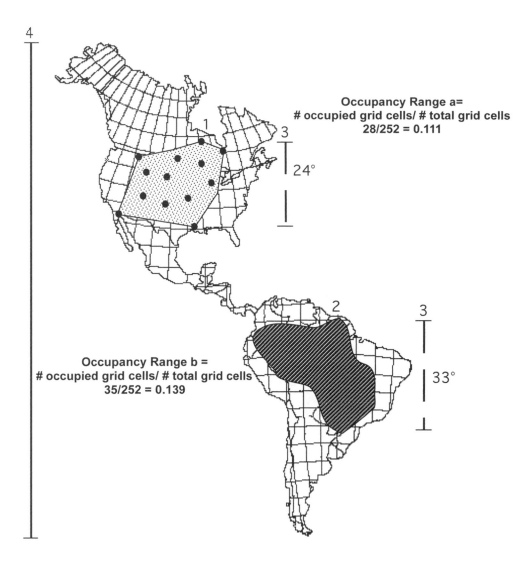

Occupancy Range a=
 # occupied grid cells/ # total grid cells
28/252 = 0.111

Occupancy Range b =
 # occupied grid cells/ # total grid cells
35/252 = 0.139

FIGURE *1*.—Schematic of the different methods for estimating geographic range size. *1*, Minimum convex polygon; *2*, Biome infilling; *3*, Latitudinal range extents; *4*, Extent of the domain used to determine the total number of grid cells for occupancy measures. Modified from Willig et al. (2009).

will have more localities spread across a larger space. This can be a difficult bias to solve because there are reasons to expect a species with a larger range to have a higher preservation potential. A species that occurs in multiple habitats or across more space will have more opportunities to be preserved. In some modern groups, there is a relationship between body size and geographic range with larger bodied species having larger geographic ranges (Brown, 1995; Gaston and Blackburn, 1996; Gaston, 2003; Madin and Lyons, 2005; Smith et al., 2008). A similar relationship exists for North American mammals in the late Pleistocene and Holocene (Lyons, unpublished data). Larger-bodied species may

be more likely to be preserved and recovered because their skeletal elements are more robust and because they are easier for workers to find. For example, studies of mammals and marine mollusks have shown that extant taxa that are recorded in the fossil record tend to have larger geographic range sizes and larger body sizes (Lyons and Smith, 2006; Valentine et al., 2006). One way to deal with variation in sampling when using minimum convex polygons is to set criteria for the degree of sampling necessary to accurately represent ranges for your system. However, sensitivity analyses should always be performed to determine the effect of the poorly sampled species on the patterns of interest

(Lyons, 2005).

A measure of geographic range size that is being increasingly used is occupancy (Ruggiero and Lawton, 1998; Gaston, 2003; Blackburn et al., 2004; Foote, 2007; Foote et al., 2007, 2008). Occupancy is the ratio of occupied sites to unoccupied sites (Fig. 1). Occupancy is positively correlated with geographic range size (Gaston, 2003). Currently there is no standard protocol for determining the number of sampled sites. As a result, sampling is usually determined by the available data. Unfortunately, there is no rigorous analysis of the effect of sampling on the accuracy of occupancy as a measure of geographic range (Willig et al., in press). However, because of the nature of fossil data, occupancy may prove to be a simple, easy, and reliable way to measure geographic range size in the fossil record. Assuming that similar taphonomic biases are operating on all species within a taxonomic group, then occupancy should give a reasonable estimate of the relative geographic range size of species in that group. In a recent application of occupancy to fossil data, Foote (2007, see also Foote et al., 2007, 2008) evaluated the trajectory of geographic range size through time and found that geographic ranges tend to increase, have a relatively short peak at mid-duration and then decline.

A final method for determining geographic range size is genetic algorithm modeling (Peterson, 2001; Martinez-Meyer et al., 2004; Peterson et al., 2004a). In this method, collection localities are combined with information about the abiotic environment such as temperature, rainfall, elevation, humidity, etc. and fed into a genetic algorithm model called GARP to produce a model of a species' requirements that best predicts its distribution. These models are often produced using part of the locality data and then tested using the rest. These models have been used quite extensively for modern species to predict distribution shifts under different models of climate warming (Peterson et al., 2001, 2002, 2004b; Thomas et al., 2004). These models are likely to be less applicable to deep time systems because of the detailed information on environmental variables that they require. However, for late Pleistocene and Holocene systems where reasonable climate models are available, these models have been used to predict the expected range changes in extinct species (Peterson, 2001; Martinez-Meyer et al., 2004; Peterson et al., 2004a). One drawback to these models is that they rely on species niches to be conservative. That is, they assume that the observed combination of abiotic variables under which a species is currently found represents the full spectrum of possibilities. If it is common for novel combinations of species to occur under novel climatic regimes (e.g., Williams et al., 2001), then predictions of past and future geographic ranges made using genetic algorithm modeling will be inaccurate, particularly for time periods of dramatic climate change.

Evaluating changes in geographic ranges over critical intervals

Predicting changes in species distributions under different scenarios of global climate change is of considerable interest in conservation ecology. Typically, these studies are limited to observing the small shifts in distributions that have already happened, or using the current distributions of species to predict what will happen to their range in the future (Peterson et al., 2002; Walker et al., 2002; Parmesan and Yohe, 2003; Thomas et al., 2004). Both have drawbacks when it comes to understanding and predicting the effects of global climate change. Analyses of current range shifts are limited to species that have extensive museum collections that can be used to define past distributions. Typically, the past is limited to the last hundred years or so. Moreover, the range shifts being observed are relatively small. For example, Parmesan and Yohe (2003) examined 1700 species and found average range shifts of 6.1 km per decade. Although their analysis determined that these were shifts of a significant distance, ecological theory predicts that the edges of species ranges are in marginal habitat and posits that a species geographic range is fluid, particularly at the edges. Therefore, some flux in a distribution is expected over time (Lomolino et al., 2006). Determining how much of a range shift represents a real change versus expected flux is difficult.

Bioclimate envelope modeling attempts to get around this problem by using the current realized niche of a species and climate models to predict what will happen to a species in the future (Foody, 2008; Jeschke and Strayer, 2008; Schweiger et al., 2008). For example, the current temperature range of a species is assumed to be the full range of temperatures under which a species can exist. By modeling what will happen to global temperatures in the future, workers can predict where the habitable area for a species is likely to be and what will happen to a species range. In some cases, the temperature range is expected to disappear and the species is predicted to become extinct (Thomas et al.,

2004). The problem with this method is that it fails to account for the possibility that species ranges are limited by something other than climate or that a species fundamental niche is greater than its realized niche. If some other aspect of its fundamental niche will be available under future climate conditions, the predictions from this type of modeling will be flawed (Willis and Birks, 2006; Willis et al., 2007a, 2007b; Beale et al., 2008; Jeschke and Strayer, 2008).

The incorporation of conservation paleobiology into these types of methods has the potential to greatly enhance our ability to understand and predict the effects of future climate change on species geographic ranges. First we have the ability to measure geographic range in the fossil record using all of the methods described above. With the increasing availability of major online databases like NEOTOMA (the current incarnation of FAUNMAP), the Paleobiology DataBase, NOW (Neogene mammals of the Old World), Miomap (Miocene Mammal Mapping Project) among others, we have the tools and data necessary to do broad scale analyses of changes in geographic ranges through time. These types of analyses have the ability to offer fundamental information that is not available using modern data alone.

Incorporating fossil data can expand our knowledge and understanding of the niches of extant species with a fossil record. For example, Pleistocene plant workers have done extensive work mapping past distributions of plant species to understand how plants shifted their distributions in response to glaciation (Overpeck et al., 1985, 1992; Jackson et al., 1997; Jackson and Overpeck, 2000; Williams et al., 2001, 2002, 2004). This work has focused extensively on identifying and understanding non-analog communities (Overpeck et al., 1992; Williams et al., 2001; Jackson and Williams, 2004). In a paper in which pollen distributions were analyzed in conjunction with climate models, Williams et al. (2001) showed that non-analog plant communities are found in areas of non-analog climate (i.e., novel combinations of temperature and precipitation). Obviously, the aspects of a species niche that are expressed in these no-analog communities are not apparent when only modern distributions are used to define a species niche space.

Our understanding of what happens to species distributions as climate changes is strongly informed by conservation paleobiology. Paleobiological studies quantifying species range shifts during the last

glaciation showed that species shift their ranges individualistically and that ecological theory predicting that communities are Clementsian superorganisms are incorrect (Graham, 1986; Graham and Mead, 1987; Webb and Barnosky, 1989; Overpeck et al., 1992; Graham et al., 1996; Davis et al., 1998; Jackson and Overpeck, 2000; Davis and Shaw, 2001; Barnosky et al., 2003; Lyons, 2003; Williams et al., 2004). This work has also shown that species did not shift their ranges in a simple north-south fashion as the glaciers expanded and contracted (Graham et al., 1996, Lyons, 2003), but that individual species are responding to changes in the environment that correspond to their niche requirements and the scale at which they perceive the environment. The widespread dissemination of this work into ecological theory and ecological text books (e.g., Lomolino et al., 2006) likely informed current ecological theory and methods used to evaluate and predict species range shifts as a result of global warming. However, there is room for the expansion of paleobiology in this area. For example, where the data exist, more detailed analyses of species past distributions and their correlation with climate models can provide a better understanding of species fundamental niche. Such information would greatly enhance the accuracy of bioclimatic envelope models in predicting species responses to global warming. As fossil databases grow, there will be more opportunities to shift these types of analyses into deeper time. Analyzing species responses to multiple climate change events can provide a baseline for how species respond under natural scenarios. Similarities and differences with the current climate change can provide insight into possible policy decisions that will mitigate anthropogenic effects.

Evaluating the role of geographic range in extinction events

Analyses of geographic ranges in the fossil record can also provide an understanding of how species geographic ranges change through time in the absence of dramatic climate change. A series of recent papers examines the patterns of geographic range expansion and contraction over a species lifetime using species occupancy as a metric (Foote, 2007; Foote et al., 2007, 2008). In general, occupancy is symmetrical with a rise, a short-lived peak, and then a decrease to extinction. As a result, species geographic ranges are already in decline when they become extinct. Moreover, the authors found no cases of subsequent increase once the decline

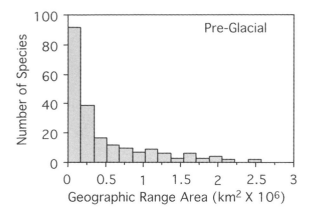

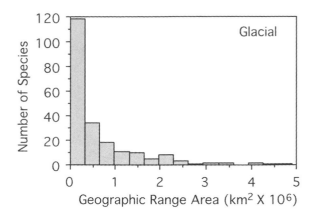

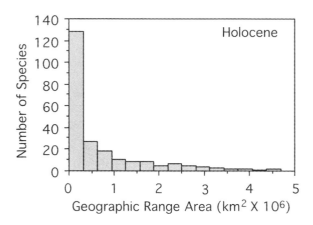

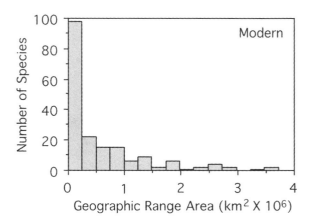

FIGURE 2.—Range size frequency distributions for late Pleistocene mammals in North America. Underlying locality data were taken from FAUNMAP and divided into four time periods. Pre-Glacial (40 kya to 20 kya), Glacial (20 kya to 10 kya), Holocene (10 kya to 500 ya) and Modern (last 500 y). Range sizes were calculated using minimum convex polygons enclosing fossil localities. Modified from Lyons (2005).

started. This implies that, in the absence of mitigating factors, species do not recover once their geographic range begins to decrease. Indeed, the species that were at the greatest risk of extinction were those whose geographic range had been declining for a substantial portion of time (Foote et al., 2007). This type of baseline information on the waxing and waning of geographic ranges is sorely missing from our understanding of species responses to climate change (Willis et al., 2007a). Moreover, it is not possible to obtain it using the modern record.

In addition to providing baseline information on the dynamics of geographic ranges over time, the fossil record allows for analyses of the role of geographic range in different types of extinction events. A classic paper by Jablonski (1986) found that broad geographic ranges enhanced the survivorship of marine mollusks

during times of background extinction, but provided little protection during mass extinctions. A similar relationship was found in an analysis of all benthic marine invertebrates across the Phanerozoic (Payne and Finnegan, 2007). Moreover, the extensive analysis of marine invertebrate genera by Foote (2007) confirmed this finding. During times of mass extinction, more of the genera that became extinct did so while their occupancy was holding steady or increasing.

This work has important policy implications for the current biodiversity crisis. It suggests that species whose geographic ranges have already declined substantially, or whose habitat is greatly degraded or destroyed, will take more effort to save. Moreover, if fears that we are in the sixth mass extinction prove true, this work argues that we cannot be complacent and assume that species with broad geographic ranges are safe.

Geographic range size distributions

Another commonly studied macroecological pattern is the frequency distribution of geographic range size. Range size frequency distributions are created by dividing species into range size categories and constructing a histogram of the number of species in each bin. They generally exhibit a characteristic shape when plotted on an arithmetic scale (e.g., Fig. 2; Brown, 1995; Gaston, 2003). Because the majority of species have small ranges and a few species have medium and large ranges, the resulting distributions typically are unimodal and right-skewed. They are typically referred to as "hollow curves" (Willis, 1922). Taking the log does not produce a normal distribution. After log-transformation, range size distributions are typically unimodal and somewhat left-skewed (Willig et al., 2003, in press).

Range size frequency distributions are potentially useful in conservation paleobiology. First, the similarity in the shape of the distribution in modern groups suggests that these patterns are the result of macroecological and evolutionary processes that have predictable effects on geographic ranges despite the multitude of factors that can affect individual ranges. If so, range size frequency distributions of fossil data should show a similar shape and can be used as a way to verify that geographic ranges of fossil taxa are likely reasonable estimates. For example Lyons (2005) used range size frequency distributions of late Pleistocene mammals derived from FAUNMAP to show that the ranges were showing similar patterns to those of modern distributions and therefore were likely to be reasonable estimates. Second, the shapes of these distributions during mass extinctions or climate change events should be evaluated. If these distributions change in predictable ways during critical intervals, then they may prove useful in evaluating the health of modern ecosystems and the impact of the current biodiversity crisis on large spatial scales.

Body size distributions

One of the most common currencies used in macroecological studies is body size. This is in part because it is the most obvious and fundamental characteristic of an organism and in part because many important biological rates and times scale predictably with body size (Calder, 1984; Peters, 1983). Body size is relatively easy to measure in most groups and methods are available for turning body size measures into biovolume so that comparisons may be made across groups with very different underlying morphologies. For example, Payne et al. (2009) used biovolume to compare the maximum size of organisms in all the major phyla since the appearance of life >3.5 billion years ago. Moreover, studies have shown that species body sizes change predictably in response to climate change (Smith et al., 1995, 1998; Hadly et al., 1998). For some species, such as pack rats (i.e., *Neotoma*) body size changes in response to climate change have been documented on the order of decades, centuries and millennial time scales.

Species body size is typically described using a single estimate in most macroecological studies. Ideally, this value should incorporate variation in body size across space, time period of interest, and sex of the species (e.g., Smith et al., 2003). For many groups mean or median body size is used. However, for indeterminate growers, it often makes more sense to use maximum body size (*sensu* Jablonski, 1997). In reality, rare species are often characterized by a single estimate representing a population or individual. Particularly for fossil species, body size *per se* is not available and surrogates must be used. Depending on the group of interest, these include body length, body area, limb measurements, shoulder height, tooth area, etc. Obviously, the best surrogates are those that correlate with body mass. For many mammalian groups, standard regression equations are available to estimate body mass from other morphological measurements such as molar area (Damuth and MacFadden, 1990). Once estimates for body size are available, body size distributions are constructed in the same way as range size distributions. Body sizes are log-transformed and then allocated into categories and the number of species or individuals in each category is tabulated and displayed as a histogram (Fig. 3).

Body size distributions are a result of evolutionary processes acting on species body sizes and ecological processes acting to sort species into communities and over longer time scales into biota. There are similarities in the shapes of body size distributions among warm-blooded vertebrates. In modern systems, body size distributions at the continental scale are unimodal and right-skewed. At smaller spatial scales, body size distributions become progressively flatter until they are nearly uniform at the community level (Brown and Nicoletto, 1991). The unimodal, right-skewed pattern seems to be limited to endotherms. Vertebrate ecotherms and invertebrate groups have unimodal, left-skewed body size distributions (Poulin and Morand, 1997; Roy and Martien, 2001; Boback and Guyer, 2003). Moreover, the pattern of progressive flattening with spatial scale differs on different continents. South American mammal communities show more peaked distributions at the local scale than North American communities (Marquet and Cofre, 1999; Bakker and Kelt, 2000) and African mammal communities are bimodal rather than uniform (Kelt and Meyer, 2009). More specialized groups of mammals such as bats have more peaked distributions at all latitudes (Willig et al., in press).

The methods used to analyze body size distributions depend upon the question being asked. If the question is simply about the shape of the distribution, the moments of the distribution (e.g., mean, median, skew and kurtosis) are often used. In particular, the kurtosis value gives the most information about the

overall shape (Alroy, 2000; Lyons, 2007). Oftentimes, the question is whether the distribution in question is significantly different from a model distribution or whether two distributions are significantly different from one another. When both distributions are known, they can be compared using a Mann-Whitney U test or a Kolmorgorov-Smirnov test (Sokal and Rholf, 1981). Randomization techniques are employed to ask whether the distribution is different from a random draw from a larger species pool. For example, Smith et al. (2004) compared continental body size distributions of modern mammals by comparing differences in the real distributions to differences between distributions generated by drawing species randomly from the global pool of species. For each randomly generated distribution, the number of species drawn was equal to the number of species on the continent. These analyses

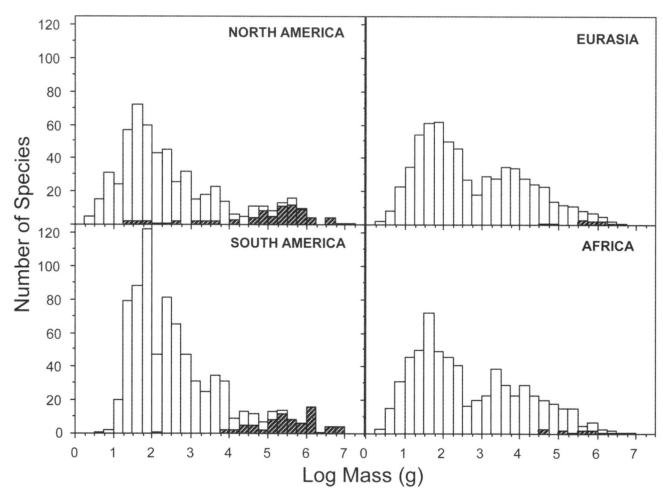

FIGURE 3.—Body size frequency distributions for late-Pleistocene mammals on four continents. Body size distribution of the surviving species are represented by white bars, those for extinct species are darkly shaded. Body sizes are taken from Smith et al. (2003).

found that continental body size distributions of mammals were significantly different from random distributions (Smith et al., 2004). All of these methods may be applied to body size distributions at any scale.

The similarity in the shapes of body size distributions at different scales suggests evolutionary and ecological factors act in predictable ways in different groups to shape these distributions. If so, then evaluating the shapes of body size distributions may be yet another tool to evaluate the effects of climate change on a biota. However, the modern systems are highly altered systems and we cannot assume that the patterns we see today are not a result of the unique anthropogenic forces acting on modern species. Indeed, analyses of mammalian body size distributions in both deep and near time suggest that modern distributions have been fundamentally altered (Alroy, 1998, 1999; Lyons et al., 2004). Shortly after the K/T extinction mammals expanded their range of body sizes into the full range we see today. Approximately 40 ma medi-

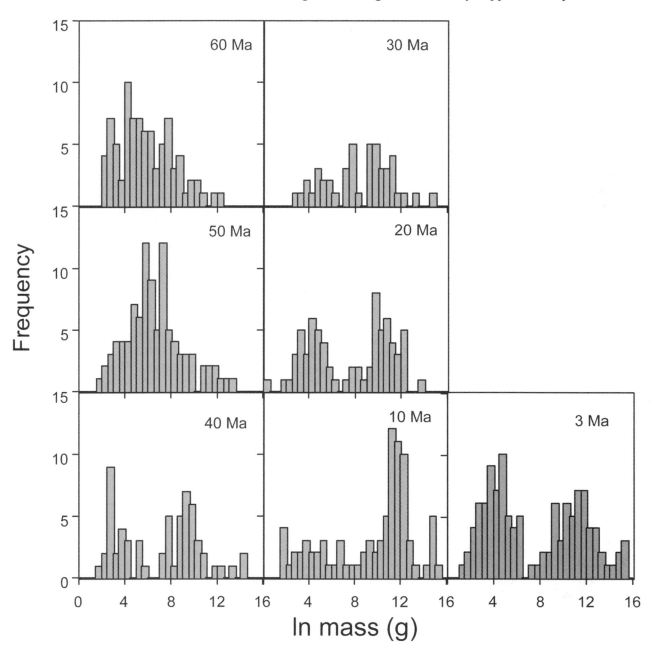

FIGURE 4.—Body size distributions of North American mammals for seven different intervals of 1 ma year each. Species and body sizes are taken from Alroy (1998, 2000).

um-size mammals became rare and a hole opened in the body size space (Alroy, 1998). Macroecological analysis of mammalian body sizes over the Cenozoic suggest that the continental body size distributions were unimodal until 40 ma at which point they became bimodal and remained that way until the extinction of the megafauna at the end of the Pleistocene (Fig. 4). Moreover, the end-Pleistocene extinctions fundamentally altered the shape of the body size distributions on all the continents on which it occurred (Fig. 3; Lyons et al., 2004). These results suggest that the natural state of mammalian body size distributions is not the unimodal, right-skewed distribution we see in the present, but the bimodal, right-skewed distribution we see for the majority of mammalian history.

Conservation paleobiology can provide insight into an additional area of modern conservation ecology, the role of body size in extinction risk. Many modern ecological studies find a strong correlation between extinction risk (as measured by inclusion on the IUCN redlist or by historical extinction) and body size (Gaston and Blackburn, 1995; Cardillo and Bromham, 2001; Jones et al., 2003), however, there is no strong signal of size selectivity in extinction in the fossil record (Jablonski and Raup, 1995; Lockwood, 2004; Jablonski, 2005). Indeed, the only factor that has consistently been associated with extinction in fossil taxa is geographic range size (Jablonski, 1986; Payne and Finnegan, 2007). The obvious difference between modern systems and the majority of history is the impact of humans. Examination of the megafaunal extinction of mammals lends support to the idea that humans are the reason why large-bodied modern species are more vulnerable to extinction, but large-bodied species in the fossil record are not. This extinction event was a highly size selective event on all the continents on which it occurred (Fig. 3, Lyons et al., 2004). Moreover the degree of size selectivity was greater than any other extinction in mammalian history (Elroy, 1999). The common denominator on all the continents that suffered an extinction was the arrival of humans. The extinction in Australia occurred earlier than in North and South America shortly after humans arrived and prior to the climate change associated with the last glaciation (e.g., Lyons et al., 2004). The lack of size selectivity in fossil extinction events combined with the strong size selectivity of the end-Pleistocene event suggests that extinction risk in modern systems is size selective because of the actions of humans and that this is not a natural

characteristic of ecosystems. These insights would not be possible without the contributions of conservation paleobiology.

MACROEVOLUTIONARY METHODS AND THEIR APPLICATIONS TO MODERN CONSERVATION BIOLOGY

Morphological disparity

Morphological disparity is the quantification of morphological diversity (Foote, 1997). Workers have used disparity largely to summarize the results of major radiations, but also to summarize the effects of extinctions (Foote, 1991; Roy, 1996; Wagner, 1997; McGowan, 2004). Roy and Foote (1997) note that it is underutilized as a conservation biology tool. In particular they offer disparity as an alternative to patristic distance or dissimilarity, i.e., summaries of phylogenetic distances among taxa (Faith, 1994, 2002). Simulation (Foote, 1996) and empirical (Wagner, 1997; Cotton, 2001) studies indicate that there is a positive correlation between morphologic disparity and either patristic distance (average numbers of branches separating taxa) or patristic dissimilarity (average sum of inferred changes separating taxa). However, whereas patristic studies require model phylogenies, disparity studies do not. As both studies require the same sort of data, this makes disparity studies a useful way to assess the effects of both past extinctions and *possible* extinctions on existing morphologic diversity.

Disparity requires a measure of how different two taxa are. When using qualitative character data (such as used in phylogenetic studies), a simple measure is:

$$D_{ij} = \frac{\text{Differences}}{\text{Comparable Characters}}$$

where D_{ij} is the difference between taxa i and j, differences is the number of characters that differ and comparable characters represents the number of characters that can be compared. The latter will be fewer than the total number of characters if there are missing data (e.g., incomplete specimens) or if there are incomparable characters (e.g., feather color characters for non-dinosaurian reptiles.) If there are "ordered" characters (e.g., state 2 is considered two units away from state 0 instead of 1), then the numerator is the sum of the differences (i.e., 1 for differing binary or unordered multistate characters, and the absolute difference between states for ordered multistate characters). For continuous (e.g., morphometric) characters, one can simply sum the absolute differences between characters. Overall disparity is simply the average among all pairwise comparisons.

We present an example here using extant lepidosaurs. The tuatara (*Sphenodon*) is well known for being the last of the sphenodontians, a formerly diverse clade of lepidosaurs. All other lepidosaur species are squamates (lizards and snakes). It is intuitively obvious that the extinction of the tuatara would eliminate far more evolutionary history than would the loss of the typical lizard or snake species. Here we will show how disparity could demonstrate this even without a complete phylogeny.

We use characters from two data sets. The squamate data are from Conrad's (2008) analysis of 221 extant and fossil lepidosaur taxa. For the purpose of this study, we limit ourselves to 88 extant taxa and 359 characters that vary among them. Conrad uses *Sphenodon* as an outgroup (i.e., an assumed closest relative of a study group thought to share primitive states with the clade of interest). Following standard convention, Conrad coded *Sphenodon* only for those characters that vary among squamates (see Kitching et al., 1998):

thus, the study omits many characters that one would recognize only if trying to discern relationships of sphenodontians, a formerly diverse lepidosaur clade now represented only by *Sphenodon*. The nature of outgroup coding would seem to reinforce this notion of the tuatara as a "living fossil." However, phylogenetic analyses of sphenodontians show that *Sphenodon* is highly derived and differs from outgroup squamates in 37 of 67 characters in one study (Apesteguia and Novas, 2003) and 33 of 49 characters in another series of studies (Reynoso, 1996, 2000). Lacking a study coding both sphenodontians and squamate lepidosaurs, we augment Conrad's matrix with the sphenodontian characters for which *Sphenodon* and lepidosaurs differ, using Apesteguia and Novas's character data. As the latter study codes squamates as polymorphic if some squamates shared states with some sphenodontians, this should not introduce redundant characters.

We calculated disparity as described above assuming unordered character states. We wish to address what the effect on lepidosaur morphological disparity would be if we lost *Sphenodon*. Foote (1993) contrasted disparity with and without whole clades to assess a similar question. Here, a whole clade is reduced to a single taxon. Therefore, we assess the effect of removing sin-

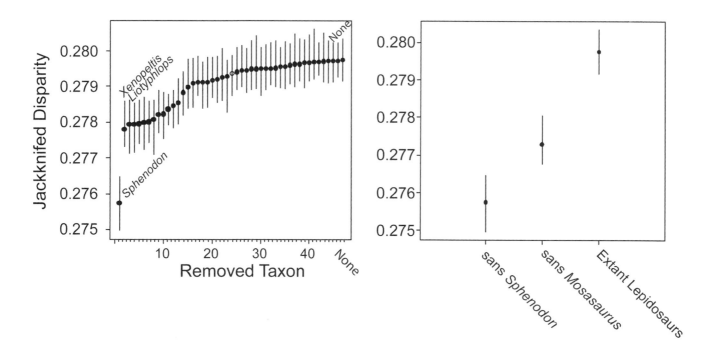

FIGURE 5.—*1*, Lepidosaur disparity following the removal of single taxa, with "none" giving disparity of all extant lepidosaurs. Error bars reflect 500 bootstrap replications; *2*-Lepidosaur disparity without *Sphenodon* vs. lepidosaur disparity without *Mosasaurus*. Data from Conrad (2008) and Apesteguia and Novas (2003).

gle taxa by jackknifing the dataset, i.e., calculating the morphospace for 89 times: once with all 88 taxa, and 88 times with one taxon removed (see Foote, 1991). Phylogenetic autocorrelation (e.g., Felsenstein, 1985) makes formal tests of differences in disparity problematic. However, bootstrapping of pairwise dissimilarities (e.g., Foote, 1992) provides an idea of how easily one could recover similar changes in disparity.

Finally, it is instructive to examine how losing another "living fossil" might affect morphological disparity. Pretend that the Maastrichtian *Mosasaurus hoffmannii* survived until the present. This presents an interesting contrast with *Sphenodon* because although mosasaurs are long extinct, they are nested high in squamate phylogeny being closely related to modern varaniforme lizards (Conrad, 2008). Thus, we have similar patristic distances between either *M. hoffmannii* or *Varanus komodoensis* (the Komodo dragon) and iguanids or gekkos. The highly derived nature of mosasaurs would elevate the patristic dissimilarity between *M. hoffmannii* and other lizards, but the majority of differences between *M. hoffmannii* and other lizards would be synapomorphies shared between mosasaurs and varaniformes.

Unsurprisingly, removing tuataras has a far greater effect on lepidosaur disparity than does removing any squamate taxon (Fig. 5.1). Removing tuataras has a far greater effect than removing the more obviously derived mosasaurs (Fig. 5.2), although losing a relic mosasaur species would reduce lepidosaur disparity greater than would losing any other lepidosaur.

There are three critical points to this analysis. One, we can easily recognize just how much morphological disparity a relict taxon such as tuatara creates only by refering to extinct taxa: without extint sphenodontians, we would have no frame of reference for describing the 30+ unique tuataran features. Second, disparity quickly recognizes that highly derived taxa closely related to other taxa (e.g., *Mosasaurus*) still represent considerable evolutionary novelty: although *Mosasaurus* and *Varanus* are equally distant from most other lizards, we lost much more morphological diversity with the loss of *M. hoffmanni* than we would with *V. komodoensis*. Finally, these disparity analyses point to other potential loses. *Sphenodon* represents an example that many non-scientists can appreciate; however, although it probably would not surprise herpetologists that losing taxa such as *Xenopeltis* and *Liotyphlops* would greatly reduce lepidosaur diversity, these taxa are nowhere as

well known to non-specialists. Disparity presents an easily repeatable and easily communicated summary of just how much diversity these few taxa actually represent.

Relative abundance distributions

Traditionally, paleontologists and conservation biologists both have equated diversity with richness (i.e., numbers of taxa; e.g., Sepkoski, 1978; Gaston and Blackburn, 2003). However, ecologists typically consider richness to be only one aspect of diversity, with the relative abundance being the other component (e.g., Hurlbert, 1971). Diversity is just as important as richness because ecological theory allows predictions about how species should allocate resources under different circumstances. This, in turn, allows predictions about different *relative abundance distributions* (RADs) describing expected proportions of species in a community. Workers have adopted two basic approaches to summarizing diversity. One is to summarize the entire RAD in a single metric, such as evenness. Another is to parameterize the RAD, usually using one of several models with theoretical implications.

Evenness metrics

Workers have devised numerous metrics to describe evenness (e.g., Smith and Wilson, 1996). In general, these metrics summarize how abundances deviate from uniform (i.e., all S taxa having abundance $1/S$). A more detailed description of several different evenness metrics is presented in Appendix 1. Studies contrasting "live" and "dead" assemblages suggest that preservational factors elevate evenness but still should permit us to discern relative evenness in assemblages (Olszewski and Kidwell, 2007). Thus, trends in evenness should be discernible in the fossil record.

Kempton (1979) suggested that there should be a positive correlation between evenness and the "health" of a community. If taxa divide resources fairly equally or when numerous taxa find "new" ways to utilize resources, then evenness should be high. Conversely, if a few taxa monopolize resources or if few taxa can find "new" ways to access resources, then evenness should be low. Most paleobiological studies have used the first proposition to assess diversification over long periods of time. For example, Cenozoic marine assemblages show significantly higher evenness than do Ordovician marine assemblages (Powell and Kowalewski, 2002). Similarly, assemblage evenness increases over the course of major radiations in both the Cambrian and the Ordovician (Peters, 2004). Conversely, evenness decreases in assemblages following rebounds from regional Ordovician extinctions (Layou, 2009).

McElwain et al. (2007) and McElwain et al. (2009) document decreasing evenness in plant communities leading up to the end-Triassic extinction. However, evenness has been under utilized as a metric for summarizing ecosystems leading up to extinctions.

Empirical examples

Here we present two empirical examples of evenness under suspected declining ecological systems. One re-examines Rhaetian (Late Triassic) plant diversity from Greenland leading up to the end-Triassic extinction (McElwain et al., 2007). The other re-examines the response of grassland diversity from 19th-20th century Sweden in response to nitrogen fertilizer (Brenchley and Warrington, 1958).

McElwain et al. (2007) use only E to note that evenness generally decreases in younger beds leading to the end-Triassic extinction, with the younger three beds showing much lower evenness than the older three beds (Fig. 6.1). All other evenness metrics repeat this, although the difference is less marked with J, F and PIE. However, Swedish grasses present an interesting contrast: all metrics show evenness decreasing from the 19th century to the 20th century, but two metrics (E and

D) suggest that the final evenness increases whereas the other three suggest that evenness plummets (Fig. 6.2). The apparent increase is an artifact of the very low sampled richness, S_o=3. S_o is the sole determinant of D_{min} and the primary determinant of E_{min}. Here, the observed value (E_o=0.343 and D_o= 0.337) are only marginally greater than the minimum possible given S_o and N=1245 (E_{min}=0.338 and D_o= 0.335). Rescaling E and D relative to the minimum possible results in evenness plummeting for the final grassland sample.

Evenness limitations

The example above illustrates additional limitations with evenness metrics. Assessing the significance of diversity change is ineloquent. Evenness metrics do not offer exact predictions about abundances. Thus, one usually must resort to bootstrapping or subsampling to contrast individual assemblages. Testing hypotheses of trends in evenness requires copious sampling, as individual assemblages are the sole data point rather than numbers of specimens. Finally, two beds might have the same evenness yet have different model RADs with very different ecological implications. Evenness clearly represents a useful exploratory tool, as it suggests patterns in examples such as the two illustrated above. However, actually examining RADs should provide far greater power for a variety of other tests.

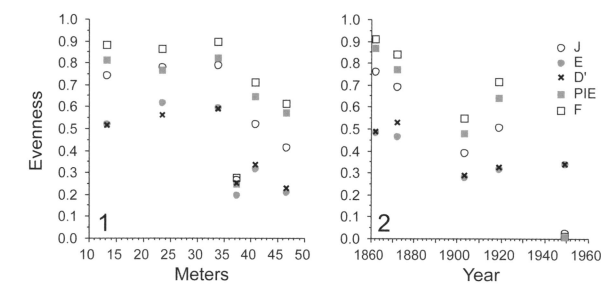

FIGURE 6.—1, Evenness over meters of sediment for Rhaetian plants, with the final bed marking the Triassic-Jurassic boundary; 2, Evenness over years in Swedish grasslands.

RAD models: theoretical expectations

Ecological models for the evolution of communities and the division of resources therein often predict relative abundance distributions (see, e.g., May, 1975; Gray, 1987; Hubbell, 2001). These models divide into two general classes. One class assumes that individuals from different species compete with each other in similar ways for resources. Examples include the geometric distribution (Motomura, 1932), the log-series distribution (Fischer et al., 1943) and the zero sum multinomial (Hubbell, 1997, 2001). These models assume that the primary controls on relative abundance are: 1) the rates at which species enter communities; 2) rates of population growth; and; 3) population size, with the different models making somewhat different assumptions about these parameters. The second class assumes that some or all species create ecological opportunities, either for themselves or for other species. Examples include the Zipf and Zipf-Mandelbrot distributions (Frontier, 1985) and the log-normal distribution (Preston, 1948). Here again, RADs reflect the order in which species colonize communities, but RADs now also reflect other biological factors such as "new" ecospace created by taxa and/or hierarchical partitioning of general niches into more specific niches. Although workers typically describe processes behind RADs in terms of the development of communities, these principles should also let us make predictions about how RADs should change as ecosystems deteriorate. In particular, we might expect changes in RAD models if ecosystems lose complexity, or we might expect changes in parameters if diversity is lost within the same RAD model. A detailed discussion of the mathematics of 4 common RAD models is

provided in Appendix 1.

Unlike evenness metrics, RADs make explicit predictions about numbers of specimens and thus allow conventional tests of hypotheses predicting changes in those numbers over time. Paleontological studies have used RADs to contrast communities over space (Buzas et al., 1977) and over time (Olszewski and Erwin, 2004; Wagner et al., 2006; Harnik, 2009; McEwlain et al., 2009). The Olszewski and Erwin and McElwain et al. studies are particularly germane here as both examine RAD shifts in response to long-term environmental change. Moreover, the likelihood framework that they employ sets up the way in which we think RAD patterns leading to extinctions (or possible extinctions) should be examined.

Testing shifts in RADs over time

Kempton (1979) noted that evenness decreased in Rothamsted grasslands over time, likely as a response to intense nitrogen-rich fertilizers. Despite the drastic decrease in the final year, there is no satisfactory test of whether the decrease is significant. Following convention (e.g., Burnham and Anderson, 2004), we accept the model with the lowest Akaike's Modified Information Criterion score as the best model (Table 1). (More exact tests can be performed using Akaike's weights; e.g., Wagner et al., 2006.) These indicate that the zero sum multinomial is the best model for the first 60+ years. However, the geometric is the best model for the final survey.

Within the first four assemblages, there is a significant decrease in θ between 1872 and 1903 (Fig. 7.1). One interpretation of this is that the realistic pool of

TABLE *1.*—Log-likelihoods and Akaike's Modified Information Criterion (AICc) scores for Rothamsted grasslands over 80 years for geometric, zero sum multinomial, Zipf and lognormal models. S_S gives sampled richness. N gives numbers of specimens, in units of biomass. AICc$= 2*\ln L + 2k + (\dfrac{N}{N-k-1})$ where k=the number of varying parameters. Data from Brenchley and Warrington (1958).

Year	S_S	N	Log-Likelihood				AICc			
			Geo	Zero	Zipf	LogN	Geo	Zero	Zipf	LogN
1862	21	3249	-128.0	-121.4	-241.4	-122.0	258.1	246.8	486.8	247.9
1872	12	2319	-72.2	-68.9	-87.7	-77.2	146.4	141.9	179.5	158.4
1903	8	1977	-76.9	-41.8	-194.2	-124.3	155.9	87.5	392.5	252.6
1919	10	4899	-72.1	-58.3	-330.8	-138.1	146.2	120.6	665.6	280.2
1949	3	1245	-7.9	-15.0	-7.3	-7.7	17.8	18.5	19.4	34.0

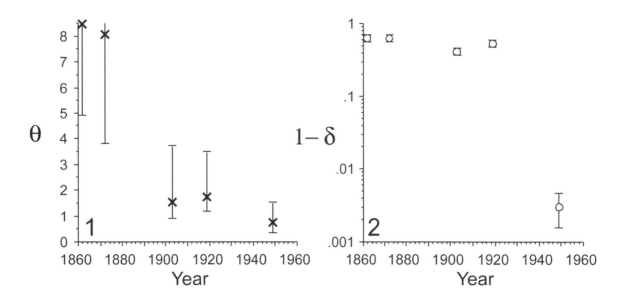

FIGURE 7.—Support bars giving diversity parameters with log-likelihoods within 1.0 of the most likely (given by x or ○). *1*, Zero sum multinomial (the best model for the first four assemblages). Each θ uses the m maximizing the θ's likelihood; *2*, Geometric (the best model for the final assemblage). If support bars do not overlap, then the log-likelihood of 2 differing parameters is significantly greater than the log-likelihood of only 1-parameter.

species that could immigrate into the Rothamsted community decreased. The local increase in nitrogen due to intense fertilizer treatments would not have affected the larger metacommunity, but it could have eliminated some species in Rothamsted and reduced the number of species within the metacommunity that could have immigrated there. Notably, the most likely *m*'s (migration rates) increase drastically, suggesting much more exchange between the local community and the larger metacommunity among those species that could migrate back and forth.

The final assemblage not only shows a shift from the zero sum to the geometric, but also to a geometric that is drastically steeper (and thus less even) than the best geometrics from prior years (Fig. 7.2). There are two issues here. One, the geometric can be thought of as a conceptual special case of the zero sum in which migration no longer is important. This would be consistent with the idea that only a few (possibly only three) species from the larger metacommunity could tolerate the heavily fertilized environment. Second, it indicates that success is very different among those few species that could tolerate the new environment.

CONCLUSIONS

Both macroecology and paleobiology concern

themselves with changes in biodiversity over longer periods of time and larger spatial scales than do traditional ecological studies. Their shared concern in loss in biodiversity (over any time scale) is shared with conservation biology. As such, pertinent methods that both fields use, whether derived independently or in tandem, should be of interest to conservation biologists. We have provided examples here from a variety of temporal and spatial scales, as well as from a variety of data types. There are, of course, several other research avenues (e.g., confidence intervals on temporal ranges or relationships between macroecological parameters) that we have not covered in this paper that clearly apply to macroecology, paleobiology and conservation biology, and which have (to varying degrees) evolved in parallel or in tandem in these different fields. Our primary point might seem obvious, but it clearly has not been properly appreciated: all three fields are dealing in similar issues and all three fields have much to offer one another.

ACKNOWLEDGMENTS

The authors would like to thank G. Dietl and K. Flessa for the invitation to participate in the 2009 Paleobiology Short Course. This manuscript was greatly improved by the comments of two anonymous reviewers. This is contribution number 183 of the Evolution of Terrestrial Ecosystems Program of the Smithsonian Natural History Museum and IMPPS RCN publication number 3 (IMPPS NSF Research Coordination Network grant DEB-0541625).

REFERENCES

ALROY, J. 1998. Cope's rule and the dynamics of body mass evolution in North American fossil mammals. Science, 280(5364):731-734.

ALROY, J. 1999. Putting North America's end-Pleistocene megafaunal extinction in context: Large scale analyses of spatial patterns, extinction rates, and size distributions, p. 105-143. *In* R. D. E. MacPhee (ed.), Extinctions in near Time: Causes, Contexts, and Consequences. Kluwer Academic/Plenum, New York.

ALROY, J. 2000. New methods for quantifying macroevolutionary patterns and processes. Paleobiology, 26(4):707-733.

APESTEGUIA, S., AND F. E. NOVAS. 2003. Large Cretaceous sphenodontian from Patagonia provides insight into lepidosaur evolution in Gondwana. Nature, 425(6958):609-612.

BAKKER, V. J., AND D. A. KELT. 2000. Scale-dependent patterns in body size distributions of neotropical mammals. Ecology, 81(12):3530-3547.

BARNOSKY, A. D., E. A. HADLY, AND C. J. BELL. 2003. Mammalian response to global warming on varied temporal scales. Journal of Mammalogy, 84(2):354-368.

BEALE, C. M., J. J. LENNON, AND A. GIMONA. 2008. Opening the climate envelope reveals no macroscale associations with climate in European birds. Proceedings of the National Academy of Sciences of the United States of America, 105(39):14908-14912.

BEGON, M. J., L. HARPER, AND C. R. TOWNSEND. 1990. Ecology: Individuals, Populations and Communities. Blackwell Scientific Publications, London, 960 p.

BLACKBURN, T. M., K. E. JONES, P. CASSEY, AND N. LOSIN. 2004. The influence of spatial resolution on macroecological patterns of range size variation: A case study using parrots (Aves: Psittaciformes) of the world. Journal of Biogeography, 31(2):285-293.

BOBACK, S. M., AND C. GUYER. 2003. Empirical evidence for an optimal body size in snakes. Evolution, 57(2):345-351.

BRENCHLEY, W. E., AND K. WARRINGTON. 1958. The Park Grass Plots at Rothamsted. Rothamsted Experimental Station, Harpenden, 144 p.

BROWN, J. H. 1995. Macroecology. University of Chicago Press, Chicago, Illinois, 269 p.

BROWN, J. H., AND B. A. MAURER. 1989. Macroecology: The division of food and space among species on continents.

Science, 243(4895):1145-1150.

BROWN, J. H., AND P. F. NICOLETTO. 1991. Spatial scaling of species composition: Body masses of North-American land mammals. American Naturalist, 138(6):1478-1512.

BURNHAM, K. P., AND D. R. ANDERSON. 2004. Multimodel inference: Understanding AIC and BIC in model selection. Sociological Methods and Research, 33(2):261-304.

BUZAS, M. A., AND T. G. GIBSON. 1969. Species diversity: Benthonic foraminifera in western North Atlantic. Science, 163:72-75.

BUZAS, M. A., R. K. SMITH, AND K. A. BEEM. 1977. Ecology and systematics of Foraminifera in two *Thalassia* habitats, Jamaica, West Indies. Smithsonian Contributions to Paleobiology, 31:1-139.

CALDER, W. A., III. 1984. Size, Function, and Life History. Harvard University Press, Cambridge, Massachusetts, 448 p.

CARDILLO, M., AND L. BROMHAM. 2001. Body size and risk of extinction in Australian mammals. Conservation Biology, 15(5):1435-1440.

COLWELL, R. K., AND G. C. HURTT. 1994. Nonbiological gradients in species richness and a spurious Rapoport effect. The American Naturalist, 144:570-595.

CONRAD, J. L. 2008. Phylogeny and systematics of Squamata (Reptilia) based on morphology. Bulletin of the American Museum of Natural History:1-182.

COTTON, T. J. 2001. The phylogeny and systematics of blind Cambrian ptychoparoid trilobites. Palaeontology, 44(1):167-207.

DAMUTH, J., AND B. J. MACFADDEN. 1990. Body Size in Mammalian Paleobiology: Estimation and Biological Implications. Cambridge University Press, New York, 442 p.

DAVIS, A. J., J. H. LAWTON, B. SHORROCKS, AND L. S. JENKINSON. 1998. Individualistic species responses invalidate simple physiological models of community dynamics under global environmental change. Journal of Animal Ecology, 67:600-612.

DAVIS, M. B., AND R. G. SHAW. 2001. Range shifts and adaptive responses to Quaternary climate change. Science, 292(5517):673-679.

DEWDNEY, A. K. 1998. A general theory of the sampling process with applications to the "Veil Line." Theoretical Population Biology, 54(2):294-302.

EDWARDS, A. W. F. 1992. Likelihood - Expanded edition. Johns Hopkins University Press, Baltimore, 275 p.

FAITH, D. P. 1994. Phylogenetic pattern and the quantification of organismal biodiversity. Philosophical Transactions of the Royal Society of London, B, 345:45-58.

FAITH, D. P. 2002. Quantifying biodiversity: A phylogenetic perspective. Conservation Biology, 16:248-252.

FELSENSTEIN, J. 1985. Phylogenies and the comparative method. The American Naturalist, 125(1):1-15.

FISHER, R. A., A. S. CORBET, AND C. B. WILLIAMS. 1943. The relation between the number of species and the number of individuals in a random sample of an animal population. Journal of Animal Ecology, 12(1):42-48.

FOODY, G. M. 2008. Refining predictions of climate change impacts on plant species distribution through the use of local statistics. Ecological Informatics, 3(3):228-236.

FOOTE, M. 1991. Morphologic patterns of diversification: Examples from trilobites. Palaeontology, 34(2):461-485.

FOOTE, M. 1992. Early morphologic diversity in blastozoan echinoderms. Fifth North American Paleontological Convention, Field Museum of Natural History, Special Publication No. l6:102.

FOOTE, M. 1993. Contributions of individual taxa to overall morphological disparity. Paleobiology, 19(4):403-419.

FOOTE, M. 1996. Models of morphologic diversification, p. 62 - 86. In D. Jablonski, D. H. Erwin, and J. H. Lipps (eds.), Evolutionary Paleobiology: Essays in honor of James W. Valentine. University of Chicago Press, Chicago.

FOOTE, M. 1997. The evolution of morphologic diversity. Annual Review of Ecology and Evolution, 28:129-152.

FOOTE, M. 2007. Symmetric waxing and waning of marine invertebrate genera. Paleobiology, 33(4):517-529.

FOOTE, M., J. S. CRAMPTON, A. G. BEU, AND R. A. COOPER. 2008. On the bidirectional relationship between geographic range and taxonomic duration. Paleobiology, 34(4):421-433.

FOOTE, M., J. S. CRAMPTON, A. G. BEU, B. A. MARSHALL, R. A. COOPER, P. A. MAXWELL, AND I. MATCHAM. 2007. Rise and fall of species occupancy in Cenozoic fossil mollusks. Science, 318(5853):1131-1134.

FRONTIER, S. 1985. Diversity and structure in aquatic ecosystems, p. 253-312. In M. Barnes (ed.), Oceanography and Marine Biology: An Annual Review. Aberdeen University Press, Aberdeen.

GASTON, K. J. 2003. The Structure and Dynamics of Geographic Ranges. Oxford University Press, Oxford, 280 p.

GASTON, K. J., AND T. M. BLACKBURN. 1995. Birds, body-size and the threat of extinction. Philosophical Transactions of the Royal Society of London, B, 347(1320):205-212.

GASTON, K. J., AND T. M. BLACKBURN. 1996. Global scale macroecology: Interactions between population size, geographic range size and body size in the Anseriformes. Journal of Animal Ecology, 65(6):701-714.

GASTON, K. J., AND T. M. BLACKBURN. 2000. Pattern and Process in Macroecology. Blackwell Science, Oxford, 377p.

GASTON, K. J., AND T. M. BLACKBURN. 2003. Macroecology and Conservation Biology, p. 345-367. In T. M. Blackburn and K. J. Gaston (eds.), Macroecology: Concepts and Consequences. Blackwell Science, Oxford.

GOSSELIN, F. 2006. An assessment of the dependence of evenness indices on species richness. Journal of Theoretical Biology, 242(3):591-597.

GOTELLI, N. J., AND G. R. GRAVES. 1996. Null Models in Ecology. Smithsonian Press, Washington, D.C., 368 p.

GRAHAM, R. W. 1986. Response of mammalian communities to environmental changes during the late Quaternary, p. 300-313. In J. Diamond and T. J. Case (eds.), Community Ecology. Harper and Row, New York.

GRAHAM, R. W., E. L. LUNDELIUS, M. A. GRAHAM, E. K. SCHROEDER, R. S. TOOMEY, E. ANDERSON, A. D. BARNOSKY, J. A. BURNS, C. S. CHURCHER, D. K. GRAYSON, R. D. GUTHRIE, C. R. HARINGTON, G. T. JEFFERSON, L. D. MARTIN, H. G. McDONALD, R. E. MORLAN, H. A. SEMKEN, S. D. WEBB, L. WERDELIN, AND M. C. WILSON. 1996. Spatial response of mammals to late Quaternary environmental fluctuations. Science, 272(5268):1601-1606.

GRAHAM, R. W., AND J. I. MEAD. 1987. Environmental fluctuations and evolution of mammalian faunas during the last deglaciation in North America, p. 372-402. In W. F. Ruddiman and H. E. Wright Jr. (eds.), North American and Adjacent Oceans during the Last Deglaciation. The Geological Society of America, Boulder, Colorado.

GRAY, J. S. 1987. Species-abundance patterns, p. 53 - 67. In J. H. R. Gee and P. S. Gillier (eds.), Organization of Communities Past and Present. Blackwell, Oxford.

HADLY, E. A., M. H. KOHN, J. A. LEONARD, AND R. K. WAYNE. 1998. A genetic record of population isolation in pocket gophers during Holocene climate change. Proceedings of the National Academy of Sciences of the United States of America, 95(12):6893-6896.

HALL, E. R. 1981. The Mammals of North America. Wiley, New York.

HARCOURT, A. H. 2006. Rarity in the tropics: Biogeography and macroecology of the primates. Journal of Biogeography, 33(12):2077-2087.

HARNIK, P. G. 2009. Unveiling rare diversity by integrating museum, literature and field data. Paleobiology, 35(2):190 - 208.

HAYEK, L.-A. C., AND M. A. BUZAS. 1997. Surveying Natural Populations. Columbia University Press, New York, 563 p.

HUBBELL, S. P. 1997. A unified theory of biogeography and relative species abundance and its application to tropical rain forests and coral reefs. Coral Reefs, 16 Supplemental:S9-S21.

HUBBELL, S. P. 2001. The Unified Neutral Theory of Biodiversity and Biogeography. Princeton University Press, Princeton, New Jersey, 375 p.

HURLBERT, S. H. 1971. The nonconcept of species diversity: A critique and alternative parameters. Ecology, 52:577-586.

JABLONSKI, D. 1986. Background and mass extinctions: The alteration of macroevolutionary regimes. Science, 231:129-133.

JABLONSKI, D. 1997. Body-size evolution in Cretaceous molluscs and the status of Cope's Rule. Nature, 385:250-252.

JABLONSKI, D. 2005. Mass extinctions and macroevolution. Paleobiology, 31(2):192-210.

JABLONSKI, D., AND D. M. RAUP. 1995. Selectivity of end-Cretaceous marine bivalve extinctions. Science, 268:389-391.

JACKSON, S. T., AND J. T. OVERPECK. 2000. Responses of plant populations and communities to environmental changes of the late Quaternary. Paleobiology, 26(4):194-220.

JACKSON, S. T., J. T. OVERPECK, T. WEBB, S. E. KEATTCH, AND K. H. ANDERSON. 1997. Mapped plant-macrofossil and pollen records of late Quaternary vegetation change in eastern North America. Quaternary Science Reviews, 16(1):1-70.

JACKSON, S. T., AND J. W. WILLIAMS. 2004. Modern analogs in Quaternary paleoecology: Here today, gone yesterday, gone tomorrow? Annual Review of Earth and Planetary Sciences, 32:495-537.

JESCHKE, J. M., AND D. L. STRAYER. 2008. Usefulness of bioclimatic models for studying climate change and invasive species, p. 1-24, In R. S. Ostfeld, and N. H. Schlesinger, Year in Ecology and Conservation Biology 2008. Wiley-Blackwell Publishing, Oxford.

JONES, K. E., A. PURVIS, AND J. L. GITTLEMAN. 2003. Biological correlates of extinction risk in bats. American Naturalist, 161(4):601-614.

KELT, D. A., AND M. D. MEYER. 2009. Body size frequency distributions in African mammals are bimodal at all spatial scales. Global Ecology and Biogeography, 18(1):19-29.

KEMPTON, R. A. 1979. Structure of species abundance and measurement of diversity. Biometrics, 35:307–322.

KITCHING, I. J., P. L. FOREY, C. HUMPHRIES, AND D. M. WILLIAMS. 1998. Cladistics. 2nd. Edition. The Theory and Practice of Parsimony Analysis. Oxford University Press, Oxford, 228 p.

KOCH, C. F. 1980. Bivalve species duration, areal extent and population size in a Cretaceous sea. Paleobiology, 6(2):184-192.

KOLEFF, P., AND K. J. GASTON. 2001. Latitudinal gradients in diversity: Real patterns and random models. Ecography, 24(3):341-351.

KRUG, A. Z., D. JABLONSKI, AND J. W. VALENTINE. 2008. Species–genus ratios reflect a global history of diversification and range expansion in marine bivalves. Proceedings of the Royal Society, B, 275(1639):1117-1123.

LAYOU, K. M. 2009. Ecological restructuring after extinction: the Late Ordovician (Mohawkian) of the Eastern United States. Palaios, 24(2):118-130.

LOCKWOOD, R. 2004. The K/T event and infaunality: Morphological and ecological patterns of extinction and recovery in veneroid bivalves. Paleobiology, 30(4):507-521.

LOMOLINO, M. V., B. R. RIDDLE, AND J. H. BROWN. 2006. Biogeography. Sinauer Associates, Inc., Sunderland, Massachusetts, 846 p.

LYONS, S. K. 1994. Areography of New World bats and marsupials. Unpublished Masters Thesis, Texas Tech University, Lubbock.

LYONS, S. K. 2003. A quantitative assessment of the range shifts of Pleistocene mammals. Journal of Mammalogy, 84(2):385-402.

LYONS, S. K. 2005. A quantitative model for assessing community dynamics of pleistocene mammals. American Naturalist, 165(6):E168-E185.

LYONS, S. K. 2007. The relationship between environmental variables and mammalian body size distributions over space and time. Annual Meeting of the Ecological Society of America. Ecological Society of America, San Jose.

LYONS, S. K., AND F. A. SMITH. 2006. Assessing biases in the mammalian fossil record using late Pleistocene mammals from North America. Abstracts with Programs, Geological Society of America, 38.

LYONS, S. K., F. A. SMITH, AND J. H. BROWN. 2004. Of mice, mastodons and men: Human-mediated extinctions on four continents. Evolutionary Ecology Research, 6(3):339-358.

LYONS, S. K., AND M. R. WILLIG. 1997. Latitudinal patterns of range size: Methodological concerns and empirical evaluations for New World bats and marsupials. Oikos, 79(3):568-580.

MADIN, J. S., AND S. K. LYONS. 2005. Incomplete sampling of geographic ranges weakens or reverses the positive relationship between an animal species' geographic range size and its body size. Evolutionary Ecology Research, 7:607-617.

MARQUET, P. A., AND H. COFRE. 1999. Large temporal and spatial scales in the structure of mammalian assemblages in South America: A macroecological approach. Oikos, 85(2):299-309.

MARTINEZ-MEYER, E., A. T. PETERSON, AND W. W. HARGROVE. 2004. Ecological niches as stable distributional constraints on mammal species, with implications for Pleistocene extinctions and climate change projections for biodiversity. Global Ecology and Biogeography, 13(4):305-314.

MAURER, B. A. 1999. Untangling Ecological Complexity: The Macroscopic Perspective. University of Chicago Press, Chicago, Illinois, 262 p.

MAY, R. M. 1975. Patterns of species abundance and diversity, p. 87-120. In M. L. Cody and J. M. Diamond (eds.), Ecology and Evolution of Communities. The Belknap Press of Harvard University Press, Cambridge.

McELWAIN, J. C., M. E. POPA, S. P. HESSELBO, M. HAWORTH, AND F. SURLYK. 2007. Macroecological responses of terrestrial vegetation to climatic and atmospheric change across the Triassic/Jurassic boundary in East Greenland. Paleobiology, 33(4):547-573.

McGILL, B. J. 2003. Does Mother Nature really prefer rare species or are log-left-skewed SADs a sampling artefact? Ecology Letters, 6(8):766-773.

McGowan, A. J. 2004. The effect of the Permo-Triassic bottleneck on Triassic ammonoid morphological evolution. Paleobiology, 30(3):369-395.

Motomura, I. 1932. A statistical treatment of associations. Zoological Magazine, Tokyo, 44:379-383.

Olszewski, T. D. 2004. A unified mathematical framework for the measurement of richness and evenness within and among multiple communities. Oikos, 104(2):377-387.

Olszewski, T. D., and D. H. Erwin. 2004. Dynamic response of Permian brachiopod communities to long-term environmental change. Nature, 428(6984):738-741.

Olszewski, T. D., and S. M. Kidwell. 2007. The preservational fidelity of evenness in molluscan death assemblages. Paleobiology, 33(1):1-23.

Overpeck, J. T., R. S. Webb, and T. Webb. 1992. Mapping Eastern North-American vegetation change of the past 18 ka: No-analogs and the future. Geology, 20(12):1071-1074.

Overpeck, J. T., T. Webb, and I. C. Prentice. 1985. Quantitative Interpretation of Fossil Pollen Spectra: Dissimilarity Coefficients and the Method of Modern Analogs. Quaternary Research, 23(1):87-108.

Parmesan, C., and G. Yohe. 2003. A globally coherent fingerprint of climate change impacts across natural systems. Nature, 421(6918):37-42.

Patterson, B. D., G. Geballos, W. Sechrest, M. Toghelli, G. T. Brooks, L. Luna, P. Ortega, I. Salazar, and B. E. Young. 2003. Digital Distribution Maps of the Mammals of the Western Hemisphere. Version 1.0. Nature Serve, Arlington, Virginia.

Payne, J. L., A. G. Boyer, J. H. Brown, S. Finnegan, M. Kowalewski, R. A. Krause, S. K. Lyons, C. R. McClain, D. W. McShea, P. M. Novack-Gottshall, F. A. Smith, J. A. Stempien, and S. C. Wang. 2009. Two-phase increase in the maximum size of life over 3.5 billion years reflects biological innovation and environmental opportunity. Proceedings of the National Academy of Sciences of the United States of America, 106(1):24-27.

Payne, J. L., and S. Finnegan. 2007. The effect of geographic range on extinction risk during background and mass extinction. Proceedings of the National Academy of Sciences of the United States of America, 104(25):10506-10511.

Peters, R. H. 1983. The Ecological Implications of Body Size. Cambridge University Press, Cambridge.

Peters, S. E. 2004. Evenness of Cambrian-Ordovician benthic marine communities in North America. Paleobiology, 30(3):325-346.

Peterson, A. T. 2001. Predicting species' geographic distributions based on ecological niche modeling. Condor, 103(3):599-605.

Peterson, A. T., E. Martinez-Meyer, and C. Gonzalez-Salazar. 2004a. Reconstructing the Pleistocene geography of the *Aphelocoma* jays (Corvidae). Diversity and Distributions, 10(4):237-246.

Peterson, A. T., E. Martinez-Meyer, C. Gonzalez-Salazar, and P. W. Hall. 2004b. Modeled climate change effects on distributions of Canadian butterfly species. Canadian Journal of Zoology-Revue Canadienne de Zoologie, 82(6):851-858.

Peterson, A. T., M. A. Ortega-Huerta, J. Bartley, V. Sanchez-Cordero, J. Soberon, R. H. Buddemeier, and D. R. B. Stockwell. 2002. Future projections for Mexican faunas under global climate change scenarios. Nature, 416(6881):626-629.

Peterson, A. T., V. Sanchez-Cordero, J. Soberon, J. Bartley, R. W. Buddemeier, and A. G. Navarro-Siguenza. 2001. Effects of global climate change on geographic distributions of Mexican Cracidae. Ecological Modelling, 144(1):21-30.

Pielou, E. C. 1966. Species diversity and pattern diversity in the study of ecological succession. Journal of Theoretical Biology, 10:370-383.

Poulin, R., and S. Morand. 1997. Parasite body size distributions: Interpreting patterns of skewness. International Journal for Parasitology, 27(8):959-964.

Powell, M. G., and M. Kowalewski. 2002. Increase in evenness and sampled alpha diversity through the Phanerozoic: Comparison of early Paleozoic and Cenozoic marine fossil assemblages. Geology, 30(4):331-334.

Preston, W. H. 1948. The commonness and rarity of species. Ecology, 29(3):254-283.

Reynoso, V.-H. 1996. A Middle Jurassic Sphenodon-Like Sphenodontian (Diapsida: Lepidosauria) from Huizachal Canyon, Tamaulipas, Mexico. Journal of Vertebrate Paleontology, 16(2):210-221.

Reynoso, V.-H. 2000. An unusual aquatic sphondontian (Reptilia: Diapsida) from the Tlayua Formation (Albian), Central Mexico. Journal of Paleontology, 74(1):133-148.

Roy, K. 1996. The roles of mass extinction and biotic interaction in large-scale replacements: A reexamination using the fossil record of stromboidean gastropods. Paleobiology, 22(3):436-452.

Roy, K., and M. Foote. 1997. Morphological approaches to measuring biodiversity. Trends in Ecology and Evolution, 12(7):277-281.

Roy, K., D. Jablonski, and J. W. Valentine. 1994. Eastern Pacific Molluscan provinces and latitudinal diversity gradient - no evidence for "Rapoport's rule". Proceedings of the National Academy of Sciences of the United States of America, 91(19):8871-8874.

Roy, K., D. Jablonski, and J. W. Valentine. 1995. Thermally anomalous assemblages revisited: Patterns in the extraprovincial latitudinal range shifts of Pleistocene marine mollusks. Geology, 23(12):1071-1074.

Roy, K., and K. K. Martien. 2001. Latitudinal distribution of body size in north-eastern Pacific marine bivalves. Journal of Biogeography, 28(4):485-493.

RUGGIERO, A., AND J. H. LAWTON. 1998. Are there latitudinal and altidudinal Rapoport effects in the geographic ranges of Andean passerine birds? Biological Journal of the Linnean Society, 63(2):283-304.

RUGGIERO, A., AND V. WERENKRAUT. 2007. One-dimensional analyses of Rapoport's rule reviewed through meta-analysis. Global Ecology And Biogeography, 16(4):401-414.

SCHWEIGER, O., J. SETTELE, O. KUDRNA, S. KLOTZ, AND I. KUHN. 2008. Climate Change Can Cause Spatial Mismatch of Trophically Interacting Species. Ecology, 89(12):3472-3479.

SEPKOSKI, J. J., JR. 1978. A kinetic model of Phanerozoic taxonomic diversity: I. Analysis of marine orders. Paleobiology, 4(2):223-251.

SHANNON, C. E. 1948. A mathematical theory of communication. Bell Systems Technology Journal, 27:379-423.

SMITH, B., AND J. B. WILSON. 1996. A consumer's guide to evenness indices. Oikos, 76(1):70-82.

SMITH, F. A., J. L. BETANCOURT, AND J. H. BROWN. 1995. Evolution of body-size in the woodrat over the past 25,000 years of climate-change. Science, 270(5244):2012-2014.

SMITH, F. A., J. H. BROWN, J. P. HASKELL, S. K. LYONS, J. ALROY, E. L. CHARNOV, T. DAYAN, B. J. ENQUIST, S. K. M. ERNEST, E. A. HADLY, K. E. JONES, D. M. KAUFMAN, P. A. MARQUET, B. A. MAURER, K. J. NIKLAS, W. P. PORTER, B. TIFFNEY, AND M. R. WILLIG. 2004. Similarity of mammalian body size across the taxonomic hierarchy and across space and time. American Naturalist, 163(5):672-691.

SMITH, F. A., H. BROWNING, AND U. L. SHEPHERD. 1998. The influence of climate change on the body mass of woodrats *Neotoma* in an arid region of New Mexico, USA. Ecography, 21(2):140 - 148.

SMITH, F. A., S. K. LYONS, S. K. M. ERNEST, AND J. H. BROWN. 2008. Macroecology: More than just the division of food and space among species on continents. Progress in Physical Geography, 32(2):115-138.

SMITH, F. A., S. K. LYONS, S. K. M. ERNEST, K. E. JONES, D. M. KAUFMAN, T. DAYAN, P. A. MARQUET, J. H. BROWN, AND J. P. HASKELL. 2003. Body mass of late quaternary mammals. Ecology, 84(12):3403-3403.

SOKAL, R. R., AND F. J. RHOLF. 1981. Biometry: The Principles and Practice of Statistics in Biological Research. W. H. Freeman and Company, New York, NY.

SUGIHARA, G. 1980. Minimal community structure: An explanation of species abundance patterns. The American Naturalist, 116:770-787.

THOMAS, C. D., A. CAMERON, R. E. GREEN, M. BAKKENES, L. J. BEAUMONT, Y. C. COLLINGHAM, B. F. N. ERASMUS, M. F. DE SIQUEIRA, A. GRAINGER, L. HANNAH, L. HUGHES, B. HUNTLEY, A. S. VAN JAARSVELD, G. F. MIDGLEY, L. MILES, M. A. ORTEGA-HUERTA, A. T. PETERSON, O. L. PHILLIPS, AND S. E. WILLIAMS. 2004. Extinction risk from climate change. Nature, 427(6970):145-148.

VALENTINE, J. W., D. JABLONSKI, S. KIDWELL, AND K. ROY. 2006. Assessing the fidelity of the fossil record by using marine bivalves. Proceedings of the National Academy of Sciences of the United States of America, 103(17):6599-6604.

VOLKOV, I., J. R. BANAVAR, S. P. HUBBELL, AND A. MARITAN. 2003. Neutral theory and relative species abundance in ecology. Nature, 424:1035-1037.

WAGNER, P. J. 1997. Patterns of morphologic diversification among the Rostroconchia. Paleobiology, 23(1):115-150.

WAGNER, P. J., M. A. KOSNIK, AND S. LIDGARD. 2006. Abundance distributions imply elevated complexity of post-Paleozoic marine ecosystems. Science, 314:1289-1292.

WALKER, K. V., M. B. DAVIS, AND S. SUGITA. 2002. Climate change and shifts in potential tree species range limits in the Great Lakes region. Journal of Great Lakes Research, 28(4):555-567.

WEBB, S. D., AND A. D. BARNOSKY. 1989. Faunal dynamics of Pleistocene mammals. Annual Review of Earth and Planetary Sciences, 17:413-438.

WILLIAMS, J. W., D. M. POST, L. C. CWYNAR, A. F. LOTTER, AND A. J. LEVESQUE. 2002. Rapid and widespread vegetation responses to past climate change in the North Atlantic region. Geology, 30(11):971-974.

WILLIAMS, J. W., B. N. SHUMAN, AND T. WEBB. 2001. Dissimilarity analyses of late-uaternary vegetation and climate in eastern North America. Ecology, 82(12):3346-3362.

WILLIAMS, J. W., B. N. SHUMAN, T. WEBB, P. J. BARTLEIN, AND P. L. LEDUC. 2004. Late-Quaternary vegetation dynamics in North America: Scaling from taxa to biomes. Ecological Monographs, 74(2):309-334.

WILLIG, M. R., AND M. R. GANNON. 1997. Gradients of species density and turnover in marsupials: A hemispheric perspective. Journal of Mammalogy, 78(3):756-765.

WILLIG, M. R., AND S. K. LYONS. 1998. An analytical model of latitudinal gradients of species richness with an empirical test for marsupials and bats in the New World. Oikos, 81(1):93-98.

WILLIG, M. R., S. K. LYONS, AND R. D. STEVENS. In press. Spatial methods for the macroecological study of bats. *In* T. H. Kunz and S. Parsons (eds.), Ecological and Behavioral Methods for the Study of Bats. Johns Hopkins University Press, Baltimore, Maryland.

WILLIG, M. R., B. D. PATTERSON, AND R. D. STEVENS. 2003. Patterns of range size, richness, and body sizes in Chiroptera., p. 580-621. *In* T. H. Kunz and M. B. Fenton (eds.), Bat Ecology. University of Chicago Press, Chicago, Illinois.

WILLIG, M. R., AND E. A. SANDLIN. 1991. Gradients of species density and turnover in New World bats: a comparison of quadrat and band methodologies, p. 81-96. *In* M. A. Mares and D. J. Schmidly (eds.), Latin American Mammals. Their Conservation, Ecology and Evolution. University of Oklahoma Press, Norman, Oklahoma.

WILLIG, M. R., AND K. W. SELCER. 1989. Bat species density gradients in the New World: A statistical assessment. Journal of Biogeography, 16(2):189-195.

WILLIS, J. C. 1922. Age and Area: A Study in Geographical Distribution and Origin of Species. Cambridge University Press, Cambridge, 259 p.

WILLIS, K. J., M. B. ARAUJO, K. D. BENNET, B. FIGUEROA-RANGEL, C. A. FROYD, AND N. MYERS. 2007a. How can a knowledge of the past help to conserve the future? Biodiversity conservation and the relevance of long-term ecological studies. Philosophical Transactions of the Royal Society of London, B, 362(1478):175-186.

WILLIS, K. J., A. KLECZKOWSKI, M. N. A., AND R. J. WHITTAKER. 2007b. Testing the impact of climate variability on European plant diversity: 320, 000 years of water-energy dynamics and its long-term influence on plant taxonomic richness. Ecology Letters, 10(8):673-679.

WILLIS, K. J., AND H. J. B. BIRKS. 2006. What is natural? The need for long-term perspective in biodiveristy conservation. Science, 314(5803):1261-1265.

APPENDIX 1

Evenness metrics

Although the basic concept of evenness, i.e., typical deviation from uniform abundance, is simple, many evenness metrics exist. When examining the logarithms of distributions, the Shannon-Wiener index, H, (Shannon, 1948) is intuitively informative. H is:

$$H = \sum_{i}^{S_o} f_i \ln(f_i)$$

where S_o is the observed richness, and f_i is the observed relative abundance (frequency) of taxon i. The maximum value for H occurs when f_i is uniform for all taxa (i.e., $f_i = \frac{1}{S_o}$) and yields $H = -\ln(S_o)$.

Two evenness metrics use this relationship: J (Pielou, 1966) and E (Buzas and Gibson, 1969). These are given as:

$$J = \frac{-H}{\ln(S_o)}$$

and:

$$E = \frac{e^{-H}}{S_o}$$

In both cases, the numerator will equal the denomina-

tor when

$$f_1 = f_2 = \ldots = f_{So} = \frac{1}{S_o}$$

and thus the evenness will equal 1.0 when sampled abundances are uniform.

Other ways of estimating evenness examines sums of squared frequencies. Hurlbert's Probability of Intraspecific Encounter (PIE) is given as:

$$PIE = 1 - \sum_{i}^{S_o} f_i^2$$

Here, the second term is simply the probability of sampling the same taxon (an interspecific encounter) twice in a row. Peters (2004) sum of squared evenness relies on the sum squared differences between observed and expected ($f = \frac{1}{S_o}$) abundances:

$$F = 1 - \frac{S_o \sum_{i}^{S_o} (f_i - \frac{1}{S_o})^2}{S_o - 1}$$

In both cases, the minimum sum of squares occurs when $f_1 = f_2 = \ldots = f_{So} = \frac{1}{S_o}$.

However, whereas F (like J and E above) is 1.0 at uniform abundance, the maximum possible PIE increases as richness (S) increases, and never can reach 1.0. PIE has a direct relationship to rarefaction curves, as the expected subsampled richness at 2 specimens is $1+PIE$ (Olszewski, 2004).

Finally, Powell and Kowalewski (2002) use the Kolmogorov-Smirnov D statistic, which is simply one minus the sum of the differences between f_i and $\frac{1}{S_o}$ for all $f_i > \frac{1}{S_o}$. Again, this is 1.0 when frequencies are uniform.

We note above that the maximum value of PIE depends on the sampled richness. For all evenness metrics, the minimum possible value depends on both sampled richness and sample size (e.g., Gosselin, 2006). For N specimens and S_o observed taxa, this would be when $f_1 = N+1-S_o$ and $f_2 = f_3 = \ldots = f_{So} = 1$. As we shall show below, this can be very important when examining trends in evenness when S_o changes.

Sample size is also important for the accuracy of evenness metrics. Small sample sizes yield accurate estimates of PIE and F (Gotelli and Graves, 1996; Peters, 2004). This is because the taxa that make the greatest contributions to PIE and F are the most common taxa. Rare taxa typically alter the sum of squared (or squared difference from expectations) only slightly whereas they alter H appreciably. Sampling standardization overcomes this problem (Layou, 2009), but then leaves all evenness metrics reflecting the most common taxa.

RAD models: theoretical expectations

Workers have proposed many more RAD models than we can review here. Therefore, we will focus on four that workers commonly use (Fig. A1). The simplest is the geometric distribution (Motomura, 1932), which relies on one parameter,

$$\delta = \frac{f_{i+1}}{f_i}$$

which reflects the rate of species entry relative to the rate of population expansion. The frequency of species rank i is approximately:

$$f_i \simeq (1-\delta)\delta^{(i-1)}$$

Although f_1 only approaches $(1-\delta)$ asymptotically, in practice only a few taxa are necessary for f_1 to approach $(1-\delta)$ to the third decimal place, especially as δ decreases. This allows us to approximate RADs using a single parameter.

The geometric is very similar to the more commonly used log series (Fisher et al., 1943), which differs in assuming that species arrival is stochastic rather than regular. However, the log series requires an iterative solution that does not lend itself so easily to calculating f_i's as the geometric does (e.g., Hayek and Buzas, 1997).

The geometric makes simplifying assumptions about both true population size and migration/origination rates. The zero-sum multinomial (Hubbell, 1997, 2001) explicitly accommodates these. The calculation is too complex to repeat here (see Volkov et al., 2003: box 1), but relies on three parameters: the local community size (N_T), the probability of immigration (m) from the larger "metacommunity" (i.e., collection of adjacent communities), and a compound parameter (θ) that is the product of the origination rate

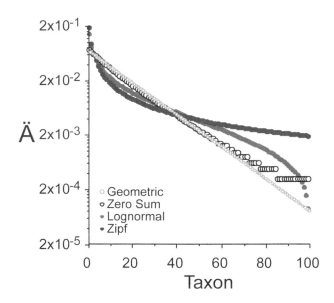

FIGURE *A1.*—Model relative abundance distributions reviewed here. Each has true richness $S_T=100$ and true $J = 0.80$.

and "metacommunity" population size (usually given as J_M). Volkov et al.'s (2003) formula allows one to estimate the exact number of species with n specimens, and thus an exact RAD. As θ increases, the probability of interspecific encounter increases, meaning a richer and more even community.

Other RAD models assume that species utilize resources differently, either by entering fundamentally different niches or by making new resources available either for themselves or for other species. The Zip-Mandelbrot (e.g., Frontier, 1985) relies on three parameters:

$$f_i = \frac{(i + \beta)^{-\gamma}}{\sum_{j=1}^{S_T}(j + \beta)^{-\gamma}}$$

Like δ for the geometric, γ reflects ecosystem stability and the regularity of immigration. However, the assumption is that new species expand the total ecospace slightly, resulting in the difference in f_i decreasing. β reflects the diversity of basic niches. The Zipf distribution is the special 2-parameter case where $\beta=0$ and yields a RAD that is linear on a log-log plot. Although f_i is proportional to the numerator regardless of true richness (S_T), it always is 1.0 for the first species. Moreover, the denominator (i.e., the sum of relative proportions) changes markedly with the addition

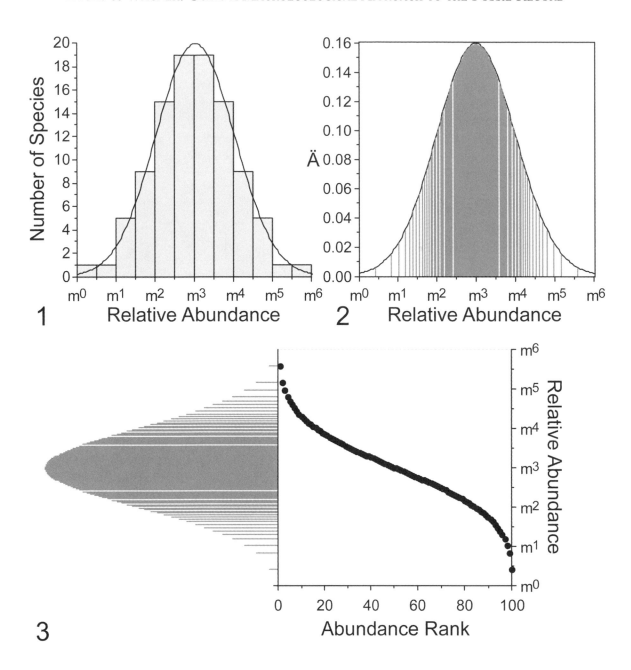

FIGURE *A2*.—Deriving lognormal RAD. *1*, Typical depiction of a lognormal distribution for 100 species. Species in each octave are m times more abundant than are species in the prior octave; *2*,The same normal curve divided by 100 partitions into 101 units of equal area, *3*, The position of each partition I on the X-axis gives the relative abundance of the ith species and becomes the Y-axis of a standard log-linear RAD plot. Modified from Wagner et al. (2006).

of new taxa even at high values of S_T. Thus, we cannot predict abundances given β and γ without specifying S_T.

RADs frequently fit lognormal distributions (Preston, 1948). This might be a simple artifact of mixing exponential distributions (May, 1975); however, it also can reflect hierarchical division of niches by incoming taxa (Sugihara, 1980). Lognormal RADs depend

on two parameters (Figure A2.1): true richness (S_T) and the magnitude difference in abundance between species separated by X standard deviations under the normal curve (m). This provides only a very general RAD: although it specifies the number of taxa in any one octave, it does not specify their relative abundances. This can be done by a simple iterative procedure where the normal curve is divided into (S_T+1) partitions of equal

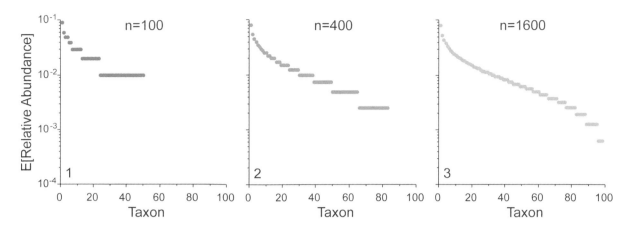

FIGURE *A3.*—Expected sampled RADs from the same model RAD (= lognormal from Fig. 7). As sample size (n) increases, the expectations become increasingly sharply delimited. This reflects both tightening of binomial error bars on expected numbers of finds given a true proportion and decreasing probabilities of sampled abundance ranks differing from true abundance ranks.

area (Fig. A2.2). The position of partition i is directly proportional to the relative abundance taxon i (Fig. A2.3).

Expected observed RADs given sampling

When comparing RADs, it is tempting to examine the goodness of fit to observed rank abundance to hypothesized curves such as shown in Figure A1. Another relevant concern is sample size (Koch, 1980; Gotelli and Graves, 1996; Dewdney, 1998; McGill, 2003). Consider the lognormal example in Figure A1. At 100 and even 400 specimens (Fig. 9.1-2), the *sampled* RADs look more like Zipf or even log series RADs than like the original model (Fig. A1). We need over 1000 specimens to recognize the characteristic lognormal sigmoidal RADs (Fig. 9.3).

Figure A3 emphasizes what we do have: predictions of expected frequencies of taxa with 1, 2, 3, etc. finds. For models predicting exact abundance frequencies given rank abundance, the expected sampled richness with abundance n given total sample N is:

$$E[S_\mathbf{n}|N] = \sum_{i=1}^{S} \binom{N}{n} \left[\left(1 - f_i\right)^{N-n} \times f_i^{n} \right]$$

where f_i is the frequency of species i implicit to the hypothesized RAD. Simple multinomial probability now gives us the likelihood of RADs that assumes that sampled RADs will differ from the original model RAD (Olszewski and Erwin, 2004). Equally important, we can use log-likelihood tests to test for changes within particular RAD models over time (e.g., Edwards, 1992); Alternatively, we can use information theory tests to test for changes in basic models over time (Burnham and Anderson, 2004).

SECTION TWO

Conservation Paleobiology in Deep Time

THE SCIENCE of conservation paleobiology also extends into the distant past. This deep-time approach is arguably "equally valuable" but is "less systematically pursued" (NRC, 2005) than the near-time approach. The deep-time approach takes advantage of the entire fossil record as a natural ecological and evolutionary laboratory. The near-time geohistorical records available to conservation paleobiology do not capture the full range of environmental conditions we are likely to encounter in the future and are also geologically too brief to encompass the full range of dynamics for some ecological and evolutionary processes. In this section, five chapters explore ways in which insights from the distant past can inform conservation biology today.

Geerat Vermeij (Chapter 9) shows that some ecological and evolutionary phenomena can behave differently over long timescales than they do over short time frames. Using examples ranging from species invasion to ocean acidification, he suggests that such stresses over the long timescales available to conservation paleobiologists do not always lead to the deleterious effects we observe today. For example, over the long run, extremely high nutrient availability may promote evolutionary diversification, opposite to the prevailing view of many conservation biologists (see also Vermeij, 2011). Geographic scale also matters because most conservation problems manifest themselves at local or regional scales or can be addressed only within political jurisdictions. Vermeij (Chapter 9), Gregory Dietl (Chapter 12), and Warren Allmon (Chapter 13) all note that the deep-time paleobiological perspective is most useful when it is employed at a local or regional scale. In this way, it does not differ from the near-time perspective or application.

Understanding exactly how—and how fast—species are likely to evolve is relevant in many conservation situations. Yet, developing practical, science-based approaches to facilitate adaptive responses that are to the benefit of improved conservation practice is difficult because of the broad geographic and temporal

scales over which evolutionary processes often work (Thompson, 2009). Dietl (Chapter 12) addresses this issue in the context of the evolution of species interactions. He shows how we can use the fossil record to help understand the conditions that limit adaptive responses to environmental change and identify which ecologically important traits are likely to evolve and keep species in the evolutionary game (see also Dietl, 2013).

Have the rules of the evolutionary game changed because of human activities? Allmon (Chapter 13) and Vermeij (Chapter 9) consider the evolutionary implications of human-induced reductions in animal abundance, increases in the fragmentation of many populations, and the globalization of others. In doing so, they question the generality of models used by conservation biologists to describe modern-species abundance relationships and the prospects for evolutionary resilience. Such human interference may make predicting evolutionary outcomes even more difficult.

If the future isn't going to look like the present day, might it look like some time in the past? Richard Aronson (Chapter 10) examines a past episode of global warming: the Eocene fossil record of benthic invertebrates in Antarctica. While the species composition of the past fauna differs from that of today's fauna, the increase of predation pressure during warm times and its decrease during cold times suggest that trait-based approaches to understanding the biotic response to climate change may prove especially fruitful. Indeed, as noted by Dietl (Chapter 12), such taxon-free approaches may be necessary for making the deep-time record even more relevant to dealing with near-time problems (see also Polly et al., 2011; Dietl, 2013).

Peter Roopnarine's (Chapter 11) primer on modeling food webs provides another tool that can be used to assess the ecological effects of top-down versus bottom-up disturbance. While the approach is amenable to application in near time as well, his deep-time applications to Permian and Miocene communities encompass the full spectrum of collapse and recovery. Today's ongoing perturbations, such as the top-down overfishing documented by Jeremy Jackson and Loren McClenachan (Chapter 5) and the bottom-up anthropogenic eutrophication documented by Susan Kidwell (Chapter 7), are not yet fully played out. Conceptual or quantitative models validated with deep-time data may help predict the future effects of present-day disturbance.

It's easy to be skeptical of the utility of deep-time conservation paleobiology: the species are unfamiliar, the temporal scale differs dramatically, and some environmental settings may have no analogs today. Nevertheless, as the chapters in this section illustrate, deep-time phenomena help us understand what is happening in near time and vice versa. Just as the present is a key to the past, the past is a key to the future.

REFERENCES

DIETL, G. P. 2013. The great opportunity to view status with an ecological lens. Palaeontology, 56:1239–1245.

NRC (NATIONAL RESEARCH COUNCIL). 2005. The Geological Record of Ecological Dynamics: Understanding the Biotic Effects of Future Environmental Change. The National Academies Press, Washington, D.C., 200 p.

POLLY, P. D., J. T. ERONEN, M. FRED, G. P. DIETL, V. MOSBRUGGER, C. SCHEIDEGGER, D. C. FRANK, J. DAMUTH, N. C. STENSETH, AND M. FORTELIUS. 2011. History matters: Ecometrics and integrative climate change biology. Proceedings of the Royal Society of London, Series B: Biological Sciences, 278:1131–1140.

THOMPSON, J. N. 2009. The coevolving web of life. The American Naturalist, 173:125–140.

VERMEIJ, G. J. 2011. The energetics of modernization: The last one hundred million years of biotic evolution. Paleontological Research, 15:54–61.

CHAPTER NINE

SEVEN VARIATIONS ON A RECENT THEME OF CONSERVATION

GEERAT J. VERMEIJ

Department of Geology, University of California at Davis, One Shields Ave., Davis, CA 95616 USA

ABSTRACT.—This essay offers a blueprint for how conservation can be informed by insights from paleontology. The fossil record offers a long-term perspective on phenomena ranging from extinction and invasion to fragmentation and ocean chemistry. I discuss seven phenomena: extinction, invasion, ecosystem recovery, the potential for adaptation, the relationship between diversity and productivity, global warming, and ocean acidification. I conclude that habitat fragmentation and the elimination of high-level consumers, especially in productive environments, exacerbate the destructive effects of invasion, global warming, and ocean acidification, and prevent or slow ecosystem recovery and the adaptation of surviving species. Paleontology therefore reinforces widely held views in conservation biology that protection of large tracts of unexploited, productive habitat and large, charismatic plants and animals is the best investment we can make in the long-term health, resilience, and adaptability of the biosphere.

INTRODUCTION

THE SCIENCE of conservation has until now been comfortably housed within the larger disciplines of ecology and genetics. Despite the rise of evolutionary biology and the increasing realization that the biosphere of today cannot be understood without reference to its history, the life sciences in general and conservation biology in particular have remained remarkably ahistorical. Those few historical insights that have crept into the field derive only from the most recent sliver of time, when humans were arguably already affecting the world's biota. A prominent review article in *Science* (Willis and Birks, 2006) enthusiastically proclaimed the virtues of historical analysis with reconstructions of conditions and communities as they existed ten thousand years ago with the stated aim of ascertaining what the world looked like before pervasive human influence changed it. But the megafaunal extinctions of North and South America in the early Holocene and of Australia during the Late Pleistocene, widely attributed to the direct and indirect effects of humans, had already taken place, and agriculture had already been invented in several parts of the world. Reconstruction of prehuman ecosystems is a worthy goal, but it can be accomplished only if conservation biologists expand their time horizons to the much more distant past, to a realm that is accessible only to those who study fossils.

How exactly can insights from the distant past enrich conservation biology? The most general answer is that the fossil record offers a long-term perspective on phenomena ranging from climate change, extinction, invasion, and fragmentation to source-sink dynamics. Nearly all change is harmful in the short term for the simple reason that individuals, species, and ecosystems work well enough that the chances of improvement are slim compared to the probability of reduced function. In the long run, however, these harmful effects are sometimes—perhaps often—overcome, or replaced by consequences that prove to be beneficial. Moreover, the fossil record demonstrates conclusively that change has been a hallmark of the biosphere for billions of years, and especially so with the evolution of multicellular animals beginning some six hundred million years ago. Time also offers a useful perspective, though not a justification or rationalization, for the changes being wrought by humans.

In this essay, I briefly outline long-term perspectives on seven phenomena of central importance in conservation biology: extinction, invasion, ecosystem recovery, the potential for adaptation, the relationship between diversity and productivity, global warming, and ocean acidification. A full treatment would fill a book. My purposes here are to highlight the often counterintuitive conclusions that emanate from paleontological studies, to urge greater acceptance of change as a necessary attribute of ecological systems, to list conservation priorities, notably the protection of top consumers and productive ecosystems, and to call upon paleontologists to emphasize ecological interactions and processes instead of abstract concepts based

In *Conservation Paleobiology: Using the Past to Manage for the Future, Paleontological Society Short Course, October 17th, 2009. The Paleontological Society Papers, Volume 15, Gregory P. Dietl and Karl W. Flessa (eds.).* Copyright © 2009 The Paleontological Society.

on unadorned lists of taxa and distributions.

1. EXTINCTION

The current high and accelerating loss of species in fresh water, on land, and increasingly in the sea has been labeled as the "sixth mass extinction," the first truly global loss of diversity since the storied end-Cretaceous crisis. The causes are many, but overexploitation, habitat destruction, invasion of foreign species to islands, damming of rivers, and perhaps pollution figure prominently in the current crisis. Causes of the previous global extinction events remain matters of intense investigation and debate. Gigantic flood-basalt eruptions coincide with the traditional "big five" events and may, in some or all cases together with collisions between Earth and celestial bodies, be the primary triggers of the ecological collapses that accompany the loss of species. As I discussed in an earlier paper (Vermeij, 2004), the extinctions of the geological past may be characterized as bottom-up, beginning with the demise of primary producers and propagating through the rest of the ecosystem, with populations of large, metabolically active and ecologically powerful species being particularly vulnerable. Modeling of food-web dynamics of communities before and after great crises shows that top-down effects, in which consumers put extra pressure on the victim species that still remain, amplify the destruction in the primary-producer guild (Roopnarine, 2006; Roopnarine et al., 2007). In modern and early Holocene systems, losses have been concentrated among large consumers—top predators, large herbivores, and large suspension-feeders like oysters—with the primary-producer guild changing in composition and becoming less productive but remaining intact functionally. With the exception of specialized parasites and symbionts of threatened hosts, these top-down losses have precipitated few collateral extinctions, in contrast to the situation during bottom-up collapses associated with the geological mass extinctions.

There have been frequent claims that the rate of extinction today greatly exceeds the rates chronicled in the fossil record. It is notoriously difficult to measure and compare rates in the geological past (Gingerich, 1983; Roopnarine, 2003), because the time interval over which rates are measured is critical and often unknown and because we lack the necessary temporal resolution; but the mass extinctions may well have proceeded extremely rapidly, and there is currently no credible evidence that modern-day extinctions are occurring at rates different from those at other times in Earth history. In many ways, the issue is irrelevant. Extinction implies an inability of a population to adapt or to move to a suitable refuge. It is far from clear whether slower deterioration would save many species.

The lesson I draw from this comparison between geological crises and the extinctions that humans have caused so far is that protection of the most productive environments and of the primary producers in them is critical if we are to avoid a wholesale ecological collapse. Geological evidence indicates that the plankton is especially vulnerable to change. As Norris (2000) points out, populations even as large as 10^{15} individuals are prone to extinction if the oceanographic conditions under which the population density of plankton is high enough to sustain the species are geographically small compared to the dispersibility of planktonic species. If these conditions change, the species as a whole is threatened, because no population is dense enough to be sustainable and to act as a reliable source.

It is in productive environments where source populations are most likely to arise. Source populations provide recruits for other populations, and adaptations that characterize the species as a whole are largely forged in them. By reducing productivity in such source regions, or by reducing the sizes of productive habitats, we are unwittingly transforming many source populations into sink populations, whose persistence depends on dispersers from elsewhere (Pulliam, 1988; Vermeij and Dietl, 2006). Paleontologists can help locate and evaluate the persistence of source regions by documenting patterns of spread of species using data from phylogeny and stratigraphic occurrences. For example, genetic and paleontological evidence indicates that the northeastern Atlantic has served as a source region since the Pliocene for many amphi-Atlantic cool-temperate marine species (Vermeij, 2005a). The identification and protection of refuges and source regions should be top priorities both for conservation biologists and paleontologists.

2. INVASION

With the globalization of commerce, accelerating in the sixteenth century, thousands of species have been transported by humans from their native geographic ranges to regions in which these species had not previously occurred. Many conservation biologists consider the spread of species by humans as unprecedented both in the number of taxa involved and in the rate at which foreign species are becoming established. Some also believe that invading species pose a serious threat to the biological diversity of recipient biotas.

The paleontological record, however, reveals dozens of large-scale biological invasions over the last twenty-five million years alone, and doubtless many more as one goes back further in time. The tectonic elimination of barriers, as well as rare dispersal events (such as the trans-Atlantic dispersal of species between Africa and South America during the Late Eocene or Early Oligocene) have enabled thousands of species to spread from one biogeographic region to another. Although rates of establishment of foreign species cannot be assessed accurately in the fossil record, the fraction of the donor biota that took part in invasion is often so high that the rate of arrival of new species in the recipient biota could well have been comparable to that seen among human-introduced species. Moreover, with the exception of endothermic mammals, few if any invading species can even plausibly be blamed for extinction of native species in the recipient biotas of continents and oceans. Instead, new arrivals enrich the biotas (Vermeij, 1991, 2005b).

There are hints in both the fossil record and in the history of human-introduced species that establishment of invaders is greatly facilitated by disruptions in the recipient biota. Island ecosystems that have remained relatively untouched by human influence have proven to be resistant to invading species that in nearby disturbed locations have become firmly established (Diamond and Veitsch, 1981). On the geological time scale, biotas that have been little affected by extinction, such as the late Neogene North Pacific, receive fewer invading species than do climatically similar ones, such as the contemporaneous North Atlantic, where extinction wiped out many incumbents (Vermeij, 1991, 2005b).

It remains possible, of course, that human-introduced species represent a different selection of native species than do species that took part in "natural" expansions, but this aspect of invasion has been little investigated. For example, mammals have in the past been poor dispersers across ocean barriers, presumably because their metabolism cannot shut down for long periods in the way that is common among lizards; whereas humans have carried mammal species to all parts of the world. Selectivity of invasion with respect

to body size, growth form (in plants), trophic level (in animals), competitive position, reproductive characteristics, type and expression of anti-consumer defenses, and even dispersibility has received very little systematic study either in human-caused or fossil episodes of species invasion.

Although conservation biologists have devoted much attention to invasion through the human transport of species, processes more akin to biotic interchange may also be in the offing as the result of global warming. We (Vermeij and Roopnarine, 2008) recently suggested that trans-Arctic movements of marine benthic species from the North Pacific to the North Atlantic, which began with the opening of Bering Strait in the latest Miocene and accelerated in the warm mid-Pliocene, will resume as Arctic Ocean ice melts and Arctic Ocean productivity rises (see also Reid et al., 2007). Most episodes of biotic interchange, on land as well as in the sea, coincide with warm periods. Thus eastward trans-tropical Pacific invasion typifies El Niño events, when ocean temperatures in the tropical eastern Pacific are warmer than normal; and trans-Beringian invasions on land during various Cenozoic episodes have been concentrated in warm time intervals, including the end-Paleocene thermal maximum (see Vermeij, 2005b, for a review).

3. ECOSYSTEM RECOVERY

Restoration ecology operates with the hope of repairing damaged ecosystems, particularly after the loss of key species. Some successes have been achieved in restoring a northern steppe community in Siberia with the reintroduction of major grazers (bison and musk ox) to Siberia from North America (Zimov, 2005). These deliberate reintroductions are problematic in that new diseases may be introduced along with them. In any case, they are necessarily incomplete, because other major consumers — mammoths and woolly rhinoceroses in the example cited — are globally extinct and therefore cannot be duplicated. In fact, the decimation of most top consumers worldwide cannot be fully compensated by second-order predators and grazers.

The extensive but still understudied geological record of ecological recovery from mass extinctions indicates that full restoration, which requires evolution of new species that replace extinct ecological equivalents, takes millions of years. The top consumers in particular are not immediately replaced. Their long generation times and long evolutionary trajectory from ancestor to top status ensure a protracted recovery phase even if small producers—herbs and phytoplankters—are ecologically reconstituted within perhaps hundreds or thousands of years (D'Hondt et al., 1998). It took some ten million years for mammalian carnivores weighing 100 kg or more to become established after the end-Cretaceous extinction (Van Valkenburgh, 1999). In temperate North America, a similar period of time was needed to re-establish the diversity of herbivorous insects and the intensity of leaf damage prevailing in Late Cretaceous plant communities (Labandeira et al., 2002).

Importantly, the new species often come from the ranks of species invading from elsewhere, not from the few local survivors. Invasion is therefore an essential process in the recovery phase. In order for them to stimulate recovery, invaders must come from geographic or ecological refuges where the causes of extinction either did not reach or were less severe. Long-term recovery in the human-dominated world will therefore also likely require reasonably intact refugial ecosystems whose species can seed and later evolve in human-disrupted systems (Vermeij, 2008).

The inherently long-term nature of ecosystem recovery makes the reconstitution of working ecosystems an unlikely target for policy-makers. Nevertheless, the historical record indicates that the likelihood of effective recovery would be raised if conservation efforts in the short term emphasized whole systems rather than individual species, and ensured that vigorous populations in large refuges remain protected. There is nothing new in such a recommendation, but the data from fossils reinforce it. The partial restoration of the Pleistocene steppe ecosystem being undertaken in Siberia through the introduction of large grazing bison and other mammals (see Zimov, 2005) offers a promising model.

4. EVOLUTION AND ADAPTATION

Whether and to what extent species can evolve and adapt under the highly restrictive conditions set by humans is an open question. With the exception of weedy species, commercially important taxa, and species well-adapted to such human environments as cities and suburbs, most remaining species are increasingly confined to small, island-like fragments separated by hostile barriers. As local extirpations of species in Sin-

gapore attest (Brook et al., 2003), many populations in these fragments are probably unsustainable in the long run. Even those that can maintain viable populations will likely be unable to expand either in numbers or in range. For them, either an adaptive status quo—conserving the adaptations that worked in the past—or the evolution of traits more consistent with island life (reduced productivity and competition imposed by small geographic extent and limited dispersibility) are the most likely evolutionary outcomes (Leigh et al., 2009). The kind of exuberant adaptive diversification and speciation observed when lineages expand geographically and ecologically when new areas are colonized (Moore and Donoghue, 2007) will be out of reach for most clades under human occupation. Thus the diversification of North American rodents in Pliocene South America, the explosion of scleractinian reef corals during and after the early Miocene in the Indo-West Pacific region, and the radiation of lemurs during the Cenozoic in Madagascar—all the result of speciation and adaptation in competitive and resource-rich environments of large extent—are unthinkable in the patchwork of ecosystem islands that human dominance has created.

These evolutionary aspects of conservation have received little attention from biologists, who have been overwhelmingly concerned with the present-day ecology and future changes in the distribution of species. Yet it is evolution that will determine whether populations, species, and clades can survive in the long run. Understanding the conditions under which adaptation, dispersal, and diversification can occur has been an important part of the research program in paleobiology and evolutionary biology, and it is high time that this perspective penetrates the field of conservation.

5. DIVERSITY AND PRODUCTIVITY

There exists a vast literature on the complex correlations between diversity (or species richness) and primary productivity (biomass made available to consumers per unit time) in present-day ecosystems. The prevailing view among ecologists is that, when a system becomes eutrophic, its diversity declines, because a few very vigorous species will overwhelm most of the rest. Yet on evolutionary time scales and on the large spatial scales of geography, the opposite pattern often seems to hold: the more productive a habitat or region, the higher is the number of coexisting species. There are, to be sure, unproductive regions where diversity is high (southwestern Australia and southern Africa, with exceptionally rich floras), and productive regions where diversity is low (marine Antarctica); but biotas that have evolved under conditions of a high, accessible, and predictable food supply contain a wealth of species, a large number of which are adapted to situations—created by the most productive members—where resources may be less prolific. Productive environments, in other words, support a wide variety of habitats, some fertile and others less so; and if there are sufficient geographic or genetic opportunities for isolation, small populations have a greater probability of becoming distinct species than in more spartan circumstances (Allmon, 1992; Vermeij, 1995).

The effects of eutrophication on diversity are thus evident when evolutionary isolation and divergence have not had time to act. The short-term consequences of eutrophication thus differ from long-term increases in resource production. Moreover, it is very likely that the observed relationship between increased productivity and lower diversity over ecological time is exacerbated by the absence of top consumers. In most present-day ecosystems, humanity has removed top predators as well as marine suspension-feeders. Without the ecological controls imposed by these consumers, species lower in the food chain are free to increase, often crowding out other species (Worm et al., 2002; Vermeij, 2004). The anoxic dead zones now spreading in the world's shallow seas have been blamed on eutrophication resulting from agricultural runoff (Diaz and Rosenberg, 2008), but I suspect strongly that the continuing elimination of the guild of top predators and large suspension-feeders is an important contributing factor.

On a more general note, many of the "natural" ecosystems that ecologists study with respect to stability and sustainability have been severely affected by humans. For example, the systems studied by Rooney et al. (2006) (including Chesapeake Bay, Minnesota grasslands, and Dutch pine forests) lack the top predators and grazers that once lived there. Likewise, the six forest types studied by Wardle et al. (2004), who claim that forests deplete their soils on time scales of thousands to hundreds of thousands of years, either grow on islands (New Zealand, the Hawaiian Islands), where top consumers are either extinct or ineffective, or at unproductive sites at high latitudes (northern Sweden, Alaska) or in dry areas (Australia), again places where megafauna has been decimated. The conclusions

reached by Wardle and colleagues may well be correct for ecosystems in general—a sobering thought in view of human attempts to create sustainable economies—but it is important either to model ecosystems with a full complement of consumers or to analyze fossil communities in which trophic relationships of a well-preserved assemblage of top consumers can be quantified, as Roopnarine (2006) and Roopnarine et al. (2007) have attempted for Permo-Triassic ecosystems in southern Africa.

The long-term and large-scale perspective offered by the fossil record and by the comparative study of biotas reinforces the recommendation made earlier in this essay: it is essential to protect—and, if possible, to expand—reasonably intact, productive ecosystems with an unexploited contingent of high-level consumers. Conservation is more than saving the complement of species we have; it is about preserving the adaptive capacity of systems that enables these systems to change and to renew themselves in the face of inevitable hazard.

6. GLOBAL WARMING

In the short term, worldwide warming will have largely deleterious effects both for humans and for the global biota. Because so many populations have been fragmented, latitudinal shifts in the thermal regimes to which species have adapted pose potentially insoluble problems for species that cannot expand to the newly favorable environments because of intervening barriers. Changes in the timing of reproduction and in the availability of critical resources, together with new relationships among predators, prey, and mutualists, will harm many species and possibly lead to extirpation and extinction.

Paleontologists, however, recognize periods of warming as times of diversification and adaptation. Cornette and colleagues (2002) provide a global documentation of this long-term relationship. This result is to be expected in view of the enduring pattern, recognized even for the Paleozoic marine world, for tropical diversity to be higher than that at cold-temperate and polar latitudes (Fischer, 1960; Leigh et al., 2004). More frequent species interactions including predation and competition, together with a greater scope for adaptation (that is, a greater range of potential adaptive states) at high temperatures create greater biological heterogeneity and more ways of life, so that when opportunities for genetic isolation present themselves, population divergence and species formation are more likely (reviewed in Vermeij, 2005c). Evolution thus reverses the short-term harm of warming as long as it is allowed to take place.

This long-term perspective on warming implies that it is not warming itself, but the human context in which warming is taking place, that causes the short-term harm of rising temperatures. Large-scale fluctuations in Pleistocene climates surely had all the disruptive effects on relationships among organisms that are observable in today's world, yet relatively few species became extinct then. The concentration of megafaunal extinctions in the Late Pleistocene of Australia, near the end of the Pleistocene in the Americas, and during the Holocene on many islands carries a distinct human signature, not the primary signature of warming of other climatically induced environmental change. Habitat fragmentation, prior elimination of top consumers, and in a few cases invasion (on islands and in lakes) are amplifying what short-term ecological damage is being wrought by the current episode of warming. Studies of past warming events and their effects on the biota may help ecologists fashion appropriate strategies for reducing these harmful effects. Particularly fruitful studies would concentrate on the Late Paleocene to Early Eocene warming, the Late Oligocene to middle Miocene expansion of warm climates, and the events leading up to the mid-Pliocene warm period just before the initiation of extensive northern glaciation at 2.7 Ma. These would provide a benchmark against which to evaluate the effects of warming in the human-dominated biosphere, and help identify and quantify the factors other than warming that are endangering the biosphere as we know it.

7. OCEAN ACIDIFICATION

With the increasing atmospheric concentration of carbon dioxide as the result of the burning of fossil fuels by humans, the pH of the oceans is declining, with observed detrimental effects on the calcification of marine organisms (Kleypas, 1999; Feely et al., 2004; Orr et al., 2005; Hoegh-Guldberg et al., 2007; Jokiel et al., 2008). Coral skeletons, echinoid tests, molluscan shells, and likely the skeletons of other organisms have become weaker, and the depth at which calcium carbonate is dissolved more than it is precipitated in the world ocean is becoming shallower. Reefs and the

calcified plankton are therefore potentially imperiled.

In the fossil record, however, times of high inferred carbon dioxide concentration are often times of exceptionally well-calcified organisms. Cretaceous rudists, thick-shelled Early Jurassic "*Liothiotis*" bivalves, and extensive Ordovician to Middle Devonian reefs existed when the partial pressure of carbon dioxide was two to ten times higher than today's. Carbonate gaps (or calcification crises) have been recognized in the aftermath of the end-Permian, end-Triassic, and end-Cretaceous extinctions and in the Valanginian (Early Cretaceous) Weissert event, among others, which have been blamed on brief, exceptionally high carbon dioxide (and low pH) (Knoll et al., 1996; Erba et al., 2004; Hautmann, 2004); but it is unclear to me that low pH is directly to blame for the extinction of calcifying species during the crises. Not only did vast numbers of other taxa become extinct as well, but very high carbon dioxide levels at the end of the Paleocene did not cause a calcification crisis among planktonic species in the upper waters (photic zone) of the ocean, where primary production takes place (Gibbs et al., 2006).

How is this contrast between present-day threats to calcifying organisms and the more complex picture presented by the fossil record to be reconciled? The answer may lie in alkalinity. Carbonate (and bicarbonate) ions, which produce alkalinity, buffer organisms against low pH (Marubini et al., 2008). They are generated by dissolution of marine carbonates, by the process of photosynthesis in calcifying marine organisms (phytoplankton, corals, and some seaweeds), and by chemical weathering on land. Weathering is enhanced at higher temperatures and higher levels of carbon dioxide. A recent discovery indicates that the digestive systems of fish are additional important sources of marine alkalinity in the form of carbonate ions (Wilson et al., 2009).

Two important human-caused changes may exacerbate the effects of warming and increased carbon dioxide on ocean acidification. First, dams have decreased the amount of sediment reaching the ocean by about 20%, from 15.5 billion metric tons in prehuman times to about 12.6 billion metric tons today (Syvitski et al., 2005). Much of this sediment today comes from lowland agricultural areas, whereas in prehuman times the bulk of sediment came from high elevations. The terrestrial source of alkalinity has thus been severely curtailed. Second, chronic overfishing worldwide may be contributing to a decline in naturally generated alkalinity. Acidification and its harmful effects on marine calcifying organisms, including many critical primary producers, may therefore have causes beyond the production of greenhouse gases in the human-controlled atmosphere.

These proposed effects remain highly speculative. In any case, remedies—the restoration of alkalinity in particular—require thousands to perhaps hundreds of thousands of years. The history of calcification in reef-like structures since the Early Cambrian indicates that calcification, especially by microbes, has never ceased despite vast fluctuations in pH and alkalinity (Webb, 1996). For animals with exoskeletons of calcium carbonate, the cost of calcification has varied according to the chemistry and temperature of ocean and fresh waters, but for tens of thousands of lineages these costs have been outweighed by the benefits of a protective house. It will be important to examine whether and to what extent organisms relying on calcification can adapt to the rapidly changing pH and alkalinity of today's oceans, and to investigate the feasibility of adding alkalinity on time scales much shorter than those prevailing in the geological past.

CONCLUDING REMARKS

This brief survey of paleontological perspectives on conservation reveals that the destructive effects of invasion, warming, and ocean acidification are amplified by habitat fragmentation and the elimination of high-energy consumers from ecosystems. It therefore shows that conserving productive environments and source populations is essential for ecosystem recovery after crises and for enabling surviving species to adapt to new circumstances. Paleontology therefore reinforces widely held views in conservation biology that protection of large tracts of unexploited, productive habitat is the best investment we can make in the long-term health, resilience, and adaptability of the biosphere.

Given the important role that top consumers play in promoting and sustaining diversity, stabilizing ecological relationships in ecosystems, and enhancing the productivity of the systems they dominate, any species-level efforts at conservation should concentrate on these charismatic organisms. Rather than expending limited resources on rare, small, unobtrusive endemics, conservationists should target such iconic species as large vertebrates, forest trees, reef-building corals and sponges, ecologically important suspension-feeders,

animals reworking marine and freshwater sediments, seagrasses and large seaweeds, and social insects. This point of view may seem retrograde, but by emphasizing these important players, we can save many less flashy species as well, not to mention the systems in which all these species live.

The contrasts often observed between short-term and long-term consequences of environmental change invite collaboration between paleontologists with a historical perspective and ecologists confronted with rapid environmental degradation, species loss, and ecosystem dysfunction. We need better modeling and more data on rates of recovery, patterns and consequences of extinction and invasion, effects of consumers on geochemical cycles, and especially on the feedbacks among producers, consumers, and climate. As we fashion environmental policy, it is essential to look to the geological and biological past for answers.

Two more controversial points also emerge. For conservation biologists, there is an enormous temptation to preserve what we have and what we know. I am as susceptible to this as anyone. But paleontology and evolutionary biology teach that the biosphere and its component life forms change. Not only will we have to live with inevitable change, but we must create and maintain conditions that allow adaptive change to do its work. For paleontologists, there has been an overwhelming temptation to document history at the global scale, with emphases placed on such properties as diversity, disparity, and the geographic and stratigraphic ranges of lineages and biotas. Engagement with conservation will require a shift in focus toward ecological relationships, selective regimes, the feedbacks between enabling factors (such as temperature and food availability) and the demand factors of consumption, the ecological causes and consequences of extinction, rates of ecological and adaptive change, and events on local and regional scales. Whether the long-term and large-scale phenomena revealed by the study of fossils can inform strategies and policies of conservation remains an open question, but without paying attention to them, we will never know.

REFERENCES

ALLMON, W. D. 1992. A causal analysis of stages in allopatric speciation. Oxford Surveys in Evolutionary Biology, 8:219-257.

BROOK, B. W., N. S. SODHI, AND P. K. L. NG. 2003. Catastrophic extinctions follow deforestation in Singapore.

Nature, 424:420-423.

CORNETTE, J. L., B. S. LIEBERMAN, AND R. H. GOLDSTEIN. 2002. Documenting a significant relationship between macroevolutionary origination and Phanerozoic pCO$_2$ levels. Proceedings of the National Academy of Sciences of the United States of America, 99:7832-7835.

D'HONDT, S., P. DONOVAN, J. C. ZACHOS, D. LUTTENBERG, AND M. LINDINGER. 1998. Organic carbon fluxes and ecological recovery from the Cretaceous-Tertiary mass extinction. Science, 282:276-279.

DIAMOND, J. M., AND C. R. VEITSCH. 1981. Extinctions and introductions in the New Zealand avifauna: Cause and effect? Science, 211:499-501.

DIAZ, R. J., AND R. ROSENBERG. 2008. Spreading dead zones and consequences for marine ecosystems. Science, 321:926-929.

ERBA, E., A. BARTOLINI, AND R. L. LARSON. 2004. Valanginian Weissert oceanic anoxic event. Geology, 32:149-152.

FEELY, R. A., C. L. SABINE, K. LEE, W. BERELSON, J. KLEYPAS, V. J. FABRY, AND F. J. MILLERO. 2004. Impact of anthropogenic CO$_2$ on the CaCO$_3$ system in the ocean. Science, 305:362-366.

FISCHER, A. G. 1960. Latitudinal variations in organic diversity. Evolution 14:64-81.

GIBBS, S. J., P. R. BOWN, J. A. SESSA, T. J. BRALOWER, AND P. A. WILSON. 2006. Nanoplankton extinction and origination across the Paleocene-Eocene thermal maximum. Science, 314:1770-1773.

GINGERICH, P. D. 1983. Rates of evolution: Effects of time and temporal scaling. Science, 222:159-161.

HAUTMANN, M. 2004. Effect of end-Triassic CO$_2$ maximum on carbonate sedimentation and marine mass extinction. Facies, 50:257-261.

HOEGH-GULDBERG, O., P. J. MUMBY, A. J. HOOTEN, R. S. STENECK, P. GREENFIELD, E. GOMEZ, C. D. HARVELL, P. F. SALE, A. J. EDWARDS, K. CALDEIRA, N. KNOWLTON, C. M. EAKIN, R. IGLESIAS-PRIETO, N. MUTHIGA, R. H. BRADBURY, A. DUBIN, AND M. E. HATZIOLOS. 2007. Coral reefs under rapid climate change and ocean acidification. Science, 318:1737-1742.

JOKIEL, P. L., K. S. RODGERS, I. B. KUFFNER, A. J. ANDERSSON, E. F. COX, AND F. T. MACKENZIE. 2008. Ocean acidification and calcifying organisms: A mesocosm investigation. Coral Reefs, 27:473-483.

KLEYPAS, J. A., R. W. BUDDEMEIER, D. ARCHER, J.-P. GATTUSO, C. LANGDON, AND B. N. OPDYKE. 1999. Geochemical consequences of increased atmospheric carbon dioxide on coral reefs. Science, 284:118-120.

KNOLL, A. H., R. K. BAMBACH, D. E. CANFIELD, AND J. P. GROTZINGER. 1996. Comparative earth history and Late Permian mass extinction. Science, 273:452-457.

LABANDEIRA, C. C., K. R. JOHNSON, AND P. WILF. 2002. Impact of the terminal Cretaceous event on plant-insect

associations. Proceedings of the National Academy of Sciences of the United States of America, 99:2061-2066.

LEIGH, E. G. JR., P. DAVIDAR, C. W. DICK, J.-P. PUYRAVAUD, J. TERBORGH, H. TER STEEGE, AND S. J. WRIGHT. 2004. Why do some tropical forests have so many species of trees? Biotropica, 36:447-473.

LEIGH, E. G., JR., G. J. VERMEIJ, AND M. WIKELSKI. 2009. What do human economies, large islands and forest fragments reveal about the factors limiting ecosystem evolution? Journal of Evolutionary Biology, 22:1-12.

MARUBINI, F., C. FERRIER-PAGÈS, P. FURLA, AND D. ALLEMAND. 2008. Coral calcification responds to seawater acidification: A working hypothesis towards a physiological mechanism. Coral Reefs, 27:491-499.

MOORE, B. R., AND M. J. DONOGHUE. 2007. Correlates of diversification in the plant clade Dipsacales: Geographic movement and evolutionary innovations. American Naturalist, 170, Supplement 2:S27-S55.

NORRIS, R. D. 2000. Pelagic species diversity, biogeography, and evolution. Paleobiology, 26, Supplement 5:236-258.

ORR, J. C., V. J. FABRY, O. AUMONT, L. BOPP, S. C. DONEY, R. A. FEELY, A. GNANADESIKAN, N. GRUBER, A. ISHIDA, F. JOOS, R. M. KEY, K. LINDSAY, E. MAIER-REIMER, R. MATEAR, P. MONFRAY, A. MOUCHET, R. G. NAJJAR, G.-K. PLATTNER, K. B. RODGERS, C. L. SABINE, J. L. SARMIENTO, R. SCHLITZER, R. D. SLATER, I. J. TOTTERDELL, M.-F. WEIRIG, Y. YAMANAKA, AND A. YOOL. 2005. Anthropogenic ocean acidification over the twenty-first century and its impact on calcifying organisms. Nature, 437:681-686.

PULLIAM, R. R. 1988. Sources, sinks, and population regulation. American Naturalist, 132:652-661.

REID, P. C., D. G. JOHNS, M. EDWARDS, M. STARR, M. POULIN, AND P. SNOEIJS. 2007. A biological consequence of reducing Arctic ice cover: Arrival of the Pacific diatom *Neodenticula seminae* in the North Atlantic for the first time in 800,000 years. Global Change Biology, 13:1910-1921.

ROONEY, N., K. McCANN, G. GELLNER, AND J. C. MOORE. 2006. Structural asymmetry and the stability of diverse food webs. Nature, 442:265-269.

ROOPNARINE, P. D. 2003. Analysis of rates of morphologic evolution. Annual Reviews of Ecology, Evolution, and Systematics, 34:605-632.

ROOPNARINE, P. D. 2006. Extinction cascades and catastrophe in ancient food webs. Paleobiology, 32:1-19.

ROOPNARINE, P. D., K. D. ANGIELCZYK, S. C. WANG, AND R. HERTOG. 2007. Trophic network models explain instability of Early Triassic terrestrial communities. Proceedings of the Royal Society, B, 274:2077-2086.

SYVITSKI, J. P. M., C. J. VÖRÖSMARTY, A. J. KETTNER, AND P. GREEN. 2005. Impact of humans on the flux of terrestrial sediment to the global coastal ocean. Science, 308:376-380.

VAN VALKENBURGH, B. 1999. Major patterns in the history of carnivorous mammals. Annual Reviews in Earth and Planetary Sciences, 27:463-493.

VERMEIJ, G. J. 1991. When biotas meet: Understanding biotic interchange. Science, 253:1099-1104.

VERMEIJ, G. J. 1995. Economics, volcanoes, and Phanerozoic revolutions. Paleobiology, 21:125-152.

VERMEIJ, G. J. 2004. Ecological avalanches and the two kinds of extinction. Evolutionary Ecology Research, 6:315-337.

VERMEIJ, G. J. 2005a. From Europe to America: Pliocene to Recent trans-Atlantic expansion of cold-water North Atlantic molluscs. Proceedings of the Royal Society, B, 272:2545-2550.

VERMEIJ, G. J. 2005b. Invasion as expectation: A historical fact of life, p. 315-339. In D. F. Sax, J. J. Stachowicz, and S. D. Gaines (eds.), Species Invasions: Insights Into Ecology, Evolution, and Biogeography. Sinauer, Sunderland, Massachusetts.

VERMEIJ, G. J. 2005c. From phenomenology to first principles: Toward a theory of diversity. Proceedings of the California Academy of Sciences, 56, Supplement I(2):12-23.

VERMEIJ, G. J. 2008. Ecosystem recovery: Lessons from the past, p. 252-258. In S. P. Carroll and C. W. Fox (eds.), Conservation Biology: Evolution In Action. Oxford University Press, Oxford, UK.

VERMEIJ, G. J., AND G. P. DIETL. 2006. Majority rule: Adaptation and the long-term dynamics of species. Paleobiology, 32:173-178.

VERMEIJ, G. J., AND P. D. ROOPNARINE. 2008. The coming Arctic invasion. Science, 321:780-781.

WARDLE, D. A., L. R. WALKER, AND R. D. BARDGETT. 2004. Ecosystem properties and forest decline in contrasting long-term chronosequences. Science, 305:509-513.

WEBB, G. E. 1996. Was Phanerozoic reef history controlled by the distribution of non-enzymatically secreted reef carbonates (microbial carbonate and biologically induced cement)? Sedimentology, 43:947-971.

WILLIS, K. J., AND H. J. B. BIRKS. 2006. The need for a long-term perspective in biodiversity conservation. Science, 314:1261-1265.

WILSON, R. W., F. J. MILLERO, J. R. TAYLOR, P. J. WALSH, V. CHRISTENSEN, S. JENNINGS, AND M. GROSELL. 2009. Contribution of fish to the marine inorganic carbon cycle. Science, 323:349-362.

WORM, B., H. K. LOTZE, H. HILLEBRAND, AND U. SOMMER. 2002. Consumer versus resource control of species diversity and ecosystem functioning. Nature, 417:848-851.

ZIMOV, S. A. 2005. Pleistocene Park: Return of the mammoth's ecosystem. Science, 308:796-798.

CHAPTER TEN

METAPHOR, INFERENCE, AND PREDICTION IN PALEOECOLOGY: CLIMATE CHANGE AND THE ANTARCTIC BOTTOM FAUNA

RICHARD B. ARONSON

Department of Biological Sciences, Florida Institute of Technology, Melbourne, FL 32901 USA

ABSTRACT—The fossil record affords us the opportunity to reconstruct the history of communities prior to human intervention, infer the causes underlying that history, and make early, accurate, and mechanistic predictions about their future as human-dominated systems. Paleoecology's unique contribution to conservation lies in providing the rationale, data, and methodological approach to view contemporary biological communities metaphorically in terms of geohistorical time-equivalence. By comparing perturbed communities of the present to paleocommunities that were prevalent in earlier time intervals, we gain insight into the causes and mechanisms of ecological degradation. Such insights are founded on testable hypotheses that ecological processes are scale independent. As an example, nearshore, shallow-benthic communities living in Antarctica today are reminiscent of Paleozoic communities, which were dominated by epifaunal suspension-feeders and lacked the functionally modern, durophagous predators that diversified in the Mesozoic. Establishing a causal link between the absence of durophagy and the retrograde community structure of the Antarctic bottom fauna requires characterizing the relevant predator–prey interactions at multiple scales using a variety of methods. Paleoecological analysis, in corroborating the postulated scale independence of those predator–prey interactions, leads us through a logical sequence of ideas with predictive power: (1) Antarctic marine communities were functionally modern before climatic cooling began 41 Ma; (2) they were forced toward a quasi-Paleozoic composition when the cooling trend reduced and ultimately eliminated durophagous predation; and (3) they will soon be re-modernized as global warming and increased ship traffic in Antarctica permit predators to reinvade. Policy recommendations follow from this paleoecological interpretation of the ecological dynamics of the Antarctic benthos.

INTRODUCTION

THE UTILITY of a scientific discipline turns on the aggregate ability of its practitioners to make predictions that are subsequently corroborated (e.g., Lakatos, 1970). So it is with paleoecology. Excoriated by Williamson (1982, p. 99) as "...a poor-man's applied ecology performed on inadequate data," paleoecology languished in purgatorial semi-legitimacy during an unnecessarily long and annoying controversy over whether there can be such a thing as historical science, in which experimentation is not possible (Cleland, 2001). This *Sturm und Drang* did not deter actual paleoecologists from conducting actual scientific studies. Mercifully, the debate has now been resolved, in large part through our collective recognition that science is about formulating and testing hypotheses, whether through observation, experimentation, or modeling. Paleoecology, like ecology, may not be amenable to strict Popperian falsificationism (Quinn and Dunham, 1983; Cleland, 2001); nevertheless, at the same time Williamson and colleagues were heaping abuse on the field, Gould (1981) and others recognized its potential as *bona fide* science, with hindcasts and forecasts that could be corroborated and had value in both theoretical and practical terms (e.g., Martin, 1998). That potential has now been fully realized.

Paleoecology is an interdisciplinary science that continues to develop rapidly in both theory and methodology. It is a motley but serviceable and ever-improving combination of geology, paleontology, evolutionary biology, paleoceanography, climatology, ecology, and latterly archeology and history. Ecology in turn draws from evolutionary biology, physiology, molecular genetics, morphology, oceanography, atmospheric science, and so on. Big questions in paleoecology are built and tackled with jury-rigged research programs, in which concepts and approaches from different fields are introduced as needed.

This paper offers a case study of interdisciplinary inference and practical application in paleoecology. It focuses on the past, present, and future implications of skeleton-breaking, or durophagous, predation for the structure of marine shallow-water, soft-bottom

In *Conservation Paleobiology: Using the Past to Manage for the Future, Paleontological Society Short Course, October 17th, 2009. The Paleontological Society Papers, Volume 15*, Gregory P. Dietl and Karl W. Flessa (eds.). Copyright © 2009 The Paleontological Society.

FIGURE *1.*—A notothenioid antifreeze fish, *Pagetopsis macropterus*, at rest in a dense population of ophiuroids, *Ophionotus victoriae* and *Ophiosparte gigas* (the two larger brittlestars to the right of the fish). Notothenioids are not durophagous, and they pose no threat to the ophiuroids. The photograph is from the Weddell Sea at Dundee Island, off the tip of the Antarctic Peninsula, at a depth of 289 m. Scenes such as this are common in nearshore habitats in Antarctica. Photo credit: Julian Gutt, copyright © AWI/Marum, University of Bremen. Used with permission.

communities. I trace the development of the idea that the geohistorical time-equivalence, or metaphorical 'flavor' of a benthic community derives from the intensity of durophagous predation: low-predation communities are structurally reminiscent of the Paleozoic, whereas high-predation communities are more Modern, in the sense of Sepkoski's (1981, 1991a) three evolutionary faunas. I then show how this framework can be used to conceptualize the history of the Antarctic benthos and predict its near-term future in response to rapid climate change. The paper concludes with a prospectus for applying metaphors of geohistorical time-equivalence in a broader sense to conservation problems that reach well beyond marine benthic communities.

I use the term paleocology to denote the multidisciplinary scientific process of drawing inferences from the fossil record about ecology in times past. Paleobiology is the multidisciplinary study of both evolutionary and ecological pattern and process in the fossil record and, therefore, subsumes paleoecology. Obviously, the demarcation between the ecological and evolutionary aspects of paleobiology is in many cases indistinct.

SCALE INDEPENDENCE

The 1980s and 1990s witnessed an explosion of literature on the issues of scale and hierarchy in ecology and paleobiology. Gould's (1985) influential essay on hierarchical disjunction was taken as marching orders by many paleobiologists, who earnestly set about demonstrating the universality of scale dependence in pattern and process. That scale dependence became the common *a priori* assumption of paleobiology was a collateral overreaction to the perceived failure of optimality models to explain empirical observations.

On the ecological side, models of optimal foraging did not adequately match the behavioral patterns of study animals, leaving investigators to search *a posteriori* for external or internal constraints to explain away departures of observation from theory (e.g., Aronson and Givnish, 1983). The situation was much the same in evolutionary biology. Gould and Lewontin (1979) famously argued that phyletic, biomechanical, and architectural constraints effectively prevent specific features of organismal design from evolving to optimality. Instead of searching for optimality *per se*, we should inquire whether the organism's complete behavioral repertoire or design package can be considered adaptive given known constraints. In arguing against optimality and for historical contingency, Gould implicitly and explicitly argued against scale independence.

Mayr (1983) reacted strongly against abandoning optimality theory. Even though natural selection does not require the perfection of design, Mayr advocated optimality as a working hypothesis, an ideal against which to compare empirical observation of the products of evolution. In a similar vein, scale independence seems the logical choice of a simple and falsifiable initial assumption about ecological/evolutionary pattern and process. Many ecological and paleoecological phenomena, including predation, diversification, and extinction are scale independent in at least some respects (Aronson, 1994; Aronson and Plotnick, 1998; Miller, 1998; Bambach et al., 2004).

THE MESOZOIC MARINE REVOLUTION

Our understanding of the scale-independent effects of predation in marine paleocommunities owes its conceptual foundation to the imaginative work of Geerat J. Vermeij. In a classic paper (Vermeij, 1977), two ground-breaking books (Vermeij, 1978, 1987), and many subsequent publications, Vermeij explored the positive correlation between durophagy (his neologism) and the defensive architecture of marine in-

vertebrates—in particular benthic gastropods—along gradients of latitude, longitude, and Phanerozoic time. Beginning in the Mesozoic, the evolutionary diversification of fast-moving, skeleton-breaking predators—teleosts (modern bony fish), neoselachians (modern sharks and rays), and reptant decapods (crabs, lobsters, and other modern, bottom-walking crustaceans)—was accompanied by, and likely drove, the evolution of larger spines, thicker outer lips, narrower apertures, and a host of other defensive adaptations in marine snails living in nearshore, shallow-water habitats. Vermeij (1977) called this macroecological trend of predator–prey escalation the Mesozoic marine revolution. In an analogous fashion, durophagous predation and defensive architecture in gastropods increase from the poles to the tropics, and in the tropics from the Atlantic to the Indo-Pacific (Vermeij, 1978; Alexander and Dietl, 2003).

Another radical change to shallow-water, soft-sediment ecosystems during the late Paleozoic and Mesozoic was an increase in the depth and intensity of bioturbation (Thayer, 1983; Bottjer and Ausich, 1986). The combination of increasing trophic pressure from diversifying predatory taxa and increasing sediment destabilization and resuspension from bioturbation led to, or at least strongly contributed to, a Mesozoic decline of epifaunal suspension-feeders on soft substrata in shallow-water environments and a trend toward infaunalization of the benthos. The effects of escalating predation and bioturbation on shallow-water taxa and the paleocommunities they comprised transcended the end-Cretaceous extinction and continued during the Cenozoic. Alexander and Dietl (2003), Harper (2003), and Aberhan et al. (2006) provide recent reviews of the Mesozoic marine revolution.

Subsequent studies have considered the morphological evolution of prey taxa other than mollusks, shell-drilling *versus* shell-crushing predation, the evolutionary implications of lethal *versus* sublethal predation, refinements to the basic scenario of a Mesozoic marine revolution, and pre-Mesozoic revolutions in predation (e.g., Signor and Brett, 1984; Vermeij, 1995; McRoberts, 2001; Kelley et al., 2003; Aberhan et al., 2006). The salient point for our purposes is that the large-scale spatial and temporal correlations of predation intensity with prey defense are scaled-up versions of phenotypic and genotypic effects in ecological space and time. Predatory introductions to modern, local communities induce the same kinds of defensive attributes in snail shells that we observe biogeographically and geologically (reviewed in Aronson, 1994; Moody and Aronson, 2007). Patterns at large scales and hierarchical levels are thus the summations of individual predator–prey interactions occurring over ecological time in local communities. Likewise, the inferred action of increasing bioturbation in radically restructuring soft-bottom paleocommunities is a scaled-up version of the trophic amensalism hypothesis that Rhoads and Young (1970) formulated for processes operating on ecological scales. The role of increasing productivity as an extrinsic cause of the Mesozoic marine revolution (Bambach, 1993; Vermeij, 1995; Martin, 1998), and the role of mass extinctions in accelerating or decelerating its macroevolutionary and macroecological effects (Dietl et al., 2002, 2004; Harper, 2003; Kelley and Hansen, 2003) remain incompletely understood.

THE ONSHORE–OFFSHORE HYPOTHESIS

Bambach (1985, 1993) viewed the Mesozoic marine revolution as part of a Phanerozoic trend of increasing utilization of ecospace: marine faunas progressively occupied major categories of marine ecosystems and major ecological roles within those ecosystems through time. Mapped onto the bathymetry of the oceans, the Mesozoic changes were part of an onshore-to-offshore trend in evolutionary innovation and community structure. The following summary of the onshore–offshore hypothesis draws from reviews by Sepkoski (1991a), Aronson (1994), and Sheehan (2001).

Sepkoski (1981, 1984) performed a factor analysis of his compilation of the stratigraphic ranges of marine families, partitioning taxa into what he termed Cambrian, Paleozoic and Modern evolutionary faunas. The Cambrian Fauna, which was dominated by trilobites and other surface deposit-feeders as well as inarticulate brachiopods and other low-lying suspension-feeders, diversified during the Cambrian Period and then declined. The brachiopod-rich Paleozoic fauna, which diversified rapidly in the Ordovician, represented the functional expansion of epifaunal suspension-feeders, which operated at multiple canopy and sub-canopy levels above the sediment-water interface (Bottjer and Ausich, 1986). Rhynchonelliform (articulate) brachiopods, stalked crinoids, ophiuroids, stromatoporoids,

stenolaemate bryozoans, rugose and tabulate corals, and a variety of other sessile and semi-mobile, epifaunal suspension-feeders characterized soft-substratum communities through much of the Paleozoic Era. Predators included asteroids, polychaete worms, ectocochleate (shelled) cephalopods, placoderms, primitive chondrichthyans, and conodonts. Bioturbation expanded downward into the sediments. The end-Permian extinction severely reduced the diversity of the Paleozoic fauna. The mollusk-rich Modern fauna, which diversified throughout the Phanerozoic, was less affected by the end-Permian extinction. In addition to gastropods, infaunal bivalves, and regular and irregular echinoids, the Modern fauna included radiations of functionally modern, durophagous predators, including teleosts, neoselachians, reptant decapods, marine reptiles, marine mammals, and coleoid cephalopods (cephalopods lacking an external shell; Packard, 1972; Aronson, 1991a).

Functional innovations (i.e., higher taxa) that are emblematic of the three faunas initially appeared in coastal and inner-shelf environments (Jablonski et al., 1983; Sepkoski and Miller, 1985; Jablonski and Bottjer, 1991; Sepkoski, 1991b). Taxa possessing key adaptations then expanded offshore to outer-shelf and deep-sea environments. The abilities of predators to break and crush heavily skeletonized prey by various means are nearshore innovations of the Modern fauna. The clade-by-clade expansion of durophagous elements of the Modern fauna progressively further offshore summed to produce an overall expansion of Modern communities offshore, and an overall displacement of the Paleozoic fauna into deeper-water environments. This is why the deep sea is popularly viewed as the biotically relaxed redoubt of living fossils and the anachronistic, Paleozoic-type communities they comprise. Aronson (1990) extended the onshore–offshore model to include human fishing activity, a trophic innovation that was recently introduced to nearshore marine ecosystems and then expanded to offshore and deep-sea environments.

Although Sepkoski's analyses, as well as subsequent family- and genus-level analyses (Benton, 1995; Foote, 2000), indicated rapid diversification of the Modern fauna during the Mesozoic and Cenozoic, reanalysis with a more complete, genus-level database did not show such a pattern (Alroy et al., 2008). Whether global marine diversity expanded dramatically after the end-Permian extinction or reached a plateau during the Mesozoic, the onshore–offshore dynamics of the predatory innovations of the Mesozoic appear robust (e.g., Jablonski and Sepkoski, 1996; Erwin, 2008).

CRINOIDS AND OPHIUROIDS

The history of stalked crinoids is perhaps the best-documented example of the onshore–offshore trend in predation and community structure. Members of the Paleozoic evolutionary fauna and abundant constituents of shallow-benthic communities in the Paleozoic and early Mesozoic, stalked crinoids persisted in near-shore shallow-water habitats through a mid-Paleozoic escalation in the diversity and function of durophagous taxa by evolving enhanced defensive architectures (Signor and Brett, 1984; Aronson, 1991b; Baumiller and Gahn, 2004). These sedentary, epifaunal suspension-feeders were then eliminated from nearshore habitats of less than ~100 m water depth during the Cretaceous and replaced by the unstalked (comatulid) crinoids (Meyer and Macurda, 1977; Oji, 1985; Bottjer and Jablonski, 1988). The comatulids, being mobile, are better able to avoid their predators in shallow water, whereas the stalked crinoids are now confined to the low-predation environments of the deep sea (Meyer, 1985; Oji, 1996).

Building on these ideas about crinoids, over the last quarter-century I have explored the hypothesis of a scale-independent trophic relationship between epifaunal, suspension-feeding ophiuroids and their durophagous predators. The overall argument fell into a logical sequence, at least in retrospect, although many of the points were addressed simultaneously.

(1) Dense populations of epifaunal, suspension-feeding ophiuroids ('brittlestar beds') are found in bathyal environments (Blaber et al., 1987; Fujita and Ohta, 1990). Dense populations also live in restricted coastal, shallow-water habitats in a number of locations, including the British Isles, California, and the northern Adriatic Sea (Warner, 1971; Morris et al., 1980; McKinney and Hageman, 2006). Population densities range in the hundreds to thousands of individuals per square meter. Observations of brittlestar beds living on soft substrata in a marine lake on Eleuthera Island, Bahamas and around the British Isles showed that these populations are restricted to habitats in which durophagous predators are uncommon or rare. Meters to kilometers away, reef habitats contained high densi-ties of durophagous predators and far fewer ophiuroids (Aronson and Harms, 1985; Aronson, 1989a).

(2) I conducted tethering experiments in the Bahamas, Scotland, and the Irish Sea to measure predation potential, which I defined as the activity and propensity of predators to consume the focal prey (Aronson, 1989a; Aronson and Heck, 1995). These experiments, in which ophiuroids were restrained and set out as 'bait,' showed that predation potential was far lower in brittlestar beds than in nearby reef habitats (Aronson and Harms, 1985; Aronson, 1989a).

(3) Observations during tethering experiments, ecological surveys, and gut-content analyses showed that the primary predators of ophiuroids in reef habitats were fish and crabs, which are fast-moving, durophagous predators of a Modern functional grade. In contrast, the primary predators of ophiuroids in brittlestar beds on nearby or adjacent soft substrata were asteroids, polychaete worms, and carnivorous ophiuroids, all of which are slow-moving, non-durophagous predators of a Paleozoic functional grade (Aronson and Harms, 1985; Aronson, 1989a; Aronson and Blake, 2001; see Blake and Guensberg [1990] on asteroids as functionally Paleozoic predators).

(4) Tethering experiments in shallow, back-reef habitats at geographically dispersed sites in the Caribbean showed that predation potential, again measured with tethering experiments, was consistently different among sites, both seasonally and on a decadal time scale. Among-site differences in predation potential correlated with differences in the abundance of predatory fish (Aronson 1992b, 1998).

(5) Predation pressure, the time-integrated encounter rate of prey with their predators, was measured as the proportion of ophiuroids in a population bearing sublethal damage in the form of one or more regenerating arms (Aronson, 1989a). Levels of sublethal arm damage were lower in brittlestar beds than in reef habitats, correlating with predation potential as measured by tethering experiments (Aronson, 1987, 1989a). Physical disturbance from storms did not appear to be a significant source of arm damage (Aronson, 1991c).

(6) Dense populations of infaunal, suspension-feeding ophiuroids (hundreds to thousands of individuals per square meter) persist in areas of high predation pressure in northern Europe, the northern Gulf of Mexico, and elsewhere. Durophagous predators, primarily teleosts, crop the brittlestar arms protruding from the

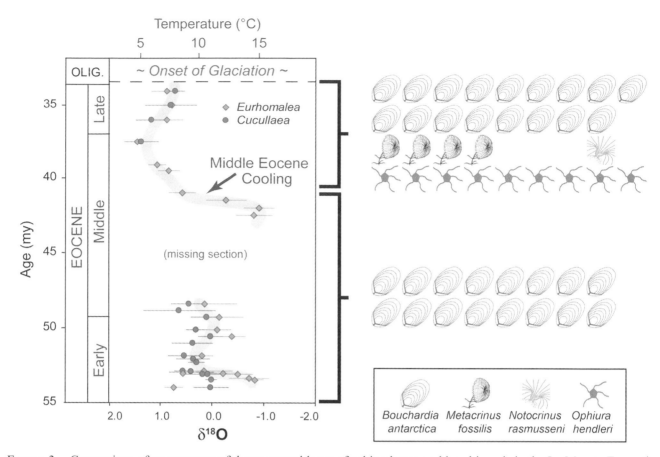

FIGURE 2.—Comparison of occurrences of dense assemblages of echinoderms and brachiopods in the La Meseta Formation at Seymour Island in response to the 41-Ma cooling step. The temperature–$\delta^{18}O$ curve for Seymour Island, left, is redrawn from Ivany et al. (2008); *Eurhomalea* and *Cucullea* were the two bivalve genera used in the isotopic analysis. The frequency distributions of dense brachiopod and echinoderm concentrations across the 41-Ma event, right, are significantly different (χ^2 = 10.291, df = 1, P = 0.001). Credit: Aronson et al. (2009), an open-access article distributed under the terms of the Creative Commons Attribution License.

sediment (Duineveld and van Noort, 1986; Munday, 1993; Sköld et al., 1994). Pilot tethering experiments with infaunal ophiuroids in the Gullmar Fjord off the west coast of Sweden indicated that predation potential was high in dense infaunal populations (R. B. Aronson, unpublished data).

(7) Historical records from the Isle of Man, corroborated through interviews with Manx fishermen, showed that an epifaunal brittlestar bed in the Irish Sea persisted continuously for more than a century (Aronson, 1989a). In contrast, data from scientific trawling surveys in the western English Channel suggested an oceanographically driven, negative relationship between epifaunal brittlestar beds and the abundance of predatory asteroids and bottom-feeding fishes (Holme, 1984). On a spatial scale of tens to hundreds of kilo-

meters and a temporal scale of decades to centuries, brittlestar beds were widespread in the western Channel when their predators were rare, and *vice versa* (see also Aronson, 1992a).

(8) On a global spatial scale and a temporal scale of tens to hundreds of millions of years, dense, autochthonous (or parautochthonous) fossil accumulations of ophiuroids from nearshore, shallow-water facies occur worldwide in Paleozoic, Mesozoic, and Cenozoic deposits (Aronson, 1989b, 1992a). Most of these dense, fossil ophiuroid assemblages are interpreted as event beds resulting from rapid burial of dense living populations, judging from the lithology of the matrix and the abundance within single bedding planes of articulated ophiuroids in life position (Aronson and Sues, 1987; Aronson, 1989b). Based on morphology, phylogenetic

relationships, and preserved bodily attitude, the ophiuroids that formed these dense paleopopulations lived epifaunally.

(9) Sublethal arm damage is a valid measure of predation pressure in fossil ophiuroid populations, as it is in living populations (Aronson, 1987, 1991c). The proportion of ophiuroids with sublethal arm damage is very low in these dense fossil accumulations, indicating that predation pressure was low (Aronson, 1992a).

(10) How are these dense, low-predation, and epifaunal brittlestar beds in shallow-water environments distributed through geological time? Brittlestar beds are uncommon in the fossil record, but sample sizes are sufficient for statistical analysis. Binomial testing of the frequencies of occurrence of brittlestar beds in adjacent stratigraphic intervals revealed that they declined precipitously after the Jurassic, around the time durophagous predators were beginning to diversity, strongly suggesting a causal connection (Aronson, 1989b, 1992a).

These results support the hypothesis of a scale-independent relationship between ophiuroids and their predators. The paleontological data suggest that dense populations of epifaunal ophiuroids in nearshore environments responded to increasing predation in the Mesozoic in a similar fashion to populations of stalked crinoids, declining in response to the radiations of Modern durophagous predators. Brittlestar beds living in nearshore, shallow-water habitats are ecological anachronisms. Low predation pressure from durophagous predators has produced a retrograde structure and function strongly reminiscent of Paleozoic shallow-water communities.

PRIMARY PRODUCTIVITY

Top-down controls clearly determine where and when brittlestar beds will persist on multiple spatiotemporal scales. Bottom-up controls are important as well. Epifaunal and infaunal brittlestar beds persist only in areas where fluxes of phytoplankton or particulate organic matter are sufficient to support the energetic requirements of the millions of suspension-feeders in the populations (Warner, 1971; Rosenberg et al., 1987; Aronson, 1989a; Zuschin and Stachowitsch, 2009).

On macroevolutionary scales, Bambach (1993) and Vermeij (1995) argued persuasively that increasing energetic inputs drove escalating predation and infau-

nalization, as discussed above. Scaling down the productivity argument, McKinney and Hageman (2006) suggested that high-nutrient conditions in the western portion of the northern Adriatic Sea have driven shallow-benthic communities away from dominance by epifaunal suspension-feeders, including foliose bryozoans and ophiuroids, and toward energetically demanding bioturbation and infaunalization. In contrast to the eutrophic northwestern Adriatic, the oligotrophic northeastern Adriatic was characterized by the Paleozoic-like dominance of epifaunal suspension-feeders. Low predation pressure was viewed as a background condition in the northern Adriatic, making possible the contrast in benthic community structure. The authors compared their results to the difference in benthic community structure on opposite sides of McMurdo Sound, Antarctica. Dayton and Oliver (1977) showed that the nutrient-rich eastern side had high infaunal densities, whereas the oligotrophic western side had "patterns of mobile epifauna and low infauna density similar to bathyal and deep-sea communities." The hypothesis of McKinney and Hageman (2006) conflicted with the idea of high energetic requirements for the establishment and persistence of dense brittlestar beds.

Zuschin and Stachowitsch (2009) argued that the greater density of epifaunal suspension-feeders observed in northwestern Adriatic reflected the greater availability of shelly and other hard substrata, rather than a difference in food availability. They concluded that nutrient concentrations were sufficient to support communities dominated by epifaunal suspension-feeders on both sides of the northern Adriatic. They also pointed out that McMurdo Sound is not comparable to the northern Adriatic: although densities of infauna are higher on the nutrient-rich eastern side of McMurdo, so too are the densities of epifaunal suspension-feeders. Finally, they suggested that levels of durophagous predation are not in general low in the northern Adriatic. This latter assertion, however, was not based on predation potential or predation pressure measured for the focal prey in the focal communities.

Sufficient nutrient concentrations and suspended-matter (i.e., energetic) flux are critical to the establishment and persistence of living, dense suspension-feeder communities on ecological scales. Increasing productivity on a macroevolutionary time scale provided the energetic driver of increased bioturbation and the resultant emphasis on infauna in Modern soft-substratum communities. There is no paradox here: unlike pre-dation, the qualitative effects of nutrient input and re-sultant primary productivity are scale-dependent. Low predation pressure from modern, durophagous taxa remains an essential background requirement for epi-faunal ophiuroids, crinoids, and presumably other Pa-leozoic-type suspension-feeders, regardless of scale.

BRACHIOPODS

Brachiopods are the most abundant marine macro-fossils from the Paleozoic. They were numerically im-portant components of Paleozoic suspension-feeding communities in shallow, nearshore environments on soft substrata as well as unconsolidated, mixed substra-ta (i.e., substrata containing small rocks and bioclasts, to which they attached). Brachiopods declined precipi-tously in the end-Permian extinction and were replaced by bivalves (Gould and Calloway, 1980).

Rhynchonelliform, or articulate, brachiopods were/are sessile, epifaunal to semi-infaunal suspension-feed-ers. Today they live on hard bottoms and unconsolidat-ed, mixed substrata. They are common components of some modern, hard- and mixed-bottom communities in nearshore and outer-shelf environments in the South-ern Hemisphere, in all latitudinal zones and under both high- and low-productivity conditions (Dell, 1972; Foster, 1974; Smith and Witman, 1999; Kowalewski et al., 2002). In Antarctica they can occur in dense ag-gregations on the continental shelf on both hard and mixed substrata (Brey et al., 1995; Peck, 1996). Dense aggregations of epifaunal brachiopods occur rarely in nearshore environments in the Northern Hemisphere (e.g., Noble et al., 1976).

Brachiopods are considered poor competitors in Modern-type benthic communities because of their low metabolism and slow growth rates (Jackson et al., 1971; Thayer, 1981; Rhodes and Thompson, 1993; Peck, 1996). Whether and how aggregations of epifau-nal brachiopods fit along the metaphorical continuum from low-predation, Paleozoic-type assemblages to high-predation, Modern-type assemblages is incom-pletely understood (James et al., 1992). Morphological evidence points to an evolutionary response to increas-ing predation in the Devonian, particularly in the paleo-tropics (Dietl and Kelley, 2001). In contrast, brachio-pods were not affected, or only weakly affected, by the Mesozoic increase in durophagous predation, possibly due to their low energy content and postulated deploy-ment of unpalatable secondary metabolites (Leighton, 2003).

THE MODERN BENTHOS
IN ANTARCTICA

The endemic, shallow-benthic fauna of Antarctica is functionally different from the nearshore benthos in tropical, temperate, and Arctic latitudes (Dell, 1972; Arntz et al., 1994; Clarke et al., 2004; Aronson et al., 2007). Dense, quasi-Paleozoic assemblages of epifau-nal suspension-feeders, including ophiuroids (Fig. 1) as well as sessile forms, are found in Antarctic shallow-water habitats where productivity is sufficiently high (Dayton and Oliver, 1977; Gili et al., 2006). A second retrograde feature is the extremely low level of du-rophagous predation in Antarctica (Aronson and Blake, 2001; Aronson et al., 2007). The top predators of shal-low-water benthos are slow-moving, non-durophagous invertebrates of a Paleozoic functional grade, including asteroids, giant nemerteans, and giant pycnogonids.

Cold sea temperatures and a generally narrow, steeply sloping continental shelf with a deep shelf-break have maintained strong ecological and evolutionary connections between the region's shallow-nearshore and deep-sea faunas (see reviews cited above, and Barnes and Peck [2008] on brachiopods). Glaciations forced the bottom fauna down the continental shelf and slope, simultaneously limiting elements of the fauna to ice-free refugia in shallow areas (Clarke et al., 2004; Thatje et al., 2008a). These strictures were then relaxed during subsequent interglacials. The continuity of deep-sea and shallow-water environments is one reason for the archaic character of the benthic fauna.

The other important reason, of course, is that du-rophagous predation is severely limited (Dell, 1972; Dayton et al., 1974; Arntz et al., 1994; Aronson and Blake, 2001; Aronson et al., 2007). Functionally mod-ern, skeleton-crushing taxa, which structure nearshore benthic communities elsewhere, are absent from Ant-arctica. There are no crabs or lobsters; sharks and rays are entirely absent as well, and skates are rare; and the limited higher-level diversity of bony fish does not in-clude durophagous taxa.

The only teleosts in Antarctica at present are the notothenioids (Fig. 1) and liparids, which have evolved antifreeze glycoproteins (AFGPs) to survive cold sea temperatures. Like antifreeze in a car, AFGPs prevent the fatal nucleation of ice crystals inside the fishes' bodies. Notothenioids and liparids are not durophagous, but there is nothing about producing AFGPs that inherently excludes durophagy in teleosts. In fact, durophagous Arctic gadids (cod) have convergently evolved AFGPs. Other durophagous taxa in the Arctic include pleuronectids (flatfish), cottids (sculpins), zoarcids (eelpout) and a squalid (dogfish, or horned shark). These fishes have a variety of physiological and behavioral adaptations to survive freezing temperatures. The Arctic also has walruses and gray whales, but there are no durophagous marine mammals in Antarctica. In summary, evolution of the Antarctic bottom fauna has been driven by the physical extremity of the polar environment and by historical contingency.

HISTORY OF THE ANTARCTIC BOTTOM FAUNA

The historical roots of the living bottom fauna date to a long-term cooling trend that began in the Eocene. By the time of the Eocene, Antarctica was close to its present position. The climate of the early Eocene was what we would describe as cool-temperate today, and the marine fauna was typical of Tertiary faunas elsewhere. Sea temperatures began to decline in the middle Eocene, eventually leading to today's extremely cold, polar environment. The first major drop in temperature, a decline of as much as 10 °C, occurred over a period of several million years beginning ~41 Ma (million years ago). The thermal step-down coincided with the initial opening of Drake Passage and establishment of the Antarctic Circumpolar Current, or ACC (Scher and Martin, 2006; Ivany et al., 2008). Permanent ice sheets were established in Antarctica at the Eocene–Oligocene boundary, ~33.5 Ma (Zachos et al., 2001; Ivany et al., 2006).

The Eocene La Meseta Formation at Seymour Island contains the best record of marine macrofossils in Antarctica, and it spans this critical time in the history of the bottom fauna. Seymour Island is located in the Weddell Sea, close to the northern tip of the Antarctic Peninsula. Geological evidence shows that the La Meseta Formation was deposited in a nearshore, shallow-water, soft-substratum setting under fully marine conditions (Ivany et al., 2008). Sometime after the 41-Ma cooling event, durophagous teleosts, reptant decapods,

and neoselachians (with the exception of a few skates) became extinct, drastically lowering predation pressure and, by their absence, restructuring the Antarctic bottom fauna. The ACC is thought to have created an oceanographic barrier that helped isolate marine life in Antarctica.

Evidence from the La Meseta Formation strongly suggests that predation on epifaunal, suspension-feeding echinoderms declined across the 41-Ma cooling event. Dense, autochthonous or parautochthonous fossil aggregations, representing dense paleopopulations of epifaunal ophiuroids (*Ophiura hendleri*) and stalked isocrinid crinoids (*Metacrinus fossilis*), as well as one dense paleopopulation of comatulid crinoids (*Notocrinus rasmusseni*), have been documented within the formation (Aronson et al., 1997, 2009). (Note that the stalk of *M. fossilis* was reduced to a short dart, making it essentially a stalkless stalked crinoid that planted itself in the soft sediment [Meyer and Oji, 1993].) Assessments of sublethal arm damage in *Ophiura* and *Metacrinus* indicated very low levels of predation pressure (Aronson et al., 1997). These low-predation paleopopulations flourished after 41 Ma but did not occur before, suggesting a decline in durophagous predation pressure associated with the cooling event (Aronson et al., 1997; Aronson and Blake, 2001).

In contrast, dense paleopopulations of epifaunal or semi-infaunal, rhynchonelliform brachiopods (*Bouchardia antarctica*) occurred commonly both before and after the 41-Ma cooling event and its associated decline in predation (Fig. 2). Of the hundreds of *Bouchardia* examined from paleopopulations that lived before and after the cooling event, none displayed signs of sublethal damage in the form of repaired shell breaks. Furthermore, none of the *Bouchardia* showed signs of predation by shell-drilling naticid gastropods. The naticids did, however, drill infaunal bivalves both before and after 41 Ma, with roughly the same frequency in both intervals (Aronson et al., 2009).

The contrasting patterns of occurrence of dense paleopopulations of epifaunal, suspension-feeding echinoderms and rhynchonelliform brachiopods are consistent with the differential responses of the two groups to escalating predation on larger scales in the Mesozoic. Supporting the hypothesis of a decline in durophagous predation in Antarctica during the Eocene, the epifaunal, suspension-feeding hiatellid bivalve *Hiatella tenuis* also formed dense populations at what is now

Seymour Island after, but not before, 41 Ma.

Conditions favorable to dense ophiuroid, crinoid, and hiatellid populations may have been enhanced after 41 Ma by upwelling events that increased productivity (references in Ivany et al., 2008). There were, however, dense populations of infaunal, suspension-feeding bivalves both before and after the cooling event (Stilwell and Zinsmeister, 1992; Aronson et al., 2009), so presumably productivity was sufficient to support dense echinoderm populations before 41 Ma, at least at certain times and places. This line of reasoning might be supported by the even distribution of dense aggregations of *B. antarctica* across the 41-Ma event, because its living congener, *B. rosea*, forms dense aggregations in upwelling zones on the outer shelf of Brazil (Kowalewski et al., 2002). Declining predation remains the best candidate for the essential condition that permitted dense ophiuroid and crinoid populations to become established and thrive after 41 Ma.

PHYSIOLOGICAL CONSTRAINTS ON DUROPHAGOUS PREDATORS

Limitations to durophagy at cold temperatures in modern Antarctic environments include: (1) thermal constraints on muscular performance; and possibly (2) the high energetic cost of depositing calcium carbonate at the cold temperatures and low saturation states of polar seas (Aronson et al., 2007). The high cost of biodeposition of calcium carbonate is inferred from the thin shells of mollusks observed in polar regions (Vermeij, 1978). If calcification is costly—and there is some doubt about that (Aronson et al., 2007)—then both the feeding apparatus of durophagous predators and the calcified defenses of their prey should be limited in polar environments.

The teleostean fauna in Antarctica is at present confined to taxa possessing AFGPs, and they happen not to be durophagous. Why there are no bottom-feeding sharks in Antarctica is more of a mystery, although some preliminary hypotheses have been offered (Aronson et al., 2007). Durophagous crustaceans are phylogenetically constrained from surviving in polar seas, and they are found in neither the Antarctic nor the high Arctic.

Reptant decapod crustaceans, including durophagous crabs and lobsters, are incapable of reducing the concentration of magnesium ions in their hemolymph, which as a result equilibrates with the surrounding seawater (Frederich et al., 2001). Magnesium is a narcotic, as anyone who has used Epsom salts—magnesium sulfate—to anesthetize marine invertebrates or ease constipation will know. At temperatures below 1 °C, corresponding to the sea temperatures in Antarctic shelf environments, magnesium narcosis amplifies the extreme torpor induced by low temperature, to the point crabs and lobsters simply stop functioning and die. Of the reptant decapods, the lithodids (anomuran king crabs) are the most cold-tolerant, being able to survive in a torpid, hypometabolic state at temperatures of 1–2 °C (Anger et al., 2003; Thatje et al., 2005). The problem of magnesium regulation does not vanish for reptant decapods at lower latitudes, but temperatures are high enough that magnesium narcosis is not debilitating. Unlike crabs and lobsters, amphipods and isopods are capable of down-regulating magnesium, and they are abundant and ecologically important in the benthic communities of Antarctica.

CLIMATE CHANGE AND THE ANTARCTIC BOTTOM FAUNA

The shallow seas off the Western Antarctic Peninsula (WAP) are among the fastest-warming in the world. Sea-surface temperatures have risen more than 1 °C in the last 50 years (Meredith and King, 2005; Clarke et al., 2007). The observations, data, and conceptual models reviewed above, taken from a variety of disciplines and from multiple scales and hierarchical levels, can be melded into a predictive scenario: durophagous predators will reinvade the nearshore-benthic communities of Antarctica within the next 50–100 years, with potentially dire consequences for the endemic bottom fauna. This unsettling prediction is already coming true, as crabs have begun to reinvade Antarctica by several routes.

First, the Antarctic Circumpolar Current is more permeable to oceanographic exchange than it was once thought to be. Like other current systems, the ACC produces eddies at a range of scales, with cold-core rings traveling north and warm-core rings traveling south (Olbers et al., 2004; Glorioso et al., 2005). Warm-core rings transport the larvae of subantarctic anomuran and brachyuran crabs into Antarctic waters (Thatje and Fuentes, 2003). Rafting on natural and anthropogenic flotsam provides a second avenue for biological invasion

(Barnes, 2002). Increasing ship traffic in Antarctica is a third mechanism for introducing alien species, which are transported in the ships' ballast tanks and on their hulls (Barnes et al., 2006). Two adult subarctic crabs, one male and one female, were recorded off the WAP in 1986, demonstrating the potential for ships to serve as vectors of predatory invasion (Tavares and De Melo, 2004).

The survival and successful recruitment of decapod larvae invading via warm-core rings, flotsam, and ballast water will require shallow-sea temperatures to rise above the critical level for magnesium narcosis. As the Antarctic seas warm, longer growing seasons for phytoplankton will combine with accelerated larval development to promote the decapod invasion (Aronson et al., 2007). On the other hand, climatically driven changes in the composition of the phytoplankton and ocean acidification could be detrimental to incoming decapod larvae (McClintock et al., 2008).

Recent explorations in deep water off the WAP have turned up large populations of adult lithodids on the continental slope (Thatje and Arntz, 2004; García Raso et al., 2005; Thatje and Lörz, 2005; McClintock et al., 2008; Thatje et al., 2008b). The oceanography of the Southern Ocean renders slope waters off the WAP slightly warmer than the shelf waters directly above, allowing the lithodids to tolerate magnesium narcosis and live, albeit at a slow pace. The origin of the lithodid populations is unknown, but it is likely their demersal larvae island-hopped along the topographic highs of the Scotia Arc, a chain of submerged peaks that connects the Andes of South America to the mountains of the Peninsula (Thatje et al., 2005).

King crabs are durophagous predators that eat echinoderms, mollusks, bryozoans, and other benthic invertebrates. At current rates of climate change, another 50–100 years of surface warming along the WAP should render nearshore waters sufficiently warm for king crabs on the slope to walk or send larvae into shallow-benthic habitats. Brachyuran crabs and durophagous fish from the subantarctic might also be able to establish viable populations within a centennial time frame.

The ecological consequences could be devastating. If we run the Eocene record of climatic cooling from Seymour Island in rapid-reverse, we can make specific predictions about particular faunal components (Aronson et al., 2009). Dense populations of ophiuroids

should be decimated, whereas shell-drilling predation on bivalves should not change significantly. According to the paleontological data, aggregations of rhynchonellid brachiopods should not be especially vulnerable to increasing predation.

Newly published ecological data potentially contradict the prediction of no strong effect on brachiopods. Harper et al. (2009) showed that unsuccccessful predatory attacks occurred much less frequently in living populations of Antarctic brachiopods than in living subantarctic and South American populations. This geographical pattern might mean that generally increasing durophagous predation in Antarctica will affect brachiopods, although issues of relative palatability (Mahon et al., 2003) and the impact of predation on population densities have yet to be resolved.

These alterations in food-web dynamics will overprint the direct, though highly variable, physiological responses to rising temperatures expected of particular taxa (Barnes and Peck, 2008). For ophiuroids and brachiopods, however, native taxa that may be excluded from Antarctica by climate change will likely be replaced by ecological equivalents from the subantarctic, leaving the predictions of altered predator–prey dynamics intact (Aronson et al., 2009). Other expected impacts of climate change include increased ice scour, which will physically disrupt benthic communities, and ocean acidification, which will inhibit calcification and, therefore, possibly de-escalate durophagous predation (Smale and Barnes, 2008).

The implications for conservation policy are clear. Preemptively disrupting the flow of invasive predators will require strengthening the Antarctic Treaty to control ship traffic in general and discharge of ballast water in particular. More difficult will be controlling emissions of greenhouse gases to slow and eventually reverse the effects of climate change (Thatje and Aronson, 2009). Both imperatives will require international cooperation that goes well beyond the usual political posturing.

QUESTIONS AND MORE QUESTIONS

The Paleozoic flavor of the living bottom fauna of Antarctica has been developing since the Eocene, apparently driven by the suppressive effect of declining sea temperatures on durophagous predation. Aronson et al. (2007) noted that the few post-Eocene records of fossil reptant decapods, which are from the early Miocene and Pliocene, correspond to times of relative warmth. Sea temperatures were warmer than at present during the previous interglacial (marine isotope stage 5e, ~125 ka), so what happened to the bottom fauna at that time? Did lithodids colonize during the Pleistocene and move down- and up-slope as the ice advanced and retreated? Did durophagous predation increase during the interglacials? The fossil record of marine life during the Pleistocene interglacials has, unfortunately, been obliterated by the scour of subsequent glacial movements (Barnes and Conlan, 2007). Ongoing genetic work on Antarctic king crabs should help fill the gaps in our understanding, by providing clues to the phylogenetic affinities of lithodids and the timing of their arrival off the WAP (Thatje et al., 2008b; Hall and Thatje, in press).

We do know the situation today is already more extreme than the Pleistocene interglacials. Atmospheric concentrations of carbon dioxide are higher than at any time during the last 2 million years and rising rapidly (IPCC, 2007; Hönisch et al., 2009). Furthermore, humans are injecting durophagous predators into Antarctic waters at a rate far exceeding what would be expected for natural, climatically mediated, latitudinal range shifts (Aronson et al., 2007, 2009).

A second, more general issue in paleobiology is whether the Mesozoic marine revolution went further in Laurasia than in Gondwana. The greater preponderance of rhynchonellid brachiopods in the nearshore and shelf waters of South America and New Zealand, and descriptions of fossil stalked crinoids from Cenozoic shallow-water deposits in South America, Australia, and New Zealand (Oji, 1996; Malumian et al., 2005), hint that living shallow-marine communities of the Southern Hemisphere may be structurally more Paleozoic than the better-studied systems to the north. The possibility of hemispheric-scale asynchrony in Phanerozoic patterns of escalation is a fascinating topic for further study.

CONCLUSION

The fossil record provides far more than a baseline against which to compare living communities threatened by degradation. It is an archive from which we can reconstruct the history of living communities and ascertain the causes of that history. The overarching method of paleoecology, if there is one, consists of conceptual jockeying between paleontology and ecology to test the hypothesis of scale independence. The rest is interdisciplinary common sense. The hypothesis is tested by moving from question to question, with technique dictated by the particulars of the case.

Shallow-benthic communities in Antarctica were functionally modern from the early Eocene until about 41 Ma. The long-term cooling trend that followed reduced durophagous predation and imparted a Paleozoic character to those communities. Ongoing reinvasions of durophagous predators could re-modernize the Antarctic benthos, accelerating the global homogenization of marine biotas. Information content has long been taken as a metaphor and a model for ecological diversity, so, metaphorically speaking, we are dumbing down global marine biodiversity by imperiling the endemic marine fauna of Antarctica.

Metaphors of geohistorical time-equivalence provide a dimension of understanding and predictive power far beyond the reach of real-time ecology. A fascinating terrestrial example is the cascading impact of human arrival and extirpation of the Pleistocene megafauna in the Americas. Adding a trophic level to modern terrestrial communities forced them into a 'post-modern' configuration, in which many tropical plants were bereft of dispersal agents (Janzen and Martin, 1982). A purely ecological perspective fails to recognize the ecological legacy of the megafauna, and instead views the plants as poorly adapted to dispersal. But there is more insight to be drawn from the metaphor. Livestock introduced following European colonization and other post-Columbian human activities have to some extent reversed the effects of the loss of megafauna (Janzen and Martin, 1982; Guimarães et al., 2008), although weak and artificial compensation through further perturbation can hardly be construed as a tidy solution of the problem. The radical policy recommendation to 're-wild' the American landscape by introducing African megafauna (Donlan et al., 2005) is paleoecology at its most imaginative. To view this proposal as the *reductio ad absurdum* of our science is to miss the essential point of how bad things really are. Environmental destruction is the absurdity, not unconventional thinking.

Assertions that marine communities are back-sliding to a Proterozoic level of organization because of climate change, nutrient loading, overfishing, and other human insults (Hallock, 2001, 2005; Jackson, 2008;

Richardson et al., 2009) are unsettling and controversial. Although the nihilistic vision of a marine biota dominated by jellyfish and cyanobacterial mats may not be correct, it is a paleoecological hypothesis about the future history of marine systems. Like the incipient predatory invasions of Antarctica and the fate of the fruits the gomphotheres ate, the Proterozoic scenario affords us the opportunity to make specific predictions and take preemptive action, a policy option far better than reacting to environmental disaster during or after the fact. Action to save the oceans could render the jellyfish–bacterial slime hypothesis untestable, a scientific irony with which I am sure we all could live. Paleoecology offers us an avenue of prediction from retrospective analysis at a time of rapid global change, when we need to know more than simply to expect nasty surprises.

ACKNOWLEDGMENTS

The work described in this paper continues to benefit from discussions with friends and colleagues. I am especially grateful to William Ausich, Martin Buzas, William DiMichele, Fred Grassle, David Jablonski, David Meyer, Rich Mooi, Tatsuo Oji, David Pawson, Roy Plotnick, Kenneth Sebens, Andrew Smith, the late Jack Sepkoski, Hans Sues, George Warner, and especially Les Kaufman for their advice and encouragement during the early stages of my career as a paleoecologist. My understanding of community dynamics in Antarctica has greatly benefited from interactions with Charles Amsler, Bill Baker, Daniel Blake, Andrew Clarke, Alistair Crame, Paul Dayton, Linda Ivany, Jere Lipps, James McClintock, Ryan Moody, Lloyd Peck, Victor Smetacek, Simon Thrush, and Sven Thatje. Mark Bush, Gregory Dietl, Jennifer Hobbs, Lauren Toth, and Robert van Woesik commented on the manuscript, and Ryan Moody drew Fig. 1. The U.S. National Science Foundation supports my research in Antarctica, currently through grant ANT-0838846. This is Contribution Number 10 from the Institute for Adaptation to Global Climate Change at the Florida Institute of Technology.

Norman Holme, who compiled the trawl data from the western English Channel, treated me with kindness and enthusiasm when I visited him at the Marine Biological Association Laboratory in Plymouth in 1985. He was one of the great pioneers of marine ecology in Britain, although he is less appreciated in the United States. This paper is dedicated to Norman's memory.

REFERENCES

ABERHAN, M., W. KIESSLING, AND F. T. FÜRISCH. 2006. Testing the role of biological interactions in the evolution of mid-Mesozoic marine benthic ecosystems. Paleobiology, 32:259-277.

ALEXANDER, R. R., AND G. P. DIETL. 2003. The fossil record of shell-breaking predation on marine bivalves and gastropods, p. 141-176. *In* P. H. Kelley, M. Kowalewski, and T. A. Hansen (eds.), Predator-Prey Interactions in the Fossil Record. Kluwer/Plenum, New York.

ANGER, K., S. THATJE, G. LOVRICH, AND J. CALCAGNO. 2003. Larval and early juvenile development of *Paralomis granulosa* reared at different temperatures: Tolerance of cold and food limitation in a lithodid crab from high latitudes. Marine Ecology Progress Series, 253:243–251.

ARNTZ, W. E., T. BREY, AND V. A. GALLARDO. 1994. Antarctic zoobenthos. Oceanography and Marine Biology: An Annual Review, 32:241-304.

ALROY, J., M. ABERHAN, D. J. BOTTJER, M. FOOTE, F. T. FURSICH, P. J. HARRIES, A. J. HENDY, S. M. HOLLAND, L. C. IVANY, W. KIESSLING, M. A. KOSNIK, C. R. MARSHALL, A. J. MCGOWAN, A. I. MILLER, T. D. OLSZEWSKI, M. E. PATZKOWSKY, S. E. PETERS, L. VILLIER, P. J. WAGNER, N. BONUSO, P. S. BORKOW, B. BRENNEIS, M. E. CLAPHAM, L. M. FALL, C. A. FERGUSON, V. L. HANSON, A. Z. KRUG, K. M. LAYOU, E. H. LECKEY, S. NURNBURG, C. M. POWERS, J. A. SESSA, C. SIMPSON, A. TOMASOVYCH, AND C. C. VISAGGI. 2008. Phanerozoic trends in the global diversity of marine invertebrates. Science, 321:97-100.

ARONSON, R. B. 1987. Predation on fossil and Recent ophiuroids. Paleobiology, 13:187-192.

ARONSON, R. B. 1989a. Brittlestar beds: Low-predation anachronisms in the British Isles. Ecology, 70:856-865.

ARONSON, R. B. 1989b. A community-level test of the Mesozoic marine revolution theory. Paleobiology 15:20-25.

ARONSON, R. B. 1990. Onshore-offshore patterns of human fishing activity. Palaios, 5:88-93.

ARONSON, R. B. 1991a. Ecology, paleobiology and evolutionary constraint in the octopus. Bulletin of Marine Science, 49:245-255.

ARONSON, R. B. 1991b. Escalating predation on crinoids in the Devonian: Negative community-level evidence. Lethaia, 24:123-128.

ARONSON, R. B. 1991c. Predation, physical disturbance, and sublethal arm damage in ophiuroids: A Jurassic–Recent comparison. Marine Ecology Progress Series, 74:91-97.

ARONSON, R. B. 1992a. Biology of a scale-independent

predator–prey interaction. Marine Ecology Progress Series, 89:1-13.

ARONSON, R. B. 1992b. The effects of geography and hurricane disturbance on a tropical predator-prey interaction. Journal of Experimental Marine Biology and Ecology, 162:15-33.

ARONSON, R. B. 1994. Scale-independent biological processes in the marine environment. Oceanography and Marine Biology: An Annual Review, 32:435-460.

ARONSON, R. B. 1998. Decadal-scale persistence of predation potential in coral reef communities. Marine Ecology Progress Series, 172:53-60.

ARONSON, R. B., AND D. B. BLAKE. 2001. Global climate change and the origin of modern benthic communities in Antarctica. American Zoologist, 41:27-39.

ARONSON, R. B., AND T. J. GIVNISH. 1983. Optimal central place foragers: A comparison with null hypotheses. Ecology, 64:395-399.

ARONSON, R. B., AND C. A. HARMS. 1985. Ophiuroids in a Bahamian saltwater lake: The ecology of a Paleozoic-like community. Ecology, 66:1472-1483.

ARONSON, R. B., AND K. L. HECK JR. 1995. Tethering experiments and hypothesis testing in ecology. Marine Ecology Progress Series, 121:307-309.

ARONSON, R. B., AND R. E. PLOTNICK. 1998. Scale-independent interpretations of macroevolutionary dynamics, p. 430-450. In M. L. McKinney and J. A. Drake (eds.), Biodiversity Dynamics: Turnover of Populations, Taxa and Communities. Columbia University Press, New York.

ARONSON, R. B., AND H.-D. SUES. 1987. The paleoecological significance of an anachronistic ophiuroid community, p. 355-366. In W. C. Kerfoot and A. Sih (eds.), Predation: Direct and Indirect Impacts on Aquatic Communities. University Press of New England, Hanover, New Hampshire.

ARONSON, R. B., D. B. BLAKE, AND T. OJI. 1997. Retrograde community structure in the late Eocene of Antarctica. Geology, 25:903-906.

ARONSON, R. B., S. THATJE, A. CLARKE, L. S. PECK, D. B. BLAKE, C. D. WILGA, AND B. A. SIEBEL. 2007. Climate change and invasibility of the Antarctic benthos. Annual Review of Ecology, Evolution, and Systematics, 38:129-154.

ARONSON, R. B., R. M. MOODY, L. C. IVANY, D. B. BLAKE, J. E. WERNER, AND A. GLASS. 2009. Climate change and trophic response of the Antarctic bottom fauna. PLoS ONE, 4:e4385.

BAMBACH, R. K. 1985. Classes and adaptive variety: The ecology of diversification in marine faunas through the Phanerozoic, p. 191-253. In J. W. Valentine (ed.), Phanerozoic Diversity Patterns: Profiles in Macroevolution. Princeton University Press, Princeton, New Jersey.

BAMBACH, R. K. 1993. Seafood through time: Changes in biomass, energetics, and productivity in the marine ecosystem. Paleobiology, 19:372-397.

BAMBACH, R. K., A. H. KNOLL, AND S. C. WANG. 2004. Origination, extinction, and mass depletions of marine diversity. Paleobiology, 30:522-542.

BARNES, D. K. A. 2002. Invasions by marine life on plastic debris. Nature, 184:203-204.

BARNES, D. K. A., AND K. E. CONLAN. 2007. Disturbance, colonization and development of Antarctic benthic communities. Philosophical Transactions of the Royal Society, B, 362:11-38.

BARNES, D. K. A., AND L. S. PECK. 2008. Vulnerability of Antarctic shelf biodiversity to predicted regional warming. Climate Research, 37:149-163.

BARNES, D. K. A., D. A. HOGDSON, P. CONVEY, C. S. ALLEN, AND A. CLARKE. 2006. Incurson and excursion of Antarctic biota: Past, present and future. Global Ecology and Biogeography, 15:121-142.

BAUMILLER, T. K., AND F. J. GAHN. 2004. Testing predator-driven evolution with Paleozoic crinoid arm regeneration. Science, 305:1453-1455.

BENTON, M. 1995. Diversification and extinction in the history of life. Science, 268:52-58.

BLABER, S. J. M., J. L. MAY, J. W. YOUNG, AND C. M. BULMAN. 1987. Population density and predators of *Ophiacantha fidelis* (Koehler, 1930) (Echinodermata: Ophiuroidea) on the continental slope of Tasmania. Australian Journal of Marine and Freshwater Research, 38:243-247.

BLAKE, D. B., AND T. E. GUENSBERG. 1990. Predatory asteroids and the fate of brachiopods—a comment. Lethaia, 23:429-430.

BOTTJER, D. J., AND W. I. AUSICH. 1986. Phanerozoic development of tiering in soft substrata suspension-feeding communities. Paleobiology, 12:400-420.

BOTTJER, D. J., AND D. JABLONSKI. 1988. Paleoenvironmental patterns in the evolution of post-Paleozoic benthic marine invertebrates. Palaios, 3:540-560.

BREY, T., L. S. PECK, J. GUTT, S. HAIN, AND W. E. ARNTZ. 1995. Population dynamics of *Magellania fragilis*, a brachiopod dominating a mixed-bottom macrobenthic assemblage on the Antarctic shelf. Journal of the Marine Biological Association of the United Kingdom, 75:857-869.

CLARKE, A., R. B. ARONSON, J. A. CRAME, J.-M. GILI, AND D. B. BLAKE. 2004. Evolution and diversity of the benthic fauna of the Southern Ocean continental shelf. Antarctic Science, 16:559-568.

CLARKE, A., E. J. MURPHY, M. P. MEREDITH, J. C. KING, L. S. PECK, D. K. A. BARNES, AND R. C. SMITH. 2007. Climate change and the marine ecosystem of the western Antarctic Peninsula. Philosophical Transactions of the Royal Society, B, 362:149-166.

Cleland, C. 2001. Historical science, experimental science, and the scientific method. Geology, 11:987-990.

Dayton, P. K., and J. S. Oliver. 1977. Antarctic soft-bottom benthos in oligotrophic and eutrophic environments. Science, 197:55-58.

Dayton, P. K., G. A. Robilliard, R. T. Paine, and L. B. Dayton. 1974. Biological accommodation in the benthic community at McMurdo Sound, Antarctica. Ecological Monographs, 44:105-128.

Dell, R. K. 1972. Antarctic benthos. Advances in Marine Biology, 10:1-216.

Dietl, G. P., and P. H. Kelley. 2001. Mid-Paleozoic latitudinal predation gradient: Distribution of brachiopod ornamentation reflects shifting Carboniferous climate. Geology, 29:111-114.

Dietl, G. P., P. H. Kelley, R. Barrick, and W. Showers. 2002. Escalation and extinction selectivity: Morphology versus isotopic reconstruction of bivalve metabolism. Evolution, 56:284-291.

Dietl, G. P., G. S. Herbert, and G. J. Vermeij. 2004. Reduced competition and altered feeding behavior among marine snails after a mass extinction. Science, 306:2229-2231.

Donlan, J., H. W. Greene, J. Berger, C. E. Bock, J. H. Bock, D. A. Burney, J. A. Estes, D. Forman, P. S. Martin, G. W. Roemer, F. A. Smith, and M. E. Soulé. 2005. Rewilding North America. Nature, 436:913-914.

Duineveld, G. C. A., and G. J. van Noort. 1986. Observations of the population dynamics of *Amphiura filiformis* (Ophiuroidea: Echinodermata) in the southern North Sea and its exploitation by the dab, *Limanda limanda*. Netherlands Journal of Sea Research, 20:85-94.

Erwin, D. 2008. Extinction: How Life Nearly Ended 250 Million Years Ago. Princeton University Press, Princeton, New Jersey, 320 p.

Foote, M. 2000. Origination and extinction components of taxonomic diversity: General problems. Paleobiology, 26(Supplement):74-102.

Foster, M. W. 1974. Recent Antarctic and Sub-Antarctic Brachiopods. American Geophysical Union, Antarctic Research Series, 22, Washington, D.C., 189 p.

Frederich, M., F. J. Sartoris, and H.-O. Pörtner. 2001. Distribution patterns of decapod crustaceans in polar areas: A result of magnesium regulation? Polar Biology, 24:719-723.

Fujita, T., and S. Ohta. 1990. Size structure of dense populations of the brittle star *Ophiura sarsii* (Ophiuroidea: Echinodermata) in the bathyal zone around Japan. Marine Ecology Progress Series 64:113-122.

García Raso, J. E., M. E. Manjón-Cabeza, A. Ramos, and I. Olasi. 2005. New record of Lithodidae (Crustacea, Decapoda, Anomura) from the Antarctic (Bellingshausen Sea). Polar Biology, 28:642-646.

Gili, J.-M., W. E. Arntz, A. Palanques, C. Orejas, A. Clarke, P. K. Dayton, E. Isla, N. Teixidó, S. Rossi, and P. J. López-González. 2006. A unique assemblage of epibenthic sessile suspension-feeders with archaic features in the high-Antarctic. Deep-Sea Research II, 53:1029-1052.

Glorioso, P. D., A. R. Piola, and R. R. Leben. 2005. Mesoscale eddies in the Subantarctic Front, southwestern Atlantic. Scientia Marina 69(Supplement 2):7-15.

Gould, S. J. 1981. Palaeontology plus ecology as palaeobiology, p. 295-317. *In* R. M. May (ed.), Theoretical Ecology: Principles and Applications (second edition). Sinauer Associates, Sunderland, Massachusetts.

Gould, S. J. 1985. The paradox of the first tier: An agenda for paleobiology. Paleobiology, 11:2-12.

Gould, S. J., and C. B. Calloway. 1980. Clams and brachiopods—ships that pass in the night. Paleobiology, 6:383-396.

Gould, S. J., and R. C. Lewontin. 1979. The spandrels of San Marco and the Panglossian paradigm: A critique of the adaptationist programme. Proceedings of the Royal Society, B, 205:581-598.

Guimarães, P. R., M. Galetti, and P. Jordano. 2008. Seed dispersal anachronisms: Rethinking the fruits the gomphotheres ate. PloS ONE, 3:e1745.

Hall, S., and S. Thatje. In press. Global bottlenecks in the distribution of marine Crustacea: Temperature constraints in the family Lithodidae. Journal of Biogeography.

Hallock, P. 2001. Coral reefs, carbonate sediments, nutrients, and global change, p. 387-427. *In* G. D. Stanley Jr. (ed.), The History and Sedimentology of Ancient Reef Systems. Kluwer/Plenum, New York.

Hallock, P. 2005. Global change and modern coral reefs: new opportunities to understand shallow-water carbonate depositional processes. Sedimentary Geology, 175:19-33.

Harper, E. M. 2003. The Mesozoic marine revolution, p. 433-455. *In* P. H. Kelley, M. Kowalewski, and T. A. Hansen (eds.), Predator-Prey Interactions in the Fossil Record. Kluwer/Plenum, New York.

Harper, E. M., L. S. Peck, and K. R. Hendry. 2009. Patterns of shell repair in articulate brachiopods indicate size constitutes a refuge from predation. Marine Biology, 156:1993-2000.

Holme, N. A. 1984. Fluctuations of *Ophiothrix fragilis* in the western English Channel. Journal of the Marine Biological Association of the United Kingdom, 64:351-378.

Hönisch, B., N. G. Hemming, D. Archer, M. Siddall, and J. F. McManus. 2009. Atmospheric carbon dioxide concentration across the mid-Pleistocene transition. Science, 324:1551-1554.

IPCC. 2007. Climate Change 2007: The Physical Science Basis: Summary for Policymakers. Intergovernmental Panel on Climate Change, United Nations World Meteorological Organization, Geneva, 18 p.

IVANY, L. C., K. C. LOHMANN, F. HASIUK, D. B. BLAKE, A. GLASS, R. B. ARONSON, AND R. M. MOODY. 2008. Eocene climate record of a high southern latitude continental shelf: Seymour Island, Antarctica. Geological Society of America Bulletin, 120:659-678.

IVANY, L. C., S. VAN SIMAEYS, E. W. DOMACK, AND S. D. SAMSON. 2006. Evidence for an earliest Oligocene ice sheet on the Antarctic Peninsula. Geology, 34:377-380.

JABLONSKI, D., AND D. J. BOTTJER. 1991. Environmental patterns in the origins of higher taxa: The post-Paleozoic fossil record. Science, 252:1831-1833.

JABLONSKI, D., AND J. J. SEPKOSKI JR. 1996. Paleobiology, community ecology, and scales of ecological pattern. Ecology, 77:1367-1378.

JABLONSKI, D., J. J. SEPKOSKI JR., D. J. BOTTJER, AND P. M. SHEEHAN. 1983. Onshore–offshore patterns in the evolution of Phanerozoic shelf communities. Science, 222:1123-1125.

JACKSON, J. B. C. 2008. Ecological extinction and evolution in the brave new ocean. Proceedings of the National Academy of Sciences of the United States of America, 105:11458-11465.

JACKSON, J. B. C., T. F. GOREAU, AND W. D. HARTMAN. 1971. Recent brachiopod–coralline sponge communities and their paleoecological significance. Science 173:623-625.

JAMES, M. A., A. D. ANSELL, M. J. COLLINS, G. B. CURRY, L. S. PECK, AND M. C. RHODES. 1992. Biology of living brachiopods. Advances in Marine Biology, 28:175-387.

JANZEN, D. H., AND P. S. MARTIN. 1982. Neotropical anachronisms: The fruits the gomphotheres ate. Science, 215:19-27.

KELLEY, P. H., AND T. A. HANSEN. 2003. The fossil record of drilling predation on bivalves and gastropods, p. 113-139. In P. H. Kelley, M. Kowalewski, and T. A. Hansen (eds.), Predator–Prey Interactions in the Fossil Record. Kluwer/Plenum, New York.

KELLEY, P. H., M. KOWALEWSKI, AND T. A. HANSEN. 2003. Predator-Prey Interactions in the Fossil Record. Kluwer/Plenum, New York, 472 p.

KOWALEWSKI, M., M. G. SIMÕES, M. CARROLL, AND D. L. RODLAND. 2002. Abundant brachiopods on a tropical, upwelling-influenced shelf (southeast Brazilian Bight, South Atlantic). Palaios, 17:277-286.

LAKATOS, I. 1970. Falsification and the methodology of scientific research programmes, p. 91-196. In I. Lakatos and A. Musgrave (eds.), Criticism and the Growth of Knowledge. Cambridge University Press, London.

LEIGHTON, L. R. 2003. Predation on brachiopods, p. 215-

237. In P. H. Kelley, M. Kowalewski, and T. A. Hansen (eds.), Predator–Prey Interactions in the Fossil Record. Kluwer/Plenum, New York.

MAHON, A. R., C. D. AMSLER, J. B. MCCLINTOCK, M. O. AMSLER, AND B. J. BAKER. 2003. Tissue-specific palatability and chemical defenses against macro-predators and pathogens in the common articulate brachiopod *Liothyrella uva* from the Antarctic Peninsula. Journal of Experimental Marine Biology and Ecology, 290:197-210.

MALUMIÁN, N., AND E. B. OLIVERO. 2005. Shallow-water late middle Eocene crinoids from Tierra del Fuego: A new southern record of a retrograde community structure. Scientia Marina, 69(Supplement 2):349-353.

MARTIN, R. E. 1998. One Long Experiment: Scale and Process in Earth History. Columbia University Press, New York, 262 p.

MAYR, E. 1983. How to carry out the adaptationist program? American Naturalist, 121:324-334.

MCCLINTOCK, J. B., H. DUCKLOW, AND W. FRASER. 2008. Ecological responses to climate change on the Antarctic Peninsula. American Scientist, 96:302-310.

MCKINNEY, F. K., AND S. J. HAGEMAN. 2006. Paleozoic to modern marine ecological shift displayed in the northern Adriatic Sea. Geology, 34:881-884.

MCROBERTS, C. A. 2001. Triassic bivalves and the initial marine Mesozoic revolution: A role for predators? Geology, 29:359-362.

MEREDITH, M. P., AND J. C. KING. 2005. Rapid climate change in the ocean west of the Antarctic Peninsula during the second half of the 20th century. Geophysical Research Letters, 32:L19604.

MEYER, D. L. 1985. Evolutionary implications of predation on Recent comatulid crinoids from the Great Barrier Reef. Paleobiology, 11:154-164.

MEYER, D. L., AND D. B. MACURDA, JR. 1977. Adaptive radiation of the comatulid crinoids. Paleobiology, 3:74-82.

MEYER, D. L., AND T. OJI. 1993. Eocene crinoids from Seymour Island, Antarctic Peninsula: Paleobiogeographic and paleoecologic implications. Journal of Paleontology, 67:250-257.

MILLER, A. I. 1998. Biotic transitions in global marine diversity. Science, 281:1157-1160.

MOODY, R. M., AND R. B. ARONSON. 2007. Trophic heterogeneity in salt marshes of the northern Gulf of Mexico. Marine Ecology Progress Series, 331:49-65.

MORRIS, R. H., D. P. ABBOTT, AND E. C. HADERLIE. 1980. Intertidal Invertebrates of California. Stanford University Press, Stanford, California, 690 p.

MUNDAY, B. W. 1993. Field survey of the occurrence and significance of regeneration in *Amphiura chiajei* (Echinodermata: Ophiuroidea) from Killary Harbour, west coast of Ireland. Marine Biology, 115:661-668.

NOBLE, J. P. A., A. LOGAN, AND R. WEBB. 1976. The Recent *Terebratulina* community in the rocky subtidal zone of the Bay of Fundy, Canada. Lethaia, 9:1-17.

OJI, T. 1985. Early Cretaceous *Isocrinus* from northeast Japan. Palaeontology, 28:629-642.

OJI, T. 1996. Is predation intensity reduced with increasing depth? Evidence from the west Atlantic stalked crinoid *Endoxocrinus parrae* (Gervais) and implications for the Mesozoic marine revolution. Paleobiology, 22:339-351.

OLBERS, D., D. BOROWSKI, C. VÖLKER, AND J.-O. WÖLFF. 2004. The dynamical balance, transport and circulation of the Antarctic Circumpolar Current. Antarctic Science, 16:439-470.

PACKARD, A. 1972. Cephalopods and fish: The limits of convergence. Biological Reviews of the Cambridge Philosophical Society, 47:241-307.

PECK, L. S. 1996. Metabolism and feeding in the Antarctic brachiopod *Liothyrella uva*: A low energy lifestyle species with restricted metabolic scope. Proceedings of the Royal Society, B, 263:223-228.

QUINN, J. F., AND A. E. DUNHAM. 1983. On hypothesis testing in ecology and evolution. American Naturalist, 122:602-617.

RHOADS, D. C., AND D. K. YOUNG. 1970. The influence of deposit-feeding organisms on sediment stability and community trophic structure. Journal of Marine Research, 28:150-178.

RHODES, M. C., AND R. J. THOMPSON. 1993. Comparative physiology of suspension-feeding in living brachiopods and bivalves: Evolutionary implications. Paleobiology, 19:322-334.

RICHARDSON, A. J., A. BAKUN, G. C. HAYS, AND M. J. GIBBONS. 2009. The jellyfish joyride: Causes, consequences and management responses to a more gelatinous future. Trends in Ecology and Evolution, 24:312-322.

ROSENBERG, R., J. S. GRAY, A. B. JOSEFSON, AND T. H. PEARSON. 1987. Petersen's benthic stations revisited: II. Is the Oslofjord and eastern Skaggerak enriched? Journal of Experimental Marine Biology and Ecology, 105:219-251.

SCHER, H. D., AND E. E. MARTIN. 2006. Timing and climatic consequences of the opening of Drake Passage. Science, 312:428-430.

SEPKOSKI, J. J., JR. 1981. A factor analytic description of the Phanerozoic marine fossil record. Paleobiology, 7:36-53.

SEPKOSKI, J. J., JR. 1984. A kinetic model of Phanerozoic taxonomic diversity. IV. Post-Paleozoic families and mass extinctions. Paleobiology, 10:246-267.

SEPKOSKI, J. J., JR. 1991a. Diversity in the Phanerozoic oceans: a partisan view, p. 210-236. *In* E. C. Dudley (ed.), The Unity of Evolutionary Biology: Proceedings of the Fourth International Congress of Systematic and Evolutionary Biology (volume 1). Dioscorides Press, Portland, Oregon.

SEPKOSKI, J. J., JR. 1991b. A model of onshore-offshore change in faunal diversity. Paleobiology, 17:58-77.

SEPKOSKI, J. J. JR., AND A. I. MILLER. 1985. Evolutionary faunas and the distribution of Paleozoic marine communities in space and time, p. 153-190. *In* J. W. Valentine (ed.), Phanerozoic Diversity Patterns: Profiles in Macroevolution. Princeton University Press, Princeton, New Jersey.

SHEEHAN, P. M. 2001. History of marine diversity. Geological Journal, 36:231-249.

SIGNOR, P. W., AND C. E. BRETT. 1984. The mid-Paleozoic precursor to the Mesozoic marine revolution. Paleobiology, 10: 229-245.

SKÖLD, M., L.-O. LOO, AND R. ROSENBERG. 1994. Production, dynamics and demography of an *Amphiura filiformis* population. Marine Ecology Progress Series, 103:81-90.

SMALE, D. A., AND D. K. A. BARNES. 2008. Likely responses of the Antarctic benthos to climate-related changes in physical disturbance during the 21st century, based primarily on evidence from the West Antarctic Peninsula region. Ecography, 31:289-305.

SMITH, F., AND J. D. WITMAN. 1999. Species diversity in subtidal landscapes: Maintenance by physical processes and larval recruitment. Ecology, 80:51-69.

STILWELL, J. D., AND W. J. ZINSMEISTER. 1992. Molluscan Systematics and Biostratigraphy: Lower Tertiary La Meseta Formation, Seymour Island, Antarctic Peninsula. American Geophysical Union, Antarctic Research Series, 55, Washington, DC, 192 p.

TAVARES, M., AND G. A. S. DE MELO. 2004. Discovery of the first known benthic invasive species in the Southern Ocean: The North Atlantic spider crab *Hyas araneus* found in the Antarctic Peninsula. Antarctic Science, 16:129-131.

THATJE, S., AND W. E. ARNTZ. 2004. Antarctic reptant decapods: More than a myth? Polar Biology, 27:195-201.

THATJE, S., AND R. B. ARONSON. 2009. No future for the Antarctic Treaty? Frontiers in Ecology and the Environment, 7:175.

THATJE, S., AND V. FUENTES. 2003. First record of anomuran and brachyuran larvae (Crustacea: Decapoda) from Antarctic waters. Polar Biology, 26:279-282.

THATJE, S., AND A. N. LÖRZ. 2005. First record of lithodid crabs from Antarctic waters off the Balleny Islands. Polar Biology, 28:334-337.

THATJE, S., K. ANGER, J. A. CALCAGNO, G. A. LOVRICH, H.-O. PÖRTNER, AND W. E. ARNTZ. 2005. Challenging the cold: Crabs reconquer the Antarctic. Ecology, 86:619-625.

THATJE, S., C.-D. HILLENBRAND, A. MACKENSEN, AND R. LARTER. 2008a. Life hung by a thread: Endurance of Antarctic fauna in glacial periods. Ecology, 89:682-692.

THATJE, S., S. HALL, C. HAUTON, C. HELD, AND P. TYLER. 2008b. Encounter of lithodid crab *Paralomis birsteini* on the continental slope off Antarctica, sampled by ROV. Polar Biology, 31:1143-1148.

THAYER, C. W. 1981. Ecology of living brachiopods, p. 110-126. *In* J. T. Dutro Jr. and R. S. Boardman (eds.), Lophophorates: Notes for a Short Course. Department of Geological Sciences, Studies in Geology, 5, University of Tennessee, Knoxville, Tennessee.

THAYER, C. W. 1983. Sediment-mediated biological disturbance and the evolution of marine benthos, p. 479-625. *In* M. J. S. Tevesz and P. L. McCall (eds.), Biotic Interactions in Recent and Fossil Benthic Communities. Plenum, New York.

VERMEIJ, G. J. 1977. The Mesozoic marine revolution: Evidence from snails, predators and grazers. Paleobiology, 3:245-258.

VERMEIJ, G. J. 1978. Biogeography and Adaptation: Patterns of Marine Life. Harvard University Press, Cambridge, Massachusetts, 332 p.

VERMEIJ, G. J. 1987. Evolution and Escalation: An Ecological History of Life. Princeton University Press, Princeton, New Jersey, 527 p.

VERMEIJ, G. J. 1995. Economics, volcanoes, and Phanerozoic revolutions. Paleobiology, 21:125-152.

WARNER, G. F. 1971. On the ecology of a dense bed of the brittle-star *Ophiothrix fragilis*. Journal of the Marine Biological Association of the United Kingdom, 51:267-282.

WILLIAMSON, P. G. 1982. Cinderella subject (book review). Nature, 296:99-100.

ZACHOS, J., M. PAGANI, I. SLOAN, E. THOMAS, AND K. BILLUPS. 2001. Trends, rhythms, and aberrations in global climate 65 Ma to present. Science, 292:686-693.

ZUSCHIN, M., AND M. STACHOWITSCH. 2009. Epifauna-dominated benthic shelf assemblages: Lessons from the modern Adriatic Sea. Palaios, 24:211-221.

CHAPTER ELEVEN

ECOLOGICAL MODELING OF PALEOCOMMUNITY FOOD WEBS

PETER D. ROOPNARINE

Department of Invertebrate Zoology and Geology, California Academy of Sciences, 55 Music Concourse Drive, Golden Gate Park, San Francisco, CA 94118 USA

ABSTRACT.—Food webs are powerful representations of some of the most significant interspecific relationships in a community. They have been used as primary tools in studies of the relationship between biodiversity and ecological stability, and the robustness and resilience of communities in the face of environmental disturbance. Those topics are central to theoretical studies in conservation biology and biodiversity ecology, and hence conservation paleobiology. Food webs can be constructed from paleocommunity data for studies relevant to conservation paleobiology and biology, given that proper attention is paid to limitations imposed by fossil preservation and the geological record, and that questions are asked at appropriate scales. This paper presents a mathematical model, CEG (Cascading Extinction on Graphs) for the reconstruction and analysis of paleocommunity food webs. The model utilizes combinatoric and network-based mathematics, combining network topological and population demographic processes. CEG is founded upon ecological first principles and effectively mimics fundamental trophic processes, such as trophic cascades, keystone predator effects, bottom-up community collapse, and parasite-mediated suppression of populations. The model is applied to two example paleocommunities, a terrestrial community from the Late Permian of the Karoo Basin in South Africa, and a Neogene coastal marine community from the Dominican Republic. Insights, assumptions and limitations of the model are explored.

INTRODUCTION

"THE COMBINATION of the experimental method with the quantitative theory is in general one of the most powerful tools in the hands of contemporary science" (Gause, 1934, p.11). Food webs summarize the trophic interactions among species in a biological community. These interactions are among a community's most important as they are the major conduits for energy transfer among species (Hairston et al., 1960). A paleontological food web couples the relative importance of trophic interactions with extensive knowledge of the general to specific trophic habits of numerous fossil taxa. This coupling is based primarily on functional morphological and uniformitarian interpretations of fossil morphologies, habitats, and interspecific associations. The ease with which much of this information can be collated or inferred for paleocommunities and fossil taxa has encouraged workers to describe paleocommunity structures and dynamics using reconstructed paleontological food webs (Butterfield, 2001; Solé et al., 2002; Angielczyk et al., 2005; Roopnarine, 2006; Roopnarine et al., 2007; Dunne et al., 2008). Conservation paleobiology is concerned with the application of paleontological ideas and data to conservation biology (Flessa, 2002), that is, understanding the circumstanc-es under which species are threatened with extinction, become extinct, or survive those circumstances. These are some of the concerns that food web models must address (May, 2009), and to be of use to conservation biology, they must make predictions of future biodiversity events on the basis of the fossil record. Food webs go beyond the consideration of individual species though, and offer the possibility of understanding the roles of community structure and ecosystem complexity in the resistance of biological communities to severe perturbation, and the circumstances that can result in community collapse (Roopnarine, 2006; May, 2009).

The importance of trophic interactions, and the extent to which food webs summarize these interactions at the community level, has also led to the development of food webs as models and tools for the analysis of modern communities (Elton, 1927; May, 1974; Pimm, 1982). Food webs subsequently figure prominently in some of the major questions of modern ecology, including whether communities comprise sets of dynamically stable populations (May, 1974; McCann, 2000; Beninca et al., 2008; Rooney et al., 2008), the roles that systematic and ecological diversity play in the resilience and resistance of communities (MacArthur, 1955; McCann, 2000; Thebault and Loreau, 2003; Okuyama and Holland, 2008), and whether general principles of community assembly exist (Bastolla et al., 2005; Azaele et al., 2006). All these questions are of central importance to conservation biology (Pimm, 1991).

Paleoecology has traditionally emphasized individual species or groups of species, but modern biodiversity crises are forcing consideration of larger groups of interacting species at the community and ecosystem scales. Furthermore, the magnitude of the problems facing natural communities today is approaching those of geological intervals associated with mass extinctions. As the deeper history of human disturbance of modern natural systems is realized, there comes the need to reconstruct and understand those systems from their fossil and sub-fossil records (Jackson et al., 2001; Flessa, 2002; Willis and Birks, 2006). The central questions of modern ecology and conservation biology remain the same when reconstructing the composition and dynamics of past communities, and food webs retain their position as important and useful community representations. Yet to utilize food webs paleobiologically, paleontologists must devise approaches that account for the nature of the geological record, as well as recognize the scales at which such representations are appropriate, or not (Bennington et al., 2009). The present paper outlines one such approach, the CEG (Cascading Extinction on Graphs) model, a combinatoric network model designed to explore the role of community structure and species interactions in the propagation of disturbances through a community (Roopnarine, 2006). The paper is intended to be instructive, and therefore includes a general discussion of mathematical model-building for paleoecology, as well as a detailed exposition of the CEG model and its application to two example paleocommunities. Finally, there is a discussion of the implications of the model for our understanding of community dynamics, its potential application to modern problems in conservation biology, and an outline of model limitations and areas for improvement.

SCIENTIFIC MODELS

A scientific model is a set of formal statements that is hypothesized to represent or explain some aspect of the natural world. The natural world is complex, and models are simplifications by necessity. They often reflect educated guesses of the relationships, or nature of relationships between known processes or observed phenomena (but they do, however, allow us to make informed predictions about the natural world). Further-more, models allow us to explore hypothetical possibilities even when our data are insufficient or incomplete.

Scientific models need not be mathematical, but in this paper I will focus on mathematical models for two reasons. First, this paper is intended to be instructional about the creation and use of food web models in conservation paleobiology. Almost any such model should contain statements of ecology and paleoecology, and mathematical modeling in ecology has matured into a rich and powerful means by which to make such statements. Therefore, "model" in this paper will always refer to a mathematical formulation. Second, when formulated mathematically, a model provides a lot of freedom to question our hypotheses and explore implications and consequences outside the bounds of empirical observations. This is particularly useful when data are unknown or incomplete, or the hypothesis itself is incomplete.

Mathematical modeling should not be taken, however, as an alternative to empiricism. If there is any order in nature, and scientific experience tells us that there is, then our models can be based on an understanding of fundamental principles, and should be expected to be congruent with empirical data to the extent that the model reflects the entirety of the processes that led to the data. Incongruence will often require a reformulation of the model, or even revision of the principles themselves, a task made easier and more precise when the model has been stated and communicated mathematically. Alternatively, one should also be mindful of the distinction between data and interpretation. While data perhaps seldom lie, scientific theories are based on histories of important and useful misinterpretations, and there are numerous examples of models that correctly predicted as yet unobserved data or phenomena while being at odds with contemporary interpretations of existing data. (For example, Laplace suggested the idea of black holes in the late 18th century, but the idea was generally ignored because of the prevailing dominant view of light consisting of massless waves.)

I will proceed to review some basic guidelines for mathematical model building and explore how such models might be applied to conservation paleobiology. Then I will present as an example the CEG model (Roopnarine, 2006). I will explore the assumptions, formulation, limitations and implications of this model, focusing on the requirements, problems and opportunities inherent in any paleoecological model.

Building a model

"...[T]he supreme goal of all theory is to make the irreducible basic elements as simple and as few as possible without having to surrender the adequate representation of a single datum of experience" (Einstein, 1934, p.165). There are numerous views and opinions on the purpose for constructing a scientific model, but they all fall between end-points on a scale of general scientific goals: problem-solving (Grimm and Railsback, 2005) and constructing an underlying theory based on observations. For example, Darwin developed the qualitative model of natural selection to explain a theory of evolution that he deduced from his observations of the natural world. In the process he *assumed* the existence of a mechanism for the inheritance of traits. Evolutionary biologists of the early 20th century were subsequently faced with reconciling the *problem* between Mendelian genetics and contemporaneous concepts of natural selection. Models of quantitative and population genetics were developed that went a long way toward that reconciliation (Fisher, 1930; Wright, 1931; Haldane, 1932). Today it is not difficult to appreciate differences between the type of model presented by Darwin, and for example Fisher, which brings us to the first guideline.

A model should be appropriate to the problem or question being addressed

Darwin set out to explain patterns which he had observed, and his work is remarkable in its interweaving of so many major themes into the theory of evolution, including the inheritance of traits, population growth and the struggle for existence, competition within and among species, and the history of life as recorded in the fossil record. Formulating his theory did not require a detailed understanding of any of these themes comparable to what is available today. More specific models were required, however, to test the theory of natural selection and search for underlying mechanisms, and that is precisely what Fisher did. Darwin presented a convincing case for modification with descent, but it was up to Fisher and others to provide adequate models of natural selection capable of explaining the data. Fisher provided a specific mathematical model explaining how natural selection can be based on the Mendelian "particulate character of the hereditary elements" (Fisher, 1930). This and related models, plus the later revolution in molecular biology, were crucial to our current understanding of natural selection as an evolutionary mechanism. Nevertheless, the details of these models are not always essential when incorporating evolutionary theory into other biological models. For example, Sepkoski (1984) did not need them for his kinetic models of Phanerozoic diversity, even though all arguments of taxon origination are based ultimately on evolution. How then does one decide on the proper ingredients for a model?

Define the domain of the model

Presumably, all things in nature are connected, yet most of these connections are not relevant to understanding the problem immediately at hand. One should define, at the outset, those connections or relationships which are causal, and those that are contingent (Bohm, 1957). To paraphrase an example from Bohm, one can quite accurately deduce a law of gravitation by dropping balls of various sizes and composition. Now drop a sheet of paper. Does the difference in motion suggest a new law, or inadequacy of the old one? The paper does eventually meet the ground, and precisely because of the same gravitational attraction to which the balls are subject. Its different motion is the result of air resistance. If your goal is to deduce a law of gravity, then you should realize that gravity is causal in the net journey of the paper, while the specific steps taken along the journey are contingent upon the sheet's air resistance. Air resistance is, therefore, irrelevant to a law of gravity. If your goal, on the other hand, is to model the motion of sheets of paper, then accounting for gravity alone is insufficient and air resistance is indeed relevant.

Constrain model specificity

Getting back to the balls for a moment, is air resistance relevant to them? If your measurements were precise enough, then you would likely discover variation based on different surface textures of the balls, perhaps variation over time corresponding to air temperature, and so on. But those details are relatively unimportant when describing the motion of a dropped ball, because mass and sphericity dominate minor differences in ball composition. Therefore, the addition of surface texture as a causal parameter is unnecessary.

There is often a temptation and tendency to strive for realism in biological models to the point at which the models become too complex. This point can be recognized either when the addition or variation of parameters fail to alter model output in a predictable manner, or variation of model output cannot be explained directly as a result of parameter variation. An accurate simulation of reality is not a useful model unless one can point to why the simulation reproduces the aspects that make it realistic. *Simulations are not necessarily good models.*

Having stated that, a major caveat must be given immediately. Ecological systems are complex (Levin, 1998), that is, they are the result of the interactions of numerous independently acting or semi-independent entities. Simple models of those entities can sometimes produce complicated results, particularly when causal relationships within the model are nonlinear (as we will see later on). The complexity of the system of entities could be modeled as a collection of simpler models, or we can ignore the lower level of entities altogether and treat the entire system as a single entity. These divergent approaches are not likely to yield the same results. The former can quickly become a hopelessly and incomprehensibly realistic simulation, while the latter simply cannot be interpreted at a level relevant to individual lower-level entities. A proper model will search for a middle path, and though the search can be as much art as it is science, that's often where the fun is.

Models in conservation paleobiology

The first step in constructing a model for application to conservation paleobiology is the same as it is for any scientific discipline: determine the question to be addressed. Ecological first principles underlying the problem then form the basis of the model to the extent to which they are known and understood. Next, one must very carefully consider the relationship between paleontological data and the ecological domain, that is, disparities of temporal and spatial scales (Bennington et al., 2009). Specifically, we must consider the nature of the paleontological data and the commensurability of those data with modern ecological dynamics. Finally, and most importantly, we must outline the assumptions of our model, the potential impact and implications of those assumptions, and the extent to which they may affect model results.

In addition to these steps, one also has to understand the resources necessary for the proper construction and implementation of the model. What type of data will be required? How will the model be constructed? Though conservation paleobiology is still a young sub-discipline, I envisage that most modeling exercises will require high quality data. Conservation biology deals with issues of immediate import, those unfolding on contemporary or generational timescales, not macroevolutionary ones. The application of chronologically (and perhaps spatially) high-resolution data, or data with well-established chronologies, will be most useful. The data must also preserve the ecological properties of interest, for example trophic interactions, symbiotic relationships or relative population sizes. If collections-based or literature research are part of the data-gathering effort, then the collections (samples and publications) must meet the necessary criteria, or broad allowance should be made in the model to incorporate the uncertainty stemming from the use of inadequate data.

Beyond the data, there are the construction and implementation of the model. There is always more than one mathematical approach to any problem, and one selects an approach suitable to both the problem and to available skills. Whatever the latter may be, any student in this discipline should understand the fundamentals of mathematics (arithmetic, geometry and basic algebra) and should have at her or his disposal more advanced skills, whether in calculus, statistics, or some other area, preferably several. Finally, abilities in computer programming are increasingly important because electronic power allows us to address questions which are beyond human computational skills. Computerized implementation of a model can take many routes, perhaps the most straightforward being the use of a high-level language or computing environment, excellent examples being Mathematica (Wolfram Research, 2008), MatLab (Mathworks, 2009), Octave (Eaton, 2007) and R (R Project, 2009). The lower the level of the language, however, the more flexibility and power available; the languages used most commonly for numeric scientific modeling are C, C++, and Fortran. Bringing all these skills and resources together, including basic field work, is a tall order for any single individual (though not impossible), and multi-disciplinary collaboration is often required.

Food web models

Food webs are depicted graphically and conventionally as species connected with arrows, a depiction referred to here as a trophic network. The arrows represent the trophic interaction, and direction moves from consumed to consumer. The size of a food web depiction depends on the diversity of the community, how much of that diversity is understood ecologically, and how much of it is actually included in the network. There are probably no complete systematic descriptions of any natural communities, and this is certainly true regarding paleocommunities. The trophic ecologies of some part of the taxonomic diversity are frequently incomplete, and therefore so are their inclusions into the network. Perhaps less commonly, however, interest lies in a specific subset of the community, and other members of the food web are omitted from the network, or are aggregated into larger units (Havens, 1992). This is possibly a reasonable way in which to model communities harboring hundreds or thousands of species, but decisions that subsequently impose an hierarchical structure on the network may have significant impacts on structural and dynamic analyses.

An added complication to the issues of diversity is the fact that food webs are dynamic; they vary over time and space. Interspecific interactions can depend upon ecological and local historical contexts, and trophic relationships can even reverse direction under different environmental conditions. Trophic relationships also vary with the life histories and relative population age structures of the interacting species. Food webs are, in reality, very complicated and complex structures (Montoya et al., 2006).

The complexity of food webs has been modeled in various ways by reducing the complexity to smaller and more manageable sets of species, or by embracing the complexity (e.g., Duffy et al., 2007; Rooney et al., 2008). The former approach is generally a classical one, using methods from population biology to study the interaction of small numbers (often five or less) of species. A serious question that arises is the extent to which the results of these models can be scaled up to describe accurately the dynamics of species embedded in communities of hundreds of species of similar and dissimilar ecologies (Teng and McCann, 2004; Brose et al., 2005). May (1972) was one of the first to attempt to bridge this gap by investigating community stability using random or Poisson graphs as models of trophic networks. Species in such networks have, on average, similar numbers of links or interactions and a uniform probability of being linked to any other species in the network. May (1972) concluded that, in general, communities become less stable as diversity and connectance increase, a conclusion in direct opposition to an older view proposing that diversity confers stability on communities (MacArthur, 1955). Subsequent work

demonstrated that random graphs are not good models for community networks, which tend to be more hierarchically structured and may be better described as non-random networks, such as small-world networks (Montoya and Solé, 2002). Today, food web complexity is often embraced with models drawn from the rapidly growing field of network theory. One issue with this approach, however, is that examinations of network topology have largely replaced population dynamics as a source of answers regarding community stability, the relationship between stability and diversity, community resilience, the nature of communities over Phanerozoic time, and so on. Ideally, our models should attempt to continue May's work, blending the mathematics of population dynamics and the mathematics of complex systems.

THE CEG MODEL

The CEG (Cascading Extinction on Graphs) model examines the extent to which the biotic interactions of an ecological community affect its resistance to secondary extinction. Secondary extinction is defined here as the propagation and perhaps perpetuation of species extinction in response to disruptions of one or more species within the community. If a major component of any species' environment is the other species with which it interacts (Van Valen, 1973), then one should expect that changes to any single species will affect others in the community. Van Valen stated this hypothesis from a macroevolutionary standpoint when formulating the Red Queen Hypothesis, but he was of course translating a basic tenet of ecology. More recently, Vermeij (2004a, b) and Roopnarine (2006) have argued that such changes have the potential to cascade through a community. This idea is reflected in concepts of food webs as non-random networks, where the topologies (patterns of linkage) of the networks are used to predict the extent to which the removal of species or links (interspecific interactions) will propagate through the community (Dunne et al., 2002a).

Basics of the model
The central assertion of the CEG model is that, all else being held constant, perturbation of one or more species within a community can result in the secondary extinction of other species, either as a result of direct or indirect interactions (Eklöf and Ebenman, 2006; Roopnarine, 2006). A network approach allows us to

evaluate the extent of the secondary and indirect inter-actions in very complicated systems. "All else being held constant" refers to the fact that the model does not allow a community to recover once a perturba-tion has been applied. Recovery could take the form of immigration-mediated replenishment of depleted or extinct (extirpated) populations, the existence of ref-uges, or evolutionary adaptation to the perturbances. While these pathways to recovery undoubtedly oper-ate under natural circumstances, their inclusion in the model would not give us insight into the role of the trophic network in the propagation and potential effects of the perturbation, nor the potential of the pathways to counteract such effects.

Another key assertion of the model is that a pa-leocommunity's trophic network can never be specified by a single topology (Roopnarine et al., 2007). There is always uncertainty associated with the biotic inter-actions of a fossil species because no one was there to observe them; this is a difficult task even for extant species. Moreover, the topology specified for a single community is expected to vary spatially and temporally (Eveleigh et al., 2007). For example, the strength and direction of interspecific interactions of extant species are known to vary according to physical conditions, the presence or absence of other species in the community (Edeline et al., 2008; Petanidou et al., 2008), relative population sizes, and the incumbency of species when addition to the community is asynchronous (Edgell and Rochette, 2008). These uncertainties will be incorpo-rated into any realistically complex model of a com-munity food web (McCann and Rooney, 2009). This raises a concern that the associated variance will then obscure any general patterns or principles, but this is not the case (as reported below). It is also evidenced by studies that have considered network uncertainty, albeit in different ways (Roopnarine, 2006; Roopnarine et al., 2007; Dunne et al., 2008; Berlow et al., 2009).

Constructing a model trophic network

The first step in constructing the CEG model of a paleocommunity is to assemble a comprehensive taxonomic database of all species known to have been members of that particular community. This is a fun-damental requirement for the construction of any food web, but CEG requires that all species be recognized individually in the web. The database may be based on field sampling specifically for the model, or the ex-amination of specimen collections, literature sources, or other databases such as the Paleobiology Database. If species-level data are not available, then one has to make assumptions of the number of species represented by higher taxa, because species' populations interact, higher taxa do not.

The second step is the ecological parameterization of the species. Ecological parameterization of a food web refers, at a minimum, to the identification of the prey and predators of each species in the web. Further parameterization may include species' demographic data, or interspecific interaction strengths, or the in-clusion of non-trophic relationships such as symbiotic partnerships. The extent to which a web is parameter-ized depends on the availability of data, the purpose for which the web is being constructed, and the ability to actually analyze and interpret the data in a meaningful manner. For example, Dunne et al. (2008) relied on the exquisite preservation of Burgess Shale and Chengji-ang faunas to infer trophic links and reconstruct highly resolved food webs of those Cambrian communities, but stopped short of inferring or modeling additional parameters such as interaction strengths.

The basic data required by CEG for ecological pa-rameterization is a description of the guild membership of each species. Here we define guilds as the trophic habits and habitats of member species, for example, the "shallow, sedentary infaunal suspension feeders" of a marine community. Guilds are further divided on the basis of all members sharing the same potential prey and predators. For example, two species of venerid bi-valves may be expected to feed from the same set of phytoplankton species, be prey to the same predators, and hence belong to the same guild. Finally, guilds are linked to each other if species within any two guilds have the potential to interact trophically; members of one may be prey or predators of species in the other (in-terguild predation), or may be both prey and predators (bi-directional interguild predation), and species within a single guild may also be predators and prey of fel-low guild members (intraguild predation). The result-ing higher guild-level representation of trophic interac-tions in the community is termed a metanetwork (Roo-pnarine et al., 2007), and summarizes the most exact information available for community interactions. An example from the Late Miocene Dominican Republic marine community is shown in Figure 1. An increase in the precision of species-level trophic data could result in further sub-division of the guilds if species become distinguished by different predators and/or prey, some-

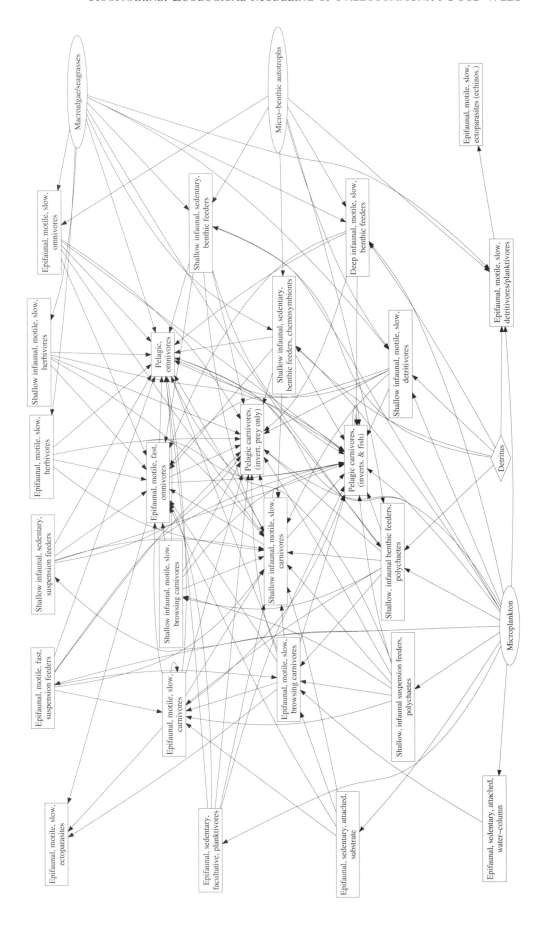

FIGURE 1.—Metanetwork of Late Miocene coastal marine communities from the Gurabo Formation of the Dominican Republic. Each compartment is an ecological guild comprising one or more species. Arrows point in the direction of energy transfer via trophic interaction.

times leading to guilds comprising a single species. A detailed example of metanetwork construction is given for the L. Permian-E. Triassic terrestrial communities of the Karoo Basin in the supplement to Roopnarine et al. (2007). Guilds represent a higher-level aggregation of specific data, and in that sense resemble trophic species, another food web aggregation strategy (Briand and Cohen, 1984). Trophic species, however, assume that the biological species which they comprise share identical resources and predators. The CEG model does not make this assumption, which is in essence untestable for fossil taxa and paleocommunities.

The final step in network construction is the generation of species-level networks from the metanetwork. A species-level network is considered a single potential pattern of community interactions in any given place at an instant of time. Species-level network generation requires the introduction of two variable parameters into the model. The first is a trophic link distribution for each guild. A trophic link distribution describes the distribution of the number of prey per species within a guild. This number ranges from one (a heterotrophic

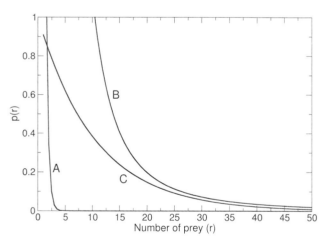

FIGURE 2.—Model trophic link distributions. X-axis indicates the number of prey species available to consumers, a maximum of 50 in this case (a higher number used to illustrate differences among the distributions sufficiently); y-axis is the probability of a consumer having a particular number of prey, as measured by the relevant distribution. A= exponential distribution of the form $p(r) = Me^{-r\gamma}$, where $p(r)$ is the probability of a particular number of prey species, M is the maximum number of prey species available, r is the number of prey species, and γ is the decay exponent of the distribution, which is 2.5 for all the distributions illustrated; B= power law distribution of the form $p(r) = M^{\gamma-1}r^{-\gamma}$; C= mixed exponential-power law distribution of the form $p(r) = e^{-r/\varepsilon}$, where $\varepsilon = e^{(\gamma-1)\ln(M)/\gamma}$ (Roopnarine et al., 2007).

species must prey upon at least one other species) up to the total species-richness of all guilds that are specified as prey of the guild in the metanetwork. For example, if guild A comprises predators of species within guilds B and C, and the species-richnesses of the latter guilds are three and five respectively, then A's trophic link distribution ranges from one (extreme specialist) to eight (extreme generalist). CEG uses distributions derived from those measured in Recent communities. Those measured distributions tend to be decay distributions, taking the form of exponential or power law functions (Camacho et al., 2002a, b) (Fig. 2). The number of measurements available from real communities is very limited, but the impact of various choices in the model can be assessed by varying those choices. Furthermore, if power law (or mixed exponential-power law) distributions are used, one also has to decide on their rates of decay (or how "fat" a distribution's tail is, fatter tailed distributions decaying more slowly). Basically, though, the distribution describes the variation of specialized to generalized dietary habits in the guild. A decay distribution means that there are more specialized than generalized feeders in a guild, with the shape of the distribution specifying *how many* more.

The second variable parameter is a minimum viable population size for each species or guild. This parameter dictates the proportion of the initial relative population size to which a species may decline before becoming extinct. Its value can vary dramatically among species based on life-history traits; for example, larger mammals generally have smaller population sizes and larger range requirements than do small mammals, and hence greater probabilities of extinction (Sibly et al., 2007). This parameter is scaled between zero and one in CEG model simulations, but its impact on simulation results appears to be minimal.

Assembling a species-level network

Once these parameters have been specified, stochastic draws from guild trophic link distributions determine the number of prey to be assigned to each species. The prey species themselves are drawn randomly from the prey guilds of the predatory species. The result is a directed graph or network, one where links are unidirectional, and in which each species in the community has been assigned prey, and many therefore also have predators (Fig. 3). Repeated stochastic generation of species-level networks accounts for the two sources of uncertainty in paleontological food webs discussed

FIGURE 3.—A single species-level network generated stochastically from the metanetwork space. This particular network is derived from the Late Miocene Dominican Republic metanetwork (Fig. 1). Lines represent interspecific trophic links, and they converge to points representing individual species.

earlier, namely uncertainty of the specific interactions of a species, and the temporal and spatial variability of a community type.

Each species-level network link is also assigned an interaction strength, calculated simply as the reciprocal of the species' number of prey (a species' incoming links are therefore all equal in strength). The issue of interaction strengths, both their measurement and distributions in real communities, remains problematic (Paine, 1992; Wootton, 1997; McCann et al., 1998; Neutel et al., 2002; Peacor and Werner, 2004; Navarette and Berlow, 2006; Novak and Wootton, 2008), but empirical and theoretical arguments suggest that most of a species' interactions are nearly equal, with a few strong ones. The CEG model uses a single-valued distribution (Dirac delta distribution), and an interesting area for extension would be the incorporation of more sophisticated models of interaction strength.

The number of species-level networks that can be derived from a metanetwork is finite, because the trophic link distributions are bounded (or truncated), and therefore the number of combinations is finite. Determination of this number is itself a useful and instructive counting and combinatoric exercise. Each species may be linked in any given instance to one or more of its potential prey, up to the maximum number of prey. The number of species-level networks that could possibly exist is therefore the product of all these arrangements, or the total number of possible combinations of all the species configurations. This number is huge for com-

munities of even several dozen species. For example, it is approximately 1.354×10^{1213} for the Late Miocene community illustrated in Figure 1, which comprises 130 heterotrophic species (Appendix 1). This means that associated with each metanetwork is a tremendous space over which community response to perturbation may vary. The significance of this variability can be estimated by sampling species-level networks from the space. This is currently done in a stochastic fashion because no solution has yet been formulated for a systematic mapping of the space (exhaustive sampling is out of the question).

Secondary extinction

Secondary extinction is defined as the extinction of a species' population in a community resulting from a primary perturbation to, or extinction of another species. For example, if carnivorous dinosaurs became extinct at the end of the Cretaceous in response to the extinction of primary herbivore consumers, themselves casualties of a reduction of primary production after the Chixulub impact, then carnivore extinctions would have been secondary extinctions. There are two paths to secondary extinction in a food web, and the CEG model utilizes both. First there is topological extinction, which occurs when all the resources to which a species is linked are lost (Amaral and Meyer, 1999). This is represented in the model simply as the removal of all a species' prey nodes. Second, there is demographic extinction, which accounts for demographic changes which must occur when a species loses *some* of its resources. That is, all of a species' prey nodes may still be extant, but if population levels are too low and the prey are functionally extinct (no longer ecologically viable), then extinction of the predator is assured.

A significant amount of recent work on complex food webs has been concerned with the potential for topological extinction and the resistance of food webs to this mechanism (Dunne et al., 2002a, 2008; Ebenman and Jonsson, 2005; Estrada, 2007). This in part stems from modern network theory, where it has been recognized that many so-called real-world networks, such as the World Wide Web and gene interaction networks, have underlying scale-free distributions (Watts and Strogatz, 1998; Dunne et al., 2002b; Williams et al., 2002; Yook et al., 2002; Newman, 2003; Doyle et al., 2005). One feature of these scale-free networks is the uneven distribution of links among nodes (for example, specialist versus generalist consumers in a guild), where a few nodes have significantly greater numbers of links, or degree, than the majority of nodes. These networks are robust against the effects of random node removal, because probability favors the removal of nodes with smaller numbers of links, and hence have limited potential for propagating effects. The networks are, however, vulnerable to the targeted removal of highly linked nodes (Albert et al., 2000; Callaway et al., 2000; Strogatz, 2001; Newman, 2003). There is supporting evidence that food webs possess scale-free structure, and hence would be similarly robust against secondary extinctions resulting from a majority of possible permutations of species removals. This idea has been tested with diverse and complex food webs, the examination of which is facilitated by the network approach (Dunne et al., 2002a; Allesina and Bodini, 2004). The general conclusion is that food webs are topologically robust against random perturbations and removals (Dunne et al., 2002a; Dunne and Williams, 2009). Topological extinction can be modeled analytically within the CEG framework if a metanetwork's underlying link distributions are known, and may be computed exactly for a species-level network (Appendix 2).

An alternative and more classical approach to examining the impact of disturbance of a species on other species in the community is to use systems of Lotka-Volterra-type equations. The classical model is

$$\frac{dN}{dt} = N|a - bP|$$

$$\frac{dP}{dt} = P|cN - d|$$

where N(t) and P(t) are prey and predator populations respectively, and a, b, c and d are positive constants or interaction terms (Murray, 1989). This model has shortcomings, and numerous modifications have been made over the almost century since the model was introduced to ecology, but the basic thesis stands, that interspecific interactions can be described using coupled differential equations. These dynamic equations model real ecological and demographic processes, such as population destabilization and extinction under conditions of resource loss (Lande et al., 2003). There are also changes in population sizes and community diversity driven by top-down cascades or the loss of keystone predators. While the topological network model does not capture these demographic dynamics (Borer et al., 2002; Eklöf

and Ebenman, 2006), it becomes mathematically intractable to scale systems of differential equations to complex communities comprising hundreds or thousands of species. The effects of community and food web complexity are therefore possibly overlooked.

Interspecific interactions in the CEG model

The interaction between two species is determined, as described above, by the stochastic assignment of one species as prey to another, within the constraints of the metanetwork guild structure. Every consumer will therefore possess a set of prey nodes. Recall that the strength of the interaction between consumer and prey is simply the reciprocal of the number of incoming links to the consumer; therefore each consumer interacts equally with each of its prey nodes. The state of a consumer species at a discrete time interval, $N_i(t)$, is a function of incoming and outgoing energy, which for a species x_i is described by a dynamic difference equation

$$N_i(t) = \frac{1}{N_i(0)} \left[\sum_{j=1}^{r_x} \frac{N_j(t)}{r_x} - \sum_{k=1}^{p_x} \frac{N_k(t)}{r_k} \right]$$

$$= \frac{1}{N_i(0)} \left[\sum_{j=1}^{r_x} s_x N_j(t) - \sum_{k=1}^{p_x} s_k N_k(t) \right]$$

where r_x is x_i's number of prey, j is a prey species, and k are predators of i, of which there are p species. "N" is initially set equal to one, and then modified as species linkages are determined. The state $N_x(0)$ of every species is therefore an equilibrium maintained by incoming and outgoing links. The contribution of each incoming link is a product of the consumer's link strength (s_x) and the state of the prey species (N_j), while the loss via each outgoing link is the product of the predator's link strength (s_k) and its state, N_k.

Perturbation (species removal) of the initial condition of the network results in the topological loss of incoming links to at least one consumer in the system. The CEG model mimics the consumer species' compensation for this loss of resources by increasing the strength of interaction with remaining resources or prey (e.g., Hebert et al., 2008). Compensation occurs by updating link strengths to equal the reciprocal of the remaining number of links (hence link strength increases as this remaining number decreases), and the difference equation is iterated until the N_x of all species stabilize. Secondary extinction occurs if a species has lost all resources ($N_x(t)<0$), or when N_x has reached a

TABLE *1*.—Paleocommunities for which CEG food webs have been constructed. Number of taxa refers to the number of consumer species in each food web.

Epoch	Location	Description	Environment	No. of taxa
Pennsylvanian	Linton, Ohio	Upper Freeport Coal	Deltaic, terrestrial	69 genera
Cisuralian (Lower Permian)	Oklahoma	Waurika I Locality	Terrestrial	128 genera
Guadalupian	Karoo Basin, South Africa	*Eodicynodon* Assemblage Zone	"	53 genera
Guadalupian	"	*Tapinocephalus* Assemblage Zone	"	101 genera
Lopingian (middle Permian)	"	*Pristerognathus* Assemblage Zone	"	65 genera
Lopingian (middle Permian)	"	*Tropidostoma* Assemblage Zone	"	73 genera
Lopingian (middle Permian)	"	*Cistecephalus* Assemblage Zone	"	92 genera
Lopingian (middle Permian)	"	*Dicynodon* Assemblage Zone	"	105 genera
Early Triassic	"	*Lystrosaurus* Assemblage Zone	"	84 genera
Early Triassic	"	*Cynognathus* Assemblage Zone	"	90 genera
Late Miocene	Dominican Republic	Four stratigraphic beds, Cercado and Gurabo Fms.	Shallow marine	113, 130, 138, 162 species
Holocene	San Francisco Bay	All invertebrate and vertebrate occurrences, including aquatic birds and marine mammals	"	~1,500 species
Holocene	Indo-Pacific	Sub-sample of invertebrates and vertebrates	Coral reef	103 species

minimum value. The minimum value for a species is a pre-determined fraction of N_x at equilibrium, and varies as one would expect a minimum viable population size to vary.

Secondary extinction in a species-level network will thus propagate after being initiated by a perturbation, both because of topological secondary extinction, as well as expected demographic consequences. Beginning in an equilibrium state, perturbation proceeds through the network, resulting in secondary extinctions, cascades of compensation, and possibly further secondary extinctions, ceasing when no further changes occur. The new network state differs from the initial one as a function of the nature and magnitude of the perturba-

tion, as well as the composition of the network, that is, systematic and ecologic diversity.

Previous results

Roopnarine (2006) raised the question that, given macroevolutionary changes to community structures through the Phanerozoic, for example changes in the numbers and types of guilds (Bambach, 1993), whether network resistance to disturbance propagation has varied or undergone secular change. The CEG numerical model was formulated to address this question specifically, and it was applied to several crude and general models of marine communities from the Cambrian, the later Paleozoic and late Mesozoic. Very slight

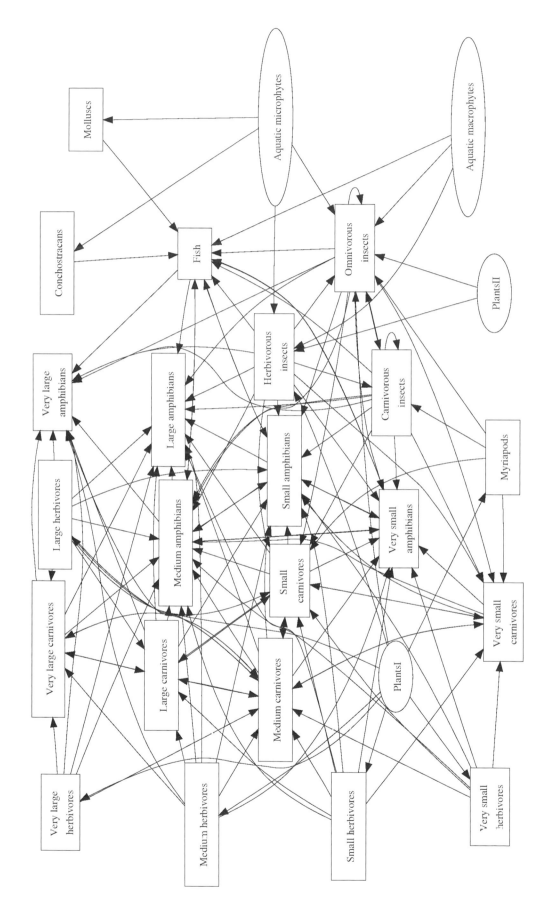

FIGURE 4.—Metanetwork of the Late Permian *Dicynodon* Assemblage Zone paleocommunity, Karoo Basin, South Africa. Producer guilds are denoted with elliptical compartments.

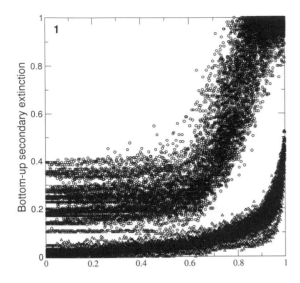

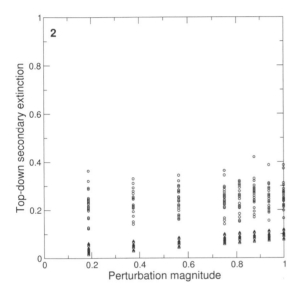

FIGURE 5.—CEG perturbation results of 100 species-level networks derived from the *Dicynodon* Assemblage Zone metanetwork (Fig. 3). *1*, Bottom-up perturbations generated by the incremental shutdown of primary productivity (x-axis). The y-axis represents the level of secondary extinction. The lower cloud of points (triangles) result from topological extinction only, while the upper cloud (circles) includes demographic extinction. *2*, Top-down perturbations generated by incremental removal of top amphibian and amniote carnivores. Symbols the same as in upper graph. There are 100 circles and 100 triangles per level of perturbation.

differences were detected among the communities, but two significant results emerged from the model.

First, when communities are perturbed by reductions of primary productivity, a threshold of perturbation exists at which levels of secondary extinction

increase nonlinearly and dramatically in each community and remain high as perturbation levels are further increased. Nonlinear threshold responses to disturbance have been noted in modern species and subsets of communities (Carpenter et al., 2008; Roopnarine, 2008), but CEG results suggest that they could be a generic feature of communities. Furthermore, a discontinuous nonlinear response is not predicted when network topology alone is considered (see below). Given the scales of extinction observed at particular intervals in the fossil record, paleontologists might well be comfortable with the possibility of such large-scale catastrophic collapses. Such patterns have not been observed in the Recent, probably because, even if communities do possess such tipping points, they are likely to reveal themselves only during rare events or under temporally-limited circumstances. For example, the Toba supereruption of 74,000 years ago may have been responsible for a significant bottleneck in the global human population (Rampino and Self, 1992; Ambrose, 1998), but the probability of such an event occurring in the next 50 years is perhaps only 0.007-0.1% (Smil, 2008). The current problem of global warming and the significant scales of predicted impacts do, however, emphasize the need to understand potential tipping points (for example, Kriegler et al., 2009).

Second, when species were ranked according to their competitive abilities within guilds, it was noted that at high levels of perturbation, competitively inferior species often had distinctly greater probabilities of survival. This result suggests that the survivors of catastrophic community collapses might not be representative of the preceding community and would have profound influences on the nature of recovery communities. The conclusion might be a valid reference to empirical observations, such as the relatively unchanged ecological structure of communities after the end-Ordovician event (Layou, 2009) compared to the dramatic changes observed after the end-Permian event (McGhee et al., 2004), as well as the numerical dominance of so-called disaster taxa in aftermath communities (Schubert and Bottjer, 1992).

These two results point to a need to understand very specific features of the CEG model, and to obtain data with better ecological and temporal resolutions. These requirements have been fulfilled with a variety of paleocommunity and Recent datasets (Table 1), two of which will be used as examples in the following sections: the Late Permian terrestrial community of the

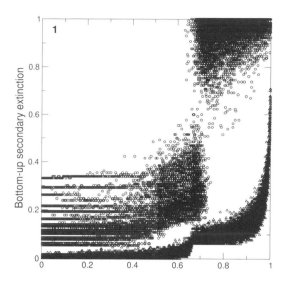

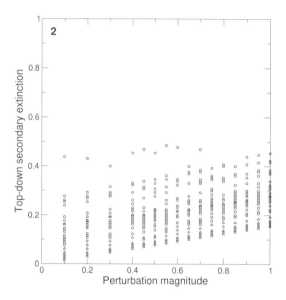

FIGURE 6.—CEG perturbation results of 100 species-level networks derived from the Late Miocene Dominican Republic marine community metanetwork. This figure may be interpreted exactly as Fig. 4, but the top-down perturbation consisted of the removal of the top pelagic carnivores.

Dicynodon Assemblage Zone from the Karoo Basin in South Africa, and a Late Miocene shallow coastal marine community from the Dominican Republic.

EXAMPLES

The following two examples serve to illustrate the model. They also serve to raise questions of interest to paleoecology, conservation paleobiology and conservation biology. The reconstructed communities are a Late Permian terrestrial community and a Late Miocene marine community. Both communities were subjected to bottom-up and top-down perturbations, comprising reductions of primary productivity and the removal of top predators respectively. Additionally, both topological secondary extinction alone, and in combination with secondary demographic effects, were simulated. For the bottom-up disruptions of primary productivity, the CEG model uses levels of primary productivity rather than individual producer species (Angielczyk et al., 2005; Roopnarine, 2006; Roopnarine et al., 2007). Utilization of actual producer taxa would require an understanding of the relationship between those taxa and the amount of autotrophic biomass available to the heterotrophic community, data which are simply not available for paleo-ecosystems.

One hundred species-level networks were generated for each paleocommunity, each one representing a stochastic draw from the total number that can be generated by the metanetworks. All simulations were performed with the CEG group of programs, descriptions of which are available upon request (Appendix 3).

Dicynodon Assemblage Zone

The *Dicynodon* Assemblage Zone (DAZ) is the Upper Permian faunal interval immediately preceding the end Permian mass extinction. The terrestrial community was dicynodont-dominated, with a diverse array of other synapsids (proto-mammals and mammalian ancestors), as well as reptilian, amphibian, insect, fish and molluscan fauna. Details of the community and metanetwork reconstruction are given elsewhere (Angielczyk et al., 2005; Roopnarine et al., 2007), though the dataset is revised continually as new systematic discoveries and revisions become available. The current dataset comprises 26 guilds, 149 interguild links, and 210 animal species (Fig. 4). Autotrophs are represented by four separate guilds based on both habitat and consumers.

Figure 5.1 illustrates the results of the bottom-up perturbation, which was implemented by an incremental decrease in the level of productivity available to the community from all producer guilds. There are three immediately obvious results. First, topological extinction underestimates the potential for secondary extinction at all levels of perturbation, because it does not incorporate further demographic effects stemming from compensation for lost or declining resources, nor top-down cascades. Second, even when productivity has been reduced to zero, topological extinction is less than 100%, implying the survival of species when no energy is being supplied to the community! This results from the presence of bidirectional interguild predation among higher trophic level guilds. Links between higher level consumers are not lost topologically when the perturbation is bottom-up, but this is of course ecologically unrealistic. Third, secondary extinction which incorporates both topological and demographic effects increases nonlinearly as a function of the magnitude of perturbation. The level of secondary extinction is relatively uniform over a broad range of perturbation ($0 \leq \omega_c \leq$ ~0.6, where ω_c is the magnitude of perturbation), and is only slightly greater than that predicted by topological effects alone. Beyond this range, however, secondary extinction increases rapidly in a critical interval (~$0.6 \leq \omega_c \leq$ ~0.8), and then returns to a uniform range, albeit at significantly higher values.

Top-down perturbations were simulated by the incremental removal of species from the largest carnivore guilds (Fig. 5.2), including Very Large Amphibians, Large Amniote Carnivores and Very Large Amniote Carnivores (Fig. 4). Topological extinction once again underestimates the potential for secondary extinction, increasing linearly since very few species are lost topologically when top consumers are removed. Total secondary extinction including demographic effects, on the other hand, has a relatively stationary mean over the total perturbation range, implying that most species lost secondarily are lost regardless of the magnitude of the perturbation. This is a result of the familiar top-down control of diversity exerted by high trophic level predators (Paine, 1966; Byrnes et al., 2006; Daskalov et al., 2007; Myers et al., 2007). Of particular interest to conservation paleobiology is the fact that top-down perturbations do not generate the catastrophically high levels of secondary extinction observed for the bottom-up perturbations. The implications of this observation are discussed below.

Late Miocene, Dominican Republic

This reconstruction is representative of the coastal marine ecosystem during the warm, productive Upper Miocene of the Caribbean Sea. The paleocommunity is based on one bulk sample out of a collection from the Cercado and Gurabo Formations of the northern Dominican Republic (DR). The sample, 525-1C (from the Gurabo Fm.), is one of several collected during an expedition in 2000, and stratigraphic setting, sampling

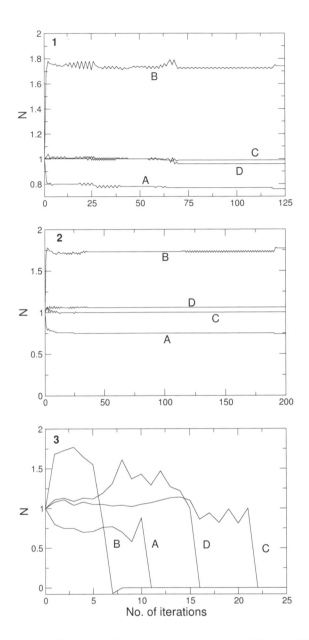

FIGURE 7.—Extended demographic dynamics of four guilds from the Late Miocene Dominican Republic community, subjected to bottom-up perturbation at three magnitudes. The y-axis indicates proportional change relative to initial population sizes (standardized to 1). Guilds and species are A= shallow infaunal, sedentary deposit and suspension feeder, for example tellinid or nuculanid bivalves; B= epifaunal motile omnivore, e.g. malacostracan crustaceans; C= pelagic carnivore, preying on invertebrates only; D= pelagic carnivore preying on invertebrates and pelagic vertebrates. *1*, Perturbation magnitude ω=0.2; *2*, Perturbation magnitude ω=0.55; *3*, Perturbation magnitude ω=0.71.

protocols and preparation are described in detail elsewhere (Tang and Pantel, 2005). The total collection comprises an estimated 250,000 specimens and while mollusc-dominated, also contains diverse echinoderm, arthropod, bryozoan, cnidarian and protistan taxa. The paleocommunity was reconstructed from both the specimen collection, as well as extensive documentation from the Dominican Republic Paleontology Project (Saunders et al., 1986) and other publications of the *Bulletins of American Paleontology*.

Species-level networks, generated stochastically from the metanetwork (Fig. 1), were subjected to bottom-up perturbations of primary production and detrital supply, and top-down perturbations of the two top pelagic carnivore guilds. Similar conclusions regarding the differences between responses to bottom-up and top-down perturbations may be drawn when compared to the *Dicynodon* Assemblage Zone community. Two differences between the communities' bottom-up responses are obvious though. First, the transition between low and high levels of secondary extinction is more abrupt and discontinuous in the DR community (Fig. 6). This can be explained largely by the second difference, apparent in the topological results, where a discontinuity coincides with the discontinuity in the demographic results. This critical interval ($0.65 \leq \omega_C \leq 0.7$) marks the complete shutdown of all benthic productivity (micro- and macroscopic), with the only remaining source of production being phytoplanktonic. These results are discussed below.

DISCUSSION

The CEG model analyzes the relationship between community composition and resistance to cascading secondary extinctions. Composition is linked to dynamics using models of the food web structure of the community. While there is considerable uncertainty associated with the reconstruction of complex food webs of fossil and modern communities, with some uncertainty stemming from incomplete data and information, uncertainty also arises because of the natural variability of food webs. The CEG model incorporates the complexity and uncertainty, modeling species-level food webs as large parameterized networks, interspecific dynamics as algorithmic and iterative difference equations, and uncertainty with the consideration of very large statistical spaces within which higher guild-level community representations may vary. Application of the CEG model to two sample paleocommunities point to several general features and raises several questions,

all of which are pertinent to almost every paleocommunity examined so far (Table 1).

Topological and demographic effects

Perturbations propagate through a community net-

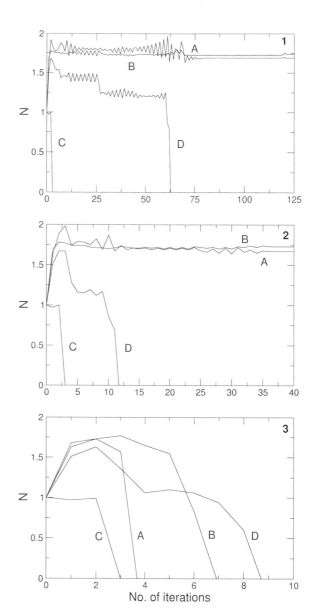

FIGURE 8.—Extended demographic dynamics of four malacostracan species from the Dominican Republic community, subjected to bottom-up perturbation at three magnitudes. A= generalist species, seven prey species; B= extreme generalist species, 59 prey species; C= extreme specialist species, one prey species; D= generalist species, 12 prey species. All species are preyed upon by pelagic carnivores, and species D is additionally infested with ectoparasites. *1*, Perturbation magnitude ω=0.2; *2*, Perturbation magnitude ω=0.55; *3*, Perturbation magnitude ω=0.71.

work both topologically and demographically. Topological losses, however, are likely to always underestimate the potential for secondary extinction because they ignore top-down cascading processes, and cannot account for the probability of extinction when resources or network links are still present. But exactly how do the demographic interactions generate secondary extinction? It is virtually impossible to answer this question analytically because of the complexity and iterative nature of the interactions, but insight is necessary because of the obvious effects of those interactions, namely: communities are very resistant to secondary extinction over a broad range of perturbation magnitude, but this resistance is lost rapidly within a narrow interval of magnitude when the perturbation is of a bottom-up type.

These nonlinear features can be explained with detailed examinations of interspecific interactions while a network is responding to a perturbation. Imagine this as being able to observe species in the community for an extended period of time after the community has been perturbed. The responses of species in four guilds from a single species-level network of the Dominican Republic community (Fig. 3) were therefore observed for 1,000 iterations after initial perturbation, at three different levels of perturbation (Fig. 7). All species are affected by the lowest and mid-level perturbations, though they respond differently. The infaunal bivalve, which as a primary consumer can be affected directly by the perturbations, declines in population size immediately after the perturbation at low and medium perturbation magnitudes (Fig. 7.1 and 7.2). Both pelagic carnivore populations oscillate about the initial value (N(0)=1), while the malacostracan predator actually increases significantly in population size. Secondary extinctions generally never occur within the community beyond step ten, but survivable demographic effects can be felt long afterward as shown by these four species. All species, however, eventually settle to new stable states within 200 iterations or less. The highest level of perturbation (Fig. 7.3), corresponding to the critical threshold of secondary extinction (ω=0.71), sees all four species become extinct. The explanation is straightforward: the drastic reduction of primary production results in a large pulse of topological extinction, the impact of which generates significantly increased top-down compensatory effects of higher-level consumers, initiating positive feedback loops of negative demographic consequences between lower and up-

per trophic levels. The malacostracan species, which at lower levels of perturbation actually benefited, is now the first to become extinct.

This particular species is one of four in a guild of malacostracans. As a guild, these species have the potential to be remarkable generalists, since their prey guilds have a total richness of 76 species, comprising numerous benthic molluscs. The malacostracan species in question actually is an extreme generalist, with 59 prey (incoming links) in the network (species B in Fig. 8). Other species in the guild range from an extreme specialist, preying on a single scaphopod species (species C), and two moderately generalist species, A and D, with seven and 12 prey species respectively. The specialist has essentially no resistance to secondary extinction, and becomes extinct at all levels of perturbation. The situation with the generalists is more complicated. The extreme generalist, species B, is very resistant to perturbation at low and mid-levels (Fig. 8.1 and 8.2), but becomes extinct at the highest level, as does species A. Their greater resistance is a function of having more prey than the specialist, and hence being more buffered from the probability of propagating topological extinction as well as having more resources for top-down compensation. Species D, however, is also a generalist, preying on the same guilds as the other generalists, and yet, after some demographic compensation (oscillations), becomes extinct at low and mid-levels of perturbation. The difference is explained by an examination of the malacostracans' predators. The four species share similar pelagic carnivore predators, but species D is additionally parasitized by a guild of ectoparasites (for example, ectoparasitic copepods or barnacles). The negative demographic consequences of such infestations is well documented (Alvarez et al., 1995; Boglio and Lucas, 1997), but the CEG model predicts that host species should have probabilities of local extinction greater than uninfected fellow guild species, when their communities are subjected to bottom-up perturbations. Lafferty and Kuris (2009) have recently documented the potentially large impact of the inclusion or exclusion of parasites on the understanding of secondary extinction in aquatic food webs.

Critical thresholds

Both the *Dicynodon* Assemblage Zone (DAZ) and Dominican Republic (DR) communities possess a critical threshold of secondary extinction when perturbed bottom-up that they do not have when perturbed top-

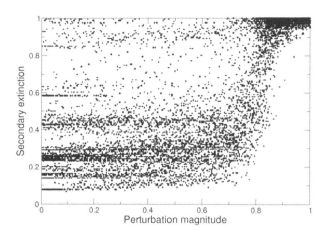

FIGURE *9.*—Bottom-up perturbation of the Early Triassic *Lystrosaurus* Assemblage Zone paleocommunity. Levels of secondary extinction are highly unpredictable and severe at low levels of perturbation (Roopnarine et al., 2007).

down. Top-down perturbations consist of the removal of top carnivores, and the resulting secondary extinctions are confined almost entirely to the secondary demographic loss of mid-level carnivores and primary consumers. They are outcompeted by other species in their guilds, and species in other guilds with which they share prey. The successful competitors are those who have been relieved of predation by the initial perturbation. The CEG model therefore successfully reproduces the keystone predator effect. Secondary extinctions occur at roughly the same levels when driven by bottom-up perturbations, until the critical threshold is reached. Up to that point, secondary extinctions are a combination of topological extinctions and the associated demographic extinctions of species with few links or resources (for example, the specialized scaphopod predator in the DR community). When the critical threshold is reached, however, the magnitude of the topological extinctions is great enough to initiate significant cascades of top-down compensation by more highly linked consumers. More species in the network are affected, and positive feedback loops are generated between predators, their extended network of prey, and species to which they are linked only indirectly. The result is a catastrophic increase in the level of secondary extinction.

The existence of these thresholds is not a function of CEG assumptions or an artifact of the model. Though thresholds exist in almost every community to which the model has been applied, there are exceptions. Most notable is the Early Triassic *Lystrosaurus* Assemblage

Zone community (LAZ), the recovery interval successor to the DAZ (Roopnarine et al., 2007) (Fig. 9). It is hypothesized that the apparently anomalous behaviour of the LAZ community arises from constraints laid upon a large fraction of the metanetwork's species-level network space, where significantly lowered diversities in many guilds lead to broadly distributed multinomial choices during network generation, a condition which may not exist in more "balanced" communities such as the DAZ and DR. This hypothesis is currently untested, but is of tremendous interest when predicting the future stability of declining modern communities.

Critical thresholds in community behavior is also not restricted to the CEG model or dynamics, but have both been observed empirically and are predicted mathematically to arise from dynamics in species populations and whole communities (Dunne et al., 2002a; Jain and Krishna, 2002; Rietkerk et al., 2004; Scheffer et al., 2004; Carpenter et al., 2008; Roopnarine, 2008; Biggs et al., 2009). The CEG model predicts that many communities have these thresholds embedded in their network dynamics. A failure to observe sudden, massive community and ecosystem collapses in the Recent is perhaps the result of the thresholds being attainable only at very high magnitudes of bottom-up perturbation. Such high energy events undoubtedly occur infrequently and on geological timescales. Moreover, the anthropogenically driven or mediated losses of species in modern times, while at a crisis level, have until recently been confined largely to higher trophic level species. However, increasing rates of habitat destruction, the shift of species overexploitation to lower trophic levels in communities ("fishing down the web") (Dobson et al., 2006), increasing human control of global photosynthetic resources, and the impacts of climate change (decreased terrestrial precipitation, increased air and water temperatures, and ocean acidification), could work in concert to move communities from CEG regimes of low bottom-up perturbation, to perturbations of increasing magnitude. The unfolding of such events would be unfortunate validation of the CEG model.

Model validation

Validation of the CEG model of any single community is not straightforward for two reasons. First, we cannot observe paleocommunity dynamics directly. Second, the most obvious feature of CEG dynamics, the nonlinear transition between low and very high levels of secondary extinction, happens only at very high levels of perturbation, and those may occur so infrequently as to be inaccessible in present times. Dynamics at low levels of perturbation are, however, easily validated circumstantially, because they reproduce the types of ecological processes commonly observed in disturbed or experimentally manipulated communities, for example top down cascades and keystone predator effects. The model could be validated, in theory, by coupling it with manipulations of experimental systems. Accuracy would be dependent though on the extent to which the manipulated populations could be modeled with the CEG difference equation, or replacement of that equation with more suitable differential counterparts. One then encounters, unfortunately, the limitations imposed by that approach, namely restriction to systems of relatively low diversity.

The real value of the CEG model, and many other food web models, lies in the resulting comparisons among communities. The comparative dynamics of multiple communities, their similarities and differences, when based on the same rules for reconstruction and parameterization, using the same underlying equations, must be the result of similarities and differences of taxonomic and ecologic diversity. The advantages of the CEG model are that it allows us to compare the relative dynamic properties of communities and observe the behaviors that emerge from these highly complex systems under similar perturbative conditions. For example, the Early Triassic *Lystrosaurus* Assemblage Zone of the Karoo Basin is unique among the CEG paleocommunities, in that it is often catastrophically susceptible to secondary extinction at even very low levels of perturbation (Roopnarine et al., 2007) (Fig. 9). Taphonomic and other potentially artifactual explanations have been ruled out, suggesting that this community of the Permo-Triassic aftermath really did function differently from other communities in the dataset. This raises the intriguing, and unanswered questions of the generality of this type of community in the aftermath of mass extinctions, and whether modern communities are being pushed toward that state by anthropogenic disturbances of increasing magnitude.

Caveat emptor

The CEG model assumes a number of conditions, but the assumptions can be tested on the basis of additional data, or their impacts can be assessed by varying the assumptions. Perhaps the most significant condition, and the one most relevant to conservation paleobiology, is that paleocommunity dynamics can be applied to an understanding of modern communities. This assumption has been tested with the CEG reconstruction of the modern marine community of San Francisco Bay (Fig. 10), comprising (at that time) a total of 1,325 invertebrate and fish taxa. Hertog et al. (2006) performed a simple experiment where all higher taxa in the community without representation in Sepkoski's genus or family-level compendia were removed from the model. The CEG dynamics remained essentially unchanged. The most noticeable difference is an increase in the variance of secondary extinctions for the reduced community. Additionally, the detail and diversities used for CEG paleocommunity reconstructions are well within the ranges available for published modern food webs (Roopnarine et al., 2007).

The other main assumptions of the CEG model lie with parameters, namely the distributions underlying interaction strengths and link number. The CEG assumptions are reasonable on the basis of current empirical observations and theory, but this is an area that

FIGURE *10*.—Metanetwork of the modern marine community of San Francisco Bay. In contrast to the paleocommunity metanetworks (Figs. 1 and 3), individual guilds cannot be visualized legibly on this network, nor can species-level networks be rendered meaningfully.

requires significantly increased ecological attention. Currently, the CEG framework allows the modeler to explore the impact of varying these assumptions.

The CEG model is also not free of the taphonomic issues that must be considered in any paleoecological study. In addition to the effects of missing data (taxonomic and ecological) discussed above, there is the issue of temporal and spatial averaging. It is extremely unlikely that any two individuals sampled from the same locality, no matter how carefully constrained stratigraphically, ever interacted during their lives. Exceptions would be actual intimate physical associations of the fossils. The careful application of sampling constraints, however, can be relied upon to infer the interaction or overlap of those species as is done for any reliable paleocommunity reconstruction. Moreover, a certain degree of temporal and spatial averaging is desirable. Such averaging smooths out short-term variability, such as seasonal, that does not contribute meaningfully to understanding processes that occur on geological and macroevolutionary scales. Suitably time-averaged samples can retain high fidelity to their parent communities (Kidwell, 2001, 2002), and in fact are likely to represent more complete depictions of the actual community, in contrast to high-resolution snapshots of interactions.

A final note

CEG is not a method. It is not a recipe for the systematic analysis of past or modern communities. It is a model, based upon ecological first principles and empirical data, of the trophic interactions of a community. The model is capable of making specific predictions of the response of a community to specific types of perturbations. Its value to conservation paleobiology lies both in its application to paleocommunities, where the predictions can be tested directly with preserved communities, and its subsequent extension to modern communities, where we do not have the luxury of observing community collapses and recovery. Additional value will be gained if: (1) it motivates paleoecological workers to adopt theoretical modeling approaches more broadly; (2) those workers subsequently extend the CEG model or supplant it altogether with improved ideas; and (3) it motivates conservation biologists to think inclusively about paleocommunities and the ecological crises of the past.

ACKNOWLEDGMENTS

I wish to thank first and foremost my excellent collaborators on this project, K. Angielczyk, R. Hertog and S. Wang. Their material contributions, ideas and discussions are woven throughout the paper. I also thank, for many helpful discussions and ideas, J. Dumbacher, D. Goodwin, B. Kraatz, R. Plotnick, C. Tang and G. Vermeij. K. Angielczyk, R. Hertog, M. Olson and E. Conel compiled significant amounts of data. The paper benefited greatly from the questions, comments and advice of C. Marshall and an anonymous reviewer. This work was supported by NSF CMG ARC-0530825 to Roopnarine and Wang, and CAS Hagey Venture Fund to Roopnarine and Dumbacher.

REFERENCES

ALBERT, R., H. JEONG, AND A. BARABASI. 2000. Error and attack tolerance of complex networks. Nature, 406:378-382.

ALLESINA, S., AND BODINI. 2004. Who dominates whom in the ecosystem? Energy flow bottlenecks and cascading extinctions. Journal of Theoretical Biology, 230:351-358.

ALVARAEZ, F., A. H. HINES, AND M. L. REAKA-KUDLA. 1995. The effects of parasitism by the barnacle *Loxothylacus panopaei* (Gissler) (Cirripedia: Rhizocephala) on growth and survival of the host crab *Rhithropanopeus harrisii* (Gould) (Brachyura: Xanthidae). Journal of Experimental Marine Biology and Ecology, 192:221-232 .

AMARAL, L. A. N., AND M. MEYER. 1999. Environmental changes, coextinction, and patterns in the fossil record. Physical Review Letters, 82:652-655.

AMBROSE, S. H. 1998. Late Pleistocene human population bottlenecks: Volcanic winter and the differentiation of modern humans. Journal of Human Evolution, 34:623-651.

ANGIELCZYK, K. D., P. D. ROOPNARINE, AND S. C. WANG. 2005. Modeling the role of primary productivity disruption in end-Permian extinctions, Karoo Basin, South Africa, p. 16-23. *In* S. G. Lucas and K. E. Zeigler (eds.), The Non-Marine Permian. New Mexico Museum of Natural History and Science Bulletin, 30.

AZAELE, S., S. PIGOLOTTI, J. R. BANAVAR, AND A. MARITAN. 2006. Dynamical evolution of ecosystems. Nature, 444:926-928.

BAMBACH, R. K. 1993. Seafood through time: Changes in biomass, energetics, and productivity in the marine ecosystem. Paleobiology, 19:372-397.

BASTOLLA, U., M. LASSIG, S. C. MANRUBIAA, AND A. VALLERIANIC. 2005. Biodiversity in model ecosystems, I: Co-existence conditions for competing species. Journal of Theoretical Biology, 235:521–530.

BENINCA, E., J. HUISMAN, R. HEERKLOOS, K. D. JOHNK, P. BRANCO, AND E. H. VAN NES. 2008. Chaos in a long-term experiment with a plankton community. Nature, 451:822-826.

BENNINGTON, J. B., W. A. DIMICHELE, C. BADGLEY, R. K. BAMBACH, P. M. BARRETT, A. K. BEHRENSMEYER, R. BOBE, R. J. BURNHAM, E. B. DAESCHLER, J. V. DAM, J. T. ERONEN, D. H. ERWIN, S. FINNEGAN, S. M. HOLLAND, G. HUNT, D. JABLONSKI, S. T. JACKSON, B. F. JACOBS, S. M. KIDWELL, P. L. KOCH, M. J. KOWALEWSKI, C. C. LABANDEIRA, C. V. LOOY, S. K. LYONS, P. M. NOVACK-GOTTSHALL, R. POTTS, P. D. ROOPNARINE, C. A. E. STRÖMBERG, H-D. SUES, P. J. WAGNER, P. WILF, AND S. L. WING. 2009. Critical issues of scale in paleoecology. Palaios, 24:1-4.

BERLOW, E. L., J. A. DUNNE, N. D. MARTINEZ, P. B. STARKE, R. J. WILLIAMS, AND U. BROSE. 2009. Simple prediction of interaction strengths in complex food webs. Proceedings of the National Academy of Sciences of the United States of America, 106:187-191.

BIGGS, R., S. R. CARPENTER, AND W. A. BROCK. 2009. Turning back from the brink: Detecting an impending regime shift in time to avert it. Proceedings of the National Academy of Sciences of the United States of America, 106:826-831.

BOGLIO, E. G., AND J. S. LUCAS. 1997. Impacts of ectoparasitic gastropods on growth, survival, and physiology of juvenile giant clams (*Tridacna gigas*), including a simulation model of mortality and reduced growth rate. Aquaculture, 150:25-43.

BOHM, D. 1957. Causality and Chance in Modern Physics. Routledge & Kegan Paul, London, 163 p.

BORER, E. T., K. ANDERSON, C. A. BLANCHETTE, B. BROITMAN, S. D. COOPER, AND B. S. HALPERN. 2002. Topological approaches to food web analyses: A few modifications-may improve our insights. Oikos, 99:397-401.

BRIAND, F., AND J. E. COHEN. 1984. Community food webs have scale-invariant structure. Nature, 307:264-267.

BROSE, U., E. L. BERLOW, AND N. D. MARTINEZ. 2005. Scaling up keystone effects from simple to complex ecological networks. Ecology Letters, 8:1317-1325.

BUTTERFIELD, N. J. 2001. Cambrian food webs, p. 40-43. *In* D. E. G. Briggs and P. R. Crowther (eds.), Paleobiology II: A Synthesis. Blackwell Scientific, Oxford.

BYRNES, J., J. J. STACHOWICZ, K. M. HULTGREN, A. R. HUGHES, S. V. OLYARNIK, AND C. S. THORNBER. 2006. Predator diversity strengthens trophic cascades in kelp forests by modifying herbivore behaviour. Ecology Letters, 9:61-71.

CALLAWAY, D. S., M. E. J. NEWMAN, S. H. STROGATZ, AND D. J. WATTS. 2000. Network robustness and fragility: Percolation on random graphs. Physical Review Letters, 85:5468-5471.

CAMACHO, J., R. GUIMERÀ, AND L. A. NUNES AMARAL. 2002a. Robust patterns in food web structure. Physical Review Letters, 188:1-4.

CAMACHO, J., R. GUIMERÀ, AND L. A. NUNES AMARAL. 2002b. Analytical solution of a model for complex food webs. Physical Review E, 65:1-4.

CARPENTER, S. R., W. A. BROCK, J. J. COLE, J. F. KITCHELL, AND M. L. PACE. 2008. Leading indicators of trophic cascades. Ecology Letters, 11:128-138.

DASKALOV, G. M., A. N. GRISHIN, S. RODIONOV, AND V. MIHNEVA. 2007. Trophic cascades triggered by overfishing reveal possible mechanisms of ecosystem regime shifts. Proceedings of the National Academy of Sciences of the United States of America, 104:10518-10523.

DOBSON, A., D. LODGE, J. ALDER, G. S. CUMMING, J. KEYMER, J. MACGLADE, H. MOONEY, J. A. RUSAK, O. SALA, V. WOLTERS, D. WALL, R. WINFREE, AND M. A. XENOPOULOS. 2006. Habitat loss, trophic collapse, and the decline of ecosystem services. Ecology, 87:1915-1924.

DOYLE, J. C., D. L. ALDERSON, L. LI, S. LOW, M. ROUGHAN, S. SHALUNOV, R. TANAKA, AND W. WILLINGER. 2005. The "robust yet fragile" nature of the Internet. Proceedings of the National Academy of Sciences of the United States of America, 102:14497-14502.

DUFFY, J. E., B. J. CARDINALE, K. E. FRANCE, P. B. MCINTYRE, E. THÉBAULT, AND M. LOREAU. 2007. The functional role of biodiversity in ecosystems: Incorporating trophic complexity. Ecology Letters, 10:522-538.

DUNNE, J. A., AND R. J. WILLIAMS. 2009. Cascading extinctions and community collapse in model food webs. Philosophical Transactions of the Royal Society, B, 364:1711-1723.

DUNNE, J. A., R. J. WILLIAMS, AND N. D. MARTINEZ. 2002a. Network structure and biodiversity loss in food webs: Robustness increases with connectance. Ecology Letters, 5:558-567.

DUNNE, J. A., R. J. WILLIAMS, AND N. D. MARTINEZ. 2002b. Food-web structure and network theory: The role of connectance and size. Proceedings of the National Academy of Sciences of the United States of America, 99:12917-12922.

DUNNE, J. A., R. J. WILLIAMS, N. D. MARTINEZ, R. A. WOOD, AND D. H. ERWIN. 2008. Compilation and network analyses of Cambrian food webs. PLoS Biology, 6:0693-0708.

EATON, J. W. 2007. GNU Octave. Available from: http://www.octave.org.

EBENMAN, B., AND T. JONSSON. 2005. Using community viability analysis to identify fragile systems and keystone species. Trends in Ecology and Evolution, 10:568-575.

EDELINE, E., T. B. ARIA, L. A. VØLLESTADA, I. J. WINFIELD, J. M. FLETCHER, J.B. JAMES, AND N. C. STENSETH. 2008. Antagonistic selection from predators and pathogens alters food-web structure. Proceedings of the National

Academy of Sciences of the United States of America, 105:19792-19796.

EDGELL, T. C., AND R. ROCHETTE. 2008. Differential snail predation by an exotic crab and the geography of shell-claw covariance in the northwest Atlantic. Evolution, 62:1216-1228.

EINSTEIN, A. 1934. On the method of theoretical physics. Philosophy of Science, 1:163-169.

EKLÖF, A., AND B. EBENMAN. 2006. Species loss and secondary extinctions in simple and complex model communities. Journal of Animal Ecology, 75:239-246.

ELTON, C. S. 1927. Animal Ecology. Sidgwick & Jackson, London, 207 p.

ESTRADA, E. 2007. Food webs robustness to biodiversity loss: The roles of connectance, expansibility and degree distribution. Journal of Theoretical Biology, 244:296-307.

EVELEIGH, E. S., K. S. McCANN, P. C. McCARTHY, S. J. POLLOCK, C. J. LUCAROTTI, B. MORIN, G. A. McDOUGALL, D. B. STRONGMAN, J. T. HUBER J. UMBANHOWAR, AND L. D. B. FARIA. 2007. Fluctuations in density of an outbreak species drive diversity cascades in food webs. Proceedings of the National Academy of Sciences of the United States of America, 104:16976-16981.

FISHER, R. A. 1930. The genetical theory of natural selection. Dover Publications, New York, 291 p.

FLESSA, K.W. 2002. Conservation paleobiology. American Paleontologist, 10: 2-5.

GAUSE, G. F. 1934. The Struggle for Existence. Courier Dover Publications, New York, 176 p.

GRIMM, V., AND S. F. RAILSBACK. 2005. Individual-Based Modeling and Ecology. Princeton University Press, Princeton, New Jersey, 428 p.

HAIRSTON, N. G., F. E. SMITH, AND L. B. SLOBODKIN. 1960. Community structure, population control, and competition. The American Naturalist, 94:421-425.

HALDANE, J. B. S. 1932. The Causes of Evolution. Princeton University Press, Princeton, New Jersey, 222 p.

HAVENS, K. 1992. Scale and structure in natural food webs. Science, 257:1107-1109.

HEBERT, C. E., D. V. C. WESELOH, A. IDRISSI, M. T. ARTS, R. O'GORMAN, O. T. GORMAN, AND B. LOCKE. 2008. Restoring piscivorous fish populations in the Laurentian Great Lakes causes seabird dietary change. Ecology, 89:891-897.

HERTOG, R., P. D. ROOPNARINE, S. C. WANG, K. D. ANGIELCZYK, AND M. OLSON. 2006. The impact of variable fossilization on reconstructions of paleocommunities and community dynamics: The San Francisco Bay community. Geological Society of America Abstracts with Programs, 38:63.

JACKSON, J. B. C., M. X. KIRBY, W. II. BERGER, K. A. BJORNDAL, L. W. BOTSFORD, B. J. BOURQUE, R. H. BRADBURY, R. COOKE, J. ERLANDSON, J. A. ESTES, T. P. HUGHES, S. KIDWELL, C. B. LANGE, H. S. LENIHAN, J. M. PANDOLFI, AND C. H. PETERSON. 2001. Historical overfishing and the recent collapse of coastal ecosystems. Science, 293:629-638.

JAIN, S., AND S. KRISHNA. 2002. Large extinctions in an evolutionary model: The role of innovation and keystone species. Proceedings of the National Academy of Sciences of the United States of America, 99:2055-2060.

KIDWELL, S. M. 2001. Preservation of species abundance in marine death assemblages. Science, 294:1091-1094.

KIDWELL, S. M. 2002. Time-averaged molluscan death assemblages: Palimpsests of richness, snapshots of abundance. Geology, 30:803-806.

KRIEGLER, E., J. W. HALL, H. HELD, R. DAWSON, AND H. J. SCHELLNHUBER. 2009. Imprecise probability assessment of tipping points in the climate system. Proceedings of the National Academy of Sciences of the United States of America, 106:5041-5046.

LAFFERTY, K. D., AND A. M. KURIS. 2009. Parasites reduce food web robustness because they are sensitive to secondary extinction as illustrated by an invasive estuarine snail. Philosophical Transactions of the Royal Society, 364:1659-1663.

LANDE, R., S. ENGEN, AND B-E. SÆTHER. 2003. Stochastic Population Dynamics in Ecology and Conservation. Oxford University Press, New York, 212 p.

LAYOU, K. M. 2009. Ecological restructuring after extinction: The Late Ordovician (Mohawkian) of the eastern United States. Palaios, 24:118-130.

LEVIN, S. A. 1998. Ecosystems and the biosphere as complex adaptive systems. Ecosystems, 1:431-436.

MacARTHUR, R. 1955. Fluctuations of animal populations and a measure of community stability. Ecology. 36:533-536.

MATHWORKS. 2009. MatLab. The Mathworks, Incorporated, Massachussetts.

MAY, R. M. 1972. Will a large complex system be stable? Nature, 238:413-414.

MAY, R. M. 1974. Stability and Complexity in Model Ecosystems. Princeton University Press, Princeton, New Jersey, 265 p.

MAY, R. M. 2009. Food web assembly and collapse: Mathematical models and implications for conservation. Philosophical Transactions of the Royal Society, B, 364:1643-1646.

McCANN, K. S. 2000. The diversity–stability debate. Nature, 405:228-233.

McCANN, K., AND N. ROONEY. 2009. The more food webs change, the more they stay the same. Philosophical Transactions of the Royal Society, B, 364:1789-1801.

McCANN, K., A. HASTINGS, AND G. R. HUXEL. 1998. Weak trophic interactions and the balance of nature. Nature, 395:794-798.

McGHEE, G. R., P. M. SHEEHAN, D. J. BOTTJER, AND M. L.

Droser. 2004. Ecological ranking of Phanerozoic biodiversity crises: Ecological and taxonomic severities are decoupled. Palaeogeography, Palaeoclimatology, Palaeoecology, 211:289-297.

Montoya, J. M., and R. V. Solé. 2002. Small world patterns in food webs. Journal of Theoretical Biology, 214:405-412.

Montoya, J. M., S. L. Pimm, and R. V. Solé. 2006. Ecological networks and their fragility. Nature, 442:259-264.

Murray, J. D. 1989. Mathematical Biology. Springer-Verlag, New York, 767 p.

Myers, R. A., J. K. Baum, T. D. Shepherd, S. P. Powers, and C. H. Peterson. 2007. Cascading effects of the loss of apex predatory sharks from a coastal ocean. Science, 315:1846-1850.

Navarette, S. A., and E. L. Berlow. 2006. Variable interaction strengths stabilize marine community pattern. Ecology Letters, 9:526-536.

Néutel, A., J. A. P. Heersterbeek, and P. C. de Ruiter. 2002. Stability in real food webs: Weak links in long loops. Science, 296:1120-1123.

Newman, M. E. J. 2003. The structure and function of complex networks. SIAM Review, 45:167-256.

Novak, M., and J. T. Wootton. 2008. Estimating nonlinear interaction strengths: An observation-based method for species-rich food webs. Ecology, 89:2083-2089.

Okuyama, T., and J. N. Holland. 2008. Network structural properties mediate the stability of mutualistic communities. Ecology Letters, 11:208-216.

Paine, R. T. 1966. Food web complexity and species diversity. The American Naturalist, 100:65-75.

Paine, R. T. 1992. Food-web analysis through field measurement of per capita interaction strength. Nature, 355:73-75.

Peacor, S. D., and E. E. Werner. 2004. How dependent are species-pair interaction strengths on other species in the food web? Ecology, 85: 2754-2763.

Petanidou, T., A. S. Kallimanis, J. Tzanopoulos, S. P. Sgardelis, and J. D. Pantis. 2008. Long-term observation of a pollination network: Fluctuation in species and interactions, relative invariance of network structure and implications for estimates of specialization. Ecology Letters, 11:1-12.

Pimm, S. L. 1982. Food Webs. University of Chicago Press, Chicago, Illinois, 219 p.

Pimm, S. L. 1991. The Balance of Nature: Eecological Issues in the Conservation of Species and Communities. University of Chicago Press, Chicago, Illinois, 434 p.

R Project. 2009. Available from: http://www.r-project.org.

Rampino, M. R., and S. Self. 1992. Volcanic winter and accelerated glaciation following the Toba super-eruption. Nature, 359:50-52.

Rietkerk, M., S. C. Dekker, P. C. de Ruiter, and J. van de Koppel. 2004. Self-organized patchiness and cata-

strophic shifts in ecosystems. Science, 305:1926-1929.

Rooney, N., K. S. McCann, and J. C. Moore. 2008. A landscape theory for food web architecture. Ecology Letters, 11: 867–881.

Roopnarine, P. D. 2006. Extinction cascades and catastrophe in ancient food webs. Paleobiology, 32:1-19.

Roopnarine, P. D. 2008. Catastrophe theory, p.531-536. In Sven Erik Jørgensen and Brian D. Fath (eds.), Ecological Informatics. Encyclopedia of Ecology. Elsevier Press, Oxford.

Roopnarine, P. D., K. D. Angielczyk, S. C. Wang, and R. Hertog. 2007. Trophic network models explain instability of Early Triassic terrestrial communities. Proceedings of the Royal Society, B, 274:2077-2086.

Saunders, J. B., P. Jung, and B. Duval. 1986. Neogene paleontology in the Northern Dominican Republic. 1. Field surveys, lithology, environment, and age. Bulletins of American Paleontology, 89:1-79.

Scheffer, M., CS. Carpenter, J. A. Foley, C. Folke, and B. Walke. 2004. Catastrophic shifts in ecosystems. Nature, 413:591-596.

Schubert, J. K., and D. J. Bottjer. 1992. Early Triassic stromatolites as post-mass extinction disaster forms. Geology, 20:883-886.

Sepkoski, J. J., Jr. 1984. A kinetic model of Phanerozoic taxonomic diversity: III, Post-Paleozoic families and mass extinctions. Paleobiology, 10:246-267.

Sibly, R. M., D. Barker, J. Hone, and M. Pagel. 2007. On the stability of populations of mammals, birds, fish and insects. Ecology Letters, 10:970-976.

Smil, V. 2008. Global Catastrophes and Trends. The Next Fifty Years. The MIT Press, Cambridge, Massachusetts, 307 p.

Solé, R. V., J. M. Montoya, and D. H. Erwin. 2002. Recovery after mass extinction: Evolutionary assembly in large-scale biosphere dynamics. Philosophical Transactions of the Royal Society of London, B, 357:697-707.

Strogatz, S. H. 2001. Exploring complex networks. Nature, 410:268-276.

Tang, C., and J. H. Pantel. 2005. Combining morphometric and paleoecological analyses: Examining small-scale dynamics in species-level and community-level evolution. Paleontologia Electronica, 8(2).

Teng, J., and K. S. McCann. 2004. Dynamics of compartmented and reticulate food webs in relation to energetic flows. The American Naturalist, 164:85-100.

Thebault, E., and M. Loreau. 2003. Food-web constraints on biodiversity–ecosystem functioning relationships. Proceedings of the National Academy of Sciences of the United States of America, 100:14949 –14954.

van Valen, L. 1973. A new evolutionary law. Evolutionary Theory, 1:1-30.

Vermeij, G. J. 2004a. Ecological avalanches and the two kinds of extinction. Evolutionary Ecology Research, 6:315–337.

Vermeij, G. J. 2004b. Nature: An Economic History. Princeton University Press, Princeton, New Jersey, 445 p.

Watts, D. J., and S. H. Strogatz. 1998. Collective dynamics of 'small-world' networks. Nature, 393:440-442.

Williams, R. J., E. L. Berlow, J. A. Dunne, A. Barabasi, and N. D. Martinez. 2002. Two degrees of separation in complex food webs. Proceedings of the National Academy of Sciences of the United States of America, 99:12913-12916.

Willis, K. J., and H. J. B. Birks. 2006. What is natural? The need for a long-term perspective in biodiversity conservation. Science, 314:1261-1265.

Wolfram Research. 2008. Mathematica, Version 6.0. Wolfram Research Incorporate, Champaign, Illinois.

Wootton, J. T. 1997. Estimates and tests of per capita interaction strength: Diet, abundance, and impact of intertidally foraging birds. Ecological Monographs, 67:45-64.

Wright, S. 1931. Evolution in Mendelian populations. Genetics, 16:97-159.

Yook, S., H. Jeong, and A. Barabasi. 2002. Modeling the Internet's large-scale topology. Proceedings of the National Academy of Sciences of the United States of America, 99:13382-13386.

Appendix *1*

A metanetwork U is a set of guilds, $\{G_1, G_2, ..., G_{|U|}\}$. $|U|$ represents the power of U, or the number of guilds, and $|G_i|$ is the species-richness or number of species in guild G_i. The number of potential prey species for each species in G_i is the sum of the species-richnesses of all guilds designated by the topology of U as prey to G_i. This number is calculated as

$$|R_i| = \sum_{j=1}^{|U|} a_{ij} |G_j|$$

where $|R_i|$ is the number of potential prey species and a_{ij} is the ij^{th} element of U's binary adjacency or connectivity matrix. The order ij designates i as a predator of j, and $a_{ij}=1$ if this is true, and zero otherwise.

If for simplicity we assume that all species in the community are extreme specialists, that is, they consume specifically a single resource, then the number of different species link configurations possible in G_i is $|R_i|^{|G_i|}$. Therefore in U there are exactly

$$\prod_{i=1}^{|U|} |R_i|^{|G_i|}$$

possible species-level networks. Given that links within each guild are described, however, by a trophic link distribution, and that species within a guild are expected to differ in their dietary breadths, the expected (average) number of species-level networks is found by first calculating the number of possible ways in which the links of a single species may be arranged among potential prey nodes, and the applying that to the above product

$$\prod_{i=1}^{|U|} \prod_{r=1}^{|R_i|} \binom{|R_i|}{r_x}^{|x_i|}$$

where $|x_i|$ is the number of species in G_i of in-degree (with number of prey) r_x.

The Late Miocene community illustrated in this paper comprises 130 species divided into 25 consumer guilds. The number of species-level networks which can be derived from this metanetwork, given the parameters outline in the main text, is 1.354×10^{1213}. Given that primary productivity is specified as units and by 1320 individual nodes in each species-level network, this number can be modified by representing primary productivity instead by the four autotrophic guilds comprising those units. The number of possible species-level networks is then reduced to 1.07×10^{237}, which is still a very large number.

Appendix *2*

x_i is a species in guild G_i (that is, $x_i \in G_i$) of in-degree r_x. The total number of prey species available to members of G_i is $|R_i|$. If a perturbation (species or node removal) of magnitude ω is applied randomly to the set of prey species R_i, the probability of x_i losing a subset n_x of its incoming links is given by the hypergeometric formula as

$$pr(n_x | \omega) = \binom{r_x}{n_x} \binom{|R_i| - r_x}{\omega - n_x} \binom{|R_i|}{\omega}^{-1}$$

which expresses the probability that of the $\omega \psi$ nodes chosen out of R_i, n_x of them will be in the set of r_x links. Topological secondary extinction of x_i will occur if $n_x = r_x$ and the above equation is then reducible to

$$pr(e, x_i | \omega) = \binom{|R_i| - r_x}{\omega - r_x} \binom{|R_i|}{\omega}^{-1} = \frac{\omega!(|R_i| - r_x)!}{|R_i|!(\omega - r_x)!}$$

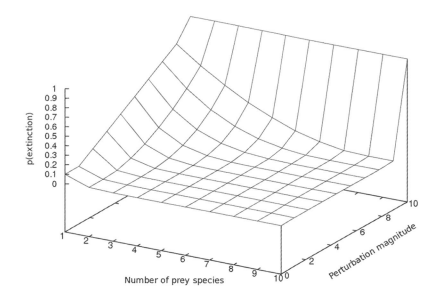

FIGURE *A1*.—Relationship between in-degree (number of prey species), the magnitude of a perturbation and the probability of extinction. In this example, there are a maximum of 10 prey species available to predators, who may therefore have in-degrees ranging from one to 10. Perturbation magnitude also ranges from one to 10. In general, the hypergeometric relationship generates a nonlinear surface, where the probability of extinction increases almost linearly at very low in-degree, but becomes strongly nonlinear as in-degree increases; for example, note that the probability of extinction of species with eight or nine prey is essentially zero unless the perturbation is of maximum magnitude.

The probability of topological extinction decreases as a function of increasing in-degree, indicating that species with more links or resources are more resistant to topological extinction (Fig. A1). The probability increases as a function of the magnitude of the perturbation.

The formulation leads us to first consider the number of species within the guild that are expected to become extinct as the result of the perturbation, and finally the total number of species in the community. First, given a guild's trophic link distribution, the number of species of a particular in-degree may be determined as

$$|x_i| = |G_i| \frac{\int_{r_x-1}^{r_x} P(r_i)}{\int_0^{|R_i|} P(r_i)}$$

where $P(r_i)$ is the trophic link distribution of guild G_i. The ratio of integrals yields the expected frequency of species of a particular in-degree in the guild. Then the expected level of topological extinction in the guild, ψ_i, is the sum over all in-degrees

$$E(\psi_i|\omega) = \sum_{r_x=1}^{|R_i|} |x_i| pr(e, x_i|\omega)$$

and the expected level for the entire community is the sum over all guilds

$$E(\Psi|\omega) = \sum_{i=1}^{|U|} E(\psi_i|\omega)$$

Topological extinction in a single species-level network, for example one drawn stochastically from the network space of the preceding metanetwork, may be computed exactly using any algorithmic method for traversing networks. A very simple matrix-based method follows (Fig. A2). Let A_0 be the binary asymmetric adjacency matrix of the species-level network. A row represents a species' incoming links, that is, the ij^{th} element, or a_{ij} is one if species i consumes species j, and zero otherwise. The initial perturbation to the network is implemented by replacing all elements of the ith row and column with zero, if the removal of species i constitutes part of the perturbation. An examination of the matrix will reveal if, as a consequence, the row elements of any other species now consists entirely of zero. If this is the case, then this species has become topologically extinct because it has lost all incoming links or resources. Its own corresponding column elements are now also replaced with zero. These steps are iterated until no more changes occur in the matrix, at which point propagation of the perturbation, and topo-

226   CHAPTER ELEVEN

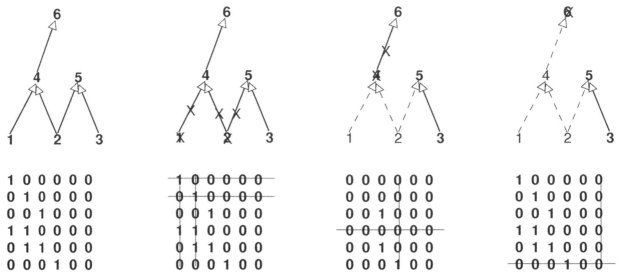

FIGURE *A2.*—Illustration of the matrix method for executing a topological extinction cascade. The upper figures show the food web or network, with species or nodes represented as numerals, and links as arrows. Nodes and links with crosses (X's) have been eliminated, either as part of the initial perturbation, or through topological secondary extinction. The cascade proceeds from left to right. Shown below the networks are the equivalent binary adjacency matrices. The initial perturbation (second network from left) is the removal of nodes 1 and 2, and therefore their out-links. The perturbation is reflected in the adjacency matrix by the conversion of rows and columns 1 and 2 to all zero values. The cascade then simply proceeds by a continued checking for all resulting row or column conversions to zero value, corresponding to the secondary loss of those species from the network.

logical extinction, have ceased.

The following example illustrates the algorithm. Figure A2 shows a simple network of six species, of which species 1-3 are producers, and species 4-6 are consumers of different trophic level. The initial perturbation consists of the removal of species 1 and 2, whose row and column elements are therefore made equal to zero. As a consequence, all row elements of species 4 also now equal zero, and hence it has become extinct. There are no further changes to the adjacency matrix, meaning that topological extinction has ceased, and iteration of the algorithm is halted.

APPENDIX *3*

CEG simulations are programmed in C++. This low-level language has several advantages, including speed, formal data types and containers, and an object-oriented interface. Computational speed is a major factor in these simulations, as run-times seem to increase polynomially (there is currently no rigorous proof of this) with the addition of taxa and more importantly, metanetwork links. Software written with C++ perform very quickly because of the language's access to low-level processor routines, and because programs are compiled to machine language. These are significant advantages when compared to interpreted or scripted languages, such as Python, Perl and R. C++ also has a large array of libraries specializing in numerical functions and data manipulation. The CEG programs make great use of the Standard Template Library (STL) for the containment and manipulation of taxon and guild properties, such as sets of predators or prey. The Boost library is also utilized to implement the matrix method for executing topological extinction. Finally, C++ is object-oriented, and all entities in the CEG programs, those are guilds and species, are implemented as individual objects, each with its own properties. This minimizes the use of large and memory-inefficient arrays, and each object may use multiple data types to describe various properties.

The CEG programs are all developed in Linux environments using GNU compilers, and the programs have also been compiled successfully for Apple OS X. Bash shell scripts have been used to implement the running of multiple simulations of a metanetwork simultaneously on both symmetric multiprocessor computers running Linux (up to 16 cores per machine), and an 88 core Apple X-Serve cluster. All programs are licensed under the GPL, and are available upon request.

CHAPTER TWELVE

PALEOBIOLOGY AND THE CONSERVATION
OF THE EVOLVING WEB OF LIFE

GREGORY P. DIETL

Paleontological Research Institution, 1259 Trumansburg Road, Ithaca, NY 14850 USA

ABSTRACT.—Predicting how society is altering the evolving web of life represents an immense scientific problem in conservation biology. An important challenge is to understand how to conserve the evolutionary processes that generate and maintain the web of life in the face of impending global environmental changes. Here I outline the role that the field of conservation paleobiology can play in addressing this challenge. I discuss why a focus on evolving ecological interactions is a vital conservation priority. I review hypotheses of adaptation of living entities to their biological surroundings, specifically coevolution and escalation, which I argue are critical processes for a full characterization of how evolving ecological interactions are linked across spatial and temporal scales. I also illustrate how paleobiological data can be applied to improve conservation "tools" developed to prioritize effort.

"It is interesting to contemplate an entangled bank, clothed with many plants of many kinds, with birds singing on the bushes, with various insects flitting about and with worms crawling through the damp earth, and to reflect that these elaborately constructed forms, so different from each other, and dependent on each other in so complex a manner, have all been produced by laws acting around us."–Darwin (1859, p. 489)

INTRODUCTION

HUMAN ACTIVITIES on a global scale have purposefully and inadvertently altered the web of life—Darwin's "entangled bank"—through the fragmentation of habitats, loss of species, overharvesting of important consumer species, and the global redistribution of species, among other changes. These influences are expected to worsen as the human population expands in numbers and global reach. Increasing evidence suggests that these anthropogenic disturbances strongly affect ecological dynamics, such as the distribution and abundance of species. However, they also do more. We are changing how organisms interact with each other and the selective environments they experience (Carroll and Fox, 2008; Jackson, 2008). Thus throughout the 21st century, by default or design, we will significantly alter the diversity and future evolutionary complexity of life on Earth (Donlan et al., 2006). We are witnessing a global evolutionary experiment with the biological diversity of the planet, upon which we ultimately depend for our own survival.

A growing list of examples—spanning a diversity of species, traits, and circumstances (Thompson, 1998; Hendry and Kinnison, 1999; Kinnison and Hendry, 2001; Reznick and Ghalambor, 2001; Stockwell et al., 2003; Ashley et al., 2003; Hairston et al., 2005; Carroll et al., 2007; Kinnison and Hairston, 2007; Carroll, 2008; Hendry et al., 2008; Darimont et al., 2009)—indicates that "natural" and human-induced evolutionary changes on contemporary timescales are evident in every corner of the globe (Palumbi, 2001; Smith and Bernatchez, 2008). For instance, size-selective harvesting of heavily exploited fishes has resulted in an evolutionary decline of age and size at sexual maturity of many species (Swain et al., 2007; Fenberg and Roy, 2008; Darimont et al., 2009). The realization that evolutionary change in ecologically important traits occurs in the here and now has also provided an impetus for the integration of community ecology and evolutionary biology, with the explicit goal of understanding how evolution shapes the ecological dynamics of communities and, in turn, how community context regulates evolutionary dynamics (Whitham et al., 2003; Johnson and Stinchcombe, 2007; Fussmann et al., 2007; Wade, 2007). Thus there is a clear and present need to develop practical, science-based approaches to managing adaptation to global environmental change. Evolutionary concerns, particularly an understanding of the conditions that limit adaptive responses to change, are just as immediate as ecological concerns, so much so that failure to acknowledge this fact may lead to unwelcome surprises in our conservation efforts (Zimmer, 2003).

In *Conservation Paleobiology: Using the Past to Manage for the Future, Paleontological Society Short Course, October 17th, 2009. The Paleontological Society Papers, Volume 15*, Gregory P. Dietl and Karl W. Flessa (eds.). Copyright © 2009 The Paleontological Society.

Despite its importance, however, an evolutionary perspective remains rare in current discussions on the effects of global environmental change (Travis and Futuyma, 1993). The canonical wisdom of conservationist thinking is that adaptation through natural selection is a glacially slow process, such that evolutionary processes are rarely thought of as a first priority for conservation. Species are thought of as fixed, non-evolving entities in a typological sense that remain unchanged over conservation-relevant periods of time. Thus an ecological rather than an evolutionary focus still dominates most academic as well as applied programs in resource management (Myers et al., 2000; Bull and Wichman, 2001).

According to Thompson (2008), a major challenge facing conservation biology today is the development of practical guidelines on how best to conserve the evolutionary processes that generate and maintain the web of life. We need to know how ecological interactions are likely to evolve given different conservation decisions (Thompson, 2008). A more complete understanding of the role of adaptation in shaping ecological interactions across a diverse range of linked spatial and temporal scales will improve management decisions to facilitate the persistence of biological diversity in the face of global environmental change (Rice and Emery, 2003; Mace and Purvis, 2008; Bell and Collins, 2008), thus securing the economic and environmental security of the planet. In other words, it is of great interest to know when adaptation can be expected to "rescue" species threatened with extinction (Bell and Collins, 2008).

In this essay, I have tried to reach into several specialized fields of study, in the hope of grasping enough common threads to weave a small fabric of diverse evolutionary phenomena that can be interrelated by a mutual, theoretical consistency to address this challenge. I highlight that an integration of approaches from the emerging broadly defined fields of evolutionary conservation biology (Ferrière et al., 2004; Thompson, 2005) and conservation paleobiology (Dietl and Flessa, 2009) will be necessary to develop practical guidelines on the management of evolutionary processes that are linked across an ecological hierarchy of spatial and temporal scales. I also outline specific hypotheses of coevolution and escalation—fundamentally hypotheses of adaptation of living entities to their biological surroundings—that can be tested with paleobiological data, point out sorely needed future research directions in the study of

evolving ecological interactions, and discuss paleobiology's role in developing future conservation tools to prioritize efforts. It is not my intent for this contribution to be exhaustive or to have an answer for all the questions that I ask; indeed, our work to apply paleobiological data and insights to conservation efforts to "manage evolution" is just beginning. I envision that this essay will help overcome past inertia in paleobiologists' training, which has resulted in a lack of exposure to applications of paleobiological data to questions relevant to society at large.

ECOLOGICAL INTERACTIONS AND CONSERVATION

Ecological interactions as the architectural fabric of biological diversity

Why is a focus on ecological interactions, such as predation, competition, and mutualism, a vital conservation priority? It is widely acknowledged that Darwin was keenly aware of the integral role ecological interactions play in driving the evolution of life (Thompson, 1994). "It is the web of interactions at the heart of an ecosystem that maintains both species and ecosystems as they are, or (more exactly) as they are evolving" (Loreau et al., 2004, p. 327; see also Thompson, 2005; Bascompte et al., 2006). Species evolve in large part by interacting with other species to gain resources and, in turn, to avoid being resources themselves (Thompson, 2005).

In an ecosystem, organisms are thus functionally interdependent (in the sense of energy flow and nutrient cycling); no organism can exist in isolation from the web of life (Worster, 1994; Lekevičius, 2002). In other words, all organisms on Earth, whether in natural or managed habitats, depend heavily on one another because of the different roles that they play and the different goods that they produce and exchange. Species are bound together in an ecosystem to varying degrees of interdependence by their differences (i.e., division of labor) (Thompson, 2005)—I thus see ecological interactions as the functional fabric of nature (Estes, 2002).

By these statements, I do not mean to imply that ecosystems—in the sense of an Aristotelian cosmology or view of nature—are perfectly balanced, static systems that are dominated by interactions between species that are tightly coevolved, continuous, and symmetric. Ecosystems in my view are dynamic, complex adaptive systems, which are structured from parts (species) that have evolved over longer time scales and broader spatial scales, and with a membership that changes adaptively as its parts interact (Levin, 1998); they are organized into panarchies (Holling, 2001; Dietl, 2008), which are hierarchically arranged, loosely structured, mutually reinforcing nested sets of interacting biotic and abiotic processes that operate at discrete spatial and temporal scales. Changes in structure with scale in ecosystems provide different templates at different scales with which species may interact (Allen and Holling, 2008). In this view, ecological interactions are varied and flexible, with species playing different roles in different places; for instance, within a pair (or group) of interacting species, interactions may be antagonistic in one place and mutualistic in another (Thompson, 1994; Thompson and Cunningham, 2002). This flexibility permits ecosystems to emerge even when species do not share a long evolutionary history (Vermeij, 2005a; see also Jablonski and Sepkoski, 1996; Jablonski, 2008; and Andersen, 1995, for useful discussion on competition in the context of the Gleasonian "individualistic" nature of communities). The implication for conservation biology is that change in many ecosystems is inevitable and thus trying to prevent it is unrealistic (Gillson and Willis, 2004).

The priority for conservation of biological diversity, however, traditionally has been from the perspective of preserving particular species threatened by extinction, with a focus on the number of species taking center stage. Perhaps we have focused on biological diversity simply because it is easier to count species than to document their myriad ecological interactions (McCann, 2007). Yet, Janzen (1974, p. 49) more than 30 years ago, insightfully remarked that "[w]hat escapes the eye…is a much more insidious kind of extinction: the extinction of ecological interactions."

Because the evolution of biological diversity is intimately linked to the evolution and maintenance of the web of species (Darwin, 1859; Vermeij, 2004; Thompson, 2005; Bascompte et al., 2006; Bascompte and Jordano, 2007), it is easy to imagine how the maintenance of biological diversity (the "parts") becomes increasingly threatened without maintenance of the interactions among species; in other words, the way the parts are interconnected (Thompson, 1996, 1997). In short, one cannot talk about biological diversity in a scientifically meaningful way without going beyond a

focus on species lists (Balmford et al., 1998; Bowen, 1999; Bohn and Amundsen, 2004).

Global environmental change and the ecology of interactions

How do global environmental change (GEC) drivers, such as climate change, overharvesting of consumer species, habitat loss and fragmentation, and alien species invasions, affect the ecology of organisms (e.g., Walther et al., 2002; MEA, 2005; Parmesan, 2006; Stiling and Cornelissen, 2007; Bell and Collins, 2008)? Such information is critical to predicting the responses of species to GEC in the future (van der Putten et al., 2004; Sutherst et al., 2007). There has been an enormous amount of effort to answer this question by ecologists and conservation biologists.

For instance, studies have shown that the loss of keystone species (Paine, 1969), ecosystem engineers (Jones et al., 1994), or foundation species (Ellison et al., 2005)—so-called "strongly interactive species" critical to the maintenance of ecological complexity at all scales *sensu* Soulé et al. (2003)—may lead to changes in the organization of ecosystems. One of the most famous examples involves the interactions between predatory sea otters (*Enhydra lustris*), their sea urchin prey, and kelp forests along the Pacific coasts of North America. As a result of intense hunting pressures in the 1800s, sea otters nearly became extinct. With one of their main predators gone (or at least no longer capable of fulfilling its ecological role), populations of sea urchins exploded to a point where they were able to heavily graze on kelp unchecked, reducing once vast forests of kelp to pavements of coralline algae. This process triggered diffuse indirect effects, decreasing ecosystem productivity and nutrient cycling (Estes, 2005).

Despite a growing list of examples, however, we still lack useful generalizations on the effects of GEC on the ecology of organisms across a wide variety of interaction types and drivers. Tylianakis et al.'s (2008) synthesis of more than 1000 terrestrial ecological interactions from 688 published studies was one of the first to try to summarize the vast literature on the effects of GEC on antagonistic and mutualistic ecological interactions. Their study posited a number of tentative generalizations including: strongly interactive species are often disproportionately affected by GEC drivers, such as habitat fragmentation; there is considerable variability in the GEC effects on the strength of plant-herbivore interactions; GEC effects are likely to reduce severely the strength of mutualistic interactions involving plants; and, antagonistic webs of interaction, such as those between predators and their prey, are more susceptible to GEC than are mutualistic webs of interaction, such as those between plants and their pollinators or seed dispersers.

Tylianakis et al. (2008, p. 6) concluded that "while some tentative generalizations can be made regarding the effects of some GEC drivers on certain types of interactions, there are many interactions for which there has been little research, or for which the effects of GEC (in terms of both magnitude and direction) are heavily dependent on environmental context and on the species involved." This statement suggests that expanding the spatial and temporal scales of studies may influence the generality of their patterns. It is also important to realize that most of the studies that they reviewed investigated the ecological effects of only a single GEC driver and did not address the possibility of interactions between the drivers themselves (see also Brook et al., 2008, for a discussion of how conservation actions targeting single-threat GEC drivers risk being inadequate in the shadow of synergistic processes between GEC drivers). Tylianakis et al. concluded that:

"If we are to reliably predict the effects of future GEC on community and ecosystem processes, then the greatest challenges lie in determining how biotic and abiotic context influences the direction and magnitude of GEC ecological effects on individual biotic interactions, and in determining how the varying responses of multiple pairwise interactions translate into altered interaction structure of entire communities" (Tylianakis et al., 2008, p. 9).

Did Tylianakis et al. (2008) leave anything out of the equation? Yes, evolution. All of the major GEC drivers of biological diversity loss today are also drivers of evolutionary change. As I pointed out above, we simply can not make the assumption that species will not adapt to altered environmental conditions, such that predictions for the future—an important goal of "ecological forecasting" (e.g., Clark et al., 2001)—based simply on the present-day ecology of a species will be incomplete (see Harmon et al., 2009, for an example that shows how ecological interactions can alter evo-

lutionary responses to external drivers such as climate change). If adaptation is itself a major component of global change (as increasing evidence suggests), "static models—those that treat the ecological players as passive bystanders in the ecological play—are now obsolete" (Carroll and Fox, 2008; p. v). We simply need a better understanding of how ecological interactions are likely to evolve across a broader spectrum of spatial and temporal scales if we are to have any hope of ever predicting the survival of species in an uncertain future; after all, it is adaptation to GEC drivers that ultimately will determine whether a species survives in the long term (see also Vermeij, 2009a).

WHY THE PAST MATTERS

Growing awareness of the inseparability of the human and natural realms (so-called social-ecological systems; Berkes and Folke, 1998), and the now widely acknowledged observation that planet earth is nowhere pristine in the sense of being free from human influence (Vitousek et al., 1997), has led to a need for a long-term historical perspective in conservation biology. As Aldo Leopold (1968, p. 196) remarked years ago in *A Sand County Almanac*, "a science of land health needs, first of all, a base datum of normality, a picture of how healthy land maintains itself as an organism." This need to determine what is "natural" was recently (Sutherland et al., 2009) proclaimed as one of the 100 questions of greatest importance to the conservation of global biological diversity. Such historical baselines—systems past or present that show no discernible human influence (NRC, 2005)—are used to define ecological patterns in the "natural" range of variability of ecological systems (Swetnam et al., 1999), differentiate between anthropogenic and non-anthropogenic change, recognize current properties—ecological legacies—that can be explained only by events or conditions that are not present in the system today, and identify phenomena outside of our experience (i.e., environmental states both like and unlike those of the present day; NRC, 2005; Willis and Birks, 2006; Jackson, 2007). Given that the linear, equilibrium-centered view of nature that has been a part of western philosophy since before Aristotle is no longer viable (Holling, 2001; Kricher, 2009), the motivation of studying past ecological patterns is not founded on the romantic vision of recreating them as much as possible as they were (but see

Pitcher, 2005). Rather, baselines help us understand how change will determine the patterns of the future, because, after all, conservation planning inevitably requires some forecasting or estimation about what will happen.

Despite the important ecological applications of this emerging "baseline industry," a missing element in studies advocating such an approach has been a focus on the preservation of evolutionary processes that produced the patterns we observe in nature in the first place (Smith et al., 1993). With a realization among community ecologists and evolutionary biologists that evolution is part of the here and now dynamics of human-dominated ecological systems, we urgently need to understand how our influences affect the adaptation of organisms to GEC. For instance, Thompson suggested that:

"For our understanding of evolution, the real tragedy accompanying the destruction of natural communities is the loss forever of specialized and highly coevolved interactions…Just as species are storehouses of information of genetic diversity, these interactions in intact communities are storehouses of information on how evolution shapes relationships between species under different ecological conditions" (Thompson,1994, p. 292).

Thompson (2005, p. 362) further suggested that environmental changes "are simultaneously creating opportunities for studying the dynamics of the early stages of coevolutionary change. It is impossible for us, however, to use these results alone to predict the long-term consequences of our actions." It is for this reason that Thompson suggested that large tracts of unexploited, productive habitat in which the dominant form of interaction is among species that have coevolved for many millennia are so critically important to our future. "The structure of these coevolved relationships can tell us much about how coevolution shapes species across space and time in complex landscapes" (*ibid.*). Thompson viewed these long-coevolved interactions as "irreplaceable free research and development for our societies" (*ibid.*).

While I agree with Thompson that we need to protect the few remaining large-scale geographic areas of relatively unmodified landscapes (*sensu* Mittermeier et al., 2003), I also think that paleobiological data have a critical part to play in understanding how ecological interactions evolve. The evolving web of life can not be fully understood without analyses at spatial and temporal scales found in the fossil record. The fossil record is the *only* place we can look to understand the process of evolution over long-time scales ($>10^3$ years). Also, although by no means perfect because of its taphonomic biases in the preservation of organisms, characters for analysis, and habitats (NRC, 2005, Jablonski, 2008), the fossil record preserves many ecological interactions that have been shaped by natural selection across varied landscapes over millions of years, each carrying information on what adaptive solutions work best in particular kinds of environments and selective regimes under every imaginable circumstance. The fossil record is full of information on how ecological interactions evolve among species given a constantly changing environmental context, if only you know how to "read it." Such historical data are necessary for a complete understanding of how variation in rates of GEC affect evolutionary outcomes (Collins et al., 2007; Kopp and Hermisson, 2007; Bell and Collins, 2008).

In this way, although historical sciences (e.g., evolutionary biology, paleobiology) are often commonly thought of as inferior to experimental sciences (see discussion in Cleland, 2001), I see life's history as replete with iterated pattern—classes of fundamentally similar processes, mechanisms, interactions, or conditions that result in a phenomenon—in other words, common causality (Gould, 1986; Eldredge, 1993). Conditions that favor or constrain evolution occur in many different places and times; thus there is a degree of independence among events (Vermeij, 2004). Although the details of place, time, and the participants involved, are matters of contingency, general principles (*sensu* Vermeij 2004, 2009b; see also Dietl, 2008)—selection and adaptation, feedback between living entities and their living and non-living surroundings, self-organization, and emergence—enable us to characterize the conditions favorable to evolutionary opportunity. Importantly, if we can apply these general principles to predict the particulars of time and place, then the long sweep of evolution's arrow cannot be ignored (contra Popper's [2002] view that there can be no predictive science of history), because it constrains the possibilities that the future holds (see Vermeij, 2004, for further discussion). I also do not make the metaphysical assumption common in many experimental sciences that a single cause can be isolated (e.g., singling out a specific GEC driver's affect on the ecology of organisms; see review in Tylianakis et al., 2008), because the fields of ecology and evolution are subject matters that are highly contingent and multi-causal (Hilborn and Stearns, 1982). Instead, discovering mechanisms in the historical and evolutionary sciences involves confirmation (Lloyd, 1987), which discriminates among mechanisms based on the amount of support each gains from independent lines of evidence; this approach is similar to Darwin's methodology (Gould, 1986) and Whewell's concept of "consilience of inference" (Ruse, 1979), which relied on multiple lines of evidence converging on a central model.

SCALE, SPECIFIC HYPOTHESES, AND THE EVOLUTION OF ECOLOGICAL INTERACTIONS

We now know that temporal and spatial patterns of landscapes are controlled by a few key abiotic and biotic structuring processes, each functioning at characteristic scales (Holling, 1992). This panarchical structure of ecological systems (see also Allen and Starr, 1982; Levin, 2000; Holling, 2001) produces discontinuous patterns on landscapes—essentially making the world uneven—across the scale over which organisms live, such as an aggregated resource distribution (e.g., Rietkerk and van de Koppel, 2008). These discontinuities, in turn, create variation in the ecological opportunities available to species; such systematic unevenness is an important property of complex systems, such as ecosystems, from the large to the small (Cumming et al., 2008) and has important implications for the adaptive process (see below). With such a wide range of spatial and temporal scales and possible interactions between them, a general understanding of how these scales are linked has been difficult to achieve. However, because these abiotic and biotic processes are critical elements in the sustained function of ecosystems, it is vital that their temporal and spatial components be maintained, and even managed. The appropriate spatial scale of conservation necessary to conserve ecological processes, however, remains incompletely understood in conservation biology (Delcourt and Delcourt, 1988; Callicott, 2002; Boyd et al., 2008); the importance of the time scale over which ecological processes oper-

ate is in no better shape (Frankham and Brook, 2004), although progress is being made (see Gunderson and Holling, 2002, and references therein).

Of even more concern are the huge gaps in our understanding of how evolutionary processes shape these patterns across linked temporal and spatial scales (Levin, 1992; Thompson, 2008) Increasing evidence suggests, however, that patterns in evolving ecological interactions observed at one spatial or temporal scale within a system are often similar to patterns occurring at larger and smaller scales, that is, there is a degree of scale independence between them (see Aronson and Plotnick, 1998, Aronson, 2009). [By scale I refer to the scale of observation, the temporal and spatial dimensions at which phenomena are observed, which should not be confused with the level of organization in a hierarchically organized system (see O'Neill and King, 1998; Jablonski, 2008).] In what follows, I briefly review several hypotheses and mechanisms of (co)-evolving ecological interactions across a variety of spatial and temporal scales that have been proposed from the fields of evolutionary biology and paleobiology. With such reciprocal illumination, we can better develop practical guidelines on how ecological interactions in the web of life are likely to evolve given different GEC drivers. Furthermore, if we appreciate that ecological interactions can evolve, are subject to feedbacks, and that their adaptation occurs in the context of interacting biotic and abiotic processes operating across linked temporal and spatial scales, this information will influence how we establish biological priorities for conservation.

[Because adaptation is and has always been a slippery concept, it is important to specify what I mean by the term. My conception is similar to Reeve and Sherman's (1993) and Vermeij's (1996) views of adaptation being fundamentally a comparative concept, which is particularly applicable to questions of why certain ecologically important traits predominate over others in nature. The discussion below concentrates on how natural selection episodically reshapes ecological interactions via the processes of coevolution and escalation across scales, but it is important to acknowledge that adaptation may also fuel speciation (Darwin, 1859; Ehrlich and Raven, 1964; Vermeij, 1987, 2005b; Smith and Grether, 2008; Schluter, 2000, 2009; Thompson, 2009). A full treatment of this latter topic in relation to conservation is beyond the scope of this essay.]

Geographic mosaic theory of coevolution

Ever since Darwin (1859), evolutionary biologists have emphasized the importance of coevolution—Darwin's "coadaptation"—in the evolution of ecological interactions. Coevolution—reciprocal adaptation between interacting species driven by natural selection—continues to be a controversial concept within evolutionary biology (and paleobiology). Thompson (1999a) suggested that studies of coevolution have waxed and waned over the last few decades between expecting every local interaction to be highly coevolved to dismissing coevolution as unimportant because of the inherent diffuse nature of ecological interactions. Increasing evidence, however, suggests that coevolution may create ever-changing geographic mosaics in how species interact with one another. For instance, Thompson's (1994, 2005) *Geographic Mosaic Theory of Coevolution* (GMTC) has provided novel insights into how ecological interactions evolve across linked space and time scales. The GMTC assumes the following: 1) variation in the strength of coevolution between populations of interacting species. In some areas, termed coevolutionary hot spots, selection is intense, whereas in coevolutionary cold spots, selection does not take place, perhaps simply because one of the interacting species is absent (Thompson, 1994, 1999b, 2005; Nash, 2008); 2) Selection mosaics are important; that is, variation exists in the direction of selection between populations. Such variability results in outcomes of coevolution that are context dependent. In other words, ecological interactions may be different in different populations, depending on their environment (*ibid.*). For instance, in "conditional mutualisms" (Cushman and Whitham, 1989), ecological interactions can be mutualistic, commensal, or parasitic depending on the ecological conditions in which the interaction is embedded; 3) Continual geographic mixing of traits, resulting, for instance, from gene flow among populations, and extinction and recolonization of local populations. Reciprocal selection therefore may be a local and episodic process. These three processes lead to predictable patterns: spatial variation in ecologically important traits, trait mismatching among interacting species, and few "species-level" coevolved traits (Thompson, 1999a, 2005; see also Gomulkiewicz et al. [2007] for a review of the fundamental components of the GMTC).

The predictions of the GMTC are supported by a variety of studies. For instance, Laine (2009) in her review of 29 studies found that they all reported spatially diverged selection that generated variable ecological outcomes among populations. This result held whether interactions were mutualisms between plants and their pollinators or antagonistic interactions, such as plants and their pathogens or herbivores. Interestingly, Laine (2009) also suggested that it is still unclear at what spatial scale the persistence and evolution of most ecological interactions occurs (see also Thompson, 1997).

In short, at its most fundamental level, Thompson's (2009, p. 130) GMTC argues that "long-term coevolution…is an inherently geographic process." This implies that studies of the local dynamics of coevolution cannot fully capture the dynamics of evolving ecological interactions (Thompson, 2009). Local coevolutionary changes are the "raw materials" for the broader coevolutionary dynamics that occur across larger spatial scales (Thompson, 1994). There are now a number of specific hypotheses about the local dynamics of the coevolutionary process: coevolving polymorphisms and coevolutionary alternation for predator and prey or parasite and host; coevolutionary complementarity and coevolutionary convergence for mutualisms; and coevolutionary displacement for competitors (see Thompson, 2005, for a detailed review). An increasing number of studies suggest that these local coevolutionary dynamics become altered in a broader geographic context (Thompson, 2009).

I will briefly discuss one specific hypothesis—coevolutionary alternation—to illustrate how consideration of the geographic mosaic of coevolution within webs of multiple species modifies expectations based solely on local coevolution between pairs of antagonistic species (or related processes such as escalation at local scales; see below). Hypotheses of pairwise predator-prey or host-parasite coevolution often predict that these interactions will lead to directional change or cycles over time in defenses or counterdefenses (see Dietl and Kelley, 2002, for review). When we consider multiple species, however, other possibilities emerge. According to Nuismer and Thompson (2006), coevolutionary alternation proceeds when natural selection favors predators (or grazers or parasites) that preferentially attack the least well-defended host or prey species among potential victims (Thompson, 2005; Nuismer and Thompson, 2006); this preference drives the evolution of increased defenses in the frequently attacked prey or host species and reduced defenses in the less frequently attacked host or prey species. Alternation occurs through evolutionary change in the preferences of parasite or predator and the relative levels of host or prey defenses (Thompson, 2005). This hypothesis may potentially explain how pairs of species and groups of species may continue to interact and coevolve over long periods of time and across broad spatial scales with no net evolutionary changes in ecologically important traits.

Although initially developed for understanding the dynamics of coevolution in extant systems, evaluating the scale independence of some of the core predictions of the GMTC in the fossil record remains an unexplored research direction (see Dietl and Kelley, 2002). Testing the predictions of the GMTC with fossil data, however, presents some challenges. Studying ecological interactions in the fossil record is difficult because direct evidence of many interactions (and their intensity) is often lacking (Jablonski, 2008). Various trace fossils left by interactions between organisms, however, offer a unique opportunity to study past ecological interactions for some systems, such as crushing and drilling predators of hard-shelled invertebrates, symbioses between various animals and skeleton-producing hosts, and plant-animal interactions (see Kowalewski and Kelley, 2002; Labandeira, 2002; Kelley et al., 2003; Miller, 2007; Tapanila, 2008; Kowalewski, 2009, for detailed treatments). The lack of data in many systems is as much a consequence of few researchers looking for it as it is in the difficulties in conducting such studies. We still have much to learn about how the spatial template of the abiotic and biotic environment affects evolving ecological interactions in the fossil record.

Hypothesis of escalation

The emphasis of evolutionary biologists on the importance of coevolution (most recently under the umbrella of Thompson's GMTC) contrasts with the view emerging from the study of ecological interactions over geological time scales. Evidence from the history of life suggests that ecological interactions often evolve in such a way that adaptation need not be reciprocal. In contrast to coevolution, Vermeij's *Escalation Hypothesis* predicts that enemies collectively exercise more intense selection on victims than victims do on their enemies (Vermeij, 1987, 1994; Dietl and Kelley, 2002). The escalation process is thus a top-down view

of adaptation. "…[A]ll forms of competition, including predation, herbivory, and parasitism, collectively create the most important agency of adaptive evolution…" (Vermeij, 2008, p. 3). A special case where escalation may involve coevolution is when prey are dangerous to their predators (Brodie and Brodie, 1999; Dietl and Kelley, 2002; Dietl, 2003). Escalation therefore implies consumer control of ecological life and change (Dietl and Kelley, 2002; Vermeij, 2004), which has resulted in long-term evolutionary trends as biological hazards have become increasingly severe over geological time. The pace, duration, and direction of escalation-related adaptation must be tracked locally or regionally in more or less physically similar environments through time at a spatial scale corresponding with the scale of interaction among entities among which selection occurs (Vermeij, 1987, 1994, 2008; Dietl and Vermeij, 2006).

Escalation is generally thought by evolutionary biologists to be of importance only in relation to "macroevolution" (Thompson, 1999c, 2005). Of course, this need not be the case. I am puzzled, therefore, why the nature of selection in the escalation process has not emerged among evolutionary biologists as a legitimate alternative mechanism to explain the ongoing evolutionary dynamics of ecological interactions. As a paleobiologist, I can only speculate that the lack of attention to this fundamental difference stems largely from the different time scale over which it has been developed and tested. As Bennington et al. (2009; p. 1) remarked, "the greatest barrier to communicating and collaborating with neoecologists is not that data collected from extant ecosystems are necessarily different or more complete than paleontological data but, rather, that these two data sets commonly represent or are collected at different scales." While the GMTC was developed in large part to counter the attitude that coevolution was too diffuse a process to develop specific hypotheses (Thompson, 1994), it seems yet another hurdle is the assumption that reciprocal change is the only game in town that drives the evolution of ecological interactions. A more pluralistic approach is sorely needed.

Escalation has been studied almost exclusively in the fossil record, with most examples discussed in reference to descriptions of large-scale ecological reorganizations in the history of life, such as the Early Paleozoic Marine Revolution (Brett and Walker, 2002;

Babcock, 2003), the mid-Paleozoic Marine Revolution (Signor and Brett, 1984; Brett and Walker, 2002; Brett, 2003), and the Mesozoic Marine Revolution (Vermeij, 1977; Walker and Brett, 2002; Harper, 2003, 2006), or competitive clade replacements over long geological time intervals, such as the replacement of cyclostome bryozoans by the competitively superior cheilostome bryozoans (McKinney et al., 1998). Far less work, however, has been done to test predictions of the escalation hypothesis at the lineage level, which usually has been conducted at the spatial scale of local stratigraphic sections (e.g., Kelley, 1989, 1992, Dietl et al., 2000, Dietl and Alexander, 2000; Dietl, 2003; Dietl et al., 2004). For instance, Kelley (1989) found that bivalve lineages of the Miocene Chesapeake Group of Maryland increased in shell defenses (thickness) in response to increased gastropod drilling predation. Lineage-level tests of the escalation hypothesis at a more regional spatially explicit scale, such as advocated in the GMTC have barely begun, but hold much promise.

Opening the black box

The view among some paleobiologists concerning the increasing evidence of ongoing coevolutionary dynamics as discussed by many evolutionary biologists studying the dynamics of the GMTC (see above) is that such changes are "ubiquitous and geologically momentary fluctuations that characterize and embellish the long-term stasis of species" (Gould, 2002, p. 800; Eldredge et al., 2005). In other words, these changes are part of the mutual adaptational stalemate occurring between adaptive breakthroughs in the history of life. While it is likely true that most of this ongoing evolutionary change does not lead to long-term directional change in the traits of species (a point evolutionary biologists seem to concede), these evolutionary "meanderings" are not just meaningless fine adjustments. Instead, as Thompson (1994, 2005) espoused, they keep the players in the evolutionary game, as populations adapt and readapt locally in an ever-changing world.

By downplaying the role of short-term evolutionary changes, however, we have unfortunately swept under the rug an important problem that unites both disciplines. Some of that short-term evolution in ecological interactions—whether the result of coevolution, escalation, or something else—must end up contributing to long-term trends. Understanding the circumstances of this process remains a black box in the study of the

evolution of ecological interactions; we have little understanding as to how it happens, such that there is no generally accepted consensus on how population processes of ecological interactions scale up into paleontological pattern. If we are to conserve the evolutionary processes that sustain the web of life we need to understand how this works.

Gould (2002), extending an argument first articulated by Futuyma (1987), believed that the speciation process played a central role in reconciling these patterns. Thompson's GMTC also seems to favor this mechanism: coevolution "begins with the coevolutionary dynamics of locally interacting populations, gathers geographic complexity…, and partitions some of that complexity more permanently over time as populations diversify into separate species" (Thompson, 2005, p. 50). In Gould's view, speciation provided individuation by "locking up" the local, ephemerally evolved changes in reproductively isolated populations. This mechanism must surely occur in some instances, especially given the increasing evidence that natural selection commonly drives the origin of species (Darwin, 1859; Schulter, 2009; Thompson, 2009). However, this explanation is by no means the only way to think about the problem. For instance, Estes and Arnold (2007) suggested that properties of Simpson's (1953) concept of an adaptive landscape accounted for the episodic nature of evolution. In their model, most of the history of a lineage is characterized by stabilizing selection at the level of the adaptive zone, within which populations can move to new local adaptive peaks. Vermeij and Dietl (2006) also proposed the "majority-rule hypothesis" to explain such patterns. This hypothesis states that: "populations in productive, effectively large environments act as sources for new or recovering populations elsewhere, and that the selective regimes to which these source populations are subjected…disproportionately influence the phenotypes and long-term evolution of species" (Vermeij and Dietl, 2006, p. 174); the majority-rule perspective emphasizes the importance of the environment in which a population is embedded for opportunities in evolution (see below).

The challenge, then, is to understand the circumstances surrounding the evolution of ecologically important traits that are "successful" enough to spread eventually to other populations and become the traits of species that can be registered in the fossil record (i.e., Thompson's [2005] "fixed traits of species"). If we are to understand fully how the ecological interactions weave the threads of biological diversity into a rich tapestry as Darwin envisioned in his description of the entangled bank, we must open the black box. As Thompson so succinctly stated:

> "We need to get the temporal and spatial scaling of ecological and evolutionary processes sorted out properly. We need to understand why the web of life is so dynamic in some respects and so static in other respects, and why there are general patterns in the organization of the web of life amid continual extinction, adaptation, and speciation" (Thompson, 2009, p. 127).

Further collaboration between evolutionary biologists and paleobiologists may lead to a better understanding of how the evolution of ecological interactions is linked across spatial and temporal scales. A search for generalities across scales has not yet been adequately conducted. An understanding of the influences of cross-scale interactions in ecosystem processes (Holling, 1992) will also be important. Cross-scale interactions have a large number of consequences for the dynamics of ecosystems, including setting the proper context within which natural selection operates (see next section). To achieve this goal will require a more comprehensive theoretical framework for the evolution of ecological interactions than is presently available (Dietl, 2008). Bridging this gap will require paleobiologists to broaden their attention to include the spatial component of evolving ecological interactions and evolutionary biologists to broaden their attention to include the temporal component of evolving ecological interactions.

EVOLUTIONARY OPPORTUNITY AND GLOBAL ENVIRONMENTAL CHANGE

Species can respond to GEC drivers in three ways (Holt, 1990; Bell and Collins, 2008): 1) species may track environmental changes and so shift their distribution; 2) species may become extinct; and, 3) species may adapt to the changed environment. For instance, Pleistocene climate change resulted in some plant species shifting their geographic range, some becoming extinct, and some adapting to the new conditions (Ge-

ber and Dawson, 1993; Davis and Shaw, 2001; Davis et al., 2005; Jump and Peñuelas, 2005).

It is the latter of these three responses that I will focus on here. Can we use our understanding of the constraints—the set of conditions that limit what is possible (Vermeij, 2004, 2005b)—on adaptation in evolving ecological interactions to guide conservation decisions to GEC? In other words, are there certain characteristics of particular environmental conditions that make adaptive responses more or less likely? The report of the Millennium Ecosystem Assessment (2005, p. vi) recommended that "to achieve greater progress toward biodiversity conservation to improve human well-being…, it will be necessary to strengthen response options that are designed with the conservation and sustainable use of biodiversity and ecosystem services as the primary goal." A key message of the report was that while "actions can be taken to reduce the drivers [of GEC] and their impacts …, some change is inevitable, and adaptation to such change will become an increasingly important component of response measures" (*ibid.*, p. 16). In short, we need to understand whether some ecological interactions are more likely than others to be constrained in their capacity for adaptive response. We need to know the environmental constraints within which natural selection operates. Such an issue is clearly important in the context of conservation biology because problems often arise when species do not adapt sufficiently to the new environments created by GEC drivers, thus increasing the risk of extinction (Holt and Gomulkiewicz, 2004; Reznick et al., 2004).

Towards achieving this goal, it is important to acknowledge that there is increasing evidence that the evolution of ecological interactions is ultimately constrained by cross-scale interactions in ecosystem processes (Holt, 1994; Holling, 2001). Feedbacks via ecosystem processes can act as constraints, channeling selection in directions different from those expected in their absence (Loreau et al., 2004; Dietl, 2008). According to Loreau et al (2004), recognition of the ecosystem as the proper context within which natural selection operates remains a major challenge, with underappreciated conservation biology implications.

Theoretical studies of the demographic aspects of environmental asymmetries (i.e., discontinuous patterns in habitats) show that there is a strong bias toward adaptation in high quality habitats—also known as source habitats—where there is an increasingly abundant, easily accessible, and predictable supply of resources (Holt, 1996a, b; Holt and Gomulkiewicz, 1997; Hochberg and van Baalen, 1998; see also Vermeij and Dietl, 2006, for an extension of this theory to the long-term evolutionary dynamics of species). Holt and Gomulkiewicz's (2004) theoretical models also suggested that large populations, which live in the most favorable parts of a species range (i.e., source habitats), are more likely to be able to adapt to environmental change. Resnick et al. (2004) further corroborated this point in their review of empirical evidence for contemporary evolution: adaptation was most likely to occur in populations moving into productive habitats, which allowed populations to expand in size. [These results have important implications for our understanding of the ecological context surrounding patterns of adaptive change discernable in the fossil record; see above.]

Vermeij's escalation hypothesis (discussed above) also posits a strong link between opportunities for adaptation and resources. The capacity to respond adaptively and to capitalize on opportunity depends on how rapidly and how reliably the environment delivers resources (Vermeij, 1987, 2002a, 2004; 2005b; Leigh et al., 2009); importantly, however, even if this environmental permissiveness is in place, without strong selection pressures, much of the potential range of adaptive possibilities will remain unrealized (Vermeij, 2005b). This underlying tenet of the escalation process typically has been ignored in tests conducted at the lineage-level (including my own cited above), which trace adaptations related to passive defenses, escape performance and feeding behavior among other energy-intensive functions within physically similar environments through time. This environmental bias, however, has played a central role in escalation tests concerned with understanding the circumstances surrounding the origins and evolutionary fates of functionally important minor adaptive innovations, which open up previously unavailable ecological possibilities. For example, Vermeij (2001) showed that the labral tooth, a structure facilitating predation, appeared in a number of Cenozoic clades of marine gastropods, with the highest concentrations originating during the early Miocene, late Miocene, and Pliocene, which were all times associated with increased resources.

We still know very little about whether the circumstances surrounding innovation and improvements that streamline or elaborate existing adaptations are always

complements, that is, similar in the details of place and time. Dietl (2007) presented a preliminary literature review of lineage-level tests of the escalation hypothesis, which suggested that the circumstances for improvement in existing adaptations are not always coincident with intervals purported to be conducive to minor innovation. In other words, the circumstances for improving existing adaptations (the most likely adaptive response of species to GEC in the short term) should be more common than are the circumstances associated with the evolution of minor innovations.

A bias toward adaptations evolving in source habitats is also evident in some studies of Thompson's GMTC. For instance, Vogwill et al. (2009) using the bacterium *Pseudomonas fluorescens* and its naturally associated phage SBW25Φ2, showed that spatial variation in the strength of selection is an important factor controlling coevolutionary dynamics within population sources or sinks. Specifically, immigration from a source population can either intensify or dampen coevolution in a recipient population. This result suggested to Vogwill et al. (2009) that source populations can act as coevolutionary "pacemakers" for recipient populations, overriding local conditions. In other words, in selection mosaics generated by productivity gradients, dispersal from hotspots is able to "warm up" coevolution in coldspots (see also Forde et al., 2007; Lopez-Pascua and Buckling, 2008). Thus hotspots have a greater evolutionary effect because they will act as net sources of immigrants, while coldspots will act as net recipients (Vogwill et al., 2009).

The kind of information described in the preceding examples is critical to conservation efforts aimed at facilitating organismal adaptation to GEC in an uncertain future, especially in an increasingly fragmented world, which leads to more and more "island-like" populations (Stockwell et al., 2003; Leigh et al., 2009; Wright et al., 2009). What emerges from this discussion is the importance of characterizing source populations and source environments. We may indirectly be able to facilitate adaptation to GEC by improving the resource predictability of habitats (Holt and Gomulkiewicz, 2004). A further implication of these findings is that, even if the genetic variation necessary for selection exists in a population (Lande and Shannon, 1996; Meyers and Bull, 2002; Charmantier and Garant, 2005; Mace and Purvis, 2008), the environment may impose constraints on the rate or direction of evolutionary change (see also

Dietl, 2008). A predictive framework in which we can identify source populations and regions is thus a crucial research direction for conservation biology and paleobiology (Vermeij and Dietl, 2006).

THE VALUE OF PALEOBIOLOGY IN THE DEVELOPMENT OF CONSERVATION TOOLS

One of the biggest challenges for conservation biology is coming up with tools that governmental and non-governmental organizations can use to prioritize effort to tackle the seemingly intractable, complex problems that confront us in the 21st century. Conservation planners continue to struggle with how best to allocate limited resources. Therefore, society's ever-increasing capacity (both intentionally and unwittingly) to alter the selective environment of organisms on a number of spatial and temporal scales—after all, we are the most coevolutionary animal of them all (Janzen, 1984)—will require big ideas, bold actions, and tough decisions about how we manage ecosystems (Hilderbrand et al., 2005). These decisions are of course not without risk, but as Vermeij (cited in Donlan, 2008, p. 13) stated: "A risk-free world is a very dull world, one from which we are apt to learn little of consequence." If we are to meet the challenges ahead, conservation tools have to be more proactive than reactive and strategic than ad hoc, with an increasing emphasis on process rather than pattern (Groves, 2003).

In what follows, three hotly-debated conservation tools—*biodiversity hotspots*, *assisted colonization*, and *rewilding of top predators*—are briefly discussed, which I believe are excellent areas where paleobiological data could prove enlightening. It is important to be realistic, however, in recognizing that conservation decisions, which are always made in a backdrop of urgency and uncertainty, will continue to be influenced by a diversity of factors, including environmental, cultural, economic and geopolitical (O'Conner et al., 2003).

Biodiversity hotspots as a conservation tool

Much attention has been focused on rank ordering of regions for conservation priorities based on species richness (i.e., biodiversity hotspots; Myers, 1988, 1990, 2002, 2003; Myers et al., 2000; Willis et al., 2007). This high-profile, "silver-bullet" conservation strategy, as it is promoted by Myers et al. (2000, p. 853), emphasizes

protection of "areas featuring exceptional concentrations of endemic species and experiencing exceptional habitat loss." Biodiversity hotspots focus on species richness patterns because obtaining the necessary information is relatively easy. Since Myers' initial hotspots study, there has been an increasing effort to identify the most important geographic areas for conservation. Most studies have approached this problem by using maps of species distributions to identify areas that contain the most species in the smallest area (see Pimm and Lawton, 1998).

Rarely, however, have the underlying evolutionary processes generating and maintaining the web of life been examined in the same framework (but see Rouget et al., 2003; Forest et al., 2007; Vandergast et al., 2008; Klein et al., 2009; Verboom et al., 2009). This lack of attention is troubling given that the maintenance of evolutionary processes is exactly the kind of problem that resource managers want answered (Balmford et al., 1998). One of the obstacles to including evolution as a key component of conservation planning has been the sizable number of unresolved questions about the spatial and temporal scales of the processes involved (see discussion above). These spatial and temporal components of evolutionary processes must first be identified in order to set targets for their effective conservation (Prendergast et al., 1999). This task will not be simple because the dynamic nature of these processes makes them difficult to quantify (Possingham et al., 2005). Thus, although the literature on evolutionary processes of evolving ecological interactions is enormous, very little can be used by conservation planners because most of the studies have failed to identify the spatial and temporal scales of the processes involved (recall Laine's [2009] conclusion that we still do not know the spatial scales at which the GMTC operates; see above).

These concerns would not matter if biodiversity hotspots always captured the relevant spatial and temporal scales of evolutionary processes. However, Vandergast et al. (2008), in a study of intraspecific genetic variation of 21 vertebrate and invertebrate taxa in one of the world's biodiversity hotspots, southern California, found that consideration of spatial patterns of genetic diversity added to the total area of land identified for conservation. A number of the areas Vandergast et al. (2008, p. 1649) identified as important for ensuring the maintenance of "at least some" of the evolutionary processes responsible for creating and maintaining the biodiversity hotspot were not consistent with the areas selected based on patterns of species richness alone. Thus there is an urgent need to incorporate the spatial components of processes into conservation planning.

New approaches should be sought that allow data on the spatial scales at which the GMTC and escalation processes operate to be mapped onto a landscape (e.g., the GIS-based approach used by Vandergast et al., 2008, to include spatial patterns of genetic diversity; see also Manel et al., 2003), which will permit different regions to be compared with regard to their potential to nurture adaptation. These data could be integrated with information on biodiversity hotspots. For instance, if future work confirms the hypothesized strong bias towards adaptation in source habitats, this information could be mapped and given disproportionate weighting in site-selection exercises. This task remains an important area for future study.

Assisted colonization as a conservation tool

A new bold conservation intervention strategy is the purposeful translocation of species to favorable habitat beyond their native range. This so-called assisted colonization (a.k.a. assisted migration and managed relocation) is rapidly emerging as a strategy in the toolbox of conservation biology. Seddon et al. (2009, p. 788) stated that the "idea of taking preemptive action to avert predicted extinction risks has been given emphasis by the recent International Union for Conservation of Nature (IUCN) assessment of species susceptibility to climate-change impacts." Proponents of assisted colonization argue that the scale of translocation includes moving populations threatened by GEC to geographic regions that share similar groups of organisms (McLachlan et al., 2007; Hoegh-Guldberg et al., 2008; Richardson et al., 2009). They also propose that translocation should only be undertaken if a species does not have the capacity to migrate on its own, respond plastically, or adapt *in situ* (Vitt et al., 2009); exactly how one goes about determining whether a species is capable of "*in situ*" adaptive response has not been outlined. Critics of assisted colonization, particularly Ricciardi and Simberloff (2009a, b), contend that such action will likely cause unintended and unpredictable consequences because we do not fully understand the ecological effects of invasive alien species.

What about the evolutionary effects of these trans-

locations (Sax et al., 2007)? There is no doubt that we are creating thousands of new interactions as we transport species around the world (see Hobbs et al., 2006, for a discussion of the management aspects of novel ecosystems resulting from this impact). From the perspective of the ongoing evolution and coevolution of ecological interactions, Thompson (2005, p. 360) stated:

"Our rearrangement of the earth's biodiversity is surely…resulting in the loss of geographic structure in some interactions…[and]…shifting the dynamics to local rather than geographic scales for some species…At the very least, these changes are likely to shift the coevolutionary dynamics of many interactions back toward earlier stages of nonequilibrium dynamics…[W]e may be creating an earth that keeps many interactions at the earlier, highly dynamic stages of coevolutionary change. We may therefore be altering coevolutionary dynamics in ways that make it increasingly difficult to predict the long-term consequences of our actions as we create new interaction networks" (Thompson, 2005, p. 360).

Evidence from the fossil record suggests that species invasion is a recurrent phenomenon. Vermeij (2005a, p. 331) concluded in his comparative study of biotic interchange in the past with the human-caused invasions of today, that purported differences between these two "either do not exist or are exaggerated, and that the lessons of past interchanges surely apply to our current predicament because the processes and consequences transcend many of the particulars of place, time, and participants." Paleobiology is thus uniquely positioned to place the debate about assisted colonization in the much needed broader context of adaptation. The study of past invasions is an opportunity to examine the circumstances of how invaders are accommodated evolutionarily by other species; that is, how they evolve and stabilize over the long term. Such information is vital if we are to reliably predict the effects of future GEC on ecological interactions and make informed decisions regarding translocations.

"Rewilding" of top predators as a conservation tool

An idea related to assisted colonization is the use

of large predators as a conservation tool to restore "big wilderness," due to their regulatory roles (Soulé and Nox, 1998). In short, the ecological argument underpinning rewilding posits that large predators are instrumental in maintaining the integrity of ecosystems and increasing the productivity of the systems in which they are embedded; that is, the entire foundation of living nature (Terborgh, 1988; Soulé and Nox, 1998; Terborgh et al., 2001; Soulé et al., 2003, 2005; Sergio et al., 2005; Myers et al., 2007; Heithaus et al., 2007; see also Sergio et al., 2008, for a more detailed review). A growing body of literature is increasingly revealing the varied ecological impacts of the global decline of top predators (Jackson et al., 2001; Jackson, 2006; Lotze and Worm, 2009). The evolutionary importance of top predators, however, is less recognized, but as equally important; high-powered top predators exercise evolutionary control by imposing intense selection on the characteristics and distribution of other species (Vermeij, 1999, 2002b, 2004).

An extreme example of rewilding is the recent proposal for "Pleistocene rewilding" (Donlan et al., 2005, 2006; Donlan, 2007; see also Janzen and Martin, 1982), which attempts to reinstitute ecological and evolutionary processes that were transformed or eliminated by megafaunal Pleistocene extinctions roughly 13,000 years ago by reintroducing large African mammals to the North American continent. The plan's idea to restore ecological and evolutionary processes is something that is generally acknowledged to be important by conservation biologists but is typically neglected (see above). Such bold ideas are not limited to the terrestrial realm. Briggs (2008) recently proposed the mass transfer of large-sized, predatory fishes from the North Pacific to the North Atlantic, where the populations of almost all of the large predatory fishes have collapsed (Essington et al., 2006). This sort of proactive vision for conservation raises some tricky questions and has generated fervid debate in the scientific literature (see Nicholls, 2006; Rubenstein et al., 2006; Caro, 2007, for critical arguments against Pleistocene rewilding).

According to Caro (2007), the evolutionary debate centers on the selection pressures rewilding might (re) introduce. There has been little research to evaluate what might happen evolutionarily after the rewilding of a top predator (that is, after the addition of a source of selection). The flip side of this coin is what happens to populations after the weakening or removal of a source

of selection—so-called "relaxed selection" (Lahti et al., 2009)—from top predators, which is what happened following the Pleistocene extinctions for many organisms. Relaxed selection pressure is a common phenomenon in nature, but remains largely understudied and misunderstood (Lahti et al., 2009). Reznick et al.'s (2008) study on the removal of predators of guppies (*Poecilia reticulata*) in Trinidadian streams provides some insight concerning the potential evolutionary consequences to prey species of predator removal. In the absence of predators, guppies have evolved morphologies that are highly susceptible to predation. These results have led critics of rewilding to suggest that the addition of strong interactors (or surrogates for them) after they have been absent for thousands or millions of years can lead to adverse ecological consequences (Reznick et al., 2008; see also Benkman et al., 2008, for a discussion of red squirrels introduced to Newfoundland 9kya). What is missing in the rewilding debate, as with assisted colonization, is the much needed broader context of adaptation that the fossil record brings to the table. The fossil record provides numerous opportunities to address the evolutionary consequences of top predators after their removal (e.g., via extinction) or addition (e.g., via invasion). A better understanding of what happens evolutionarily after the addition or removal of an important source of selection, such as a top predator, should facilitate more informed debate on Pleistocene rewilding and lead to better policy and practice of predator restoration in the future.

A "sea change" in paleobiology

These three examples are meant to be illustrative (but certainly not exhaustive) of the potential applications of paleobiological data derived from the study of evolving ecological interactions, which can be used to inform strategies and policies governing how we interact with the web of life. An emphasis on ecological interactions will generate critical data on how best to preserve evolutionary processes, which are linked across multiple spatial and temporal scales of ecological systems.

To achieve this goal, however, we need to reconsider the ways we view our science. Present discourse in theoretical paleobiology is dominated by a focus on broader (e.g., global) scale patterns. I suggest, however, that any efforts to bring the lessons derived from the past within the purview of evolutionary conservation biology based on this approach will fail due to obstacles represented primarily by modes of procedure and methodological conceptions. The global scale is simply too large to be relevant to the lives of organisms (Dietl and Vermeij, 2006). Variation in context—environment, interaction, functional role, adaptive syndrome, and geographic origin—in which organisms live and evolve makes any global-scale analysis of evolutionary processes, such as escalation or coevolution, an amalgamation that integrates heterogeneous signals. To have an impact on the infusion of evolution into conservation biology (an agenda that is itself still fighting an uphill battle of its own; see Ashley et al., 2003; Kinnison et al., 2007), I believe we need to shift focus from such broad scale approaches to spatially explicit local and regional studies of evolving lineages in the fossil record, in which ecological context is not sacrificed (see above). Such studies, which inevitably involve an intimate understanding of the ecology of organisms beyond their abundances and distributions, are sorely needed and have great potential to inform conservation planning, and perhaps fundamentally change our views on large-scale evolutionary processes in the fossil record.

The theory of punctuated equilibrium (Eldredge and Gould, 1972) challenged a generation of paleobiologists to collect detailed lineage-level data on morphological change in the fossil record; it is my hope that the unfortunate growing GEC problems that we face in the 21st century will sow the seeds of a similarly impassioned community effort. It is high time to dissect the geologically momentary fluctuations in trait values that characterize the long-term stasis of species with an ecological lens. We need to know which ecologically important traits are likely to evolve and keep species in the ecological game. We currently lack enough data across a broad range of traits and taxa to address this problem. In a big way, we have a great and current need to return to a Darwinian appreciation of the entangled bank.

CONCLUDING STATEMENTS

The world's ecosystems provide us with goods and services that support our society (Daily, 1997; MEA, 2005), including all the technological inventions that make our lives more comfortable. By this statement, I mean that the present alteration of the interdependent,

web of life matters, not because, as, in my view, Gould (cited in Allmon et al., 2009, p. 161) correctly espoused, "we fear for global stability in a distant future not likely to include us," but rather, from a philosophical stance of enlightened self-interest, because of the still largely imperfectly known extent to which human well-being is vitally dependent on the many benefits and services that ecological systems provide us.

While deciding how society should best manage this support system remains a scientific challenge of paramount importance, I see conservation paleobiology in the coming decades playing an increasingly important role in the debate. We need to understand how to turn "the potential of species to adapt to a changing world to the benefit of improved conservation practices" (Smith and Bernatchez, 2008, p. 1; Mace and Purvis, 2008). In other words, beyond the simple fact of knowing that we are changing the evolutionary course of many ecological interactions in the web of life, we need to apply our knowledge to facilitate and perhaps even direct the adaptive responses of species to GEC—after all, change is ubiquitous in nature. We need an "evolutionarily enlightened management" (Ashley et al., 2003; Stockwell and Ashley, 2004), in which the evolutionary consequences of resource management decisions are considered. However, the transformation of conservation biology from a discipline dominated by phenomenological descriptions to one that is informed by theory and capable of making generalizations remains a major challenge (see also Ferrière et al, 2004).

I have discussed a number of ideas that, when combined, help build the foundation of the theoretical evolutionary approaches that I believe will ultimately provide answers to some of society's most pressing questions, such as how can we best avoid foreclosing the evolutionary options of species (Ehrlich, 2001). To ignore such questions risks our own future and renders all our conservation efforts as temporary only. As Willis et al. (2007, p. 169) eloquently stated: "Lateral thinking, across knowledge systems and with open-mindedness about bridging data gaps, will be necessary for our accumulating knowledge about our planet's past to be brought to bear on our attempts to conserve it in the future."

Often the greatest impediment to a new way of thinking is an unwillingness to imagine something different. I hope that this paper will make a small contribution towards developing a research program in *evolutionary conservation paleobiology.* Such a research program offers stimulating intellectual challenges, opportunities for new collaborative, integrative research efforts across traditional academic boundaries, and is of the greatest societal relevance. In short, I see evolutionary conservation paleobiology as having a central role to play in ongoing efforts to conserve Darwin's entangled bank.

ACKNOWLEDGMENTS

I am thankful to Tricia Kelley for her helpful comments on the manuscript. The ideas presented here were developed during work on NSF grants EAR-0719130 and EAR-0755109.

REFERENCES

ANDERSEN, A. N. 1995. Palaeontology, adaptation, and community ecology: A response to Walter and Paterson (1994). Australian Journal of Ecology, 20:458-462.

ALLEN, C. R., AND C. S. HOLLING. 2008. Cross-scale structure and the generation of innovation and novelty in discontinuous complex systems, p. 219-233. *In* C. R. Allen, and C. S. Holling (eds.), Discontinuities in Ecosystems and Other Complex Systems. Columbia University Press, New York.

ALLEN, T. F. H., AND T. B. STARR. 1982. Hierarchy: Perspectives for Ecological Complexity. University of Chicago Press, Chicago, Illinois, 310 p.

ALLMON, W. D., P. J. MORRIS, AND L. C. IVANY. 2009. A tree grows in Queens: Stephen Jay Gould and ecology, p. 146-170. *In* W. D. Allmon, P. H. Kelley, and R. M. Ross (eds.), Stephen Jay Gould: Reflections on His View of Life. Oxford University Press, Oxford.

ARONSON, R. B. 2009. Metaphor, inference and prediction in paleoecology: Climate change and the Antarctic bottom fauna. *In* G. P. Dietl, and K. W. Flessa (eds.), Conservation Paleobiology: Using the Past to Manage for the Future. The Paleontological Society Papers, 15 (this volume).

ARONSON, R. B., AND R. E. PLOTNICK. 1998. Scale-independent interpretations of macroevolutionary dynamics, p. 430-450. *In* M. L. McKinney, and J. A. Drake (eds.), Biodiversity Dynamics: Turnover of Populations, Taxa, and Communities. Columbia University Press, New York.

ASHLEY, M. V., M. F. WILLSON, O. R. W. PERGAMS, D. J. O'DOWD, S. M. GENDE, AND J. S. BROWN. 2003. Evolu-

tionarily enlightened management. Biological Conservation, 111:115-123.

BABCOCK, L. E. 2003. Trilobites in Paleozoic predator-prey systems, and their role in reorganization of Early Paleozoic ecosystems, p. 55-92. *In* P. H. Kelley, M. Kowalewski, and T. A. Hansen (eds), Predator-Prey Interactions in the Fossil Record. Kluwer Academic/Plenum Publishers, New York.

BALMFORD, A., G. M. MACE, AND J. R. GINSBERG. 1998. The challenges to conservation in a changing world: Putting processes on the map, p. 1-28. *In* G. M. Mace, A. Balmford, and J. R. Ginsberg (eds.), Conservation in a Changing World. Cambridge University Press, Cambridge.

BASCOMPTE, J., AND P. JORDANO. 2007. Plant-animal mutualistic networks: The architecture of biodiversity. The Annual Review of Ecology, Evolution and Systematics, 38:567-593.

BASCOMPTE, J., P. JORDANO, AND J. M. OLESEN. 2006. Asymmetric coevolutionary networks facilitate biodiversity maintenance. Science, 312:431-433.

BELL, G., AND S. COLLINS. 2008. Adaptation, extinction and global change. Evolutionary Applications, 1:3-16.

BENKMAN, C. W., A. M. SEIPIELSKI, AND T. L. PARCHMAN. 2008. The local introduction of strongly interacting species and the loss of geographic variation in species and species interactions. Molecular Ecology, 17:395-404.

BENNINGTON, J. B., W. A. DIMICHELE, C. BADGLEY, R. K. BAMBACH, J. M. BARRETT, A. K. BEHRENSMEYER, R. BOBE, R. J. BURNHAM, E. B. DAESCHLER, J. VAN DAM, J. T. ERONEN, D. H. ERWIN, S. FINNEGAN, S. M. HOLLAND, G. HUNT, D. JABLONSKI, S. JACKSON, B. F. JACOBS, S. M. KIDWELL, P. L. KOCH, M. J. KOWALEWSKI, C. C. LABANDEIRA, C. V. LOOY, S. K. LYONS, P. M. NOVACK-GOTTSHALL, R. POTTS, P. D. ROOPNARINE, C. A. E. STRÖMBERG, H. D. SUES, P. J. WAGNER, P. WILF, AND S. L. WING. 2009. Critical issues of scale in paleoecology. Palaios, 24:1-4.

BERKES, F., AND C. FOLKE. 1998. Linking Social and Ecological Systems: Management Practices and Social Mechanisms for Building Resilience. Cambridge University Press, Cambridge, 476p.

BØHN, T., AND P. AMUNDSEN. 2004. Ecological interactions and evolution: Forgotten parts of biodiversity? Bioscience, 54:804-805.

BOWEN, B. W. 1999. Preserving genes, species, or ecosystems? Healing the fractured foundations of conservation policy. Molecular Ecology, 8:S5-S10.

BOYD, C., T. M. BROOKS, S. H. M. BUTCHART, G. J. EDGAR, G. A. B. DA FONSECA, F. HAWKINS, M. HOFFMANN, W. SECHREST, S. N. STUART, AND P. P. VAN DIJK. 2000. Spatial scale and the conservation of threatened species. Conservation Letters, 1:37-43.

BRETT, C. E. 2003. Durophagous predation in Paleozoic marine benthic assemblages, p. 401-432. *In* P. H. Kelley, M. Kowalewski, and T. A. Hansen (eds), Predator-Prey Interactions in the Fossil Record. Kluwer Academic/Plenum Publishers, New York.

Brett, C. E., and S. E. Walker. 2002. Predators and predation in Paleozoic marine environments, p. 93-118. *In* M. Kowalewski and P. H. Kelley (eds.), The Fossil Record of Predation. The Paleontological Society Papers, 8.

Briggs, J. C. 2008. The North Atlantic Ocean, need for proactive management. Fisheries, 33:180-185.

Brodie, E. D., III, and E. D. Brodie, Jr. 1999. Predator-prey arms races: Asymmetrical selection on predators and prey may be reduced when prey are dangerous. Bioscience, 49:557-568.

Brook, B. W., N. S. Sodhi, and C. J. A. Bradshaw. 2008. Synergies among extinction drivers under global change. Trends in Ecology and Evolution, 28:453-460.

Bull, J. J., and H. A. Wichman. 2001. Applied evolution. Annual Review of Ecology and Systematics, 32:183-217.

Callicott, J. B. 2002. Choosing appropriate temporal and spatial scales for ecological restoration. Journal of Biosciences, 27:409-420.

Caro, T. 2007. The Pleistocene re-wilding gambit. Trends in Ecology and Evolution, 22:281-283.

Carroll, S. P. 2008. Facing change: Forms and foundations of contemporary adaptation to biotic invasions. Molecular Ecology, 17:361-372.

Carroll, S. P., and C. W. Fox. 2008. Preface, p. v-vi. *In* S. P. Carroll, and C. W. Fox (eds.), Conservation Biology: Evolution in Action. Oxford University Press, Oxford.

Carroll, S. P., A. P. Hendry, D. N. Reznick, and C. W. Fox. 2007. Evolution on ecological time-scales. Functional Ecology, 21:387-393.

Charmantier, A., and D. Garant. 2005. Environmental quality and evolutionary potential: Lessons from wild populations. Proceedings of the Royal Society, B, 272:1415-1425.

Clark, J. S., S. R. Carpenter, M. Barber, S. Collins, A. Dobson, J. A. Foley, D. M. Lodge, M. Pascual, R. Pielke, Jr., W. Pizer, C. Pringle, W. V. Reid, K. A. Rose, O. Sala, W. H. Schleisinger, D. H. Wall, and D. Wear. 2001. Ecological forecasts: An emerging imperative. Science, 293:657-660.

Cleland, C. E. 2001. Historical science, experimental science, and the scientific method. Geology, 29:987-990.

Collins, S., J. de Meaux, and C. Acquisti. 2007. Adaptive walks toward a moving optimum. Genetics, 176:1089-1099.

Cumming, S., G. Barnes, and J. Southworth. 2008. Environmental asymmetries, p. 15-39. *In* J. Norberg, and G. S. Cumming (eds.), Complexity Theory for a Sustainable Future. Columbia University Press, New York.

Cushman, J. H., and T. G. Whitham. 1989. Condition-al mutualism in a membracid-ant association: Temporal, age-specific, and density-dependent effects. Ecology, 70:1040-1047.

Daily, G. C. 1997. Nature's Services: Societal Dependence on Natural Ecosystems. Island Press, Washington D.C., 392 p.

Darimont, C. T., S. M. Carlson, M. T. Kinnison, P. C. Paquet, T. E. Reimchen, and C. C. Wilmers. 2009. Human predators outpace other agents of trait change in the wild. Proceedings of the National Academy of Sciences of the United States of America, 106:952-954.

Darwin, C. 1859 [1964]. On the Origin of Species. Harvard University Press, Cambridge, Massachusetts, 513 p.

Davis, M. B., and R. G. Shaw. 2001. Range shifts and adaptive responses to Quaternary climate change. Science, 292:673-679.

Davis, M. B., R. G. Shaw, and J. R. Etterson. 2005. Evolutionary responses to changing climate. Ecology, 86:1704-1714.

Delcourt, H. R., and P. A. Delcourt. 1988. Quaternary landscape ecology: Relevant scales in space and time. Landscape Ecology, 2:23-44.

Dietl, G. P. 2003. Coevolution of marine gastropod predator and its dangerous bivalve prey. Biological Journal of the Linnean Society, 80:409-436.

Dietl, G. P. 2007. On the circumstances of opportunity in escalation: The complementarity principle. Geological Society of America Abstracts with Programs, 39(6):611.

Dietl, G. P. 2008. On the adaptive cycle of transformational change: A proposal for a panarchical expansion of escalation theory, p. 335-355. *In* P.H. Kelley, and R. K. Bambach (eds.), From Evolution to Geobiology: Research Questions Driving Paleontology at the Start of a New Century. Paleontological Society Papers, 14.

Dietl, G. P., and R. R. Alexander. 2000. Post-Miocene shift in stereotypic naticid predation on confamilial prey from the mid-Atlantic shelf: Coevolution with dangerous prey. Palaios, 15:414-429.

Dietl, G. P., and K. W. Flessa. 2009. Conservation Paleobiology: Using the Past to Manage for the Future. The Paleontological Society Papers, 15 (this volume).

Dietl, G. P., and P. H. Kelley. 2002. The fossil record of predator-prey arms races: Coevolution and escalation hypothesis, p. 353-374. *In* M. Kowalewski, and P. H. Kelley (eds.), The Fossil Record of Predation. The Paleontological Society Papers, 8.

Dietl, G. P., and G. J. Vermeij. 2006. Comment on "Statistical independence of escalatory ecological trends in Phanerozoic marine invertebrates". Science, 314:925e.

Dietl, G. P., R. A. Alexander, and W. F. Bien. 2000. Escalation in Late Cretaceous-early Paleocene oysters (Gryphaeidae) from the Atlantic Coastal Plain. Paleobiology, 26:215-237.

DIETL, G. P., G. S. HERBERT, AND G. J. VERMEIJ. 2004. Reduced competition and altered feeding behavior among marine snails after a mass extinction. Science, 306:2229-2231.

DONLAN, C. J. 2007. Restoring America's big, wild animals. Scientific American, 296:70-77.

DONLAN, C. J. 2008. Pleistocene dreams: It's time to bring back North America's charismatic megafauna. Orion, 24:13.

DONLAN, C. J., H. W. GREENE, J. BERGER, C. E. BOCK, J. H. BOCK, D. A. BURNEY, J. A. ESTES, D. FOREMAN, P. S. MARTIN, G. W. ROEMER, F. A. SMITH, AND M. E. SOULÉ. 2005. Re-wilding North America. Nature, 436:913-914.

DONLAN, C. J., J. BERGER, C. E. BOCK, J. H. BOCK, D. A. BURNEY, J. A. ESTES, D. FOREMAN, P. S. MARTIN, G. W. ROEMER, F. A. SMITH, M. E. SOULÉ, AND H. W. GREENE. 2006. Pleistocene rewilding: An optimistic agenda for twenty-first century conservation. The American Naturalist, 168:660-679.

EHRLICH, P. R. 2001. Intervening in evolution: Ethics and actions. Proceedings of the National Academy of Sciences of the United States of America, 98:5477-5480.

EHRLICH, P. R., AND P. H. RAVEN. 1964. Butterflies and plants: A study in coevolution. Evolution, 18:586-608.

ELDREDGE, N. 1993. History, function, and evolutionary biology, p. 33-50. In M. K. Hecht, R. J. MacIntyre, M. T. Clegg (eds.), Evolutionary Biology, 27. Plenum Press, New York.

ELDREDGE, N., AND S. J. GOULD. 1972. Punctuated equilibria: An alternative to phyletic gradualism, p. 82-115. In T. J. M. Schopf (ed.), Models in Paleobiology. Freeman, Cooper, San Francisco, California.

ELDREDGE, N., J. N. THOMPSON, P. M. BRAKEFIELD, S. GAVRILETS, D. JABLONSKI, J. B. C. JACKSON, R. E. LENSKI, B. S. LEIBERMAN, M. A. MCPEEK, AND W. MILLER, III. 2005. The dynamics of evolutionary stasis. Paleobiology, 31:133-145.

ELLISON, A. M., M. S. BANK, B. D. CLINTON, E. A. COLBURN, K. ELLIOTT, C. R. FORD, D. R. FOSTER, B. D. KLOEPPEL, J. D. KNOEEPP, G. M. LOVETT, J. MOHAN, D. A. ORWIG, N. L. RODENHOUSE, W. V. SOBCZAK, K. A. STINSON, J. K. STONE, C. M. SWAN, J. THOMPSON, B. VON HOLLE, AND J. R. WEBSTER. 2005. Loss of foundation species: Consequences for structure and dynamics of forested ecosystems. Frontiers in Ecology and the Environment, 3:479-486.

ESSINGTON, T. E., A. H. BEAUDREAU, AND J. WIEDENMANN. 2006. Fishing through marine food webs. Proceedings of the National Academy of Sciences of the United States of America, 103:3171-3175.

ESTES, J. A. 2002. Then and now, p. 61-71. In R. L. Knight, and S. Riedel (eds.), Aldo Leopold and the Ecological Conscience. Oxford University Press, Oxford.

ESTES, J. A. 2005. Carnivory and trophic connectivity in kelp forests, p. 61-81. In J. C. Ray, K. H. Redford, R. S. Steneck, and J. Berger (eds.), Large Carnivores and the Conservation of Biodiversity. Island Press, Washington, D.C..

ESTES, S. AND S. J. ARNOLD. 2007. Resolving the paradox of stasis: Models with stabilizing selection explain evolutionary divergence on all timescales. The American Naturalist, 169:227-244.

FENBERG, P. B., AND K. ROY. 2008. Ecological and evolutionary consequences of size-selective harvesting: How much do we know? Molecular Ecology, 17:209-220.

FERRIÈRE, R., U. DIECKMANN, AND D. COUVET. 2004. Evolutionary Conservation Biology. Cambridge University Press, Cambridge, 448 p.

FORDE, S. E., J. N. THOMPSON, AND B. J. M. BOHANNAN. 2007. Gene flow reverses an adaptive cline in a coevolving host-parasitoid interaction. The American Naturalist, 169:794-801.

FOREST, F., R. GRENYER, M. ROUGET, T. J. DAVIES, R. M. COWLING, D. P. FAITH, A. BALMFORD, J. C. MANNING, S. PROCHES, M. VAN DER BANK, G. REEVES, T. A. J. HEDDERSON, AND V. SAVOLAINEN. 2007. Preserving the evolutionary potential of floras in biodiversity hotspots. Nature, 445:757-760.

FRANKHAM, R., AND B. W. BROOK. 2004. The importance of time scale in conservation biology and ecology. Annales Zoologici Fennici, 41:459-463.

FUSSMANN, G. F., M. LOREAU, AND P. A. ABRAMS. 2007. Eco-evolutionary dynamics of communities and ecosystems. Functional Ecology, 21:465-477.

FUTUYMA, D. J. 1987. On the role of species in anagenesis. The American Naturalist, 130:465-473.

GEBER, M. A., AND T. E. DAWSON. 1993. Evolutionary responses of plants to global change, p. 179-197. In. P. M. Kareiva, J. G. Kingsolver, and R. B. Huey (eds.), Biotic Interactions and Global Change, Sinauer Associates, Sunderland, Massachusetts.

GILLSON, L., AND K. J. WILLIS. 2004. 'As Earth's testimonies tell': Wilderness conservation in a changing world. Ecology Letters, 7:990-998.

GOMULKIEWICZ, R., D. M. DROWN, M. F. DYBDAHL, W. GODSOE, S. L. NUISMER, K. M. PEPIN, B. J. RIDENHOUR, C. I. SMITH, AND J. B. YODER. 2007. Dos and don'ts of testing the geographic mosaic theory of coevolution. Heredity, 98:248-258.

GOULD, S. J. 1986. Evolution and the triumph of homology, or why history matters. American Scientist, 74:60-69.

GOULD, S. J. 2002. The Structure of Evolutionary Theory. The Belknap Press of Harvard University Press, Cambridge, Massachusetts, 1433 p.

GROVES, C. R. 2003. Drafting a Conservation Blueprint. Island Press, Washington, D.C., 457 p.

GUNDERSON, L. H., AND C. S. HOLLING. 2002. Panarchy: Understanding Transformations in Human and Natural Systems. Island Press, Washington, D. C., 507 p.

HAIRSTON, N. G., JR., S. P. ELLNER, M. A. GEBER, T. YOSHIDA, AND J. A. FOX. 2005. Rapid evolution and the convergence of ecological and evolutionary time. Ecology Letters, 8:1114-1127.

HARMON, J. P., N. A. MORAN, AND A. R. IVES. 2009. Species response to environmental change: Impacts of food web interactions and evolution. Science, 323:1347-1350.

HARPER, E. M. 2003. The Mesozoic marine revolution, p. 433-455. In P. H. Kelley, M. Kowalewski, and T. A. Hansen (eds), Predator-Prey Interactions in the Fossil Record. Kluwer Academic/Plenum Publishers, New York.

HARPER, E. M. 2006. Dissecting post-Palaeozoic arms races. Palaeogeography, Palaeoclimatology, Palaeoecology, 232:322-343.

HEITHAUS, M. R., A. FRID, A. J. WIRSING, AND B. WORM. 2007. Predicting ecological consequences of marine top predator declines. Trends in Ecology and Evolution, 23:202-210.

HENDRY, A. P., AND M. T. KINNISON. 1999. The pace of modern life: Measuring rates of contemporary microevolution. Evolution, 53:1637-1653.

HENDRY, A. P., T. J. FARRUGIA, AND M. T. KINNISON. 2008. Human influences on rates of phenotypic change in wild animal populations. Molecular Ecology, 17:20-29.

HILBORN, R., AND S. C. STEARNS. 1982. On inference in ecology and evolutionary biology: The problem of multiple causes. Acta Biotheoretica, 31: 145-164.

HILDERBRAND, R. H., A. C. WATTS, AND A. M. RANDLE. 2005. The myths of restoration ecology. Ecology and Society, 10:19.

HOBBS, R. J., S. ARICO, J. ARONSON, J. S. BARON, P. BRIDGEWATER, V. A. CRAMER, P. R. EPSTEIN, J. J. EWEL, C. A. KLINK, A. E. LUGO, D. NORTON, D. OJIMA, D. M. RICHARDSON, E. W. SANDERSON, F. VALLADARES, M. VILÀ, R. ZAMORA, AND M. ZOBEL. 2006. Novel ecosystems: Theoretical and management aspects of the new ecological world order. Global Ecology and Biogeography, 15:1-7.

HOCHBERG, M. E., AND M. VAN BAALEN. 1998. Antagonistic coevolution over productivity gradients. The American Naturalist, 152:620-632.

HOEGH-GULDBERG, O., L. HUGHES, S. MCINTYRE, D. B. LINDENMAYER, C. PARMESAN, H. P. POSSINGHAM, AND C. D. THOMAS. 2008. Assisted colonization and rapid climate change. Science, 321:345-346.

HOLLING, C. S. 1992. Cross-scale morphology, geometry, and dynamics of ecosystems. Ecological Monographs, 62:447-502.

HOLLING, C. S. 2001. Understanding the complexity of economic, ecological, and social systems. Ecosystems,

4:390-405.

HOLT, R. D. 1990. The microevolutionary consequences of climate change. Trends in Ecology and Evolution, 5:311-315.

HOLT, R. D. 1994. Linking species and ecosystems: Where's Darwin? p. 273-279. In C. Jones, and J. Lawton (eds.), Linking Species and Ecosystems. Chapman and Hall, London.

HOLT, R. D. 1996a. Demographic constraints in evolution: Towards unifying the evolutionary theories of senescence and niche conservatism. Evolutionary Ecology, 10:1-11.

HOLT, R. D. 1996b. Adaptive evolution in source-sink environments: Direct and indirect effects of density-dependence on niche evolution. Oikos, 75:182-192.

HOLT, R. D., AND R. GOMULKIEWICZ. 1997. How does immigration influence local adaptation? A reexamination of a familiar paradigm. The American Naturalist, 149:563-572.

HOLT, R. D., AND R. GOMULKIEWICZ. 2004. Conservation implications of niche conservatism and evolution in heterogeneous environments, p. 244-264. In R. Ferrière, U. Dieckmann, and D. Couvet (eds.), Evolutionary Conservation Biology. Cambridge University Press, Cambridge.

JABLONSKI, D. 2008. Biotic interactions and macroevolution: Extensions and mismatches across scales and levels. Evolution, 62:715-739.

JABLONSKI, D., AND J. J. SEPKOSKI., JR. 1996. Paleobiology, community ecology and scales of ecological pattern. Ecology, 77:1367-1378.

JACKSON, J. B. C. 2006. When ecological pyramids were upside down, p. 27-37. In J. A. Estes, D. P. DeMaster, D. F. Doak, T. M. Williams, and R. L. Brownell, Jr. (eds.), Whales, Whaling, and Ocean Ecosystems. University of California Press, Berkeley, California.

JACKSON, J. B. C. 2008. Ecological extinction and evolution in the brave new ocean. Proceedings of the National Academy of Sciences of the United States of America, 105:11458-11465.

JACKSON, J. B. C., M. X. KIRBY, W. H. BERGER, K. A. BJORNDAL, L. W. BOTSFORD, B. J. BORQUE, R. H. BRADBURY, R. COOKE, J. ERLANDSON, J. A. ESTES, T. P. HUGHES, S. KIDWELL, C. B. LANGE, H. S. LENIHAN, J. M. PANDOLFI, C. H. PETERSON, R. S. STENECK, M. J. TEGNER, AND R. R. WARNER. 2001. Historical overfishing and the recent collapse of coastal ecosystems. Science, 293:629-638.

JACKSON, S. T. 2007. Looking forward from the past: history, ecology, and conservation. Frontiers in Ecology and the Environment, 5:455.

JANZEN, D. H. 1974. The deflowering of Central America. Natural History, 83:48-53.

JANZEN, D. H. 1984. The most coevolutionary animal of

them all. Crafoord Lectures of the Royal Swedish Academy of Sciences, 3:2-20.

JANZEN, D. H., AND P. S. MARTIN. 1982. Neotropical anachronisms: The fruits the gomphotheres ate. Science, 215:19-27.

JOHNSON, M. T. J., AND J. R. STINCHCOMBE. 2007. An emerging synthesis between community ecology and evolutionary biology. Trends in Ecology and Evolution, 22:250-257.

JONES, C. G., J. H. LAWTON, AND M. SHACHAK. 1994. Organisms as ecoystem engineers. Oikos, 69:373-386.

JUMP, A. S., AND J. PEÑUELAS. 2005. Running to stand still: Adaptation and the response of plants to rapid climate change. Ecology Letters, 8:1010-1020.

KELLEY, P. H. 1989. Evolutionary trends within bivalve prey of Chesapeake group naticid gastropods. Historical Biology, 2:139-156.

KELLEY, P. H. 1992. Evolutionary patterns of naticid gastropods of the Chesapeake Group: An example of coevolution? Journal of Paleontology, 66:794-800.

KELLEY, P. H., M. KOWALEWSKI, AND T. A. HANSEN. 2003. Predator-Prey Interactions in the Fossil Record. Kluwer Academic/Plenum Publishers, New York, 464 p.

KINNISON, M. T., AND N. G. HAIRSTON, JR. 2007. Eco-evolutionary conservation biology: Contemporary evolution and the dynamics of persistence. Functional Ecology, 21:444-454.

KINNISON, M. T., AND A. P. HENDRY. 2001. The pace of modern life II: From rates of contemporary microevolution to patterns and process. Genetica, 112-113:145-164.

KINNISON, M. T., A. P. HENDRY, AND C. A. STOCKWELL. 2007. Contemporary evolution meets conservation biology II: Impediments to integration and application. Ecological Research, 22:947-954.

KLEIN, C., K. WILSON, M. WATTS, J. STEIN, S. BERRY, J. CARWARDINE, M. S. SMITH, B. MACKEY, AND H. POSSINGHAM. 2009. Incorporating ecological and evolutionary processes into continental-scale conservation planning. Ecological Applications, 19:206-217.

KOPP, M., AND J. HERMISSON. 2007. Adaptation of a quantitative trait to a moving optimum. Genetics, 176:715-719.

KOWALEWSKI, M. 2009. The youngest fossil record and conservation biology: Holocene shells as eco-environmental recorders. In G. P. Dietl, and K. W. Flessa (eds.), Conservation Paleobiology: Using the Past to Manage for the Future. The Paleontological Society Papers, 15 (this volume).

KOWALEWSKI, M., AND P. H. KELLEY. 2002. The Fossil Record of Predation. The Paleontological Society Papers, 8.

KRICHER, J. 2009. The Balance of Nature: Ecology's Enduring Myth. Princeton University Press, Princeton, New Jersey, 237 p.

LABANDIERA, C. C. 2002. The history of associations between plants and animals, p. 26-74. In C. M. Herrera, and O. Pellmyr (eds.), Plant-Animal Interactions: An Evolutionary Approach. Blackwell Publishing, Oxford.

LAHTI, D. C., N. A. JOHNSON, B. C. AJIE, S. P. OTTO, A. P. HENDRY, D. T. BLUMSTEIN, R. G. COSS, K. DONOHUE, AND S. A. FOSTER. 2009. Relaxed selection in the wild. Trends in Ecology and Evolution, 24:487-496.

LAINE, A. L. 2009. Role of coevolution in generating biological diversity: Spatially divergent selection trajectories. Journal of Experimental Botany, 60:2957-2970.

LANDE, R., AND S. SHANNON. 1996. The role of genetic variation in adaptation and population persistence in a changing environment. Evolution, 50:434-437.

LEIGH, E. G., JR., G. J. VERMEIJ, AND M. WIKELSKI. 2009. What do human economies, large islands and forest fragments reveal about the factors limiting ecosystem evolution? Journal of Evolutionary Biology, 22:1-12.

LEKEVIČIUS, E. 2002. The Origin of Ecosystems by Means of Natural Selection. Institute of Ecology, Lithuanian Academy of Sciences, Vilnius, Lithuania, 88 p.

LEOPOLD, A. 1968. A Sand County Almanac: And Sketches Here and There. Oxford University Press, London, 226 p.

LEVIN, S. A. 1992. The problem of pattern and scale in ecology. Ecology, 73:1943-1967.

LEVIN, S. A. 1998. Ecosystems and the biosphere as complex adaptive systems. Ecosystems, 1:431-436.

LEVIN, S. A. 2000. Multiple scales and the maintenance of biodiversity. Ecosystems, 3:498-506.

LLOYD, E. A. 1987. Confirmation of ecological and evolutionary models. Biology and Philosophy, 2:277-293.

LOPEZ-PASCUA, L. D. C., AND A. BUCKLING. 2008. Increasing productivity accelerates host-parasite coevolution. The Journal of Evolutionary Biology, 21:853-860.

LOREAU, M., C. DE MAZANCOURT, AND R. D. HOLT. 2004. Ecosystem evolution and conservation, p. 327-343. In R. Ferrière, U. Dieckmann, and D. Couvet (eds.), Evolutionary Conservation Biology. Cambridge University Press, Cambridge.

LOTZE, H. K., AND B. WORM. 2009. Historical baselines for large marine animals. Trends in Ecology and Evolution, 24:254-262.

MACE, G. M., AND A. PURVIS. 2008. Evolutionary biology and practical conservation: Bridging a widening gap. Molecular Ecology, 17:9-19.

MANEL, S., M. K. SCHWARTZ, G. LUIKART, AND P. TABERLET. 2003. Landscape genetics: Combining landscape ecology and population genetics. Trends in Ecology and Evolution, 18:189-197.

MCCANN, K. 2007. Protecting biostructure. Nature, 446:29.

MCKINNEY, F. K., S. LIDGARD, J. J. SEPKOSKI, JR., AND P. D.

TAYLOR. 1998. Decoupled temporal patterns of evolution and ecology in two post-Paleozoic clades. Science, 281:807-809.

McLACHLAN, J. S., J. J. HELLMAN, AND M. W. SCHWARTZ. 2007. A framework for debate of assisted migration in an era of climate change. Conservation Biology, 21:297-302.

MEYERS, L. A., AND J. J. BULL. 2002. Fighting change with change: Adaptive variation in an uncertain world. Trends in Ecology and Evolution, 17:551-557.

MEA (MILLENNIUM ECOSYSTEM ASSESSMENT). 2005. Ecosystems and Human Well-Being: Biodiversity Synthesis. World Resources Institute, Washington, D.C., 100 p.

MILLER, W., III. 2007. Trace Fossils: Concepts, Problems, Prospects. Elsevier, Amsterdam, 611 p.

MITTERMEIER, R. A., C. G. MITTERMEIER, T. M. BROOKS, J. D. PILGRIM, W. R. KONSTANT, G. A. B. DA FONSECA, AND C. KORMOS. 2003. Wilderness and biodiversity conservation. Proceedings of the National Academy of Sciences of the United States of America, 100:10309-10313.

MYERS, N. 1988. Threatened biotas: "Hot spots" in tropical forests. The Environmentalist, 8:187-208.

MYERS, N. 1990. The biodiversity challenge: Expanded hotspots analysis. The Environmentalist, 10:243-256.

MYERS, N. 2002. Biodiversity hotspots for conservation priorities. Nature, 403:853-858.

MYERS, N. 2003. Biodiversity hotspots revisited. Bioscience, 53:916-917.

MYERS, N., R. A. MITTERMEIER, C. G. MITTERMEIER, G. A. B. DA FONSECA, AND J. KENTS. 2000. Biodiversity hotspots for conservation priorities. Nature, 403:853-858.

MYERS, R. A., J. K. BAUM, T. D. SHEPHERD, S. P. POWERS, AND C. H. PETERSON. 2007. Cascading effects of the loss of apex predatory sharks from a coastal ocean. Science, 315:1846-1850.

NASH, D. R. 2008. Process rather than pattern: Finding pine needles in the coevolutionary haystack. Journal of Biology, 7:14.

NICHOLLS, H. 2006. Restoring nature's backbone. PLoS Biology, 4:893-896.

NRC (NATIONAL RESEARCH COUNCIL). 2005. The Geological Record of Ecological Dynamics: Understanding the Biotic Effects of Future Environmental Change. National Academies Press, Washington, D.C., 200 p.

NUISMER, S. L., AND J. N. THOMPSON. 2006. Coevolutionary alternation in antagonistic interactions. Evolution, 60:2207-2217.

O'CONNOR, C., M. MARVIER, AND P. KAREIVA. 2003. Biological vs. social, economic and political priority-setting in conservation. Ecology Letters, 6:706-711.

O'NEILL, R. V., AND A. W. KING. 1998. Homage to St. Michael; or, why are there so many books on scale? p. 3-15. In D. L. Peterson, and V. T. Parker (eds.), Ecological Scale: Theory and Applications. Columbia University Press, New York.

PAINE, R. T. 1969. A note on trophic complexity and species diversity. The American Naturalist, 1103:91-93.

PALUMBI, S. R. 2001. Humans as the world's greatest evolutionary force. Science, 293:1786-1790.

PARMESAN, C. 2006. Ecological and evolutionary responses to recent climate change. The Annual Review of Ecology, Evolution and Systematics, 37:637-669.

PIMM, S. L., AND J. H. LAWTON. 1998. Planning for biodiversity. Science, 279:2068-2069.

PITCHER, T. J. 2005. Back-to-the-future: A fresh policy initiative for fisheries and a restoration ecology for ocean ecosystems. Philosophical Transactions of the Royal Society, B, 360, 107-121.

POPPER, K. 2002 [1957]. The Poverty of Historicism. Routledge, New York, 156 p.

POSSINGHAM, H. P., J. FRANKLIN, K. WILSON, AND T. J. REGAN. 2005. The roles of spatial heterogeneity and ecological processes in conservation planning, p. 386-406. In G. M. Lovett, C. G. Jones, M. G. Turner, and K. C. Weathers (eds.), Ecosystem Function in Heterogeneous Landscapes. Springer-Verlag, New York.

PRENDERGAST, J. R., R. M. QUINN, AND J. H. LAWTON. 1999. The gaps between theory and practice in selecting nature reserves. Conservation Biology, 13:484-492.

REEVE, H. K., AND P. W. SHERMAN. 1993. Adaptation and the goals of evolutionary research. The Quarterly Review of Biology, 68:1-32.

REZNICK, D. N., AND C. K. GHALAMBOR. 2001. The population ecology of contemporary adaptations: What empirical studies reveal about the conditions that promote adaptive evolution. Genetica, 112-113:183-198.

REZNICK, D., H. RODD, AND L. NUNNEY. 2004. Empirical evidence for rapid evolution, p. 100-118. In R. Ferrière, U. Dieckmann, and D. Couvet (eds.), Evolutionary Conservation Biology. Cambridge University Press, Cambridge.

REZNICK, D. N., C. K. GHALAMBOR, AND K. CROOKS. 2008. Experimental studies of evolution in guppies: A model for understanding the evolutionary consequences of predator removal in natural communities. Molecular Ecology, 17:97-107.

RICCIARDI, A., AND D. SIMBERLOFF. 2009a. Assisted colonization is not a viable conservation strategy. Trends in Ecology and Evolution, 24:248-253.

RICCARDI, A., AND D. SIMBERLOFF. 2009b. Assisted colonization: Good intentions and dubious risk assessment. Trends in Ecology and Evolution, 24:476-477.

RICE, K. J., AND N. C. EMERY. 2003. Managing microevolution: Restoration in the face of global change. Frontiers in Ecology and the Environment, 1:469-478.

RICHARDSON, D. M., J. J. HELLMAN, J. S. McLACHLAN, D. F. SAX, M. W. SCHWARTZ, P. GONZALEZ, E. J. BRENNAN,

A. CAMACHO, T. L. ROOT, O. E. SALA, S. H. SCHNEIDER, D. M. ASHE, S. POLASKY, H. D. SAFFORD, A. R. THOMPSON, AND M. VELLEND. 2009. Multidimensional evaluation of managed relocation. Proceedings of the National Academy of Sciences of the United States of America, 106:9721-9724.

RIETKERK, M., AND J. VAN DE KOPPEL. 2008. Regular pattern formation in real ecosystems. Trends in Ecology and Evolution, 23:169-175.

ROUGET, M., R. M. COWLING, R. L. PRESSEY, AND D. M. RICHARDSON. 2003. Identifying spatial components of ecological and evolutionary processes for regional conservation planning in the Cape Floristic Region, South Africa. Diversity and Distributions, 9:191-210.

RUSE, M. 1979. Falsifiability, consilience, and synthesis. Systematic Zoology, 28:530-536.

RUBENSTEIN, D. R., D. I. RUBENSTEIN, P. W. SHERMAN, AND T. A. GAVIN. 2006. Pleistocene park: Does re-wilding North America represent sound conservation for the 21st century? Biological Conservation, 132:232-238.

SAX, D. F., J. J. STACHOWICZ, J. H. BROWN, J. F. BRUNO, M. N. DAWSON, S. D. GAINES, R. K. GROSBERG, A. HASTINGS, R. D. HOLT, M. M. MAYFIELD, M. I. O'CONNOR, AND W. R. RICE. 2007. Ecological and evolutionary insights from species invasions. Trends in Ecology and Evolution, 22:465-471.

SCHLUTER, D. 2000. The Ecology of Adaptive Radiation. Oxford University Press, Oxford, 296 p.

SCHLUTER, D. 2009. Evidence for ecological speciation and its alternative. Science, 323:737-741.

SEDDON, P. J., D. P. ARMSTRONG, P. SOORAE, F. LAUNAY, S. WALKER, C. R. RUIZ-MIRANDA, S. MOLUR, H. KOLDEWEY, AND D. G. KLEINMAN. 2009. The risks of assisted colonization. Conservation Biology, 23:788-789.

SERGIO, F., I. NEWTON, AND L. MARCHESI. 2005. Top predators and biodiversity. Nature, 436:192.

SERGIO, F., T. CARO, D. BROWN, B. CLUCAS, J. HUNTER, J. KETCHUM, K. MCHUGH, AND F. HIRALDO. 2008. Top predators as conservation tools: Ecological rationale, assumptions, and efficacy. The Annual Review of Ecology, Evolution and Systematics, 39:1-19.

SIGNOR, P. W., III, AND C. E. BRETT. 1984. The mid-Paleozoic precursor to the Mesozoic marine revolution. Paleobiology, 10:229-245.

SIMPSON, G. G. 1953. The Major Features of Evolution. Columbia University Press, New York, 434 p.

SMITH, T. B., AND L. BERNATCHEZ. 2008. Evolutionary change in human-altered environments. Molecular Ecology, 17:1-8.

SMITH, T. B., AND G. F. GRETHER. 2008. The importance of conserving evolutionary processes, p. 85-98. In S. P. Carroll, and C. W. Fox (eds.), Conservation Biology: Evolution in Action. Oxford University Press, Oxford.

SMITH, T. B., M. W. BRUFORD, AND R. K. WAYNE. 1993. The preservation of process: The missing elements of conservation programs. Biodiversity Letters, 1:164-167.

SOULÉ, M., AND R. NOSS. 1998. Rewilding and biodiversity: Complementary goals for continental conservation. Wild Earth, 8:18-28.

SOULÉ, M. E., J. A. ESTES, J. BERGER, AND C. M. DEL RIO. 2003. Ecological effectiveness: Conservation goals for interactive species. Conservation Biology, 17:1238-1250.

SOULÉ, M. E., J. A. ESTES, B. MILLER, AND D. L. HONNOLD. 2005. Strongly interacting species: Conservation policy, management, and ethics. Bioscience, 55:168-176.

STILING, P., AND T. CORNELISSEN. 2007. How does elevated carbon dioxide (CO_2) affect plant-herbivore interactions? A field experiment and meta-analysis of CO_2-mediated changes on plant chemistry and herbivore performance. Global Change Biology, 13:1823-1842.

STOCKWELL, C. A., AND M. V. ASHLEY. 2004. Rapid adaptation and conservation. Conservation Biology, 18:272-273.

STOCKWELL, C. A., A. P. HENDRY, AND M. T. KINNISON. 2003. Contemporary evolution meets conservation biology. Trends in Ecology and Evolution, 18:94-101.

SUTHERST, R. W., G. F. MAYWALD, AND A. S. BORNE. 2007. Including species interactions in risk assessments for global change. Global Change Biology, 13:1843-1859.

SUTHERLAND, W. J., W. M. ADAMS, R. B. ARONSON, R. AVELING, T. M. BLACKBURN, S. BROAD, G. CEBALLOS, I. M. CÔTÉ, R. M. COWING, G. A. B. DA FONSECA, E. DINERSTEIN, P.J. FERRARO, E. FLEISHMAN, C. GASCON, M. HUNTER, JR., J. HUTTON, P. HAREIVA, A. HURIA, D. W. MACDONALD, K. MACKINNON, F. J. MADGWICK, M. B. MASCIA, J. MCNEELY, E. J. MILNER-GULLAND, S. MOON, C. G. MORLEY, S. NELSON, D. OSBORN, M. PAI, E. C. M. PARSONS, L. S. PECK, H. POSSINGHAM, S. V. PRIOR, A. S. PULLIN, M. R. W. RANDS, J. RANGANATHAN, K. H. REDFORD, J. P. RODRIGUEZ, F. SEYMOUR, J. SOBEL, N. S. SODHI, A. STOTT, K. VANCE-BORLAND, AND A. R. WATKINSON. 2009. One hundred questions of importance to the conservation of global biological diversity. Conservation Biology, 23:557-567.

SWAIN, D. P., A. F. SINCLAIR, AND J. M. HANSON. 2007. Evolutionary response to size-selective mortality in an exploited fish population. Proceedings of the Royal Society, B, 274:1015-1022.

SWETNAM, T. W., C. D. ALLEN, AND J. L. BETANCOURT. 1999. Applied historical ecology: Using the past to manage for the future. Ecological Applications, 9:1189-1206.

TAPANILA, L. 2008. Direct evidence of ancient symbiosis using trace fossils, p. 271-287 In P. H. Kelley, and R. K. Bambach (eds.), From Evolution to Geobiology: Research Questions Driving Paleontology at the Start of a

New Century. Paleontological Society Papers, 14.

TERBORGH, J. 1988. The big things that run the world: A sequel to E. O. Wilson. Conservation Biology, 2:402-403.

TERBORGH, J., L. LOPEZ, P. NUÑEZ V., M. RAO, G. SHAHABUDDIN, G. ORIGUELA, M. RIVEROS, R. ASCANIO, G. H. ADLER, T. D. LAMBERT, AND L. BALBAS. 2001. Ecological meltdown in predator-free forest fragments. Science, 294:1923-1926.

THOMPSON, J. N. 1994. The Coevolutionary Process. The University of Chicago Press, Chicago, Illinois, 376 p.

THOMPSON, J. N. 1996. Evolutionary ecology and the conservation of biodiversity. Trends in Ecology and Evolution, 11:300-303.

THOMPSON, J. N. 1997. Conserving interaction biodiversity, p. 285-293. In S. T. A. Pickett, R. S. Ostfeld, M. Shachak, G. E. Likens (eds.), The Ecological Basis of Conservation: Heterogeneity, Ecosystems, and Biodiversity. Chapman and Hall, New York.

THOMPSON, J. N. 1998. Rapid evolution as an ecological process. Trends in Ecology and Evolution, 13:329-332.

THOMPSON, J. N. 1999a. Specific hypotheses on the geographic mosaic of coevolution. The American Naturalist, 153:S1-S4.

THOMPSON, J. N. 1999b. The evolution of species interactions. Science, 284:2116-2118.

THOMPSON, J. N. 1999c. Coevolution and escalation: Are ongoing coevolutionary meanderings important? The American Naturalist, 153:S92-S93.

THOMPSON, J. N. 2005. The Geographic Mosaic of Coevolution. The University of Chicago Press, Chicago, Illinois, 443 p.

THOMPSON, J. N. 2008. Conservation of the coevolving web of life, p. 221-223. In S. P. Carroll, and C. W. Fox (eds.), Conservation Biology: Evolution in Action. Oxford University Press, Oxford.

THOMPSON, J. N. 2009. The coevolving web of life. The American Naturalist, 173:125-140.

THOMPSON, J. N., AND B. M. CUNNINGHAM. 2002. Geographic structure and dynamics of coevolutionary selection. Nature, 417:735-738.

TRAVIS, J., AND D. J. FUTUYMA. 1993. Global change: Lessons from and for evolutionary biology, p. 251-263. In P. M. Kareiva, J. G. Kingsolver, and R. B. Huey (eds.), Biotic Interactions and Global Change. Sinauer Associates, Sunderland, Massachusetts.

TYLIANAKIS, J. M., R. K. DIDHAM, J. BASCOMPTE, AND D. A. WARDLE. 2008. Global change and species interactions in terrestrial ecosystems. Ecology Letters, 11:1-13.

VANDERGAST, A. G., A. J. BOHONAK, S. A. HATHAWAY, J. BOYS, AND R. N. FISHER. 2008. Are hotspots of evolutionary potential adequately protected in southern California? Biological Conservation, 141:1648-1664.

VAN DER PUTTEN, W. H., P. C. DE RUITER, T. M. BEZEMER, J. A. HARVEY, M. WASSEN, AND V. WOLTERS. 2004. Trophic interactions in a changing world. Basic and Applied Ecology, 5:487-494.

VERBOOM, G. A., L. L. DREYER, AND V. SAVOLAINEN. 2009. Understanding the origins and evolution of the world's biodiversity hotspots: The biota of the African 'Cape Floristic Region' as a case study. Molecular Phylogenetics and Evolution, 51:1-4.

VERMEIJ, G. J. 1977. The Mesozoic marine revolution: Evidence from snails, predators, and grazers. Paleobiology, 3:245-258.

VERMEIJ, G. J. 1987. Evolution and Escalation: An Ecological History of Life. Princeton University Press, Princeton, New Jersey, 526 p.

VERMEIJ, G. J. 1994. The evolutionary interaction among species: Selection, escalation, and coevolution. Annual Review of Ecology and Systematics, 25:219-236.

VERMEIJ, G. J. 1996. Adaptations of clades: Resistance and response, p. 363–380. In M. R. Rose and G. V. Lauder (eds.), Adaptation. Academic Press, San Diego, California.

VERMEIJ, G. J. 1999. Inequality and the directionality of history. The American Naturalist, 153:243-253.

VERMEIJ, G. J. 2001. Innovation and evolution at the edge: Origins and fates of gastropods with a labral tooth. Biological Journal of the Linnean Society, 72:461-508.

VERMEIJ, G. J. 2002a. The geography of evolutionary opportunity: Hypothesis and two cases in gastropods. Integrative and Comparative Biology, 42:935-940.

VERMEIJ, G. J. 2002b. Evolution in the consumer age: Predators and the history of life, p. 375-393. In M. Kowalewski, and P. H. Kelley (eds.), The Fossil Record of Predation. The Paleontological Society Papers, 8.

VERMEIJ, G. J. 2004. Nature: An Economic History. Princeton University Press, Princeton, New Jersey, 445 p.

VERMEIJ, G. J. 2005a. Invasion as expectation: a historical fact of life, p. 315-339. In D. F. Sax, J. J. Stachowicz, and S. Gaines (eds.), Species Invasions: Insight into Ecology, Evolution and Biogeography. Sinauer Associates, Inc., Sunderland, Massachusetts.

VERMEIJ, G. J. 2005b. From phenomenology to first principles: Toward a theory of diversity. Proceedings of the California Academy of Sciences, 56 (Supplement 1(2)):12-23.

VERMEIJ, G. J. 2008. Escalation and its role in Jurassic biotic history. Palaeogeography, Palaeoclimatology, Palaeoecology, 263:3-8.

VERMEIJ, G. J. 2009a. Seven variations on a recent theme of conservation. In G. P. Dietl, and K. W. Flessa (eds.), Conservation Paleobiology: Using the Past to Manage for the Future. The Paleontological Society Papers, 15 (this volume).

VERMEIJ, G. J. 2009b. Comparative economics: Evolution

and the modern economy. Journal of Bioeconomics, 11:105-134.

VERMEIJ, G. J., AND G. P. DIETL. 2006. Majority rule: Adaptation and the long-term dynamics of species. Paleobiology, 32:173-178.

VITOUSEK, P. M., H. A. MOONEY, J. LUBCHENCO, AND J. M. MELILLO. 1997. Human domination of earth's ecosystems. Science, 277:494-499.

VITT, P., K. HAVENS, AND O. HOEGH-GULDBERG. 2009. Assisted migration: Part of an integrated conservation strategy. Trends in Ecology and Evolution, 24:473-474.

VOGWILL, T., A. FENTON, A. BUCKING, M. E. HOCHBERG, AND M. A. BROCKHURST. 2009. Source populations act as co-evolutionary pacemakers in experimental selection mosaics containing hotspots and coldspots. The American Naturalist, 173:E171-E176.

WADE, M. J. 2007. The co-evolutionary genetics of ecological communities. Nature Reviews, Genetics, 8:185-195.

WALKER, S. E., AND C. E. BRETT. 2002. Post-Paleozoic patterns in marine predation: Was there a Mesozoic and Cenozoic marine predatory revolution? p. 119-193. In M. Kowalewski, and P. H. Kelley (eds.), The Fossil Record of Predation. The Paleontological Society Papers, 8.

WALTHER, G., E. POST, P. CONVEY, A. MENZEL, C. PARMESAN, T. J. C. BEEBEE, J. FROMENTIN, O. HOEGH-GULDBERG, AND F. BAIRLEIN. 2002. Ecological responses to recent climate change. Nature, 416:389-395.

WHITHAM, T. G., W. P. YOUNG, G. D. MARTINSEN, C. A. GHERING, J. A. SCHWEITZER, S. M. SHUSTER, G. M. WIMP, D. G. FISCHER, J. K. BAILEY, R. L. LINDROTH, S. WOLBRIGHT, AND C. R. KUSKE. 2003. Community and ecosystem genetics: A consequence of the extended phenotype. Ecology, 84:559-573.

WILLIS, K. J., AND H. J. B. BIRKS. 2006. What is natural? The need for a long-term perspective in biodiversity conservation. Science, 314:1261-1265.

WILLIS, K. J., L. GILLSON, AND S. KNAPP. 2007. Biodiversity hotspots through time: An introduction. Philosophical Transactions of the Royal Society, B, 362:169-174.

WORSTER, D. 1994. Nature's Economy: A History of Ecological Ideas. Cambridge University Press. Cambridge, 423 p.

WRIGHT, S. D., L. N. GILLMAN, H. A. ROSS, AND D. J. KEELING. 2009. Slower tempo of microevolution in island birds: Implications for conservation biology. Evolution, 63:2275-2287.

ZIMMER, C. 2003. Rapid evolution can foil even the best-laid plans. Science, 300:895.

CHAPTER THIRTEEN

SPECIATION AND SHIFTING BASELINES: PROSPECTS FOR RECIPROCAL ILLUMINATION BETWEEN EVOLUTIONARY PALEOBIOLOGY AND CONSERVATION BIOLOGY

WARREN D. ALLMON

Paleontological Research Institution and Department of Earth and Atmospheric Sciences, Cornell University, 1259 Trumansburg Road, Ithaca, NY 14850 USA

ABSTRACT.—Humans have caused not only extinction of species; they have dramatically reduced the abundance of many species, especially over the past 500 years. These reductions have come not just in already rare species, but in many formerly abundant species. Modern abundances of these species are therefore not representative of pre-human nature. This "shifting baselines" view suggests that it may be "natural" not just for a few species to be abundant and many rare, but for many species to be abundant and a few to be superabundant. In addition to having implications for conservation, this conclusion has significance for paleontology: many dense fossil assemblages may have formed under conditions very different than at present, and we may therefore have to rethink the shape of common species-abundance curves. Humans have also significantly changed the distribution of global biomass and biological production, not just by reducing the abundance of formerly numerous species, but also by habitat modification and resource use.

These human-induced changes in biomass, production, and abundance are likely altering future patterns of speciation in many clades. An explicit model for the stages of allopatric speciation allows for predictions about which species will see increased speciation in the future, and which are likely to have speciation suppressed. If the present is not an entirely adequate key to past levels of abundance, then the fossil record is one of the best places to test these ideas. Because of the incompleteness of that record, paleontologists are sometimes reluctant to claim reliable estimates of absolute abundance for fossil species. While not unfounded, this reluctance should be balanced with more serious attempts to assess population sizes of fossil species when preservation allows, and to apply these to individual species lineages in an effort to determine their effects on the processes of cladogenesis.

"We should not take it for granted that the way things are now is the way they have always been. ... We usually infer that a system as we observe it represents the normal structure and functioning of the system. If we think about it for a moment, we realize it doesn't make much sense to assume that a system that has suffered heavy fishing ... is 'normal'..."–Apollonio (2002, p. 45)

"Those early reports suggest extraordinary, almost unbelievable numbers of fish, though such numbers are rarely taken into account when assessing what a healthy stock should look like. And even if we use what hard data can be mustered to build up a picture of a pristine ecosystem, we may still be grossly underestimating what nature should look like."–Nicholls (2009, p. 42-43; emphasis in original)

"A historical ecologist studies man's effects on a past landscape. The discipline requires an aptitude for science but also the ability to imagine things differently from the way they appear, which is not an ordinary habit of mind."–Wilkinson (2006, p. 56)

INTRODUCTION

THE CREATION story presented in the Book of Genesis addresses two of the most conspicuous features of life: diversity and abundance.

"And God said, Let the waters bring forth abundantly the moving creature that hath life...And God created great whales, and every living creature that moveth, which the waters brought forth abundantly, after their kind... and God saw that it was good... And God said,

In *Conservation Paleobiology: Using the Past to Manage for the Future, Paleontological Society Short Course, October 17th, 2009. The Paleontological Society Papers, Volume 15, Gregory P. Dietl and Karl W. Flessa (eds.). Copyright © 2009 The Paleontological Society.*

Table *1.*—Various human influences on evolutionary processes.

Human impact	Evolutionary consequence(s)	References
Decreased abundance	Change in speciation rates[1] Extinction Genetic impoverishment Ecological changes[2]	Myers and Knoll (2001)
Habitat/population fragmentation	Change in speciation rates Decreased abundance	Lawton (1993); Maurer and Heywood (1993); Templeton et al. (2001)
Migration obstruction	Decreased abundance Geographic range changes Anagenesis[3]	Wilcove (2007); Waples et al. (2008)
Habitat loss	Decreased abundance Ecological changes	Eldredge (1991); Wilcove et al. (1998)
Habitat alteration	Decreased abundance Geographic range changes Anagenesis	Singer et al. (1993); Tilman and Lehman (2001); Western (2001)
Climate change	Abundance changes Anagenesis Geographic range changes	Hughes (2000); Lovejoy and Hannah (2005); Parmesan (2006); Bradshaw and Holzapfel (2008); Bell and Collins (2008); Gienapp et al. (2008)
Introduced/invasive species	Abundance changes Ecological changes Anagenesis	Carlton (1999); Mooney and Cleland (2001); Cox (2004); Lockwood et al. (2006); Suarez and Tsutsui (2008); McKinney and Lockwood (2001)
Removal of predators, competitors, or food sources	Increase in abundance Ecological changes Anagenesis	Smith et al. (1995); Reznick et al. (2008)
Genetic modification	Anagenesis[4] Ecological changes	Palumbi (2001a, b)
Pesticides, antibiotics	Anagenesis	Palumbi (2001a, b)
Size selection	Changes in population structure Abundance changes Anagenesis	Fenberg and Roy (2008)

Notes:
    1. See discussion below.
    2. Ecological changes including change in local diversity, competition, predator-prey relationships, and species rank abundance and dominance.
    3. Anagenesis (evolutionary change within lineages), as opposed to cladogenesis (speciation).
    4. Biotechnological insertion of exogenous genes into domesticated plants and animals have the potential to "effectively increase the rate of generation of new traits—akin to increasing the rate of macromutation. When these traits cross from domesticated into wild species, they can add to the fuel of evolution and allow rapid spread of the traits in natural populations." (Palumbi, 2001a, p. 1787).

Let the earth bring forth the living creature after his kind, cattle, and creeping thing, and beast of the earth after his kind: and it was so" (Genesis, 1:20-24).

These words express what is known as the principle of "plentitude", the idea that for the universe to be as perfect as possible it must be as full as possible, containing as many individuals of as many kinds of things as it possibly could contain. This has long been part of the ontological argument for God's existence, that existence is perfect, and that there can be no possibilities that remain unrealized throughout eternity (Lovejoy, 1936). In a secular context, plentitude also expresses the Malthusian-Darwinian observation of life's enormous capacity for increase, its seeming ability to fill all available space. Left unsaid in most such ruminations, however, is exactly what the connection might be between these two phenomena, between the number of living things and the number of kinds of living things.

After 1859, questions about the diversity and abundance of species continued to be important questions in the new fields of ecology and evolutionary biology, yet the relationship between them has remained incompletely explored, and this has had implications for the consideration of these two phenomena in both evolutionary and conservation biology. On one hand, the evolutionary/ecological connection between abundance and species extinction seems clear: because it must precede extinction, a decrease in abundance is often taken as an indicator of increased risk of species extinction (e.g., Diamond, 1984; Pimm et al., 1993; Mace and Kershaw, 1997; Abrams, 2002). On the other hand, the connection between abundance and the origin or transformation of species is less straightforward. Darwin (1859) suggested that abundant, widespread species had the greatest potential to give rise to new descendant forms; yet the abundance of species in the tropics (where diversity is highest) is usually low (e.g., Fedorov, 1966; Hubbell and Foster, 1986; Symonds et al., 2006, and references therein), and some evolutionary theory suggested that smaller populations might evolve more rapidly (Mayr, 1963). Conservation biology, furthermore, has only recently begun to focus on evolutionary processes other than extinction.

The idea that present levels of abundance and diversity in many communities are not what they were even a few hundred years ago—what has come to be called the phenomenon of "shifting baselines"—has gained increasing support among ecologists and conservation biologists over the past decade or so (e.g., Pauly, 1995; Jackson, 1997, 2001; Jackson et al., 2001; Baum and Myers, 2004; Sáenz-Arroyo et al., 2005; Knowlton and Jackson, 2008; Lotze and Worm, 2009; Dietl, 2009; Jackson, 2009). These authors argue that we need to use the approaches of historical ecology (e.g., Swetnam et al., 1999), and not just observations of the present, as a basis for comparison with the current condition of the natural world. Exploration of the evolutionary implications of these changes in abundance, however, has not gone much beyond dire (but unfortunately plausible) predictions of a future ocean dominated by bacteria (e.g., Pandolfi et al., 2005; Jackson, 2008).

In this chapter I try to expand on this perspective, by considering the potential evolutionary consequences of anthropogenic changes in animal abundance from a more mechanistic point of view. If the reductions in population size implied by shifting baselines—especially in formerly abundant or "superabundant" species—are a widespread feature of modern, human-dominated ecosystems, this in turn implies that modern abundances of many species are poor bases for comparison with many times in the geological past, which may have been characterized by higher abundances, higher levels of productivity, and differently shaped species-abundance distributions than conventional ecological wisdom has led us to expect. If this is true, then we might look to the fossil record for data on past abundances, and try to use this information to test hypotheses about the causal connection between abundance and evolution—especially speciation—in the geological past as well as in the future. Such a collaborative focus by paleobiologists and conservation biologists can provide a perspective that neither discipline can produce on its own. Such a collaboration can bring to paleobiology an improved understanding of the evolutionary and taphonomic consequences of changes in biological productivity, and to conservation an improved understanding of the long-term impact of the changes that humans are wreaking on the modern biosphere.

HUMAN EFFECTS ON EVOLUTION

After a slow start in the 1970s, a consensus finally emerged that humans are significantly altering evolutionary processes and the course of future evolution,

TABLE 2.—Examples of species that were once abundant and have declined significantly over the past 100-1000 years.

Species	Common name	Reference
PLANTS		
Castanea dentate	American chestnut	Rosenzweig (1995)
Swietenia macrophylla	Big-leaved mahogany	Gaston and Fuller (2007)
MOLLUSKS		
Haliotis sorenseni	White abalone	Tegner et al. (1996)
Strombus gigas	Queen conch	Robertson (2005)
Tridacna gigas	Giant clam	Wells (1997)
Pinctada mazatlanica	Pearl oyster	Sáenz-Arroyo et al. (2006)
Crassostrea virginiana	American oyster	Jackson (2001)
INSECTS		
Melanopus spretus	Rocky Mountain grasshopper	Gaston and Fuller (2007)
Nicrophorus americanus	American burying beetle	"
FISH [1]		
Oncorhynchus spp.	Pacific salmon	Montgomery (2003)
Salmo salar	Atlantic salmon	"
Engraulis ringens	Peruvian anchoveta	Gaston and Fuller (2007)
Gadus morhua	Northern cod	Hutchings and Myers (1994); Kurlansky (1997)
Totoaba macdonaldi	Totoaba	Roberts and Hawkins (1999)
Hippoglossus hippoglossus	Atlantic halibut	Gaston and Fuller (2007); Grasso (2008)
Dipturus batis	Common skate	Gaston and Fuller (2007)
Raja laevis	Barndoor skate	Casey and Myers (1998)
Pristis pectinata	Smalltooth sawfish	"
Echinorhinus brucus	Bramble shark	"
Squatina squatina	Angel shark	"
Carcharhinus longimanus [2]	Whitetip shark	Baum and Myers (2004); Myers and Worm (2005)
REPTILES		
Chelonia mydas	Green turtle	Jackson (1997)
Eretmochelys imbricate	Hawksbill turtle	McClenachan et al. (2006)
Dermochelys coriacea	Leatherback turtle	Safina (2006)
BIRDS		
Ectopistes migratorius	Passenger pigeon	Schorger (1955)
Conuropsis carolinensis	Carolina parakeet	Fuller (2001)
Alca impennis	Great auk	Nicholls (2009); Fuller (2001)
Numenius borealis	Eskimo curlew	Fuller (2001)
Tympanuchus cupido	greater prairie chicken	Bouzat et al. (1998)
Songbirds (many species)[3]		
MAMMALS		
Cynomys ludovicianus	Black-tailed prairie dog	Nicholls (2009)
Chinchilla spp.	Chinchilla	Gaston and Fuller (2008)
Spilogale putorius	Eastern spotted skunk	"
Bison bison	American bison	Isenberg (2000); Lott (2002)
Bison bonasus	European bison	Gaston and Fuller (2007)
Saiga tatarica	saiga antelope	"
Antilope cervicapra	Blackbuck	Gaston and Fuller (2008)
Dugong dugon	Dugong	"
Loxodonta africana	African elephant	"
Phocoena phocoena	Harbor porpoise	Gaston and Fuller (2007)
Eubalaena glacialis [4]	North Atlantic right whale	Kenney (2002); Allmon (2004)

and that this fact should be taken into account in conservation biology (e.g., Frankel, 1974; Soulé and Wilcox, 1980; Frankel and Soulé, 1981; Myers, 1985; Allmon, 1990; Ward, 1994; Bowen, 1999; Crandall et al., 2000; Myers and Knoll, 2001; Palumbi, 2001a,b; Western, 2001; Woodruff, 2001; Ashley et al., 2003; Faith et al., 2004; Ferrière et al., 2004; Young, 2004; Redding and Mooers, 2006; Kinnison and Hairston, 2007; Carroll and Fox, 2008; Smith and Bernatchez, 2008). Consideration of this topic has been particularly widespread in fisheries science, where terms such as "fisheries-induced evolution" have become common (e.g., Policansky, 1993; Stokes and Law, 2000; Conover and Munch, 2002; Law and Stokes, 2005; Reznick and Ghalambor, 2005; Hutchings and Fraser, 2008; ICES, 2007; Law, 2007; Hard et al., 2008, and references therein; Heino and Deickmann, 2008; McClure et al., 2008; Waples et al., 2008).

It is now clear that humans are affecting evolution in a variety of ways (Table 1). Most often discussed are: 1) changes in selection pressure (via habitat or ecological changes, pesticides, antibiotics, etc.) leading to anagenetic change in novel directions or at greater rates (e.g., Hendry and Kinnison, 1999; Cowling and Pressey, 2001; Palumbi, 2001a, b); 2) extinction of species (e.g., Myers, 1985, 1986, 1987; Wilson, 1992; Nee and May, 1997; Pimm and Raven, 2000); and 3) decrease in genetic variation within species or populations (e.g., Franklin, 1980; Frankel and Soulé, 1981; Allendorf and Leary, 1986; Ellstrand and Elam, 1993; Avise, 2008).

Less often discussed are potential anthropogenic effects on the origin of species, the process(es) of speciation. Several authors have considered this topic (e.g., Soulé, 1980; Frankel and Soulé, 1981; Webb and Gaston, 2000; Webb et al., 2000; Barraclough and Davies, 2005), but conclusions have been hesitant and contradictory. It has been suggested, for example, that human influences might lead to an increase in speciation, due to some combination of: a) fragmentation of species ranges; b) founder effects associated with species invasions; c) empty niches left by extinction of incumbent species; and d) expansion of species particularly able to thrive in human-dominated environments (Myers and Knoll, 2001, and references therein). But it has also been suggested that speciation might decrease, due to some combination of reduction of species geographic ranges, especially for large terrestrial vertebrates, and disruption of incipient speciation, by extinction or hybridization of isolated populations, including espe-

---

←

Table 2, notes:

1. Globally, marine fish populations are now estimated to have declined by 80-90% (Jackson, 2008; Lotze and Worm, 2008), although there are some recent signs of improvement (Worm et al., 2009).

2. Global shark populations have decreased 50-80% since the 1970s (Manire and Gruber, 1990; Baum and Myers, 2004; Helfman, 2007; Jackson, 2008).

3. Reliable estimates of present total population sizes for birds are extremely few. Peterson (1941) suggested that the population of all breeding birds in the United States and Canada is around 6 billion. More recent estimates that have been done for individual species (e.g., www.rmbo.org/pif_db/laped/ plus major ecological groups (e.g., waterfowl and shorebirds) suggest that this number may be closer to 7 billion breeding birds. What proportion this may be of pre-human or pre-European songbird abundance is impossible to say with any confidence. Most authors agree that many species of North American birds have declined in abundance over the past century (e.g., Robbins et al., 1989; Terborgh, 1989; Askins et al., 1990; Franzreb and Rosenberg, 1997; Rodriguez, 2002), probably mostly due to habitat destruction, much of which may be occurring in regions of highest bird abundance (Rodriguez, 2002). At least some portion of this decline, however, is due to anthropogenic causes such as domestic cats, towers, and glass windows; studies of mortality from these causes are not very precise, but suggest that perhaps 500 million to 1 billion birds are killed annually (e.g., Calver et al., 2007, and references therein). This is probably less than natural annual mortality of adults, which might be as high as 20-30% (Gill, 2006). If this direct anthropogenic mortality is distributed nonrandomly, however, and species of conservation concern with lower total populations are affected mainly at migration concentration points, then this mortality could be even more significant (K. Rosenberg, pers. comm., 2009). It might therefore be safe to say that anthropogenic mortality in the form of "indirect taking", as described above, effectively approximately doubles total annual adult mortality, with mortality or reduction in reproduction due to habitat loss being sources of additional abundance decrease.

4. Although it has long been known that the population sizes of all large whales were once much higher than they are today, it is only recently that studies of genetic diversity have indicated that they were much, much higher. As pointed out by Jackson (2006, p. 27), the previous abundance of whales is a "question more controversial than it should be… but the total for all species combined was certainly many millions". The waters off Labrador and Newfoundland were known to early European colonists as the "Sea of Whales" (Nicholls, 2009, p. 46) and, indeed, pre-whaling abundance of all baleen whale species was higher than present by at least a factor of two, and in some cases by more than an order of magnitude (Roman and Palumbi, 2003; Alter et al., 2007).

cially putative evolutionary "hotspots," which appear to contain multiple incipient species (e.g., Myers and Knoll, 2001, and references therein; also Ehrlich and Daily, 1993; Hughes et al., 1997; Hobbs and Mooney, 2008; Rosenzweig, 2001; Ceballos and Ehrlich, 2002; Barraclough and Davies, 2005; Gow et al., 2006; Hendry et al., 2006; Seehausen, 2006; Davis et al., 2008; Seehausen et al., 2008).

How can these apparently contradictory conclusions be resolved? In the following discussion I suggest that we need to consider the effects of abundance on speciation within an explicit framework of the stages by which speciation occurs (Allmon, 1992), and that by doing so we can also gain insights into the degree to which modern species-abundance relationships are adequate models for such relationships in the geological past.

WHAT ABUNDANCES ARE "NATURAL"?

What is "natural"?

What was the living world like prior to humans? At one level, this is one of the central questions of paleontology, and the field has provided abundant answers to it over the past two centuries. At another level, however, paleontologists have always struggled with this question because so much of what we do depends on comparison with the present, which may or may not be representative of the past. It is obvious that humans have had major effects on the Earth's flora and fauna. Until relatively recently, however, both scientists and society at large believed that—outside of areas of dense human habitation or activity (and even in some enclaves within them)—"nature" was still much as it had been prior to human presence. It is now clear that few if any such "pristine" places exist; that is, there are essentially no places left on Earth that do not show some measurable effect of human presence (e.g., Gallagher and Carpenter, 1997; Jackson, 2001; Western, 2001; Halpern et al., 2008). The effects of humans, furthermore, are not just a recent phenomenon; there is widespread recognition that even prehistoric humans had major impact on the landscape and biota in many regions of the world, through fire, hunting, deforestation, and agriculture (e.g., Delcourt and Delcourt, 2008). Within this recognition, however, controversy persists about the magnitude of these effects. Two questions, in particular, are relevant to the present discussion. First, what,

if any, role did humans play in the extinction of large mammals in the Late Pleistocene, especially in North America? Second, how much impact did humans have on the North American biota between the Pleistocene megafaunal extinctions and European contact? These questions are connected; if humans were the primary cause of the Pleistocene extinction, then it seems reasonable that they could have significantly modified the subsequent Holocene fauna. If, on the other hand, Native Americans were not a significant contributor to the demise of the megafauna, they could still have had a major effect on the biota prior to the arrival of Europeans.

Despite a quarter-century of intense discussion, the role of human hunting (the "Overkill" or "Blitzkrieg" hypothesis) versus climate change in the Pleistocene extinction remains unclear (e.g., Martin, 1984, 2002; Haynes, 2009; Meltzer, 2009). The impact of humans on the North American fauna of the Holocene is also controversial. It was long thought that the Americas in 1492 were relatively sparsely populated by humans, that Native Americans had relatively little impact on animal populations, and that non-human nature in the New World was essentially as it had been prior to human arrival. Critics of this "pristine myth", however, have argued that it owed more to Western cultural biases than to science, that at the time of European contact Native American populations were substantial and much or most of the biota had been significantly altered by human activities (e.g., Denevan, 1992; Kay, 1994; Grayson, 2001; Broughton, 2002; Mann, 2005; Delcourt and Delcourt, 2008). When Europeans first arrived, for example, there may have been as many as 100 million Native Americans living in the hemisphere, andthrough fire, agriculture, hunting, and landscape modification—these people had already had considerable impact on the non-human environment. Yet by 1800, if not before, perhaps 90% of these people had succumbed to European diseases (Dobyns, 1983). Some, but not all, of these authors have argued that this post-contact decline of Native Americans allowed the abundance of many game animals to abruptly increase, so that by the eighteenth and nineteenth centuries (when many natural history observations were made and recorded by Euro-Americans) many animal populations had assumed much higher levels than they likely had in pre-Columbian times (e.g., Broughton, 2002; Mann, 2005).

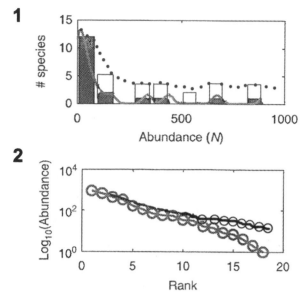

FIGURE 1.—Two of the many different forms of species-abundance distributions. *1*, Simple histogram (shaded boxes) of number of species vs. abundance on an arithmetic scale. A smoothed line is added to highlight the overall shape. Open histogram boxes and dotted line show what such a curve might look like if higher overall abundances were the norm (i.e., a larger number of abundant species). *2*, Rank-abundance diagram (sometimes called an RAD), in which log abundance is plotted against the rank. Upper line of open circles shows what such a curve might look like if higher overall abundances were the norm (i.e., a larger number of abundant species). Modified from McGill et al. (2007).

While this view is clearly at least partly correct, there are also significant reasons for concluding that at least some post-contact, pre-twentieth century animal abundances are indicative of those that might have existed prior to or without humans, and that early European travelers' and explorers' accounts can provide potentially valid evidence for historical ecology (e.g., Jackson, 1997, 2001; Jackson and Sala, 2001; Jackson et al., 2001; Sáenz-Arroyo et al., 2006). First, many of the animal species showing very high pre-twentieth century abundance appear to be adapted to such abundance, suggesting that they existed at near these levels prior to human effects (the "Allee effect"; see discussion below). Second, although pre-Columbians did have some significant effects on many marine animal populations (e.g., Hildebrandt and Jones, 2002; Spotila, 2004; Rick and Erlandson, 2008), these effects were clearly not as large as those they probably had on terrestrial animals, and yet many marine species show the same patterns

of high pre-modern abundance as many of these terrestrial species (see below). Finally, even if we accept that post-contact pre-modern abundances is a result of the decimation of Native American populations, and not indicative of pre-contact Holocene abundance, they might still be representative of faunal abundance prior to the arrival of any humans in the New World in the Late Pleistocene.

It is therefore reasonable to conclude that the generally high abundances of many species recorded in the nineteenth and early twentieth centuries are at least somewhat representative of "natural" (i.e., pre-human) abundances, and that pre-human nature (in the Americas and elsewhere) was in general much more abundant than we can now easily envision (e.g., Pearson, 1972; Mowat, 1984; Jackson, 1997, 2001, 2008; Kurlansky, 1997; Jackson et al., 2001; Apollonio, 2002; Pauly and MacLean, 2003; Crowder, 2005; Roberts, 2007; Grasso, 2008; Nicholls, 2009). Documenting individual examples of such changes, however, is frequently difficult. Estimating modern abundances is complex enough (e.g., Witt et al., 2009); estimating past abundances—in historical, archeological, or paleontological records—is even more challenging. There are nevertheless a sufficient number of examples to support the conclusion of an overall pattern (Table 2).

Several generalizations arise from consideration of these and other similar species histories:

1) Natural patterns of and controls on rarity and abundance are poorly understood.—Determining why some species are rare and others abundant may be the most basic question in ecology (e.g., Andrewartha and Birch, 1954; MacArthur, 1972; Whittaker, 1975; Colinvaux, 1978; Gaston, 1994), but we still do not fully understand many aspects of it. What, for example, is the "natural" (i.e., pre-human) frequency of rare versus abundant species? What factors affect this frequency? Do all or most "abundant" and "rare" species share important characteristics? We do know that rarity is a relative, and highly scale-dependent, condition (e.g., depending on the size of the area in which they are censused). It is also frequently variable in time and space (Gaston, 1994, 1997), with many, perhaps most, natural populations fluctuating considerably in abundance across their ranges and over time (e.g., May, 1979; Ariño and Pimm, 1995). Thus a very useful distinction is between species that are rare only in some places or times ("somewhere-abundant" or "sometimes-abundant" species) and those that are always rare through-out their range and duration ("everywhere-sparse", "always rare" or "naturally rare" species) (Murray et al., 1999; Murray and Lepschi, 2004; Harrison et al., 2008). The definitions used will therefore determine to a considerable degree which and how many species are considered "rare". While it is common in ecology to say that "most species are rare", most studies conclude that only 20-40% of species in a community are "rare" (Gaston, 1997, p. 36). "Everywhere-sparse" species often share a number of characteristics (Gaston and Kunin, 1997), including: breeding systems biased away from outcrossing and sexual reproduction; lower repro-ductive investment; poorer dispersal ability; higher ho-mozygosity; using less common or a narrower range of resources; occupying higher trophic levels; and having larger body size.

An underappreciated factor in the frequency of abundant versus rare species may be the "Allee ef-fect", in which individuals of a species benefit from the presence of conspecifics (Stephens et al., 1999). The phenomenon is named for ecologist W.C. Allee, who noted that many species suffer a decrease in per capita rate of increase as their populations reach small sizes or low densities (Allee, 1931; Allee et al., 1949). Al-though it was originally referred to as "Allee's Prin-ciple" (Odum, 1971), it later came to be known as the Allee effect; it is also known as "inverse density dependence" (Courchamp et al., 1999) or, in fisheries science, as "depensation" (e.g., Liermann and Hilborn, 1997). The effect can be manifest at either low or high densities. In the former, populations are too small to survive not because there are insufficient resources, but because of lack of some form of interaction. In the latter, populations are large in order to maintain such interactions. Stephens et al. (1999) distinguished be-tween two forms of Allee effect: 1) *component Allee ef-fects* (positive density dependence in some component of individual fitness), and 2) *demographic Allee effects* (positive density dependence in the per capita popula-tion growth rate). A range of mechanisms may lead to Allee effects, including: "predator dilution or satura-tion; antipredator vigilance or aggression; cooperative predation or resource defense; social thermoregulation; collective modification or amelioration of the environ-ment; increased availability of mates; increased polli-nation or fertilization success; conspecific enhancement of reproduction; and reduction of inbreeding, genetic

drift, or loss of integrity by hybridization" (Stephens et al., 1999, p. 185). Roberts and Hawkins (1999, p. 244) have also noted that Allee effects at reproduction may render marine species vulnerable to population decline due to overexploitation.

The Allee effect has been cited as a contributing factor in the decline of several of the formerly very abundant species listed in Table 2 (e.g., bison, Isenberg, 2000, p. 28; passenger pigeon, Petersen and Levitan, 2001; and whales, Jackson et al., 2008). Yet, although the effect has been identified in a number of taxa, and "the logic for a potential Allee effect exists in many cases", there is not always strong empirical support for it (Petersen and Levitan, 2001, p. 289; Gascoigne and Lipcius, 2004). There are at least three possible reasons for this apparent lack of demonstrable cases (Petersen and Levitan, 2001): 1) it may actually be rare in nature, "perhaps because the magnitude of intraspecific com-petition is too large relative to countering demographic Allee effects to produce possible density dependence at the population level" (*ibid.*, p. 289); 2) demograph-ic Allee effects might be difficult to observe because populations in which they can occur quickly become extinct; and 3) demographic Allee effects may typi-cally occur in populations where human exploitation and habitat degradation also occur, making it difficult to partition Allee effects from other negative effects on population growth. Petersen and Levitan conclude that: "Allee effects may be particularly important in species that typically have occurrence at relatively high density and only recently had their densities dramatically re-duced by exploitation… It is just these types of species that may be unsuited for success at low densities… Pas-senger pigeons probably fell into this category" (2001, p. 295; see also Halliday, 1980).

Thus we do not know what proportion of species in the pre-human world were "rare" and what proportion "abundant" or why. It is possible that our perceptions are skewed by the existence of a few "high-abundance biased" (HAB) species (Rodriguez, 2002). It may also be, however, that there were many more "abundant" or "superabundant" species in pre-human times (because of Allee effects or some other intrinsic aspects of their biology), and that this has been obscured by what we see around us today.

2) Humans have significantly reduced the abun-dance of many, and many different, species.—Different human impacts have different effects on different species; for some, direct taking (e.g., hunting and fishing) is the largest source of mortality; for others, it is habitat alteration. In any case, species showing significant anthropogenic declines in abundance (Table 2) are many and varied—marine and terrestrial, large and small bodied, and occurring across a range of higher taxa. These declines range from the historic to the prosaic: from the grand sweep of the American frontier (e.g., bison, passenger pigeons), to codfish and birds hitting windows. They include species that at their peaks, or even after population declines were well along, were widely considered impervious to major insult. Yet, as historian Hanna Rose Shell has put it, stories such as those of the bison remind us "that an environment's most plentiful and massive organisms may be those at highest risk of disappearing" (2002, p. viii).

3) Humans have significantly altered many species abundance distributions.—Species abundance distributions (SADs) follow a "hollow" or right-skewed curve: on an arithmetic scale, a large number of species show low abundance and a smaller number (represented by the long "tail") show higher abundance (Fig. 1). This phenomenon has been called "one of ecology's oldest and most universal laws" (McGill et al., 2007, p. 995). Although within this pattern it is acknowledged that "there is a great deal of variation in the details, especially as highlighted on a log-scale", the general "most species are rare" pattern is psychologically dominant (*ibid.*).

Yet there is evidence that this "universal" pattern has not always been obtained. Wagner et al. (2006), for example, showed that SADs of marine taxa were different in the Paleozoic compared to the post-Paleozoic, as a result of the effects of the massive end-Permian mass extinction. What if something similar is happening in the Holocene? What if the modern populations, on the observations of which modern ecology is based, are not representative of the history of most groups? Modern ecosystems may, in other words, contain fewer abundant species, and fewer species at high trophic levels, than there are "supposed" to be. As Jackson and Sala (2001, p. 278) suggest:

"Colinvaux ...[1978] explained '*Why Big Fierce Animals are Rare*' in terms of basic ecological principles, including most importantly the constraints on size imposed by the mechanics of feeding relationships and by the efficiency of assimilation of energy among trophic levels. Another explanation, however, is that we ate all the big animals before we studied them so therefore they are rare... [T]op and middle-level carnivores have been and continue to be a major component of our food from the oceans, which is perhaps the strongest evidence available that these large animals comprised a much larger proportion of the total pristine animal biomass than has been generally assumed."

Similarly, much of ecology is dominated by density-dependent thinking (e.g., White, 2008), "that if fish are abundant they are most likely undernourished, thin, and perhaps in poor condition and stressed because they must be crowded and competing and pressing upon their food sources... [and] that if one species is abundant then another species must be scarce" (Appollonio, 2002, p. 50-51). But what if this is yet another uniformitarian bias? What if this hasn't always been true? What if many species of fish are "supposed" to be abundant and healthy, and, as Appollonio puts it:

"...that in fact there was remarkable constancy of population sizes over a long period of time; certainly much longer than there has been scientific study of fish dynamics... that there was then an abundance of large, old fish within numerous species, and that there was a significant presence then of species whose abundances are now negligible...We might say that there was there a quality of fish ... that is now greatly changed" (Appollonio, 2002, p. 50-51).

4) Humans have significantly reduced and/or rearranged global biomass and primary productivity, and altered trophic pyramids.—Anthropogenic reduction in abundance goes beyond direct taking by hunting or fishing, or even habitat alteration; humans may also be causing a significant shift in the overall levels and/or behavior of Earth's primary productivity. Total net primary production (NPP; the net amount of biomass produced each year by plants) on the modern Earth has been estimated at approximately 170×10^9 dry tons / yr (Whittaker, 1975) (= 170,000 million metric tons, MMT). Several recent estimates suggest that 20-40% of terrestrial NPP now passes through humans (Vi-

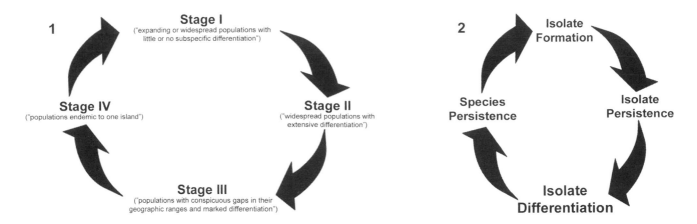

FIGURE 2.—*1*, Schematic representation of the Taxon Cycle, based on Ricklefs and Cox (1978); *2,* The Speciation Cycle (*sensu* Grant and Grant, 1997), in the same format.

tousek et al., 1997; Haberl et al., 2008). This has been referred to as "human appropriation of net primary production" (HANPP) (Luck, 2007, and references therein), defined as "the difference between the amount of NPP that would be available in an ecosystem in the absence of human activities (NPP_0) and the amount of NPP which actually remains in the ecosystem, or in the ecosystem that replaced it under current management practices (NPP_t). NPP_t can be calculated by quantifying the NPP of the actual vegetation (NPP_{act}) and subtracting the amount of NPP harvested by humans (NPP_h)" (Haberl et al., 2008). The human share of marine production may be comparable. Pauly and Christensen (1995) found that fisheries catches from oceanic upwellings and shelves require 24-35% of total primary productivity in these ecosystems to be sustained. Theoretical total fisheries productivity of the world's oceans is probably 80-100 MMT, and present total catch levels are around 80-85 MMT (Helfman, 2007). Humans are thus taking "close to what the ocean can produce" (Helfman, 2007, p. 261).

At the same time there are signs that overall marine primary productivity may be declining, probably because of climate change and associated shifts in oceanic circulation. For example, zooplankton biomass off Southern California decreased 80% in the second half of the twentieth century (Roemmich and McGowan, 1995; Veit et al., 1997; Helfman, 2007, p. 261). Similar declines have been noted in productivity in the deep-sea benthos, the northeastern Atlantic, the Baltic, and sub-Antarctic regions (Druffel and Robison, 1999; Flinkman et al., 1998; Hunt et al., 2001; Sims and Reid,

2002; Helfman, 2007), although plankton and fish production in the south Atlantic has increased (Verheye, 2000; Helfman, 2007).

Humans have thus removed a huge quantity of organic material from the biosphere and locked it away or redistributed it in various forms (e.g., human biomass, landfills, increase in abundance of "weedy" and domesticated species), and may also be reducing overall production, at least in the seas, via climate change. We do not know what the long-term consequences of this sequestration and reduction will be for the functioning or evolution of the biosphere. As Apollonio notes:

"Some might say that the energy that previously went into the growth of whales now appears elsewhere as other kinds of marine organisms. That very likely is true, but the fact that energy is not present as whales ... suggests that certain things or functions that used to happen when whales were abundant are not happening now... While it is true that other species may have increased to replace that lost biomass, the system itself has changed qualitatively; the dynamics must be different. ... Specifically, the qualities of hierarchical structure, internal constraints, naturally evolved stabilizing mechanisms, self-organization, and information inherent in some species and age classes have been removed and not replaced" (Apollonio, 2002, p. 176).

Ecology texts teach that in almost all biological

communities, the "trophic pyramid" consists of a broad base of primary producers, with narrower upper levels of consumers. In human-dominated communities, however, such pyramids are flattened by diminished abundance at higher levels (Jackson, 1997, 2001; Pauly and Maclean, 2003). It may be, however, that what ecology has taken for "natural" was already significantly altered by humans, and that pre-human trophic pyramids were "upside down"—that is, with much larger higher layers than textbooks have traditionally shown (Jackson, 2006; Sáenz-Arroyo et al., 2006).

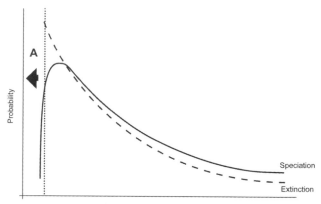

Abundance, range size, dispersal ability

FIGURE 3.—Chown's (1997) modification of Stanley's (1986) "fission effect" model of speciation. As Chown describes the model: "geographical range size, abundance and dispersal ability are presumed to be positively interrelated... The area marked A represents narrow geographical range and low abundance. Species in this area are fated with extinction... As abundance, range size and dispersal ability increase, extinction rate declines due to the rescue effect from other populations... and the distribution of populations over a wider geographic range, which facilitates escape from inimical environmental conditions... At the same time, the distribution of populations over a wider geographical area increases the likelihood of discontinuous, multifarious, divergent selection at a rate faster than that which can be overcome by gene-flow via dispersal..., resulting in an increased probability of speciation... As dispersal ability continues to rise and populations become more panmictic, the probability of speciation declines rapidly. However, speciation probability remains higher than that of extinction, due to the possibility of vicariance, resulting from extrinsic factors and/or the formation of peripheral isolates" (Chown, 1997, p. 100-101). Figure modified from Chown (1997).

ABUNDANCE, DIVERSITY, AND SPECIATION

If some or all of the changes noted in the preceding section are in fact taking place, what might be their potential evolutionary consequences, especially with respect to speciation? This is a difficult question to answer because we lack a clear understanding of the relationship between abundance and species diversity, even in modern communities (Rosenzweig and Abramsky, 1993; Abrams, 1995; Rosenzweig, 1995; Waide et al., 1999; Mittelbach et al., 2001; Whittaker and Heegaard, 2003; Bailey et al., 2004; Evans et al., 2005a, b, 2006; Martin and Allmon, in press). Furthermore, because diversity is a product of both speciation and extinction, its relationship to abundance may tell us only indirectly anything about speciation. Nevertheless, despite the complex and even contradictory results of previous studies of abundance and diversity, there are some suggestions about the connection between abundance and the origin of that diversity in the process(es) of speciation.

General patterns

Darwin's view—that abundant, widespread species had more "evolutionary potential"—represented one of what Chown (1997, p. 95) calls the two "remarkably divergent points of view" on the relationship between geographical range size, abundance, and speciation. On one end of this spectrum are authors who have adopted Darwin's view, that species with broad geographic range and high dispersal ability should have relatively high rates of speciation (e.g., Vermeij, 1987; Brooks and McLennan, 1993). On the other are those who have argued that such widespread and abundant

species should have lower rates of speciation than narrowly distributed species with lower abundance and dispersal ability (e.g., Mayr, 1963; Boucot, 1975; Stanley, 1979).

The frequently positive correlation between geographical variation in species richness and energy availability was an early observation of biogeography (e.g., Wallace, 1878). Separately, energy came to be viewed as an important organizing feature of ecosystems early in the history of modern ecology (e.g., Lotka, 1925; Elton, 1927; Lindeman, 1942; Evans, 1956; Odum, 1971). These two lines of thought stayed largely separate during much of the development of modern community ecology, but eventually converged in suggestions that spatial variation in energy availability might exert significant influence on species richness (e.g., Hutchinson, 1959; Connell and Orias, 1964; Leigh, 1965; Brown, 1981). Wright (1983) built upon such studies to propose the "species–energy theory", that species richness is positively correlated with environmental energy, which ultimately sets its upper limit. Subsequent work (see reviews by Waide et al., 1999; Evans et al., 2005a, b) has noted the complexity of the relationship between energy and species richness, but has generally supported the hypothesis that the two are frequently positively correlated. A mechanistic understanding for these patterns, however, has proven elusive (see general discussions by Brown, 1981, 2004).

Empirical species–energy relationships in modern biotas generally fall into one of two categories. Species richness may vary positively with energy, along a *monotonically* increasing curve, or it may show a more complex relationship, at first increasing but then declining at higher energy levels, forming a *unimodal*, or "hump-shaped", species-energy relationship. It remains unclear, however, which of these patterns is more common or why (Rosenzweig and Abramsky, 1993; Waide et al., 1999; Mittelbach et al., 2001; Whittaker and Heegaard, 2003; Evans et al., 2005a, b, 2006).

Positive relationships

Two general explanations for positive or monotonic relationships between abundance and diversity have been proposed: 1) the More Individuals Hypothesis (MIH), and 2) the Environmental Heterogeneity Hypothesis (EHH). The MIH (a term proposed by Wright et al., 1993) argues that the more energy available in usable form, the more organisms and hence, the more species the environment can support (Brown, 1981). More individuals might lead to more species by one or more of several mechanisms: a) More individuals may provide more loci for mutation and provide greater genetic variation. (Darwin, for example, believed that it was the abundant and wide-spread species that were likely to be the sources of most evolutionary innovation and new species: "We can so far take a prophetic glance into futurity as to foretell that it will be the common and widely-spread species, belonging to the larger and dominant groups, which will ultimately prevail and procreate new and dominant species" (1859, p. 489).) b) Higher population sizes may reduce extinction risk (Hutchinson, 1959; Leigh, 1981; Lande, 1998); c) in-

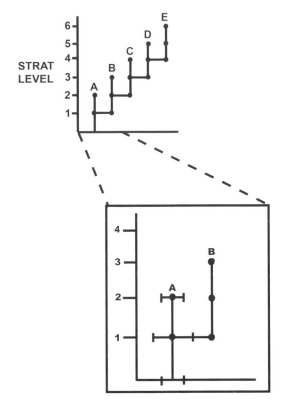

FIGURE 4.—Schematic illustration showing how estimates of abundance of fossil species might be plotted on a phylogeny to examine the relative timing of changes in abundance compared to the occurrence of events of cladogenesis.

creased population size may correlate positively with range size (Blackburn and Gaston, 1997; Gaston and Blackburn, 2000; Gaston, 2003), which may buffer species from extinction (e.g., Cardillo et al., 2005).

From the perspective of the fossil record, a number of authors have argued that overall levels of biological productivity—in the seas and perhaps also on land—have increased during parts or all of the Phanerozoic, and that this may have had some causal connection to either overall diversity levels or the ascendancy (or decline) of particular clades (e.g., Valentine, 1971; Bambach, 1993, 1999; Vermeij, 1987, 1995; Martin, 1996, 2003; Allmon and Ross, 2001; Martin and Allmon, in press).

More complex relationships

Unimodal or "hump-shaped" relationships between abundance and diversity have been interpreted as resulting from low habitat fertility, which reduces diversity through nutrient stress, while high fertility removes the constraints imposed by such stress, resulting in simplified (lower-diversity) communities as the outcome of competitive exclusion (Rosenzweig and Abramsky, 1993). Gaston (2003, p. 82) suggests that there is significant evidence for a unimodal ("hump-shaped") relationship between geographical range size (and by implication its frequent correlate, abundance) and the probability of speciation, with species with intermediate range sizes having the greatest probability of speciation (Gaston and Chown, 1999; Chown and Gaston, 2000).

In a more general context, the connection between evolution and population size has long been controversial within evolutionary biology. Sewell Wright suggested in the 1930s that small populations would be especially subject to change via genetic drift (Wright, 1931). Wright's work was the basis for two separate theoretical developments: neutralism and the founder principle. The neutral hypothesis suggested that small populations could change more quickly due to genetic drift (Ohta, 1972; Kimura, 1983; Woolfit and Bromham, 2005). The founder principle was proposed by Mayr (1954, 1963). It held that a new population established by only a few original founders (in the extreme case a single fertilized female) would carry only a small fraction of the total genetic variation of the parental population, and might therefore be subject to a relatively large and rapid genetic change (Mayr, 1963, p. 211).

The founder effect was widely accepted and elaborated on among evolutionary biologists and paleobiologists (e.g., Eldredge and Gould, 1972; Boucot, 1975; Giddings et al., 1989). Since the 1980s, however, it has come under increasing theoretical criticism (e.g., Barton and Charlesworth, 1984; Barton, 1989, 1996; Coyne and Orr, 2004), and the claim that there are few data that support it (Grant, 2001; Florin and Ödeen, 2002; McKinnon and Rundle, 2002; Grant and Grant, 2008, p. 44), although a significant number of authors continue to defend some or all of the components of the theory (e.g., Gavrilets and Hastings, 1996; Slatkin, 1996; Paulay and Meyer, 2002; Newton, 2003; Futuyma, 2005; Uyeda et al., 2009).

Large populations almost always contain greater genetic variation than smaller populations, and can therefore respond more quickly than smaller populations to changed selection pressures (Ridley, 1996; Charlesworth, 2009). The occurrence of smaller populations, however, may be higher than it seems. The number of individuals in a population that are actually breeding—referred to as the "effective population size" (N_e)—is always smaller than the total or census size of the population (Charlesworth, 2009). Furthermore, many large populations are geographically subdivided into metapopulations, which means that the effective population size is smaller still (Gavrilets, 2004; Sambatti et al., 2008). There is in any case still considerable empirical and theoretical support for the conclusion that genetic drift can overpower selection in such small populations, and that this can lead to more rapid fixation of alleles and so to more rapid genetic divergence in such populations (Ridley, 1996; Grant and Grant, 1997; Newton, 2003; Charlesworth, 2009). There is also still considerable support for allopatric speciation by peripheral isolate (peripatry), which requires daughter populations of smaller size than the parental population (e.g., Frey, 1993; Chown, 1997; Maurer and Nott, 1998; Futuyma, 2005). In sum, although the "genetic revolutions" that Mayr envisioned in such small populations may not occur very frequently (if at all), it still appears that smaller populations can in many instances change more quickly than large populations.

THE SPECIATION CYCLE

What then are the potential effects of abundance on probability and/or rate of speciation? Is it possible to combine and resolve all of these conflicting observations and conclusions into one causal model? I believe that it is, and that by doing so we can gain useful insights into the potential evolutionary consequences of human-mediated HAB species. The answer depends on where individual species are in the speciation process.

Steven Chown (1997; Gaston and Chown, 1999) has suggested that the concept of the "taxon cycle" can be modified and applied to this problem. The "taxon cycle" (Fig. 2.1) was introduced by E.O. Wilson (1961), and developed further by Robert Ricklefs and colleagues (Ricklefs, 1970; Ricklefs and Cox, 1972, 1978; Ricklefs and Bermingham, 1999, 2002; see also Roughgarden and Pacala, 1989; Webb and Gaston, 2000). It describes the series of processes that occur as species disperse from islands, differentiate into new species, and invade or reinvade islands. Grant and Grant (1997, 2008) used a version of the same idea, calling it the "speciation cycle," to explain the history of the diversification of Galapagos finches. Using the multi-stage framework for allopatric speciation (Allmon, 1992), such a cycle can be generalized even further to explain the relationship of abundance and speciation in a wider variety of cases (Allmon, 2005; Martin and Allmon, in press; Fig. 2.2).

For allopatric speciation in general, all events of speciation must involve at least three stages (Allmon, 1992): formation of an isolated population, persistence of the isolate, and differentiation of the isolate. If we ask what will be the most likely population sizes during each of these stages, we can see that there will probably be a cyclic series of changes. In the first stage, a relatively abundant parental population gives rise to a smaller daughter population—by either vicariance or dispersal. The daughter population either persists in isolation, merges with the parental population, or becomes extinct. If it persists long enough, it will become sufficiently genetically differentiated so that if/when it comes into contact with the parent it will be reproductively isolated. During this interval of persistence, the size of the isolate will likely increase. Once it becomes a fully differentiated species, it may give rise to its own daughter populations, and the cycle begins again.

Widespread (and therefore usually abundant) parental species are more likely to form more isolated daughter populations (usually at their margins), and so should give rise to more daughter species (Boucot, 1975; Rosenzweig, 1975; Stanley, 1986; Chown, 1997). More narrowly distributed (and usually less abundant) populations and species may change more quickly, under the influence of genetic drift, but are more susceptible to extinction. For an initially small population, probability of speciation therefore should increase with increasing abundance, but only up to a point, after which it declines rapidly (Fig. 3). Rare species, but not the rarest species, thus have a higher speciation probability than more common species (Lawton, 1993; Chown, 1997; Martin and Allmon, in press).

It is noteworthy that many of the species listed in Table 2 belong to clades (e.g., genera) that are currently relatively species-poor. This might therefore argue for low rates of speciation in super-abundant species. Yet all that these species accounts show is that these particular species—and perhaps many others—*could* achieve these spectacular population sizes under certain conditions. Had they been allowed to persist unmolested by humans, some of them might well have passed into the Speciation Cycle and given rise to numerous descendant species.

Testing the Speciation Cycle in the fossil record

The "speciation cycle" shown in Figure 2 is nothing more than an interesting thought unless it can be tested empirically. Doing so in the fossil record, however, requires an ability to obtain biologically meaningful estimates of abundance of fossil species.

1) The problem of fossil abundance.—Although it is also frequently acknowledged that paleontological consideration of abundance must confront the ubiquitous challenge of taphonomic bias in the fossil record (e.g., McKinney, 1997a; Vermeij and Herbert, 2004; Finnegan and Droser, 2005; Blob and Badgley, 2007; Brinkman et al., 2007; Pyenson et al., 2009), including incomplete or non-preservation, and postmortem transport, and (especially) time-averaging of biogenic hardparts, many paleontological studies nevertheless take the ability to obtain at least some kind of biologically relevant measure of abundance for granted (e.g., Boucot, 1975; Reboulet et al., 2005; Jackson and Erwin, 2006; Deline, 2009). In practice, one can either accept a biological interpretation of fossil data only after the necessary taphonomic analysis (e.g., Kidwell et al., 1986) has been completed, or one can assume that the processes adding and removing individuals from a fossil assemblage will more or less cancel each other out, and it will "all come out in the wash". It may be the increasing unacceptability of the latter that has limited the number of paleontological studies that have tried to examine the role of abundance in evolution (e.g., Boucot, 1975; Enquist et al., 1995; Jernvall and Fortelius, 2002, 2004; Tuite, 2009).

2) Use relative abundance.—Paleontologists have long reported abundance of different fossil taxa as a proportion of all individuals in a sample or set of samples (e.g., Ager, 1963; Boucot, 1981; Dodd and Stanton, 1990), and assigned such data various biological (instead of or in addition to taphonomic) explanations. As noted by Finnegan and Droser (2005), such relative abundance measures are fairly easily collected, robust to variation in sample size, and yield ecologically useful measures such as evenness. Most significantly, Kidwell (2001) found that many modern time-averaged death assemblages, especially if very small individuals are excluded, retain a strong signal of their component species' original rank order abundance. Relative abundance cannot, however, tell us everything we want to know because, as Finnegan and Droser (2005, p. 481) put it: "It is impossible to resolve the ecological meaning of abundance trends on the basis of relative abundance alone ... [and] normalizing to the number of individuals in the sample necessarily couples variables that may be entirely uncorrelated", such as the rise and decline in absolute abundance of two different species. In the context of this discussion, relative abundance estimates might allow for comparing the evolutionary behavior of two taxa in a single assemblage, but might not allow for very precise analysis of anything resembling effective population size as measured by a population geneticist. There is therefore also a need to generate absolute abundance estimates from fossil assemblages, if and when it is possible.

3) Use absolute abundance, cautiously.—As in all attempts to extract biological information from fossil assemblages, taphonomic analysis must come first; in this case, the role of abundance or productivity of living individuals in creating fossil abundance must be assessed relative to that of sedimentological processes (Kidwell et al., 1986). Beyond this essential preliminary, the abundances of different species can be treated differently. As discussed above, almost all biological communities show a "hollow curve" pattern of species abundances. In fossil assemblages, the smaller number of more numerous species (the long tail on the right of such curves) tend to also be those that occur repeatedly in samples and at more localities (CoBabe and Allmon, 1994); indeed the fossil record is likely highly biased toward sampling such relatively abundant species (McKinney, 1997a, b). This suggests an approach to examining species abundance in fossils. If a species is among the most abundant, widespread, and regularly occurring species in an assemblage, then when this same species is found to be rare in other samples or localities, this may represent actual rarity rather than just a taphonomic artifact. This could be tested further by use of "taphonomic control" taxa in the same assemblage (*sensu* Jablonski and Bottjer, 1990).

4) Use occurrence or geographic range.—Because local abundance and total geographic range are so often closely linked (e.g., Gaston, 2003, and references therein), geographic range can be used as a proxy for local abundance and compared to evolutionary behavior (e.g., Jernvall and Fortelius, 2004). Jablonski and Roy (2003), for example, found that Late Cretaceous gastropod genera show a strong negative relation between geographic range of constituent species and speciation rate per species, per million years. It should

also in principle be possible to examine the role of abundance in particular speciation events in the fossil record by plotting changes in absolute or relative abundance or geographic range through a phylogeny (Fig. 4; Allmon, in press).

DISCUSSION/CONCLUSIONS

The future of speciation

Serious, scientifically-informed predictions about the future of evolution are in their infancy, but a few conclusions seem indisputable. In the short term, it is clear that we are dramatically accelerating anagenetic change in many taxa—from plants to bacteria to insects to salmon (Palumbi, 2001a, b). We are also rapidly "homogenizing" the world's biota, by facilitating the replacement of local biotas by non-indigenous and locally expanding species that can co-exist with humans (McKinney and Lockwood, 1999, 2001; Olden and Rooney, 2006). On longer time scales, the "Anthropocene mass extinction" (Jackson, 2008, p. 11,458) is clearly underway and may rival its famous K-T precursor, leading to comparable long-term evolutionary consequences (e.g., Jablonski, 2001). If, as discussed above, we are also altering patterns of speciation among those taxa that remain, then our macroevolutionary importance is even greater.

The question, however, is whether such a conclusion about speciation can go beyond the simple statement that humans are changing the course of future evolution? Does it provide any insights that might allow or encourage us to alter the situation? It is reasonable to expect that anthropogenic effects on speciation will not be evenly distributed among surviving lineages, but will be concentrated in particular clades or ecological types (Myers and Knoll, 2001, p. 5389). Does this mean we can predict which lineages, clades and types will be more affected and which less? The Speciation Cycle discussed here offers, if not clear predictions on particular taxa, then at least a potential roadmap for deriving such predictions.

Decreasing population size and geographic fragmentation of formerly abundant and more widely distributed forms will increase the rate of isolate formation, and so might—all things being equal—be expected to lead to higher rates of speciation (cf. Allmon, 1992; Allmon et al., 1998). All things, however, are not equal in this case, and the key is what happens *after*

TABLE 3.—Some specific features ("ecological attributes") that might cause species to have increased (or at least not decreased) rates of speciation in a human-dominated world.

Evolutionary attribute	Ecological attribute
1) High ability to survive in small patches/small populations	Broad ecological range ("generalists")
2) Ability to survive in human-dominated landscapes	Broad ecological range ("generalists")
3) Rapid genetic differentiation	Short generation time High mutation rate
4) Potential to expand in a difficult environment	High fecundity High dispersal capability

isolation. If population reduction continues (through continued direct taking, such as fishing of hunting, or due to continued habitat degradation), isolates will have lower probabilities of persisting long enough to differentiate sufficiently to become new species (Frankel and Soulé, 1981, p. 94). As discussed above, this already seems to be occurring, with high rates of disappearance of populations (e.g., Ehrlich and Daily, 1993; Ceballos and Ehrlich, 2002). Even if populations do survive long enough to diverge, however, success as a new species depends in most cases on subsequent expansion of population size (and usually also geographic range) (e.g., Jablonski and Roy, 2003; Liow and Stenseth, 2007; Martin et al., 2007). If many current trends of abundance and habitat alteration continue, this may also be unlikely, for example because habitat islands (such as parks and preserves) are unlikely to expand in size any time soon.

Thus, the species that might be expected to show increased (or at least not declining) rates of speciation in an increasingly human-dominated world might be those that can: 1) survive well in human-dominated environments, in small habitat patches, and/or at low population sizes; 2) differentiate genetically at rapid rates; and 3) retain the potential to expand post-divergence in a difficult environment. Some specific features that might allow species to fulfill these conditions are listed in Table 3. Conversely, species expected to show declining rates of speciation (and so declining evolutionary rates and futures) might be those with the reverse attributes (Table 4). As noted by McKinney and Lockwood (1999), species that benefit from human activities (and therefore have increasing abundances) are fewer (5-29%) than species that decline from human activities (>50%). A similarly detailed survey might indicate the relative proportions of species with these other characteristics.

At an even larger scale, human-caused reductions and redirections of global productivity and biomass may have significant effects on speciation in general. If less biological production is available for conversion into individual organisms, it seems likely that many species will experience reduced population growth even if they have not been specific victims of other anthropogenic changes. In this respect, the human-dominated world may be similar to the broad productivity declines that may have characterized post-mass extinction ecosystems (e.g., Zachos et al., 1989; Caldeira and Rampino, 1993; D'Hondt et al., 1998; Martin, 1998). This in turn suggests that we might look at the early diversification history of mass extinction recovery biotas for indications of not just which taxa will survive, but also which are most likely to show various levels of speciation in the near-term evolutionary future (e.g., Hansen et al., 1993; Hart, 1996; Erwin, 2001).

Other paleontological implications

The main argument of this chapter is that application of the Speciation Cycle to fossil lineages can yield insights of potential utility to mitigating human effects on future speciation. There are, however, potential insights flowing in the other direction that may assist paleontologists in interpreting fossil assemblages. The biological interpretation of fossil abundance needs more careful attention. Many high-density fossil assemblages (e.g., bone or shell beds) may have formed under conditions very different than those that exist at present, that is, in conditions of higher productiv-

TABLE 4.—Some specific features that might cause species to have decreased rates of speciation in a human-dominated world.

Evolutionary attribute	Ecological attribute
1) Low ability to survive in small patches/small populations	Narrow ecological range ("specialists") Subject to Allee effect Permanently small population size ("natural rarity")
2) Slow genetic differentiation	Long generation time
3) Low potential to expand in a difficult environment	Low fecundity Low dispersal capability

ity and higher abundance (e.g., Allmon, 1993, 2007), meaning that attempts to reconstruct their processes of formation that reference only modern conditions and processes may be misleading.

In addition, the conventional ecological wisdom of the ubiquity of the "hollow curve" species-abundance relationship (Fig. 1) that is commonly applied to analysis of fossil assemblages (e.g., Dodd and Stanton, 1990; CoBabe and Allmon, 1994; Zuschin et al., 2006) may have to be investigated more carefully, on a case-by-case basis. It may not be reasonable to assume that species-abundance distributions have always looked the same (cf., Wagner et al., 2006; McGill et al., 2007).

Directions for future research

As the other papers in this volume amply testify, exploration of the practical utility of the fossil record for conservation purposes is in only its beginning stages. It is therefore of particular importance to examine what avenues of research might be fruitful in the near future on the subject of the relationship between abundance and speciation.

1) Comparison of fossil and Recent species abundance distributions.—The hypothesis that late Holocene anthropogenic reductions in abundance are responsible for "shifting baselines" can be tested by comparing species abundance distributions of modern and taphonomically-appropriate fossil assemblages taken from comparable environments.

2) In living species, which ones are declining and which not?—This information is commonly collected for modern species, but mainly in the context of assessing their status with respect to the threat of extinction. Species abundance changes could also be compiled

and compared to characteristics such as those listed in Tables 3 and 4 in order to make preliminary predictions of likelihood of future speciation potential.

3) In fossils, measures of abundance, plotted against phylogenies.—The paleontological literature is increasingly filled with detailed, species-level phylogenies, and the Paleobiology Database is increasingly filled with abundance data. This would seem to present excellent opportunities for large-scale comparisons of cladogenesis and population sizes (Fig. 4), while giving due caution to the familiar issues of bias and incompleteness of the record.

FINAL THOUGHTS

Just because we know that we might be significantly affecting future evolution does not mean that we can or will change our collective behavior, or even that we should. Humans were well aware of the extermination of the bison (Isenberg, 2000) and the passenger pigeon (Schorger, 1955) while they were happening. Recognition of the ethical issues of our potential effects on evolution (e.g., Ehrlich, 2001) does not necessitate our altering course. Part of the problem, of course, is temporal scale. Even the fastest rates of speciation (ca. 12,000 years; Fryer, 2001) mean that the actual effects are incomprehensibly far in the future on the time scale of our sociopolitical decision-making (e.g., Barraclough and Davies, 2005). Why protect an evolutionary future that even our great-great-great grandchildren will never see?

In any case, the data and theory available, as discussed above, strongly support the conclusion that humans are already altering the patterns of species formation that will create the diversity of life in the future. The broader implications of this conclusion are at least two-fold. First, if society decides that it is a priority to be concerned about this, then science might be called upon to predict the likely direction(s) of these alterations in speciation, which we are currently not able to do. Further research would therefore be required. More importantly, however, society must engage in the conversation necessary to make such a decision on prioritization, and the chances of this seem even more remote. Yet unless and until such discussion and decisions take place, science's predictions of the future of evolutionary change on Earth will be little more than academic exercises. It therefore would seem that the highest priority for evolutionary biologists (including paleobiologists) interested in this issue is to do all we can to include consideration of long-term human influences on evolution in the ongoing (and increasingly urgent) discussion about our impact on the Earth's biota and environment.

ACKNOWLEDGMENTS

I am grateful to many people who have stimulated and endured my thinking on these topics over the years, especially Ernest Williams and Tom Lovejoy. For advice, discussion, and/or comments on previous drafts of the manuscript I am grateful to John Alroy, Greg Dietl, John Fitzpatrick, Austin Hendy, Jeremy Jackson, Tom Olszewski, Ken Rosenberg, and Jennifer Tegan. I also thank Kelly Cronin for help with the references, and Greg Dietl and Karl Flessa for allowing me to participate in the short course. This work has been supported in part by NSF grant EAR-0719642.

REFERENCES

ABRAMS, P. A. 1995. Monotonic or unimodal diversity-productivity gradients: What does competition theory predict? Ecology, 76(7):2019-2027.

ABRAMS, P. A. 2002. Will small population sizes warn us of impending extinctions? American Naturalist, 160(3):293-305.

AGER, D. V. 1963. Principles of Paleoecology. McGraw-Hill, New York, 371 p.

ALLEE, W. C. 1931. Animal Aggregations: A Study in General Sociology. University of Chicago Press, Chicago, Illinois, 431 p.

ALLEE, W. C., A. E. EMERSON, O. PARK, T. PARK, AND K. P. SCHMIDT. 1949. Principles of Animal Ecology. Saunders, Philadelphia, Pennsylvania, 837 p.

ALLENDORF, F. W., AND R.F. LEARY. 1986. Heterozygosity and fitness in natural populations of animals, p. 57-76. In M. Soulé (ed.), Conservation Biology. The Science of Scarcity and Diversity. Sinauer Associates, Sunderland, Massachusetts.

ALLMON, W. D. 1990. What are we doing to evolution? The World and I, 5(12):337-343.

ALLMON, W. D. 1992. A causal analysis of stages in allopatric speciation. Oxford Surveys in Evolutionary Biology, 8:219-257.

ALLMON, W. D. 1993. Age, environment and mode of deposition of the densely fossiliferous Pinecrest Sand (Pliocene of Florida): Implications for the role of biological productivity in shell bed formation. Palaios, 8:183-201.

ALLMON, W. D. 2004. A Leviathan of Our Own. The Tragic and Amazing Story of North Atlantic Right Whale #2030. Special Publication No. 25, Paleontological Research Institution, Ithaca, New York, 71 p.

ALLMON, W. D. 2005. Profusion and plenitude: what is the connection between abundance and speciation? North American Paleontology Convention, Abstracts. PaleoBios, 25(Supplement to no. 2):12-13.

ALLMON, W. D. 2007. Cretaceous marine nutrients, greenhouse carbonates, and the abundance of turritelline gastropods. Journal of Geology, 115(5):509-524.

ALLMON, W.D. In press. Testing the evolutionary effects of abundance in the fossil record. Geological Society of America, Abstracts with Program.

ALLMON, W. D., AND R. M. ROSS. 2001. Nutrients and evolution in the marine realm, p. 105-148. In W. D. Allmon and D. J. Bottjer (eds.), Evolutionary Paleoecology: The Ecological Context of Macroevolutionary Change. Columbia University Press, New York.

ALLMON, W. D., P. J. MORRIS, AND M. L. MCKINNEY. 1998. An intermediate disturbance hypothesis of maximal speciation, p. 349-376. In M. L. McKinney and J.A. Drake (eds.), Biodiversity Dynamics. Turnover of Populations, Taxa, and Communities. Columbia University Press, New York.

ALTER, S. E., E. RYNES, AND S. R. PALUMBI. 2007. DNA evidence for historic population size and past ecosystem impacts of gray whales. Proceedings of the National Academy of Sciences of the United States of America, 104(38):15162-15167.

ANDREWARTHA, H. G., AND L. C. BIRCH. 1954. The Distribution and Abundance of Animals. University of Chicago Press, Chicago, Illinois, 782 p.

APOLLONIO, S. 2002. Hierarchical Perspectives on Marine Complexities. Searching for Systems in the Gulf of Maine. Columbia University Press, New York, 229 p.

ARIÑO, A., AND S. L. PIMM. 1995. On the nature of population extremes. Evolutionary Ecology, 9:429-443.

ASHLEY, M. V., M. F. WILLSON, O. R. W. PERGAMS, D. J. O'DOWD, S. M. GENDE, AND J. S. BROWN. 2003. Evolutionarily enlightened management. Biological Conservation, 111:115-123.

ASKINS, R. A., J. F. LYNCH, AND R. GREENBERG. 1990. Population declines in migratory birds in eastern North America. Current Ornithology, 7:1-57.

AVISE, J. C. 2008. The history, purview, and future of conservation genetics, p. 5-15. In S. P. Carroll and C. W. Fox (eds.), Conservation Biology: Evolution in Action. Oxford University Press, New York.

BAILEY, S.-A., M. C. HORNER-DEVINE, G. LUCK, L. A. MOORE, K. M. CARNEY, S. ANDERSON, C. BETRUS, AND E. FLEISHMAN. 2004. Primary productivity and species richness: Relationships among functional guilds, residency groups and vagility classes at multiple spatial scales. Ecography, 27:207-217.

BAMBACH, R. K. 1993. Seafood through time: changes in biomass, energetics, and productivity in the marine ecosystem. Paleobiology, 19(3):372-397.

BAMBACH, R. K. 1999. Energetics in the global marine fauna: A connection between terrestrial diversification and change in the marine biosphere. Geobios, 32(2):131-144.

BARRACLOUGH, T. G., AND T. J. DAVIES. 2005. Predicting future speciation, p. 400-418. In A. Purvis, J. L. Gittleman, and T. M. Brooks (eds.), Phylogeny and Conservation. Cambridge University Press, Cambridge, UK.

BARTON, N. 1989. Founder effect speciation, p. 229-256. In D. Otte and J. A. Endler (eds.), Speciation and Its Consequences. Sinauer Associates, Sunderland, Massachusetts.

BARTON, N. 1996. Natural selection and random genetic drift as causes of evolution on islands. Philosophical Transactions of the Royal Society of London, B, 351:785-795.

BARTON, N., AND B. CHARLESWORTH. 1984. Genetic revolutions, founder effects and speciation. Annual Review of Ecology and Systematics, 15:133-164.

BAUM, J. K., AND R. A. MYERS. 2004. Shifting baselines and the decline of pelagic sharks in the Gulf of Mexico. Ecology Letters, 7:135-145.

BELL, G., AND S. COLLINS. 2008. Adaptation, extinction and global change. Evolutionary Applications, 1:3-16.

BLACKBURN, T. M., AND K. J. GASTON. 1997. Who is rare? Artefacts and complexities of rarity determination. In W. E. Kunin, W. E., and K. J. Gaston (eds.), The Biology of Rarity: Causes and Consequences of Rare-Common Differences. Chapman & Hall, London, UK.

BLOB, R. W., AND C. BADGLEY. 2007. Numerical methods for bonebed analysis, p. 333-396. In R. R. Rogers, D. A. Eberth, and A. R. Fiorillo (eds.), Bonebeds: Genesis, Analysis, and Paleobiological Significance. University of Chicago Press, Chicago, Illinois.

BOUCOT, A. J. 1975. Evolution and Extinction Rate Controls. Elsevier, Amsterdam, 427 p.

BOUCOT, A. J. 1981. Principles of Benthic Marine Paleoecology. Academic Press, New York, 463 p.

BOUZAT, J. L., H. A. LEWIN, AND K. N. PAIGE. 1998. The ghost of genetic diversity past: Historical DNA analysis of the greater prairie chicken. American Naturalist, 152:1-6.

BOWEN, B.W. 1999. Preserving genes, species, or ecosystems? Healing the fractured foundations of conservation policy. Molecular Ecology, 8:S5-S10.

BRADSHAW, W. E., AND C. M. HOLZAPFEL. 2008. Genetic response to rapid climate change: It's seasonal timing that matters. Molecular Ecology, 17:157-166.

BRINKMAN, D. B., D. A. EBERTH, AND P. J. CURRIE. 2007. From bonebeds to paleobiology: Applications of bonebed data, p. 221-264. In R. R. Rogers, D. A. Eberth, and A. R. Fiorillo (eds.), Bonebeds: Genesis, Analysis, and Paleobiological Significance. University of Chicago Press, Chicago, Illinois.

BROOKS, D. R., AND D. A. MCLENNAN. 1993. The Nature of Diversity. An Evolutionary Voyage of Discovery. University of Chicago Press, Chicago, Illinois, 668 p.

BROUGHTON, J. M. 2002. Pre-Columbian human impact on California vertebrates: Evidence from old bones and implications for wilderness policy, p. 44-71. In C. E. Kay and R. T. Simmons (eds.), Wilderness and Political Ecology. Aboriginal Influences and the Original State of Nature. University of Utah Press, Salt Lake City, Utah.

BROWN, J. H. 1981. Two decades of homage to Santa Rosalia: Toward a general theory of diversity. American Zoologist, 21:877-888.

BROWN, J. H. 2004. Toward a metabolic theory of ecology. Ecology, 85(7):1771-1789.

CALDEIRA, K., AND M. R. RAMPINO. 1993. Aftermath of the end-Cretaceous mass extinction: Possible biogeochemical stabilization of the carbon cycle and climate. Paleoceanography, 8(4):515-525.

CALVER, M., S. THOMAS, S. BRADLEY, AND H. MCCUTCHEON. 2007. Reducing the rate of predation on wildlife by pet cats: The efficacy and practicability of collar-mounted pounce protectors. Biological Conservation, 137:341-348.

CARDILLO, M., C. D. L. ORME, AND I. P. F. OWENS. 2005. Testing for latitudinal bias in diversification rates: An example using New World birds. Ecology, 86:2278-2287.

CARLTON, J. T. 1999. The scale and ecological consequences of biological invasions in the world's oceans. Population and Community Biology, 24:195-212

CARROLL, S. P., AND C. W. FOX, 2008. Conservation Biology. Evolution in Action. Oxford University Press, New

York, 380 p.

CASEY, J. M., AND R. A. MYERS. 1998. Near extinction of a large, widely distributed fish. Science, 281:690-692.

CEBALLOS, G., AND P. R. EHRLICH. 2002. Mammal population losses and the extinction crisis. Science, 296:904-907.

CHARLESWORTH, B. 2009. Effective population size and patterns of molecular evolution and variation. Nature Reviews Genetics, 10:195-205.

CHOWN, S. L. 1997. Speciation and rarity: Separating cause from consequence, p. 91-109. *In* W. E. Kunin, W. E., and K. J. Gaston (eds.), The Biology of Rarity: Causes and Consequences of Rare-Common Differences. Chapman & Hall, London, UK.

CHOWN, S. L., AND K. J. GASTON. 2000. Areas, cradles, and museums: The latitudinal diversity gradient in species richness. Trends in Ecology and Evolution, 15:311-315.

COBABE, E. A., AND W. D. ALLMON. 1994. Effects of sampling on paleoecologic and taphonomic analyses in high diversity fossil accumulations: An example from the Gosport Sand, Middle Eocene, Alabama. Lethaia, 27:167-178.

COLINVAUX, P. 1978. Why Big Fierce Animals Are Rare. An Ecologist's Perspective. Princeton University Press, Princeton, New Jersey, 256 p.

CONNELL, J. H., AND E. ORIAS. 1964. The ecological regulation of species diversity. American Naturalist, 98:399-414.

CONOVER, D. O., AND S. B. MUNCH. 2002. Sustaining fisheries yields over evolutionary time scales. Science, 297:94-96.

COURCHAMP, F., T. CLUTTON-BROCK, AND B. GRENFELL. 1999. Inverse density dependence and the Allee effect. Trends in Ecology and Evolution, 14(10):405-410.

COWLING, R. M., AND R. L. PRESSEY. 2001. Rapid plant diversification: Planning for an evolutionary future. Proceedings of the National Academy of Sciences of the United States of America, 98(10):5452-5457.

COX, G. W. 2004. Alien Species and Evolution: The Evolutionary Ecology of Exotic Plants, Animals, Microbes, and Interacting Native Species. Island Press, Washington, D.C., 400 p.

COYNE, J., AND A. ORR. 2004. Speciation. Sinauer Associates, Sunderland, Massachusetts, 545 p.

CRANDALL, K. A., O. R. P. BININDA-EMONDS, G. M. MACE, AND R. K. WAYNE. 2000. Considering evolutionary processes in conservation biology. Trends in Ecology and Evolution, 15(7):290-295.

CROWDER, L. B. 2005. Back to the future in marine conservation, p. 19-29. *In* E. A. Norse and L. B. Crowder (eds.), Marine Conservation Biology: The Science of Maintaining the Sea's Biodiversity. Island Press, Washington, D.C.

DARWIN, C. 1859. On the Origin of Species. John Murray,

London, 513 p.

DAVIS, E. B., M. S. KOO, C. CONROY, J. L. PATTON, AND C. MORITZ. 2008. The California Hotspots Project: Identifying regions of rapid diversification of mammals. Molecular Ecology, 17:120-138.

DELCOURT, P. A., AND H. R. DELCOURT. 2008. Prehistoric Native Americans and Ecological Change: Human Ecosystems in Eastern North America Since the Pleistocene. Cambridge University Press, New York, 216 p.

DELINE, B. 2009. The effects of rarity and abundance distributions on measurements of local morphological disparity. Paleobiology, 35(2):175-189.

DENEVAN, W. M. 1992. The pristine myth: The landscape of the Americas in 1492. Annals of the Association of American Geographers, 82(3):369-385.

D'HONDT, S., P. DONAGHAY, J. C. ZACHOS, D. LUTTENBERG, AND M. LINDINGER. 1998. Organic carbon fluxes and ecological recovery from the Cretaceous-Tertiary mass extinction. Science, 282:276-279.

DIAMOND, J. M. 1984. "Normal" extinctions of isolated populations, p. 191-246. *In* M. Nitecki, (ed.), Extinctions. University of Chicago Press, Chicago.

DIETL, G. P. 2009. Paleobiology and the conservation of the evolving web of life. *In* G. P. Dietl, and K. W. Flessa, (eds.), Conservation Paleobiology: Using the Past to Manage for the Future. The Paleontological Society Papers, 15 (this volume).

DOBYNS, H. F. 1983. Their Number Become Thinned: Native American Population Dynamics in Eastern North America. University of Tennessee Press, Knoxville, Tennessee, 378 p.

DODD, J. R., AND R. J. STANTON, JR. 1990. Paleoecology. Concepts and Applications. 2nd edition, Wiley, New York, 502 p.

DRUFFEL, E. R. M., AND B. H. ROBISON. 1999. Oceanography: Is the deep sea on a diet? Science, 284:1139-1140.

EHRLICH, P. R. 2001. Intervening in evolution: Ethics and actions. Proceedings of the National Academy of Sciences of the United States of America, 98(10):5477-5480.

EHRLICH, P. R., AND G. DAILY. 1993. Population extinction and saving biodiversity. Ambio, 22(2-3):64-68.

ELDREDGE, N. 1991. The Miner's Canary. Unraveling the Mysteries of Extinction. Prentice Hall, Englewood Cliffs, New Jersey, 246 p.

ELDREDGE, N., AND S. J. GOULD. 1972. Punctuated equilibria: An alternative to phyletic gradualism, p. 82-115. *In* T. J. M. Schopf, (ed.), Models in Paleobiology. Freeman, Cooper & Co., San Francisco, California.

ELLSTRAND, N. C., AND D. R. ELAM. 1993. Population genetic consequences of small population size: Implications for plant conservation. Annual Review of Ecology and Systematics, 24:217-242.

ELTON, C. 1927. Animal Ecology. Macmillan, New York,

207 p.

ENQUIST, B. J., M. A. JORDAN, AND J. H. BROWN. 1995. Connections between ecology, biogeography, and paleobiology: Relationship between local abundance and geographic distribution in fossil and recent molluscs. Evolutionary Ecology, 9:586-604.

ERWIN, D. H. 2001. Lessons from the past: Biotic recoveries from mass extinctions. Proceedings of the National Academy of Sciences of the United States of America, 98(10):5399-5403.

EVANS, F. C. 1956. Ecosystem as the basic unit in ecology. Science, 123:1127-1128.

EVANS, K. L., P. H. WARREN, AND K. J. GASTON. 2005a. Species-energy relationships at the macroecological scale: A review of the mechanisms. Biological Reviews, 80:1–25.

EVANS, K. L., J. J. D. GREENWOOD, AND K. J. GASTON. 2005b. Dissecting the species-energy relationship. Proceedings of the Royal Society of London, B, 272:2155-2163.

EVANS, K. L., S. F. JACKSON, J. J. D. GREENWOOD, AND K. J. GASTON. 2006. Species traits and the form of individual species-energy relationships. Proceedings of the Royal Society of London, B, 273:1779-1787.

FAITH D. P., C. A. M. REID, AND J. HUNTER. 2004. Integrating phylogenetic diversity, complementarity and endemism for conservation assessment. Conservation Biology, 18(1):255-261.

FEDOROV, A. A. 1966. The structure of the tropical rain forest and speciation in the humid tropics. Journal of Ecology, 54:1-11.

FENBERG, P. B., AND K. ROY. 2008. Ecological and evolutionary consequences of size-selective harvesting: How much do we know? Molecular Ecology, 17:209-220.

FERRIÈRE, R., U. DIECKMANN, AND D. COUVERT, 2004. Evolutionary Conservation Biology. Cambridge University Press, Cambridge, UK, 448 p.

FINNEGAN, S., AND M. DROSER. 2005. Relative and absolute abundance of trilobites and rhynchonelliform brachiopods across the Lower/Middle Ordovician boundary, eastern Basin and Range. Paleobiology, 31(3):480-502.

FLINKMAN, J., E. ARO, I. VUORINEN, AND M. VIITASALO. 1998. Changes in northern Baltic zooplankton and herring nutrition from 1980s to 1990s: Top down and bottom up processes at work. Marine Ecology Progress Series, 165:127-136.

FLORIN, A. -B., AND A. ÖDEEN. 2002. Laboratory environments are not conducive for allopatric speciation. Journal of Evolutionary Biology, 15:10-19.

FRANKEL, O. 1974. Genetic conservation: Our evolutionary responsibility. Genetics, 78:53-65.

FRANKEL, O., AND M. E. SOULÉ. 1981. Conservation and Evolution. Cambridge University Press, New York, 327 p.

FRANKLIN, I. R. 1980. Evolutionary change in small populations, p. 135-150. In M. E. Soulé and B. A. Wilcox (eds.), Conservation Biology. An Evolutionary-Ecological Perspective. Sinauer Associates, Sunderland, Massachusetts.

FRANZREB, K. E., AND K. V. ROSENBERG. 1997. Are forest songbirds declining? Status assessment from the southern Appalachians and northeastern forests. Transactions of the 62nd North American Wildlife and Natural Resources Conference, p. 264-279.

FREY, J. K. 1993. Modes of peripheral isolate formation and speciation. Systematic Biology, 42:373-381.

FRYER, G. 2001. On the age and origin of the species flock of haplochromine cichlid fishes of Lake Victoria. Proceedings of the Royal Society of London, B, 268:1147-1152.

FULLER, E. 2001. Extinct Birds. Revised edition, Comstock Publishing Associates/Cornell University Press, Ithaca, New York, 398 p.

FUTUYMA, D. J. 2005. Evolution. Sinauer Associates, Sunderland, Massachusetts, 603 p.

GALLAGHER, R., AND B. CARPENTER. 1997. Human dominated ecosystems. Science, 277:485.

GASCOIGNE, J., AND R. N. LIPCIUS. 2004. Allee effects in marine systems. Marine Ecology Progress Series, 269:49-59.

GASTON, K. J. 1994. Rarity. Chapman & Hall, London, 224 p.

GASTON, K. J. 1997. What is rarity?, p. 30-47. In Kunin, W. E., and K. J. Gaston (eds.), The Biology of Rarity: Causes and Consequences of Rare-Common Differences. Chapman & Hall, London, UK.

GASTON, K. J. 2003. The Structure and Dynamics of Geographic Ranges. Oxford University Press, Oxford, 280 p.

GASTON, K. J., AND T. M. BLACKBURN, 2000. Patterns and Process in Macroecology. Blackwell Science, Oxford, 377 p.

GASTON, K. J., AND S. CHOWN. 1999. Geographic range size and speciation, p. 236-272. In A. E. Magurran and R. M. May (eds.), Evolution of Biological Diversity. Oxford University Press, Oxford, UK.

GASTON, K. J., AND R. A. FULLER. 2007. Biodiversity and extinction: Losing the common and the widespread. Progress in Physical Geography, 31:213-225.

GASTON, K. J., AND R. A. FULLER. 2008. Commonness, population depletion and conservation biology. Trends in Ecology and Evolution 23(1):14-19.

GASTON, K. J., AND W. E. KUNIN. 1997. Rare-common differences: An overview, p. 13-29. In W. E. Kunin and K. J. Gaston (eds.), The Biology of Rarity: Causes and Consequences of Rare-Common Differences. Chapman & Hall, London, UK.

GAVRILETS, S. 2004. Fitness Landscapes and the Origin of

Species. Princeton University Press, Princeton, New Jersey, 476 p.

GAVRILETS, S., AND A. HASTINGS. 1996. Founder effect speciation: A theoretical reassessment. American Naturalist, 147(3):466-491.

GIDDINGS, L. V., K. Y. KANESHIRO, AND W. W. ANDERSON, 1989. Genetics, Speciation and the Founder Principle. Oxford University Press, Oxford, UK, 373 p.

GIENAPP, P., C. TEPLITSKY, J. S. ALHO, J. A. MILLS, AND J. MERILA. 2008. Climate change and evolution: Disentangling environmental and genetic responses. Molecular Ecology, 17:167-178.

GILL, F. B. 2006. Ornithology. 3rd ed. W.H. Freeman, New York, 720 p.

GOW, J. L., C. L. PEICHEL, AND E. B. TAYLOR. 2006. Contrasting hybridization rates between sympatric threespine sticklebacks highlight the fragility of reproductive barriers between evolutionarily young species. Molecular Ecology, 15:739-752.

GRANT, P.R. 2001. Reconstructing the evolution of birds on islands: 100 years of research. Oikos, 92:385-403.

GRANT, P. R., AND B. R. GRANT. 1997. Genetics and the origin of bird species. Proceedings of the National Academy of Sciences of the United States of America, 94:7768-7775.

GRANT, P. R., AND R. GRANT. 2008. How and Why Species Multiply: The Radiation of Darwin's Finches. Princeton University Press, Princeton, New Jersey, 272 p.

GRASSO, G. 2008. What appeared limitless plenty: The rise and fall of the nineteenth-century Atlantic halibut fishery. Environmental History, 13(1):66-91.

GRAYSON, D. K. 2001. The archaeological record of human impacts on animal populations. Journal of World Prehistory, 15(1):1-68.

HABERL, H., K. -H. ERB, AND F. KRAUSMANN. 2008. Global human appropriation of net primary production (HANPP). In C. J. Cleveland (ed.), Encyclopedia of Earth. Environmental Information Coalition, National Council for Science and the Environment, Washington, D.C. Available from: www.eoearth.org/article/Global_human_appropriation_of_net_primary_production_(HANPP).

HALLIDAY, T. R. 1980. The extinction of the passenger pigeon Ectopistes migratorius and its relevance to contemporary conservation. Biological Conservation, 17:157-167.

HALPERN, B. S., S. WALBRIDGE, K. A. SELKOE, C. V. KAPPEL, F. MICHELI, C. D'AGROSA, J. F. BRUNO, K. S. CASEY, C. EBERT, H. E. FOX, R FUJITA, D. HEINEMANN, H. S. LENIHAN, E. M. P. MADIN, M. T. PERRY, E. R. SELIG, M. SPALDING, R. STENECK, AND R. WATSON. 2008. A global map of human impact on marine ecosystems. Science, 319:948–952.

HANSEN, T. A., B. R. FARRELL, AND B. UPSHAW, III. 1993. The first two million years after the Cretaceous-Tertiary boundary in east Texas: Rate and paleoecology of the molluscan recovery. Paleobiology, 19:251-265.

HARD, J. J., M. R. GROSS, M. HEINO, R.HILBORN, R. G. KOPE, R. LAW, AND J. D. REYNOLDS. 2008. Evolutionary consequences of fishing and their implications for salmon. Evolutionary Applications, 1:388-408.

HARRISON, S., J. H. VIERS, J. H. THORNE, AND J. B. GRACE. 2008. Favorable environments and the persistence of naturally rare species. Conservation Letters, 1:65-74.

HART, M. B. 1996. Biotic Recovery from Mass Extinction Events. Geological Society Special Publication 102, 392 p.

HAYNES, G. 2009 American Megafaunal Extinctions at the End of the Pleistocene. Springer, Berlin, 202 p.

HEINO, M., AND U. DEICKMANN 2008. Detecting fisheries-induced life-history evolution: An overview of the reaction-norm approach. Bulletin of Marine Science, 83(1):69-93.

HELFMAN, G. S. 2007. Fish Conservation. A Guide to Understanding and Restoring Global Aquatic Biodiversity and Fishery Resources. Island Press, Washington, D.C., 584 p.

HENDRY, A. P., AND M. T. KINNISON. 1999. The pace of modern life: Measuring rates of contemporary microevolution. Evolution, 53(6):1637-1653.

HENDRY, A. P., P. R. GRANT, B. R. GRANT, H. A FORD, M. J. BREWER, AND J. PODOS. 2006. Possible human impacts on adaptive radiation: Beak size bimodality in Darwin's finches. Proceedings of the Royal Society of London, B, 273:1887-1894.

HILDEBRANDT, W. R., AND T. L. JONES. 2002. Depletion of prehistoric pinniped populations along the California and Oregon coasts: Were humans the cause? p. 72-110. In C. E. Kay and R. T. Simmons (eds.), Wilderness and Political Ecology: Aboriginal Influences and the Original State of Nature. University of Utah Press, Salt Lake City, Utah.

HOBBS, R. J., AND H. A. MOONEY. 2008. Broadening the extinction debate: Population deletions and additions in California and Western Australia. Conservation Biology, 12:271-283.

HUBBELL, S. P., AND R. B. FOSTER. 1986. Biology, chance, and history and the structure of tropical rain forest tree communities, p. 314-329. In J. Diamond and T. J. Case (eds.), Community Ecology. Harper and Row, New York.

HUGHES, L. 2000. Biological consequences of global warming: Is the signal already apparent? Trends in Ecology and Evolution, 15(2):56-61.

HUGHES, J. B., G. C. DAILY, AND P. R. EHRLICH. 1997. Population diversity: Its extent and extinction. Science, 278:689-692.

HUNT, B. P. V., E. A. PAKHOMOV, AND C. D. MCQUAID. 2001. Short-term variation and long-term changes in the oceanographic environment and zooplankton community in the vicinity of a sub-Antarctic archipelago. Marine Biology, 138:369-381.

HUTCHINGS, J. A., AND D. J. FRASER. 2008. The nature of fisheries- and farming-induced evolution. Molecular Ecology, 17:294-313.

HUTCHINGS, J. A., AND R. A. MYERS. 1994. What can be learned from the collapse of a renewable resource? Atlantic cod, *Gadus morhua*, of Newfoundland and Labrador. Canadian Journal of Fisheries and Aquatic Sciences, 51:2126-2146.

HUTCHINSON, G. E. 1959. Homage to Santa Rosalia: Why are there so many kinds of animals? American Naturalist, 93:145-159.

ICES (INTERNATIONAL COUNCIL FOR THE EXPLORATION OF THE SEA). 2007. Report of the Study Group on Fisheries Induced Adaptive Change (SGFIAC), 26 February-2 March 2007, Lisbon, Portugal. ICES CM 2007/RMC:03, 25 p.

ISENBERG, A. C. 2000. The Destruction of the Bison. An Environmental History, 1750-1920. Cambridge University Press, New York, 206 p.

JABLONSKI, D. 2001. Lessons from the past: Evolutionary impacts of mass extinctions. Proceedings of the National Academy of Sciences of the United States of America, 98(10):5393-5398.

JABLONSKI, D., AND D. J. BOTTJER. 1990. Onshore-offshore trends in marine invertebrate evolution, p. 21-75. *In* R. M. Ross and W. D. Allmon (eds.), Causes of Evolution. A Paleontological Perspective. University of Chicago Press, Chicago, Illinois.

JABLONSKI, D., AND K. ROY. 2003. Geographical range and speciation in fossil and living molluscs. Proceedings of the Royal Society of London, B, 270:401-406.

JACKSON, J. A., N. J. PATENAUDE, E. L. CARROLL, AND C. S. BAKER. 2008. How few whales were there after whaling? Inference from contemporary mtDNA diversity. Molecular Ecology, 17:236-251.

JACKSON, J. B. C. 1997. Reefs since Columbus. Coral Reefs, 16(Supplement):S23-S32.

JACKSON, J. B. C. 2001. What was natural in the coastal oceans? Proceedings of the National Academy of Sciences of the United States of America, 98(10):5411-5418

JACKSON, J. B. C. 2006. When ecological pyramids were upside down, p. 27-37. *In* J. A. Estes, D. P. Demaster, D. F. Doak, T. M. Williams, and R. L. Brownell, Jr. (eds.), Whales, Whaling, and Ocean Ecosystems. University of California Press, Berkeley, California.

JACKSON, J. B. C. 2008. Ecological extinction and evolution in the brave new ocean. Proceedings of the National Academy of Sciences of the United States of America, 105:11458-11465.

JACKSON, J. B. C. AND L. MCCLENACHAN. 2009. Historical ecology for the paleontologist. *In* G. P. Dietl, and K. W. Flessa, (eds.), Conservation Paleobiology: Using the Past to Manage for the Future. The Paleontological Society Papers, 15 (this volume).

JACKSON, J. B. C., AND D. H. ERWIN. 2006. What can we learn about ecology and evolution from the fossil record? Trends in Ecology and Evolution, 21(6):322-328.

JACKSON, J. B. C., AND E. SALA. 2001. Unnatural oceans. Scientia Marina, 65(Suppl. 2):273-281.

JACKSON, J. B. C., M. X. KIRBY, W. H. BERGER, K. A. BJORNDAL, L. W. BOTSFORD, B. J. BOURQUE, R. H. BRADBURY, R. COOKE, J. ERLANDSON, J. A. ESTES, T. P. HUGHES, S. KIDWELL, C. B. LANGE, H. S. LENIHAN, J. M. PANDOLFI, C. H. PETERSON, R. S. STENECK, M. J. TEGNER, AND R. R. WARNER. 2001. Historical overfishing and the recent collapse of coastal ecosystems. Science, 293:629-638.

JERNVALL, J., AND M. FORTELIUS. 2002. Common mammals drive the evolutionary increase of hypsodonty in the Neogene. Nature, 417:538-540.

JERNVALL, J., AND M. FORTELIUS. 2004. Maintenance of trophic structure in fossil mammal communities: Site occupancy and taxon resilience. American Naturalist, 164(5):614-624.

KAY, C. E. 1994. Aboriginal overkill: The role of Native Americans in structuring western ecosystems. Human Nature, 5(4):359-398.

KENNEY, R. D. 2002. North Atlantic, North Pacific, and Southern Right Whales, p. 806-813. *In* W. F. Perrin, B. Wursig, and J. G. M. Thewissen (eds.), Encyclopedia of Marine Mammals. Academic Press, San Diego, California.

KIDWELL, S. M. 2001. Preservation of species abundance in marine death assemblages. Science, 294:1091-1094.

KIDWELL, S. M., F. T. FURSICH, AND T. AIGNER. 1986. Conceptual framework for the analysis and classification of fossil concentrations. Palaios, 1:228-238.

KIMURA, M. 1983. The Neutral Theory of Molecular Evolution. Cambridge University Press, Cambridge, UK, 384 p.

KINNISON, M. T., AND N. G. HAIRSTON, JR. 2007. Eco-evolutionary conservation biology: Contemporary evolution and the dynamics of persistence. Functional Ecology, 21:444-454.

KNOWLTON, N., AND J. B. C. JACKSON. 2008. Shifting baselines, local impacts, and climate change on coral reefs. PLoS Biology, 6:e54.

KURLANSKY, M. 1997. Cod. A Biography of the Fish That Changed the World. Knopf, New York, 294 p.

LANDE, R. 1998. Anthropogenic, ecological and genetic factors in extinction, p. 29-52. *In* G. M. Mace, A. Balmford, and J. R. Ginsberg (eds.), Conservation in a Changing World. Cambridge University Press, Cambridge, UK.

LAW, R. 2007. Fisheries-induced evolution: Present status and future directions. Marine Ecology Progress Series, 335:271-277.

LAW, R., AND K. STOKES. 2005. Evolutionary impacts of fishing on target populations, p. 232-246. *In* E. A. Norse and L. B. Crowder (eds.), Marine Conservation Biology. The Science of Maintaining the Sea's Biodiversity. Island Press, Washington, D.C.

LAWTON, J. H. 1993. Range, population abundance and conservation. Trends in Ecology and Evolution, 8:409-413.

LEIGH, E. G. 1965. On the relation between the productivity, biomass, diversity, and stability of a community. Proceedings of the National Academy of Sciences of the United States of America, 53:777-783.

LEIGH, E. G. 1981. The average lifetime of a population in a varying environment. Journal of Theoretical Biology, 90(22):213-239.

LIERMANN, M., AND R. HILBORN. 1997. Depensation in fish stocks: A hierarchic Bayesian meta-analysis. Canadian Journal of Fisheries and Aquatic Sciences, 54:1976-1984.

LINDEMAN, R. 1942. The trophic-dynamic aspect of ecology. Ecology, 23: 399-418.

LIOW, L. H., AND N. C. STENSETH. 2007. The rise and fall of species: Implications for macroevolutionary and macroecological studies. Proceedings of the Royal Society, London, B, 274:2745-2752.

LOCKWOOD, J., M. HOOPES, AND M. MARCHETTI. 2006. Invasion Ecology. University of Arizona Press, Tucson, Arizona, 312 p.

LOTKA, A. J. 1925. Elements of Physical Biology. Williams and Wilkins, Baltimore, Maryland, 495 p.

LOTT, D. F. 2002. American Bison. A Natural History. University of California Press, Berkeley, California, 229 p.

LOTZE, H. K., AND B. WORM. 2009. Historical baselines for large marine animals. Trends in Ecology and Evolution, 24:254-262.

LOVEJOY, A. O. 1936, The Great Chain of Being. Harvard University Press, Cambridge, Massachusetts, 382 p.

LOVEJOY, T. E., AND L. HANNAH. 2005. Climate Change and Biodiversity. Yale University Press, New Haven, Connecticut, 440 p.

LUCK, G. W. 2007. The relationship between net primary productivity, human population density and species conservation. Journal of Biogeography, 34:201-212.

MACARTHUR, R. H. 1972. Geographical ecology. Harper & Row, New York, 269 p.

MACE, G., AND M. KERSHAW. 1997. Extinction risk and rarity on an ecological timescale, p. 130-149. *In* W. Kunin and K. J. Gaston, (eds.), The Biology of Rarity, Chapman and Hall, London.

MANIRE, C. A., AND S. H. GRUBER. 1990. Many sharks may be headed toward extinction. Conservation Biology, 4(1):10-11.

MANN, C. C. 2005. 1491: New Revelations of the Americas Before Columbus. Alfred A. Knopf, New York, 480 p.

MARTIN, P. R., F. BONIER, AND J. J. TEWKSBURY. 2007. Revisiting Jablonski (1993): Cladogenesis and range expansion explain latitudinal variation in taxonomic richness. Journal of Evolutionary Biology, 20:930-936.

MARTIN, P. S. 1984. Prehistoric overkill: The global model, p. 354-403. *In* P. S. Martin and R. G. Klein (eds.), Quaternary Extinctions: A Prehistoric Revolution. University of Arizona Press, Tucson, Arizona.

MARTIN, P. S. 2002. Prehistoric extinctions: In the shadow of man, p. 1-27. *In* C. E. Kay and R. T. Simmons (eds.), Wilderness and Political Ecology. Aboriginal Influences and the Original State of Nature. University of Utah Press, Salt Lake City, Utah.

MARTIN, R. E. 1996. Secular increase in nutrient levels through the Phanerozoic: Implications for productivity, biomass, and diversity of the marine biosphere. Palaios, 11:209-220.

MARTIN, R. E. 1998. Catastrophic fluctuations in nutrient levels as an agent of mass extinction: Upward scaling of biological processes?, p. 405-429. *In* M. L. McKinney and J. A. Drake (eds.), Biodiversity Dynamics: Turnover of Populations, Taxa, and Communities. Columbia University Press, New York.

MARTIN, R. E. 2003. The fossil record of biodiversity: Nutrients, productivity, habitat area and differential preservation. Lethaia, 36:179-194.

MARTIN, R. E., AND W. D. ALLMON. In press. Seafood through time revisited: The Phanerozoic increase in marine trophic resources and its macroevolutionary consequences. Geological Society of America, Abstracts with Program.

MAURER, B. A., AND S. G. HEYWOOD. 1993. Geographic range fragmentation and abundance in Neotropical migratory birds. Conservation Biology, 7(3):501-509.

MAURER, B. A., AND M. P. NOTT. 1998. Geographic range fragmentation and the evolution of biodiversity, p. 31-50. *In* M. L. McKinney and J. A. Drake (eds.), Biodiversity Dynamics: Turnover, Populations, Taxa, and Communities. Columbia University Press, New York.

MAY, R. M. 1979. Fluctuations in abundance of tropical insects. Nature, 278:505-507.

MAYR, E. 1954. Change of genetic environment and evolution, p. 157-180. *In* J. Huxley, A. C. Hardy, and E. B. Ford (eds.), Evolution as a Process. Allen and Unwin, London, UK.

MAYR, E. 1963. Animal Species and Evolution. Harvard University Press, Cambridge, Massachusetts, 797 p.

MCCLENACHAN, L., J. B. C. JACKSON, AND M. J. H. NEWMAN.

2006. Conservation implications of historic sea turtle nesting beach loss. Frontiers in Ecology and Environment, 4:290-296.

McCLURE, M. M., S. M. CARLSON, T. J. BEECHIE, G. R. PESS, J. C. JORGENSEN, S. M. SOGARD, S. E. SULTAN, D. M. HOLZER, J. TRAVIS, B. L. SANDERSON, M. E. POWER, AND R. W. CARMICHAEL. 2008. Evolutionary consequences of habitat loss for Pacific anadromous salmonids. Evolutionary Applications, 1:300-318.

McGILL, B. J., R. S. ETIENNE, J. S. GRAY, D. ALONSO, M. J. ANDERSON, H. K. BENECHA, M. DORNELAS, B. J. ENQUIST, J. L. GREEN, F. HE, A. H. HURLBERT, A. E. MAGURRAN, P. A. MARQUET, B.A. MAURER, A. OSTING, C. U. SOYCAN, K. I. UGLAND, AND E. P. WHITE. 2007. Species abundance distributions: Moving beyond single prediction theories to integration within an ecological framework. Ecology Letters, 10:995-1015.

McKINNEY, M. L. 1997a. The biology of fossil abundance. Revista Espanola de Paleontologia, 11(2):125-133.

McKINNEY, M. L. 1997b. How do rare species avoid extinction? A paleontological view, p. 110-129. In W. E., Kunin and K. J. Gaston (eds.), The Biology of Rarity: Causes and Consequences of Rare-Common Differences. Chapman & Hall, London, UK.

McKINNEY, M. L., AND J. L. LOCKWOOD. 1999. Biotic homogenization: A few winners replacing many losers in the next mass extinction. Trends in Ecology and Evolution, 14:450-453.

McKINNEY, M. L., AND J. L LOCKWOOD. 2001. Biotic Homogenization. Kluwer Academic/Plenum Publishers, New York, 289 p.

McKINNON, J. S., AND H. D. RUNDLE. 2002. Speciation in nature: The threespine stickleback model systems. Trends in Ecology and Evolution, 17(10):480-488.

MELTZER, D. J. 2009. First Peoples in a New World. Colonizing Ice Age America. University of California Press, Berkeley, California, 446 p.

MITTLEBACH, G. G., C. F. STEINER, S. M. SCHEINER, K. L. GROSS, H. L. REYNOLDS, R. B. WAIDE, M. R. WILLIG, S. I. DODSON, AND L. GOUGH. 2001. What is the observed relationship between species richness and productivity? Ecology, 82(9):2381-2396.

MONTGOMERY, D. R. 2003. King of Fish. The Thousand-Year Run of Salmon. Westview Press, Boulder, Colorado, 290 p.

MOONEY, H. A., AND E. E. CLELAND. 2001. The evolutionary impact of invasive species. Proceedings of the National Academy of Sciences of the United States of America, 98:5446-5451.

MOWAT, F. 1984. Sea of Slaughter. Atlantic Monthly Press, Boston, Massachusetts, 438 p.

MURRAY, B. R., AND B. J. LEPSCHI. 2004. Are locally rare species abundant elsewhere in their geographical range? Austral Ecology, 29:287-293.

MURRAY, B. R., B. L. RICE, D. A. KEITH, P. J. MYERSCOUGH, J. HOWELL, A. G. FLOYD, K. MILLS, AND M. WESTOBY. 1999. Species in the tail of rank-abundance curves. Ecology, 80:1806-1816.

MYERS, N. 1985. A look at the present extinction spasm and what it means for the future evolution of species, p. 47-58. In R. J. Hoage (ed.), Animal Extinctions. What Everyone Should Know. Smithsonian Institution Press, Washington, D.C.

MYERS, N. 1986. Tropical deforestation and a mega-extinction spasm, p. 394-409. In M. E. Soule (ed.), Conservation Biology: The Science of Scarcity and Diversity. Sinauer Associates, Sunderland, Massachusetts, 394-409.

MYERS, N. 1987. The extinction spasm impending: Synergisms at work. Conservation Biology, 1(1):14-21.

MYERS, N., AND A. H. KNOLL. 2001. The biotic crisis and the future of evolution. Proceedings of the National Academy of Sciences of the United States of America, 98(10):5389-5392.

MYERS, R. A., AND B. WORM. 2005. Extinction, survival or recovery of large predatory fishes. Philosophical Transactions of the Royal Society of London, B, 360:13-20.

NEE, S., AND R. M. MAY. 1997. Extinction and the loss of evolutionary history. Science, 278:692-694.

NEWTON, I. 2003. The Speciation and Biogeography of Birds. Academic Press, Amsterdam, 668 p.

NICHOLLS, S. 2009. Paradise Found: Nature in America at the Time of Discovery. University of Chicago Press, Chicago, Illinois, 524 p.

ODUM, E. 1971. Fundamentals of Ecology. 3rd edition, Saunders, Philadelphia, Pennsylvania, 574 p.

OHTA, T. 1972. Population size and the rate of evolution. Journal of Molecular Evolution, 1:305-314.

OLDEN, J. D., AND T. P. ROONEY. 2006. On defining and quantifying biotic homogenization. Global Ecology and Biogeography, 15:113-120.

PALUMBI, S. R. 2001a. Humans as the world's greatest evolutionary force. Science, 293:1786-1790.

PALUMBI, S. R. 2001b. The Evolution Explosion. How Humans Cause Rapid Evolutionary Change. W. W. Norton, New York, 277 p.

PANDOLFI, J. M., J. B. C. JACKSON, N. BARON, R. H. BRADBURY, H. M. GUZMAN, T. P. HUGHES, C. V. KAPPEL, F. MICHELI, J. C. OGDEN, H. P. POSSINGHAM, AND E. SALA. 2005. Are U.S. coral reefs on the slippery slope to slime? Science, 307:1725-1726.

PARMESAN, C. 2006. Ecological and evolutionary responses to recent climate change. Annual Review of Ecology and Systematics, 37:637-669.

PAULAY, G., AND C. MEYER. 2002. Diversification in the tropical Pacific: Comparisons between marine and terrestrial systems and the importance of founder speciation. Integrative and Comparative Biology, 42:922-934.

PAULY, D. 1995. Anecdotes and the shifting baseline syndrome of fisheries. Trends in Ecology and Evolution, 10:430.

PAULY, D., AND V. CHRISTENSEN. 1995. Primary production required to sustain global fisheries. Nature, 374:255-257.

PAULY, D., AND J. MACLEAN. 2003. In a Perfect Ocean: The State of Fisheries and Ecosystems in the North Atlantic Ocean. Island Press, Washington, D.C., 175 p.

PEARSON, J. C. 1972. The Fish and Fisheries of Colonial North America. A Documentary History of the Fishery Resources of the United States and Canada. National Technical Information Service, U.S. Dept. of Commerce, Rockville, Maryland, 9 vols., 3090 p.

PETERSEN, C. W., AND D. R. LEVITAN. 2001. The Allee effect: A barrier to recovery by exploited species, p. 281-300. In J. D. Reynolds, G. M. Mace, K. H. Redford, and J. G. Robinson (eds.), Conservation of exploited species. Cambridge University Press, Cambridge, UK.

PETERSON, R. T. 1941. How many birds are there? Audubon Magazine, 43:179-187.

PIMM, S., AND P. RAVEN. 2000. Biodiversity: Extinction by numbers. Nature, 403:843-845.

PIMM, S. L., J. M. DIAMOND, T. M. REED, G. J. RUSSELL, AND J. VERNER. 1993. Times to extinction for small populations of large birds. Proceedings of the National Academy of Sciences of the United States of America, 90:10871-10875.

POLICANSKY, D. 1993. Fishing as a cause of evolution in fishes, p. 2-18. In T. K. Stokes, J. M. McGlade, and R. Law (eds.), The Exploitation of Evolving Resources. Sinauer Associates, Sunderland, Massachusetts.

PYENSON, N. D., R. B. IRMIS, J. H. LIPPS, L. G. BARNES, E. D. MITCHELL, JR., AND S. A. MCLEOD. 2009. Origin of a widespread marine bonebed deposited during the middle Miocene Climatic Optimum. Geology, 37(6):519-522.

REBOULET, S., F. GIRAUD, AND O. PROUX. 2005. Ammonoid abundance variations related to changes in trophic conditions across the Oceanic Anoxic Event 1d (latest Albian, SE France). Palaios, 20:121-141.

REDDING, D. W., AND A. Ø. MOOERS. 2006. Incorporating evolutionary measures into conservation prioritization. Conservation Biology, 20(6):1670-1678.

REZNICK, D. N., AND C. K. GHALAMBOR. 2005. Can commercial fishing cause evolution? Answers from guppies (Poecilia reticulata). Canadian Journal of Fisheries and Aquatic Sciences, 62:791-801.

REZNICK, D.N., C. K. GHALAMBOR, AND K. CROOKS. 2008. Experimental studies of evolution in guppies: A model for understanding the evolutionary consequences of

predator removal in natural communities. Molecular Ecology, 17:97-107.

RICK, T. C., AND J. M. ERLANDSON. 2008. Human Impacts on Ancient Marine Ecosystems: A Global Perspective. University of California Press, Berkeley, California, 320 p.

RICKLEFS, R. E. 1970. Stage of taxon cycle and distribution of birds on Jamaica, Greater Antilles. Evolution, 24:475-477.

RICKLEFS, R. E., AND E. BERMINGHAM. 1999. Taxon cycles in the Lesser Antillean avifauna. Ostrich, 70:49-59.

RICKLEFS, R. E., AND E. BERMINGHAM. 2002. The concept of the taxon cycle in biogeography. Global Ecology and Biogeography 11:353-361.

RICKLEFS, R. E., AND G. W. COX. 1972. Taxon cycles in the West Indian avifauna. American Naturalist 106:195-219.

RICKLEFS, R. E., AND G. W. COX. 1978. Stage of taxon cycle, habitat distribution, and population density in the avifauna of the West Indies. American Naturalist, 112:875-895.

RIDLEY, M. 1996. Evolution. 2nd edition, Blackwell Science, Cambridge, Massachusetts, 670 p.

ROBERTS, C. 2007. The Unnatural History of the Sea. Island Press, Washington, D.C., 435 p.

ROBERTS, C. M., AND J. P. HAWKINS. 1999. Extinction risk in the sea. Trends in Ecology and Evolution, 14(6):241-246.

ROBERTSON, R. 2005. Large conchs (Strombus) are endangered herbivores having many predators and needing dense populations of adults to reproduce successfully. American Conchologist, 33(3):3-7.

ROBBINS, C. S., J. R. SAUER, R. GREENBERG, AND S. DROEGE. 1989. Population declines in North American birds that migrate to the tropics. Proceedings of the National Academy of Sciences of the United States of America, 86:7658-7662.

RODRIGUEZ, J. P. 2002. Range contraction in declining North American bird populations. Ecological Applications, 12(1): 238-248.

ROEMMICH, D., AND J. MCGOWAN. 1995. Climate warming and the decline of zooplankton in the California current. Science, 267:1324-1326.

ROMAN, J., AND S. R. PALUMBI. 2003. Whales before whaling in the North Atlantic. Science, 301:508-510.

ROSENZWEIG, M. L. 1975. On continental steady states of species diversity, p. 121-140. In M. L. Cody and J. M. Diamond (eds.), The Ecology of Species Communities. Harvard University Press, Cambridge, Massachusetts.

ROSENZWEIG, M. L. 1995. Species Diversity in Space and Time. Cambridge University Press, New York, 436 p.

ROSENZWEIG, M. 2001. Loss of speciation rate will impoverish future diversity. Proceedings of the National Academy of Sciences of the United States of America,

98(10):5404-5410.

Rosenzweig, M. L., and Z. Abramsky. 1993. How are diversity and productivity related?, p. 52-65. In R. Ricklefs and D. Schluter (eds.), Species Diversity in Ecological Communities: Historical and Geographical Perspectives. University of Chicago Press, Chicago, Illinois.

Roughgarden, J., and S. Pacala. 1989. Taxon cycle among *Anolis* lizard populations: Review of evidence, p. 403-432. In D. Otte and J. A. Endler (eds.), Speciation and Its Consequences. Sinauer Associates, Sunderland, Massachusetts.

Sáenz-Arroyo, A., C. M. Roberts, J. Torre, M. M. Cariño-Olivera, and R. Enríquez-Andrade. 2005. Rapidly shifting environmental baselines among fishers of the Gulf of California. Proceedings of the Royal Society of London, B, 272:1957-1962.

Sáenz-Arroyo, A., C.M. Roberts, J. Torre, M. Cariño-Olivera, and J. P. Hawkins. 2006. The value of evidence about past abundance: Marine fauna of the Gulf of California through the eyes of 16th to 19th century travelers. Fish and Fisheries, 7:128-146.

Safina, C. 2007. Voyage of the Turtle: In Pursuit of the Earth's Last Dinosaur. Henry Holt, New York, 400 p.

Sambatti, J. B. M., E. Stahl, and S. Harrison. 2008. Metapopulation structure and the conservation consequences of population fragmentation, p. 50-67. In S. P. Carroll and C. W. Fox (eds.), Conservation Biology: Evolution in Action. Oxford University Press, New York.

Schorger, A. W. 1955. The Passenger Pigeon: Its Natural History and Extinction. University of Wisconsin Press, Madison, Wisconsin, 424 p.

Seehausen, O. 2006. Conservation: Losing biodiversity by reverse speciation. Current Biology, 16(9):R334-R337.

Seehausen, O., G. Takimoto, D. Roy, and J. Jokela. 2008. Speciation reversal and biodiversity dynamics with hybridization in changing environments. Molecular Ecology, 17:30-44.

Shell, H. R. 2002. Introduction: Finding the soul in the skin, p. viii-xxiii. In W. T. Hornaday, The Extermination of the American Bison. Smithsonian Institution Press, Washington, D.C.

Sims, D. W., and P. C. Reid. 2002. Congruent trends in long-term zooplankton decline in the north-east Atlantic and basking shark fishery catches off west Ireland. Fisheries Oceanography, 11:59-63.

Singer, M. C., C. D. Thomas, and C. Parmesan. 1993. Rapid human-induced evolution of insect-host associations. Nature, 366:681-683.

Slatkin, M. 1996. In defense of founder-flush theories of speciation. American Naturalist, 147(4):493-505.

Smith, T. B., and L. Bernatchez. 2008. Evolutionary change in human-altered environments. Molecular Ecology, 17(1):1-8.

Smith, T. B., L. A. Freed, J. Kaimanu Lepson, J. H. Carothers. 1995. Evolutionary consequences of extinctions in populations of a Hawaiian honeycreeper. Conservation Biology, 9(1):107-113.

Soulé, M. E. 1980. Thresholds for survival: Maintaining fitness and evolutionary potential, p. 151-170. In M. E. Soulé and B. A. Wilcox (eds.), Conservation Biology: An Evolutionary-Ecological Perspective. Sinauer Associates, Sunderland, Massachusetts.

Soulé, M. E., and B. A. Wilcox. 1980. Conservation Biology: An Evolutionary-Ecological Perspective. Sinauer Associates, Sunderland, Massachusetts, 395 p.

Spotila, J. R. 2004. Sea Turtles: A Complete Guide to Their Biology, Behavior, and Conservation. The Johns Hopkins University Press, Baltimore, Maryland, 227 p.

Stanley, S. M. 1979. Macroevolution. W.H. Freeman, San Francisco, California, 332 p.

Stanley, S. M. 1986. Population size, extinction, and speciation: The fission effect in Neogene Bivalvia. Paleobiology 12:89-110.

Stephens, P. A., W. J. Sutherland, and R. P. Freckleton. 1999. What is the Allee effect? Oikos, 87:185-190.

Stokes, T. K., and R. Law. 2000. Fishing as an evolutionary force. Marine Ecology Progress Series, 208:307-309.

Suarez, A. V., and N. D. Tsutsui. 2008. The evolutionary consequences of biological invasions. Molecular Ecology, 17:351-360.

Swetnam, T. W., C. D. Allen, J. L. Betancourt. 1999. Applied historical ecology: Using the past to manage for the future. Ecological Applications, 9(4):1189-1206.

Symonds, M. R. E., L. Christidis, and C. N. Johnson. 2006. Latitudinal gradients in abundance, and the causes of rarity in the tropics: A test using Australian honeyeaters (Aves: Meliphagidae). Oecologia, 149:406-417.

Tegner, M. J., L. V. Basch, and P. K. Dayton. 1996. Near extinction of an exploited marine invertebrate. Trends in Ecology and Evolution, 11:278-280.

Templeton, A. R., R. J. Robertson, J. Brisson, and J. Strasburg. 2001. Disrupting evolutionary processes: The effect of habitat fragmentation on collared lizards in the Missouri Ozarks. Proceedings of the National Academy of Sciences of the United States of America, 98(10):5426-5432.

Terborgh, J. 1989. Where Have All the Birds Gone? Princeton University Press, Princeton, New Jersey, 207 p.

Tilman, D., and C. Lehman. 2001. Human-caused environmental change: Impacts on plant diversity and evolution. Proceedings of the National Academy of Sciences of the United States of America, 98(10):5433-5440.

Tuite, M. L., Jr. 2009. Linking terrestrial biogeochemistry to declining rates of origination in Middle and Late Devonian seas. 9th North American Paleontological Convention Abstracts. Cincinnati Museum Center Scientific Contributions, 3:5.

UYEDA, J. C., S. J. ARNOLD, P. A. HOHENLOHE, AND L. S. MEAD. 2009. Drift promotes speciation by sexual selection. Evolution, 63(3):583-594.

VALENTINE, J. W. 1971. Resource supply and species diversity patterns. Lethaia, 4:51-61.

VEIT, R. R., J. A. McGOWAN, D. G. AINLEY, T. WAHL, AND P. PYLE. 1997. Apex marine predator declines ninety percent in association with changing oceanic climate. Global Change Biology, 3:23-28.

VERHEYE, H. M. 2000. Decadal scale trends across several marine trophic levels in the southern Benguela upwelling system off South Africa. Ambio, 29:30-34.

VERMEIJ, G. J. 1987. Evolution and Escalation: An Ecological History of Life. Princeton University Press, Princeton, New Jersey, 544 p.

VERMEIJ, G. J. 1995. Economics, volcanoes, and Phanerozoic revolutions. Paleobiology 21(2):125-152.

VERMEIJ, G. J., AND G. HERBERT. 2004. Measuring relative abundance in fossil and living assemblages. Paleobiology, 30(1):1-4.

VITOUSEK, P. M., H. A. MOONEY, J. LUBCHENCO, AND J. M. MELILLO. 1997. Human domination of earth's ecosystems. Science, 277:494-499.

WAGNER, P. J., M. A. KOSNIK, AND S. LIDGARD. 2006. Abundance distributions imply elevated complexity of post-Paleozoic marine ecosystems. Science, 314:1289-1291.

WAIDE, R. B., M. R. WILLIG, C. F. STEINER, G. MITTELBACH, L. GOUGH, S. I. DODSON, G. P. JUDAY, AND R. PARMENTER. 1999. The relationship between productivity and species diversity. Annual Review of Ecology and Systematics, 30:257-300.

WALLACE, A. R. 1878. Tropical Nature, and Other Essays. Macmillan & Co., London, 356 p.

WAPLES, R. S., R. W. ZABEL, M. D. SCHEUERELL, AND B. L. SANDERSON. 2008. Evolutionary responses by native species to major anthropogenic changes to their ecosystems: Pacific salmon in the Columbia River hydropower system. Molecular Ecology, 17:84-96.

WARD, P. D. 1994. The End of Evolution. On Mass Extinctions and the Preservation of Biodiversity. Bantam, New York, 301 p.

WEBB, T. J., AND K. J. GASTON. 2000. Geographic range size and evolutionary age in birds. Proceedings of the Royal Society of London, B, 267:1843-1850.

WEBB, T .J., M. KERSHAW, AND K. J. GASTON. 2000. Rarity and phylogeny in birds, p. 61-84. In J. L. Lockwood and M. L. McKinney (eds.), Biological Homogenization. Kluwer Academic/Plenum Publishers, New York.

WELLS, S. M. 1997. Giant Clams: Status, Trades and Mariculture, and the Role of CITES in Management. World Conservation Union (IUCN), ix+77 p.

WESTERN, D. 2001. Human-modified ecosystems and fu-ture evolution. Proceedings of the National Academy of Sciences of the United States of America, 98(10):5458-5465.

WHITE, T. C. R. 2008. The role of food, weather and climate in limiting the abundance of animals. Biological Reviews, 83:227-248.

WHITTAKER, R. H. 1975. Communities and Ecosystems. 2nd edition, MacMillan, New York, 385 p.

WHITTAKER, R. H., AND E. HEEGAARD. 2003. What is the observed relationship between species richness and productivity? Comment. Ecology, 84(12):3384-3390.

WILCOVE, D. 2007. No Way Home: The Decline of the World's Great Animal Migrations. Island Press, Washington, D.C., 256 p.

WILCOVE, D., D. ROTHSTEIN, J. DUBOW, A. PHILLIPS, AND E. LOSOS. 1998. Quantifying threats to imperiled species in the United States. Bioscience, 48:607-615.

WILKINSON, A. 2006. The lobsterman: How Ted Ames turned oral history into science. The New Yorker, 82(23):56-65.

WILSON, E. O. 1961. The nature of the taxon cycle in the Melanesian ant fauna. American Naturalist, 95:169-193.

WILSON, E. O. 1992. The Diversity of Life. Harvard University Press, Cambridge, Massachusetts, 424 p.

WITT, M. J., B. BAERT, A. C. BRODERICK, A. FORMIA, J. FRETEY, A. GIBUDI, C. MOUSSOUNDA, G. A. M. MOUNGUENGUI, S. NGOUESSONO, R. J. PARNELL, D. ROUMET, G. -P. SOUNGUET, B. VERHAGE, A. ZOGO, AND B. J. GODLEY. 2009. Aerial surveying of the world's largest leatherback turtle rookery: A more effective methodology for large-scale monitoring. Biological Conservation, 142(8):1719-1727.

WOODRUFF, D. S. 2001. Declines of biomes and biotas and the future of evolution. Proceedings of the National Academy of Sciences of the United States of America, 98(10):5471-5476.

WOOLFIT, M., AND L. BROMHAM. 2005. Population size and molecular evolution on islands. Proceedings of the Royal Society, B, 272:2277-2282.

WORM, B., R. HILBORN, J. K. BAUM, T. A. BRANCH, J. S. COLLIE, C. COSTELLO, M. J. FOGARTY, E. A. FULTON, J.A. HUTCHINGS, S. JENNINGS, O. P. JENSEN, H. K. LOTZE, P. M. MACE, T. R. McCLANAHAN, C. MINTO, S. R. PALUMBI, A. M. PARMA, D. RICARD, A. A. ROSENBERG, R. WATSON, AND D. ZELLER. 2009. Rebuilding global fisheries. Science, 325:578-585.

WRIGHT, D. H. 1983. Species-energy theory: An extension of species-area theory. Oikos, 41:496-506.

WRIGHT, D. H., D. J. CURRIE, AND B. A. MAURER. 1993. Energy supply and patterns of species richness on local and regional scales, p. 66-74. In R. Ricklefs and D. Schluter (eds.), Species Diversity in Ecological Communities. University of Chicago Press, Chicago, Illinois.

WRIGHT, S. 1931. Evolution in Mendelian populations. Genetics, 16:97-159.

YOUNG, K. A. 2004. Toward evolutionary management: Lessons from salmonids, p. 359-376. *In* A. P. Hendry and S. C. Stearns, (eds.), Evolution Illuminated. Salmon and Their Relatives. Oxford University Press, Oxford.

ZACHOS, J., M. ARTHUR, AND W. DEAN. 1989. Geochemical evidence for suppression of marine pelagic productivity at the Cretaceous/Tertiary boundary. Nature, 337:61-64.

ZUSCHIN, M., M. HARZHAUSER, AND K. SAUERMOSER. 2006. Patchiness of local species richness and its implication for large-scale diversity patterns: An example from the Middle Miocene of the Paratethys. Lethaia, 39:65-78.

SECTION THREE

Conservation Paleobiology at Work

IN THIS section, we explore conservation paleobiology at work. As the discipline has developed its academic underpinnings and approaches, it has begun to put those approaches to work.

Like conservation biology, one of our parent disciplines, conservation paleobiology is also an applied science—we seek to use geohistorical information to inform decisions about the preservation of biodiversity, habitats, and ecosystem services (Dietl, 2016). And, as in other sciences, the application of our approaches depends on the body of theory, knowledge, and techniques developed both by basic research and in practice. The first two sections of this book explore the foundations of conservation paleobiology: the many approaches to extracting and understanding the conservation-relevant information that exists in the geohistorical record.

Of course, many of the chapters in the previous sections also draw on real-world applications to illustrate the power of the past in solving present-day problems. John Smol (Chapter 2) demonstrates how geohistorical information in lake deposits has proven especially effective in conservation. Similarly, Stephen Jackson, Stephen Gray, and Bryan Shuman (Chapter 4) show how the tree-ring, packrat midden, and lake records in western North America are directly pertinent to resource management. This final section of the book, however, provides more local and more personal accounts of conservation paleobiology at work.

If it is sometimes said that all politics are local, the same might be said of conservation issues. The problems are local not so much because of their geographic extent—just consider the area covered by Australia's Great Barrier Reef—but because political jurisdictions define the sort of conservation solutions that are available. Consider that changes in land use and changes in climate are probably the two major threats to biodiversity, habitats, and ecosystem services. Changes in land use (and associated effects of economic development and water management) are often obvious at the local and regional scale and are often amenable to mitigation and management under existing regulatory policies. The global nature of climate change has made mitigation and adaptation difficult because of the many jurisdictions involved.

Chapter 14, in this section, recounts Karl Flessa's personal scientific journey from taphonomy to conservation paleobiology—a journey that took place in one location: the Colorado River delta. He shows how many of the standard tools in the paleontologist's tool kit were brought to bear on documenting the impact of human activities, deciphering the biology of now-threatened species, and providing reference conditions for estuarine restoration. His story is also a lesson in serendipity—of being in the right place, at the right time, and among the right people. Since this chapter was written, a new agreement between the United States and Mexico has resulted in the first environmental flow to the delta, the biotic and hydrologic effects of which are being monitored (Flessa et al., 2013). While the studies of pre-dam estuarine mollusks and fish that Flessa reviews played only a minor role in this recent development, they do demonstrate how conservation paleobiology can inform policy by working with non-governmental organizations and agencies.

Chapter 15 in this section is a roundtable discussion by five practicing conservation paleobiologists: Alison Boyer, Mark Brenner, David Burney, John Pandolfi, and Michael Savarese. The five panelists in this virtual roundtable share some of their experiences and in doing so may provide some practical advice for the student or professional interested in the applications of conservation paleobiology. The panelists are at different points in their careers and work on different systems: two (Boyer and Burney) decipher the cave record of island biota, two (Pandolfi and Savarese) look at marine or estuarine systems, and one (Brenner) examines the record preserved in lake sediments.

We then reflect on the similarities and differences expressed in the panelists' responses to our fifteen questions.

REFERENCES

DIETL, G. P. 2016. Brave new world of conservation pale-obiology. Frontiers in Ecology and Evolution, 4:21, doi:10.3389/fevo.2016.00021.

FLESSA, K. W., E. P. GLENN, O. HINOJOSA-HUERTA, C. A. DE LA PARRA-RENTERIA, J. RAMIREZ-HERNANDEZ, J. C. SCHMIDT, AND F. ZAMORA-ARROYO. 2013. Flooding the Colorado River Delta: A landscape-scale experiment. EOS, 94:485–486.

PUTTING THE DEAD TO WORK: TRANSLATIONAL PALEOECOLOGY

KARL W. FLESSA

Department of Geosciences, University of Arizona, Tucson, AZ 85721 USA

ABSTRACT.—Paleontological and paleoecological techniques can be enormously effective in understanding the biology of species at risk and in reconstructing their habitats. In the case of the Colorado River delta and its species, the toolkit is a familiar one: taphonomy, taxonomy, geochronology, sclerochronology and geochemistry. The accumulations of skeletal remains and sediments in the delta and estuary convincingly demonstrate that habitats, species composition, relative abundances and individual growth rates in the delta and estuary have been profoundly affected by the nearly complete anthropogenic diversion of the river's water. Establishing a pre-diversion baseline documents an environmental impact not otherwise demonstrable because of the lack of an observational record. Such a baseline also provides goals for species recovery and habitat restoration. Translating paleoecological information into conservation action requires collaborations with agencies and environmental organizations. Paleontologists who want to put the dead to work need to expand their usual tool kit to include environmental law and policy, need to respond to the concerns of agencies and environmental groups, and need to work with the media.

INTRODUCTION

PALEONTOLOGISTS ARE famous for being able to reconstruct ancient life forms, communities and environments with the scrappiest bits of information imaginable—a toe bone here, a broken shell there, a few grams of carbon here, a few stable isotopes there. And voila, the dead are brought back to life. Habitats are restored, in sketches or museum dioramas, at any rate.

How can paleontologists help to bring nearly dead species back to real life? How can they aid in the restoration of habitats? That's the focus of Conservation Paleobiology. The objects of study are the remains of species at risk of extinction and the sedimentary and geochemical record of their habitats. The techniques of study are familiar (or nearly so to most of us in this room): taxonomy, geochronology, sclerochronology, geochemistry, trace fossil analysis, paleoecology, and on and on. And the objectives of study are the recovery of species populations now at risk and the restoration of their critical habitats.

The objects of study—at risk species and habitats—are familiar. We see them and read and hear about them often in the public media and, if you belong to certain scientific societies or environmental NGOs (non-government organizations), in professional journals and requests for donations. Some of those at-risk species have left a record of their existence—in museum collections, in explorers' notebooks, in archaeological middens, and in natural accumulations of pollen, shells and bones.

And while we sometimes bemoan the difficulties of working with dead materials, they have two real advantages in the current context:

1. Many remains date from before the species were at risk. Thus, the biology and habitat requirements of those individuals reflect pre-impact conditions. Live individuals, on the other hand, may no longer be typical of the once-thriving population and may be living in marginal conditions.

2. The difficulty of work with endangered species is compounded by the fact that they are endangered. This is because the research itself may add to the risk of extinction if it disturbs the behavior, biology or habitat of the species in any way. Working with already-dead or long-dead specimens is certainly non-invasive and does not threaten any live individuals. The dead are both convenient and safe.

The techniques of study are also familiar—though paleontologists usually think of their application in so-called "deep time", rather than the "near time" that is more pertinent to the preservation and restoration of species at risk today.

That leaves us with the objectives of the effort—the recovery of species at risk and the restoration of their habitats. How do we best translate what we do in the field and in the lab into policy and into action? How do we put the dead to work? How do we become—

In *Conservation Paleobiology: Using the Past to Manage for the Future, Paleontological Society Short Course, October 17th, 2009. The Paleontological Society Papers, Volume 15*, Gregory P. Dietl and Karl W. Flessa (eds.). Copyright © 2009 The Paleontological Society.

should we become—advocates for a particular cause? We don't just want to study nature, we want to save it. How can we be effective?

FROM TAPHONOMY TO CONSERVATION PALEOBIOLOGY

Here's my story of how I got into this game. It may be typical in its serendipity. The story is intended to be instructive.

After many years and many students working on the tidal flats near a rapidly developing coastal town in Sonora, Mexico, I was getting bored with the same old place—and the same old place just wasn't the same anymore: roads were getting closed off, camping became more restricted and there were just too many people around. So I was looking for new places to explore. So I asked a faculty colleague who had just gotten her pilot's license—Judy Parrish—if she could fly me over the Colorado River delta. She was up for the adventure so up we went.

I had my socks knocked off—miles and miles of mudflats; white, empty beaches and a sparkling ocean without a jet ski in sight. There was so little vegetation that I could imagine ourselves flying across an Ordovician coastline. The flight convinced me that I had to see this place on the ground.

So the next means of transport was a Chevy Suburban (Flessa et al., 1992), together with Sally Walker and Michal Kowalewski. On that trip we learned a lot: the beaches—islands at high tide—were made of shells—billions of shells, there was no shame in getting stuck in the mud, and you could get around fairly easily by paying local fishermen to take you in their boats. We founded the Centro de Estudios de Almejas Muertas during that trip.

So here it was, a great new setting in which to do some more taphonomic research: the old live-dead comparison game. I knew how to do that. But I'm not sure exactly when we realized that we had a problem. Remember, we had beaches made up of billions of shells—mostly of one species, as it turns out. The problem was that we hadn't found any live specimens.

Archimedes is supposed to have exclaimed "Eureka!" (Greek for "I found it!") upon discovering the principle of displacement while in the bath. He then ran naked through the streets of Syracuse (not the city in upstate New York) proclaiming his insight.

My personal "Eureka" moment was in fact more of a "Well, duh!" moment. Lots of dead, but no live? Okay, the Colorado delta environment had changed. The river was no longer delivering 16 billion cubic meters of fresh water and tons of sediment and nutrients to the Gulf of California. It hadn't done so since about 1960. I did not run naked down the beach.

Those billons of dead shells were from an earlier time, from another era: the Predambrian. The era before dams and diversions prevented the Colorado River from reaching the sea.

We could use those dead shells (and other remains) to tell us what the delta was like before the construction of upstream dams and water diversions. In a part of the world where there were few if any scientific observations made before large scale human alteration of the environment, this was going to be the best way to establish an environmental baseline—a baseline to estimate the environmental impact of human activity. And maybe we could figure out how much river water it would take to restore some of these habitats.

This task is easy to say, more difficult to do. We needed to make the case.

LEARNING EXPERIENCES

Geochronology

How old were the shells? If we were right, the shells would date from before the era of large-scale water diversions on the Colorado River. That era began with the completion of Hoover Dam in 1935. Thompson's (1968) heroic dissertation on the delta reported raw radiocarbon dates based on bulk samples of shells. But those were problematic for two reasons: 1. dating more than one shell, when the shells are likely time-averaged, produces something like an average; 2. Thompson didn't have any estimate of the reservoir effect for the Gulf of California—the effect of the upwelling of old carbon from deep water, resulting then in an apparent age potentially much greater than the actual age of the shell.

Lucky for us, the technology of radiocarbon dating had advanced significantly since the 1960's. The advent of accelerator mass spectrometry (AMS)-dating allowed dating small portions of single shells. Bowman (1990) is a wonderful primer on radiocarbon dating in general and most AMS labs provide what you need to know about sampling, sample preparation and interpretation on their websites (e.g., http://www.physics.arizona.edu/ams/, http://www.physics.purdue.edu/primelab/, http://www.radiocarbon.com/international.htm) and converting radiocarbon dates to calendar-year dates (i.e., calibration), is easily done via software packages available on the Web (e.g., http://calib.qub.ac.uk/calib/, or, quickly online http://www.calpal-online.de/).

We estimated the reservoir effect for Gulf of California shells by either dating shells of known collection age, such as a shell collected in Steinbeck and Rickett's expedition to the Sea of Cortez (Steinbeck, 1951) and shells and associated charcoal from Indian middens (Goodfriend and Flessa, 1997). The difference between the collection age and the radiocarbon age equals the reservoir effect. The difference between the age of the charcoal (atmospheric radiocarbon) and the age of the associated shell (marine radiocarbon) equals the reservoir effect. Reservoir effects vary throughout the ocean—your area may already have a known reservoir effect. See http://intcal.qub.ac.uk/marine/ for a very useful database of marine reservoir ages.

But a problem remained: AMS dates are costly. While there's a substantial discount for those lucky enough to be supported by NSF, the list price for a single date for a non-profit academic institution or museum is $475 (http://www.physics.arizona.edu/ams/service/fee.htm). What can be done when facing the prospect of dating many, many shells?

Radiocarbon-calibrated amino acid racemization has proven very useful because it's much less expensive (about $35/sample (http://jan.ucc.nau.edu/~dsk5/AAGL/services/price.html)) and, for most purposes, just as accurate as radiocarbon dates within the past 1,000 years. See Kowalewski et al. (1998, 2000), Kidwell et al. (2005), Best et al. (2007), and Kosnik et al. (2007, 2008) for examples and applications. Darrell Kaufman's lab at Northern Arizona University (http://jan.ucc.nau.edu/~dsk5/AAGL/) will run samples for you.

Live: dead comparisons

We needed to prove that the live fauna differed significantly from the dead. While we had a good understanding of the composition, distribution and relative abundance of the shelly remains on the delta (Kowalewski et al., 1994; Kowalewski and Flessa, 1995), we did not have a comparable understanding of the live

fauna. Early results from the surveys of Avila-Serrano et al. (2006) enabled a comparison with the dead fauna (Kowalewski et al., 2000). The fauna had indeed changed dramatically. *Mulinia coloradoeneis*, the once dominant bivalve, endemic to the delta, was now rare. Indeed, abundance may have decreased by as much as 90%, with no other shelly species taking its place. Circumstantial evidence—the great abundance of *M. colorodoensis* in the time before water diversions and its rarity since then—suggested that the lack of river flow was responsible for the decreased benthic productivity. Could we make a better case with more direct forensic evidence?

Stable isotopes and sclerochronology

Most applications of stable isotopes of oxygen in paleoecology concern the use of $\delta^{18}O$ values in carbonates as indicators of temperature change. See, for example Norris and Corfield (1998). Indeed, when the isotopic composition of the ambient water is known, then the growth temperature of the shell can be inferred from the $\delta^{18}O$ signature of the shell's $CaCO_3$. It's an extraordinarily useful technique.

So consider than when the growth temperature and the $\delta^{18}O$ of the shell are known, the isotopic composition of the water can be inferred. And in the case of the Colorado River delta and estuary, the isotopic composition of the water is a function of the amount and isotopic composition of the Colorado River water entering the mixing zone of seawater and river water.

Or think of it like this: As water evaporates from the surface of the ocean (the source of most of the atmosphere's water vapor), relatively more of the isotopically "light" $H_2^{16}O$ becomes water vapor; slightly more of the isotopically "heavy" $H_2^{18}O$ is left behind. As water vapor moves inland, some of it rains out, with the $H_2^{18}O$ preferentially becoming precipitation. Water vapor—and the resulting precipitation—deep in continental interiors is relatively depleted in $H_2^{18}O$, and relatively enriched in $H_2^{16}O$.

Because most of the Colorado River's water comes from snowpack on the western slope of the Rocky Mountains, its water is relatively depleted in $H_2^{18}O$, and relatively enriched in $H_2^{16}O$. While still true today, its $\delta^{18}O$ was even lower before dams and reservoirs—as demonstrated by the very low $\delta^{18}O$ values in the shell of a freshwater mussel collected from the river in the 1880s and reposited to the Smithsonian Institution

(Dettman et al., 2004).

Arriving at the mouth of the river, this isotopically distinct river water mixes with normal seawater of the Gulf of California. Mollusks, and other carbonate-secreting organisms record this mix. Portions of shells strongly depleted in ^{18}O—those with a very low δ^{18}O signature—indicate the presence of Colorado River water during growth, assuming, of course, the same growth temperature.

δ^{18}O in biogenic aragonite also varies with temperature. And temperature varies seasonally. Thus, the need for sclerochronology (Jones and Quitmyer, 1996; Schöne et al., 2002). Micro-sampling a shell through a year of its growth in the absence of any river water—the condition today—reveals the temperature-driven cycle of δ^{18}O (Goodwin et al., 2001). Variation in δ^{18}O values below these values indicate the influx of Colorado River water. δ^{18}O values below the lowest temperature-driven values are unambiguous indicators of some Colorado River water in the mixing zone.

Shells of pre-dam *M. coloradoensis* are characterized by much lower values of δ^{18}O than shells of live-collected *M. coloradoensis* or live-collected *Chione* (*Chionista*) *cortezi*, shells that grew in the absence of any Colorado River water. When *M. coloradoensis* was abundant, it grew in brackish water. Today, when *M. coloradoensis* is rare, it grows in 100% seawater (Rodriguez et al., 2001a, b)—more evidence for implicating the lack of river flow as a factor in the dramatic drop in productivity and change in relative abundance.

Note, however, that not all rivers are as isotopically distinct as the Colorado River. Low-lying coastal rivers that derive their water from coastal storms may not differ as strongly from the adjacent seawater. Rivers that drain continental interiors are best suited for this approach. For information on isotopic variation in precipitation, see http://www-naweb.iaea.org/napc/ih/GNIP/IHS_GNIP.html#.

Clams lack charisma

While some are economically important, *M. coloradoensis* is not. So when asked by a reporter: "Who cares about clams?", I struggled to point out that clams once constituted an important part of the delta's food web. Maybe people would care about the animals that ate the clams. So to prove that the clams were on the diet of other, perhaps more charismatic species, we demonstrated that a large percentage of shells were ei-

ther drilled by predatory gastropods or nibbled to death by predatory crabs (Cintra-Buenrostro et al., 2005).

A fish out of water

The same isotopic and sclerochronologic techniques can be applied to otoliths—the aragonite "ear bones" of fish. Fish biologists have long used the increments in otoliths to estimate age (Campana and Moksness, 1991) and are also employing geochemical techniques to infer habitat (e.g., Ingram and Weber, 1999; Campana and Thorrold, 2001).

The Colorado River estuary is spawning and nursery habitat for *Totoaba macdonaldi*, an endangered species (listed in both Mexico and the U.S.A.). The fish was once highly prized by sport fishers and commercial fishing operations. Its population crashed in the 1960s and 1970s, just as Colorado River flow diminished greatly. Fishing is now prohibited but the species has yet to recover.

Rowell et al. (2008) compared growth rates and δ^{18}O values in five otoliths from live-caught *Tototaba macdonaldi* (no Colorado River flow) and to those in five otoliths from archaeological middens dating from well before upstream dams and diversions (Colorado River flow). They found that pre-dam Totoaba grew faster in their first year than do present-day individuals. Because sexual maturity is size-dependent in this fish, pre-dam Tototaba reached sexual maturity from two to three years earlier than fish today. This difference in life history would have resulted in larger populations before upstream dams and diversions. This analysis suggests that fisheries management alone (i.e., the ban on catches) will not suffice in restoration of the species. Habitat restoration through river management (i.e., release of water to the estuary) is also necessary.

Next steps

While documenting environmental damage is important—especially when no baseline had been previously established—it is not enough. We are now working to prescribe the amount of water that would be needed to restore some of the delta and estuary's habitats. This is not an easy number to estimate and we want to get it right. Negotiating for water from an already over-allocated, bi-national river is difficult enough. Doing so when you don't know how much you need is impossible.

FROM CONSERVATION PALEOBIOLOGY TO CONSERVATION ACTION

Training as a paleontologist doesn't include training in law, policy or the rough-and-tumble world of western water. So what's the best way to get the word out and to influence the future?

Agency scientists and people at environmental NGOs don't read the journals that we usually publish in. What works in getting their attention is media coverage. A press release and subsequent media coverage of Kowalewski et al.'s (2000) *Geology* paper got us the attention that we needed. Smart people at NGOs will have a Google News Alert that keys on locations or species that are of particular interest to them. That's how our work got noticed by Environmental Defense.

Agency officials tend to be overworked and well-entrained in the sort of data that they are willing to consider. The value of paleobiological information can be a hard sell to an agency biologist. What they might have learned about the fossil record in their college and graduate school classes is that the record is incomplete and biased, and that fossils are incomplete and transported, etc., etc. They probably haven't heard that the fossil record is, under the right circumstances, a pretty-good archive of the species, populations and habitats of the past (Kidwell and Flessa, 1996; Kidwell, 2001; NRC, 2005) and that differences between live and dead shelly assemblages are reliable indicators of anthropogenic impact (Kidwell, 2007).

In my experience, the organizations who will be most interested in such work will be the environmental NGOs. They may share the same skepticism about the reliability of skeletal remains, but they tend to be less-invested in the status quo and more likely to be willing to take a chance. They also know how to best use the information you provide. This may make you nervous, so think carefully about how much you would like to engage in the activities of environmental NGOs. And keep in mind that they are not all the same. Some are mainstream and middle of the road (e.g., The Nature Conservancy), some are engaged in so-called community-level projects (e.g., The Sonoran Institute in the western U.S. and many local and regional groups elsewhere), some lobby the government or work with agencies to change policies (e.g., Environmental Defense), and some are aggressively litigious (e.g., The Center for Biological Diversity).

You need to find your comfort zone. Do you want to work behind the scenes supplying data or advocate for a particular outcome? A thoughtful treatment of the ways in which scientists can participate in policy is Pielke's (2007) *The Honest Broker*. Peilke identifies four potential roles for scientists: Pure Scientist, Science Arbiter, Honest Broker of Policy Alternatives, and Issue Advocate, in increasing order of active engagement of issues.

What about scientific objectivity? Being in the pocket of an environmental organization is just as bad as being in the pocket of an oil company when it comes to speaking out on environmental issues. We do indeed need to be advocates for the facts before we can advocate for a particular cause. But I am also cognizant of Howard Zinn's (2004) point that you can't be neutral on a moving train. If you believe that society is moving toward greater environmental degradation, deciding to take no action (so-called "neutrality") means that you have accepted the outcome. Objectivity in science does not require neutrality in policy decisions.

Things I've learned (that I should have known already):

1. Associate with smart, skilled people. Our delta work required diverse talents—from the help of fishermen in the field to isotope geochemists in the lab; from specialists in remote sensing to professors of water law; and from smart students.

2. Museum collections can be extraordinarily valuable and curators are willing to help, even if some part of the specimen will be destroyed. We depended on specimens from the U.S. Museum of Natural History to help us figure out reservoir effects, provide a proof of concept for the use of oxygen isotopes as indicators of river flow, and for figuring out what the isotopic composition of the Colorado River was before upstream dams and diversions. We depended on the Scripps Institution fo Oceanography's collections to get pre-diversion otoliths so that we could reconstruct growth rates in an endangered fish.

3. Engage with agencies and NGOs. To be effective, academic scientists need to step back from telling other people what it is they need to know. Take the time to find out what scientific information they say they need. They probably know. Go to their meetings—don't expect them to come to ours. Spend part of your next sabbatical at an environmental group or agency. Get internships for your students at NGOs.

4. Learn the law and policy. Agency scientists and NGOs work within an existing framework of environmental law and policy. Do you know your NEPA? Does an EA and FONSI mean anything to you? Do riparian rights or prior appropriation apply in your state? Sit in on classes at your institution. Find out for yourself how science could inform law and policy. And find out how law and policy can inform science. You don't need to become an attorney or a policy wonk, but you do need to keep from saying or doing something really stupid. But the more you know, the more useful you can be.

5. Learn how to deal with the media. Progress on a lot of environmental problems requires media attention. If the issue is in the news, it's hard to ignore. If you've published something that's relevant to some environmental issue, chances are your institution's public information office will want to write a press release about it. Cooperate with them, and chances are they will get the story right. Then reply to queries from the media. Distill your message down to a few sentences, practice your sound bites and don't go beyond the data.

6. Know who you are. Academic scientists get a lot of respect—for their expertise and for their objectivity. Peer review ensures that our conclusions are supported by the data. Academic scientists have other advantages too—we have the luxury of being able to follow problems wherever they lead, to learn new techniques, to get funding for projects, and to enlist enthusiastic students. Our obligation to be honest brokers for the facts gives us the opportunity to work with both advocacy groups and their adversaries.

REFERENCES

AVILA-SERRANO, G. E., K.W. FLESSA, M. A. TÉLLEZ-DUARTE, AND C. E. CINTRA-BUENROSTRO. 2006. Distribución de la macrofauna intermareal del Delta del Río Colorado, norte del Golfo de California, México. Ciencias Marinas, 32:649-661.

BEST, M. M. R., T. C. W. KU, S. M. KIDWELL, AND L. M. WALTER. 2007. Carbonate preservation in shallow marine environments: Unexpected role of tropical siliciclastics. Journal of Geology, 115:437-456.

BOWMAN, S. 1990. Radiocarbon dating. Interpreting the Past. British Museum Press, London, 64 p.

CAMPANA, S. E., AND E. MOKSNESS. 1991. Accuracy and precision of age and hatch date estimates from otolith microstructure examination. ICES Journal of Marine Science, 48:303–316.

CAMPANA. S. E., AND S.R. THORROLD. 2001. Otoliths, incre-

ments and elements: Keys to a comprehensive understanding of fish populations. Canadian Journal of Fisheries and Aquatic Science, 58:30–38.

CINTRA-BUENROSTRO, C. E., K. W. FLESSA, AND G. E. AVILA-SERRANO. 2005. Who cares about a vanishing clam? Trophic importance of *Mulinia coloradoensis* inferred from predatory damage. Palaios, 20:296-302.

DETTMAN, D. L., K. W. FLESSA, P. D. ROOPNARINE, B. R. SCHÖNE, AND D. H. GOODWIN. 2004. The use of oxygen isotope variation in shells of estuarine mollusks as a quantitative record of seasonal and annual Colorado River discharge. Geochimica et Cosmochimica Acta, 68:1253-1263.

FLESSA, K. W., M. KOWALEWSKI, AND S. E. WALKER. 1992. Post-collection taphonomy: Shell destruction and the Chevrolet. Palaios, 7:553-554.

GOODFRIEND, G. A., AND K. W. FLESSA. 1997. Radiocarbon reservoir ages in the Gulf of California: Roles of upwelling and flow from the Colorado River. Radiocarbon, 39:139-148.

GOODWIN, D. H., K. W. FLESSA, B. R. SCHÖNE, AND D. L DETTMAN. 2001. Cross-calibration of daily growth increments, stable isotope variation, and temperature in the Gulf of California bivalve mollusk *Chione* (*Chionista*) *cortezi*: Implications for paleoenvironmental analysis. Palaios, 16:387-398.

INGRAM, B.L., AND P. K. WEBER. 1999. Salmon origin in California's Sacramento-San Joaquin River system as determined by otolith strontium isotope compositions. Geology, 27: 851-854.

JONES, D. S., AND I. R. QUITMYER. 1996. Marking time with bivalve shells: Oxygen isotopes and season of annual increment formation. Palaios, 11:340-346.

KIDWELL, S. M. 2001. Preservation of species abundance in marine death assemblages. Science, 294:1091-1094.

KIDWELL, S. M. 2007. Discordance between living and death assemblages as evidence for anthropogenic ecological change. Proceedings of the National Academy of Sciences of the United States of America, 104:17701-17706.

KIDWELL, S. M., AND K. W. FLESSA. 1996. The quality of the fossil record: Populations, species and communities. Annual Review of Earth and Planetary Sciences, 24:433-464.

KIDWELL, S. M., M. R. BEST, AND D. S. KAUFMAN. 2005. Taphonomic tradeoffs in tropical marine death assemblages: Differential time-averaging, shell loss, and probable bias in siliciclastic versus carbonate facies. Geology, 33:729-732.

KOSNIK, M. A., Q. HUA, G. E. JACABSEN, D. S. KAUFMAN, AND R. A. WÜST. 2007. Sediment mixing and stratigraphic disorder revealed by the age-structure of *Tellina* shells in Great Barrier Reef sediment. Geology, 35:811-814.

KOSNIK, M. A., D. S. KAUFMAN, AND H. QUAN. 2008. Identifying outliers and assessing the accuracy of amino acid racemization measurements for geochronology: I. Age calibration curves. Quaternary Geochronology, 3/4, 308-327.

KOWALEWSKI, M., AND K. W. FLESSA. 1995. Comparative taphonomy and faunal composition of shelly cheniers from northeastern Baja California, Mexico. Ciencias Marinas 21:155-177.

KOWALEWSKI, M., K. W. FLESSA, AND J. A. AGGEN. 1994. Taphofacies analysis of Recent shelly cheniers (beach ridges), northeastern Baja California, Mexico. Facies, 31:209-242.

KOWALEWSKI, M., G. A. GOODFRIEND, AND K. W. FLESSA. 1998. High-resolution estimates of temporal mixing within shell beds: The evils and virtues of time-averaging. Paleobiology, 24:287-304.

KOWALEWSKI, M., G. E. AVILA-SERRANO. K. W. FLESSA, AND G. A. GOODFRIEND. 2000. Dead delta's former productivity: Two trillion shells at the mouth of the Colorado River. Geology, 28:1059-1062.

NRC (NATIONAL RESEARCH COUNCIL). 2005. The Geological Record of Ecological Dynamics: Understanding the Biotic Effects of Future Environmental Change. National Academies Press, Washington, D.C., 200 p.

NORRIS, R.D., AND R.M. CORFIELD. 1998. Isotope Paleobiology and Paleoecology. Paleontological Society Papers, No. 4., 285 p.

PIELKE, R.A., JR. 2007. The Honest Broker: Making Sense of Science in Policy and Politics. Cambridge University Press, Cambridge, 188 p.

RODRIGUEZ, C.A., D. L. DETTMAN AND K. W. FLESSA. 2001a. Effects of upstream diversion of Colorado River water on the estuarine bivalve mollusk *Mulinia coloradoensis*. Conservation Biology, 15:249-258.

RODRIGUEZ, C. A., K. W. FLESSA, M. A. TELLEZ-DUARTE, D. L. DETTMAN, AND G. E. AVILA-SERRANO. 2001b. Macrofaunal and isotopic estimates of the former extent of the Colorado River estuary, upper Gulf of California, Mexico. Journal of Arid Environments, 49:185-195.

ROWELL K., K. W. FLESSA, D. L. DETTMAN, M. J. ROMÁN, L. R. GERBER, AND L. T. FINDLEY. 2008. Diverting the Colorado River leads to a dramatic life history shift in an endangered marine fish. Biological Conservation 141:1138–1148.

SCHÖNE, B. R., D. H. GOODWIN, K. W. FLESSA, D. L. DETTMAN, AND P. D. ROOPNARINE. 2002. Sclerochronology and growth of the bivalve mollusks, *Chione* (*Chionista*) *fluctifraga* and *Chione* (*Chionista*) *cortezi* in the Gulf of California, Mexico. The Veliger, 45:45-54.

STEINBECK, J. 1951. The Log from the Sea of Cortez. Reprint 1986. Penguin Books, New York, 282 p.

THOMPSON, R.W. 1968. Tidal flat sedimentation on the Colorado River delta, northwestern Gulf of California. Geological Society of America Memoir, 107, 133 p.

ZINN, H. 1994. You Can't be Neutral on a Moving Train. Beacon Press, Boston, Massachusetts, 224 p.

CHAPTER FIFTEEN

CONSERVATION PALEOBIOLOGY ROUNDTABLE: FROM PROMISE TO APPLICATION

ALISON G. BOYER[1,*], MARK BRENNER[2,*], DAVID A. BURNEY[3,*], JOHN M. PANDOLFI[4,*], MICHAEL SAVARESE[5,*], GREGORY P. DIETL[6], KARL W. FLESSA[7]

[1]Department of Ecology and Evolutionary Biology, University of Tennessee, 569 Dabney Hall, Knoxville, TN 37996 USA
[2]Department of Geological Sciences, University of Florida, Gainesville, FL 32611 USA
[3]National Tropical Botanical Garden, 3530 Papalina Road, Kalaheo, HI 96741 USA
[4]School of Biological Sciences, ARC Centre of Excellence for Coral Reef Studies, The University of Queensland, Brisbane, Queensland 4072 Australia
[5]Department of Marine & Ecological Sciences, Florida Gulf Coast University, Ft. Myers, FL 33965 USA
[6]Paleontological Research Institution, 1259 Trumansburg Road, Ithaca, NY 14850 USA, and Department of Earth and Atmospheric Sciences, Cornell University, Ithaca, NY 14853 USA
[7]Department of Geosciences, University of Arizona, Tucson, AZ 85721 USA
*authors contributed equally and listed alphabetically.

ABSTRACT.—This conversation among five conservation paleobiologists highlights some of the personal, practical, and institutional factors affecting the application of conservation paleobiology. The five practitioners work with island faunas and floras, lakes, coral reefs, and estuaries. They all highlight the importance of personal initiative in engaging with conservation practice and policy, and patience in seeing the results of academic work put into application. They note that their work is most effective when they make the effort to learn what the conservation-oriented non-governmental organizations and governmental agencies need to know. They agree that applied work provides intellectual challenges, opportunities for students, and funding. All the participants in the roundtable came to conservation paleobiology from geological, paleontological, or ecological backgrounds. The hybrid nature of the field made their skills transferable.

INTRODUCTION

IN THE conservation biology community, there is growing awareness of an "implementation" gap, whereby conservation research often fails to be translated into meaningful conservation policy and practice (Nature, 2007; Hulme, 2014). Several reasons have been put forward to explain the gap. Part of the problem is that what conservation scientists find interesting may not always be needed by practitioners. There are many incentives in the scientific community for publishing research and obtaining funding but few for helping to implement conservation actions (Arlettaz et al., 2010). Then there is the issue of academic training in conservation biology, which often fails to teach the practical skills required to engage with the management community.

Conservation paleobiology is no different (Dietl and Flessa, 2011). Many conservation paleobiologists perceive the translation of their research into the practice of conservation and restoration as a one-way flow of information and advice from scientists to policy makers, managers, or stakeholders. This pathway is legitimate, valuable, and within the comfort zone of most scientists. Further steps, such as advocating and implementing conservation action, require deliberate engagement and occur not by accident but by intent. However, first you must decide what you need to do to get involved.

Therefore, we invited five practicing conservation paleobiologists to participate in a virtual roundtable to offer practical advice to those wishing to develop management-relevant conservation paleobiology. The roundtable panelists are all academic scientists who are also applying conservation paleobiology in search of solutions to site-specific problems. They are all based at academic or research institutions and work with government agencies or non-governmental organizations. Panelists include

- Alison Boyer, an ecologist who uses fossil and modern data to study extinction risk and biogeography in island birds. Her research is motivated by a desire to understand and prevent human-caused losses of biodiversity. She has pursued paleoecological fieldwork in both Hawaii and New Caledonia and works with both local governmental agencies and non-governmental organizations (NGOs). See Boyer and Jetz (2014) for a recent example of her work.

- Mark Brenner, who was trained as a biologist and has special interests in tropical limnology and paleolimnology. Most of his "paleo" work has focused on inferring late Quaternary climate change and evaluating human impacts on lakes and watersheds. See Mueller et al. (2010) for a recent example of his work.

- David Burney, who has been both a conservation biologist and a paleoecologist by training and practice for more than three decades. His research has focused on the human role in extinction and environmental change, and his applied work seeks to restore damaged ecosystems. He has worked on islands in all three tropical oceans, as well as in Africa and the Americas. He now holds the title of "Professor of Conservation Paleobiology"—likely the first such professorial title. See Burney (2010) for a recent example of his work.

- John Pandolfi, who has broad research interests in marine paleoecology, with an emphasis on the effects of anthropogenic impacts and climate change on the recent-past history of modern coral reefs. He works with the Great Barrier Reef Marine Park Authority and other Australian governmental departments and with NGOs and academic partners in both Australia and the Pacific. See Pandolfi et al. (2011) for a recent example of his work.

- Michael Savarese, who was trained as a traditional paleobiologist and geologist. Now, his research and teaching are principally focused on the application of geological principles and methods to problems in environmental management and restoration. See Savarese and Pierce (2013) for a recent example of his work.

We asked the panelists to respond to fifteen questions about how they applied conservation paleobiology in practice and how they managed to integrate their applied efforts into the teaching and research missions of their home institutions. After their replies, we reflect on the similarities and differences between them and offer recommendations to help conservation paleobiologists bridge the implementation gap and turn their science into conservation action.

ROUNDTABLE

1. How did you become involved in the application of some of your work to the actual practice of conservation biology?

Alison Boyer: It has been one of my goals as an academic scientist to bridge the fields of paleoecology and conservation biology. My first major postdoctoral field project was on a Holocene cave site on the Pacific island of New Caledonia. This French territory has tremendous biodiversity but faces many serious environmental challenges, including widespread strip mining, deforestation, and wildfires. One goal of our project was to understand the changes that have taken place within the island's birds (including extinction) due to these destructive human activities. We realized that information from our cave studies could represent valuable baseline data and could really help inform restoration efforts.

Mark Brenner: I first became involved in applied paleolimnology as a postdoctoral associate in the 1980s working on agency contracts held by regular faculty members at the University of Florida. The early work generally focused on large, hypereutrophic lakes in the state, like Okeechobee and Apopka, with the goal of better understanding baseline conditions in the water bodies, as well as the timeline for changes that occurred. In many respects, applied paleolimnology work for agencies became my "bread and butter" for nearly two decades, during which I was self-funded.

David Burney: I started my career as a professional conservation biologist. After undergraduate studies in zoology, I worked for six years as a regional naturalist for the North Carolina Division of State Parks. I obtained an MS in conservation biology from the University of Nairobi, where I studied human effects on cheetahs in the Mara region of Kenya. During my African years I became interested in deeper time perspectives on our conservation research and applications, and that led me to eventually earn a PhD at Duke University with Daniel Livingstone, the noted palynologist, paleolimnologist, and coring-techniques pioneer best known for his work in African paleoecology. For my dissertation project I chose to study the late Quaternary paleoenvironments of Madagascar, in hopes of shedding some light from a past perspective on the current environmental dilemma there. Over the subsequent nearly three decades I have conducted multidisciplinary paleoecological studies in Africa, Madagascar, the Americas, and remote oceanic islands in all three tropical oceans. Although I have always followed my intellectual curiosity in choosing paleoecological research topics, my curiosity has always turned to topics that centered around interactions between humans and nature on longer timescales, a focus that I would argue is equally interesting from both theoretical and applied perspectives.

John Pandolfi: Back in the early 1990s an issue of considerable concern for the Great Barrier Reef Marine Park Authority was the wave of outbreaks of the crown-of-thorns starfish, which was decimating coral cover on the Great Barrier Reef. A team of scientists from James Cook University gathered some cores from reef sediments and found skeletal elements of crown-of-thorns starfish in their cores, and recommended to the Park Authority that crown-of-thorns starfish outbreaks were natural as they had been found in the geological past. My work and familiarity with the field of taphonomy allowed me to take exception to such a literal interpretation of skeletal spines in reef sediment, so I became embroiled in a controversy about the utility of spines in reef sediments as being able to depict "outbreak events." We were able to convince the Park Authority that just because some spines were found in the sediment did not mean that the frequency and intensity of any potential previous outbreaks or fluctuations in population sizes were preserved in the sedimentary record. Because of this work, the Park Authority has followed our work on corals and their history on the Great Barrier Reef.

Michael Savarese: Florida Gulf Coast University hired me when the university first opened in 1997. A group of interdisciplinary scientists were brought on board to help facilitate environmental management and restoration in southwest Florida. The university became southwest Florida's center for Greater Everglades restoration. As the first geoscientist to join the faculty my role was to provide the geological perspective to environmental science and restoration. As the lead paleobiologist/sedimentary geologist, I made conservation paleobiology a key component of the university's involvement and curriculum in environmental science.

2. Are you working primarily with agencies or with environmental NGOs?

Alison Boyer: We are working mostly with an NGO, the main ornithological society in New Caledonia—the Société Calédonienne d'Ornithologie—although the governmental agency Direction de l'Environnement has also expressed interest in our work.

Mark Brenner: All the applied work that I have done has been with state, federal, county, and even city agencies. The majority has been funded by Florida's water management districts, agencies entrusted with management of state waters. But I have also done work funded by the United States Geological Survey, the Florida Department of Environmental Protection, the Florida Department of Transportation, and private consulting firms.

David Burney: About half and half, and consistently so for the last thirty years. Increasingly, my projects in both research and conservation are funded by partnerships that generally incorporate governmental, NGO, and private-sector funding sources and collaborators.

John Pandolfi: I work with both. The Great Barrier Reef Marine Park Authority is the primary agency I work with, though there are others. I liaise with environmental NGOs (Pew, WWF, Conservation International, The Nature Conservancy) mostly for providing advice on conservation initiatives (e.g., the Coral Sea marine park) or in writing documents (letters or white papers) for influencing government policy.

Michael Savarese: Southwest Florida has a strong network of collaborative agencies, NGOs, and academic groups. The nature of Greater Everglades restoration projects requires the development of teams to plan and implement restoration projects. At the heart of these efforts is a federal/state partnership between the U.S. Army Corps of Engineers (USACE) and the South Florida Water Management District (SFWMD). In recent years, while federally appropriated dollars have dwindled, the SFWMD has funded much of the restoration science and implementation. (These dollars unfortunately have also become limited as the state budget has tightened and a more conservative political agenda has arisen.) Nonetheless, NGOs, as well as an assortment of governmental agencies, are involved or interested in our conservation paleobiology research. Groups from the governmental side include: USGS, US Fish & Wildlife, Everglades National Park, Ten Thousand Islands & Florida Panther National Wildlife Refuges, Big Cypress National Preserve, Florida Sea Grant, SFWMD, City of Naples, City of Ft. Myers, Collier and Lee Counties. NGOs include the Conservancy of Southwest Florida, the Audubon Society, Caloosahatchee River Watch, and the Southwest Florida Watershed Council.

3. Did the agencies or NGOs become aware of your work (and if so, how) or did you reach out to them?

Alison Boyer: We set out immediately to develop a working relationship with both groups, in the hopes that we would mutually benefit. We had to initiate all collaborations, and we faced many barriers, like language differences, cultural differences, and overworked civil servants. Just setting up our first meetings was a challenge, and they might not have happened at all if we had not set up a wide network of contacts both in country and abroad.

Mark Brenner: I worked as part of a group that soon became known for our research approach, and we generally were contacted by agencies when they had questions they thought we might be able to address. In some cases they would simply alert us that a call for proposals was to be issued, whereas in other cases they would contact us and sole-source projects to us. Sometimes when they did reach out to us, they had a general notion of the questions they needed answered, but they would work cooperatively with us to develop a scope of work.

David Burney: Both situations have occurred, but one common scenario is that one of my publications or a talk that I gave led to someone contacting me for advice on technical matters relating to conservation or helping design a research project to clarify a question from conservation managers.

John Pandolfi: The coral reef community is a rather small one, so word spreads about who is doing what. My work on historical ecology of coral reefs is well-known to the coral reef community, from scientists to managers. Having said that, I do take frequent opportunities to keep letting these organizations know what my work is, as there are always new people and the turnover at these organizations can be high.

Michael Savarese: Because of my university's youth and environmental stewardship mission and as a consequence of the collaborative team approach to restoration planning, strong partnerships and communications already existed. Perhaps the secret to

success for someone or some program hoping for these kinds of relationships is time spent providing service to local environmental agencies and NGOs. Comparable environmental restoration teams or collaborative groups should exist everywhere.

4. What results of yours did the practitioners find to be especially useful?

Alison Boyer: The study is still ongoing, but our initial findings documenting substantial changes in the dry-forest avifauna over the past approximately 1200 years are being used now by the ornithology society. They are involved in efforts to establish several new dry-forest reserves, and in a major restoration project on an offshore island.

Mark Brenner: Often the water management districts are under pressure to "restore" eutrophic lakes to a lower trophic status, and they feel intense public pressure to do so. Our ability to infer past trophic status using diatoms, nutrient (e.g., P) accumulation rates, pigments, and stable isotopes ($\delta^{13}C$ and $\delta^{15}N$) of organic matter in ^{210}Pb-dated cores provides agencies with the information they require to make informed decisions about investing in "restoration." They rely on our "paleo" data because direct measures of water quality were not made in most Florida lakes at least until the 1980s, by which time many had already experienced change. In some cases, we have been able to demonstrate that a lake is naturally eutrophic, perhaps because it lies atop phosphate-rich deposits. In such cases, it is clear that large and often costly efforts to mitigate eutrophic conditions would be futile. In other cases in which we can demonstrate that trophic status changed in the recent past, restoration may be feasible if nutrient sources can be identified and controlled.

David Burney: My collaborators and I have been able, through paleoecological techniques, to provide practitioners with detailed reconstructions of the species composition and environmental dynamics of ecosystems in Madagascar, Hawaii, and other areas prior to humans and during prehistoric occupation. This information has been widely applied in ecological restoration activity as an ecological baseline and potential reference community for developing restoration goals and methods. For instance, my research results are often used to specify what type of plant assemblages are an appropriate target for a particular restoration, including a list of species to reintroduce.

John Pandolfi: One was the Pandolfi et al. (2003) paper in *Science*, which documented coral-reef decline globally using historical information on a number of different reefs. NGOs especially found the work to be useful for understanding the enormity of reef decline throughout the world, and it became a clarion call for more and better reef conservation throughout the world.

The other work is work I am now involved in—coring the Great Barrier Reef for evidence of historical decline associated with European colonization from the mid-1800s onward. The Park Authority and other federal agencies are very interested to see the role of land-use changes on the nearshore reefs of the Great Barrier Reef, which have suffered recent degradation due to water-quality issues.

Michael Savarese: A good example of a well-applied and successful paleobiologic approach to a problem in southwest Florida concerns the management of the Caloosahatchee River. This river serves as a major storm-water management flow-way for southwest Florida to avoid flooding of the Lake Okeechobee basin in central south Florida. Freshwater is shunted through the Caloosahatchee to avoid flooding during times of extreme rainfall events or is retained to ensure an adequate drinking- and irrigation-water supply during the dry season. As a consequence, the estuary at the river's downstream terminus can experience extreme freshets during times of excessive release or high marine salinities when too much freshwater is withheld. Estuarine scientists documented the effects these water management practices are having on existing oyster reef faunas. My students and I used a comparison of life and death assemblages of mollusks to document the pre- and post-alteration effects of these practices. The district has used the results from both approaches to alter its practice of freshwater releases.

5. What has been your biggest success in the application of your results to conservation practice?

Alison Boyer: Really I am just getting started in the New Caledonian work, and I plan to develop a long-term research program there, but I am pleased that our first study was so warmly received by the ornithology NGO and that local officials are also showing interest in our work. I would love to see some of the extirpated species reintroduced to the new forest reserve on Léprédour Island, once restoration efforts are further along.

Mark Brenner: I prefer speaking in terms of "management," as opposed to "conservation" *per se*, but I think the ability to demonstrate that some lakes are naturally eutrophic, whereas others display changes in trophic status in response to human activities, has been very helpful. I also think that some of our findings serve as cautionary tales for managers who are contemplating strategies to manage lakes. For instance, we have been in a long-term drought in north Florida, and there has been recent discussion about pumping groundwater into local lake basins that have shown sharp stage declines. Our demonstration that such pumping loaded systems with ^{226}Ra that was subsequently bioaccumulated by mussels is at least being considered as managers think about alternatives.

David Burney: Makauwahi Cave Reserve on Kaua'i, in the Hawaiian Islands, is a prototype for an approach that I have strived to demonstrate in my work. In the last two decades, with the aid of a host of accomplished scientists in paleontology, archaeology, paleoecology, geology, and other disciplines, my wife Lida Pigott Burney and I have developed this site from our initial discovery of its tremendous potential for fossil and artifact recovery to what is now generally acknowledged as the richest fossil site in the Hawaiian Islands and perhaps the Pacific region. As summarized in a book I wrote for general readers entitled *Back to the Future in the Caves of Kaua'i: A Scientist's Adventures in the Dark* (Yale University Press, 2010), over subsequent years we have taken the exciting fossil discoveries, which provide a detailed picture of what the area was like before humans and how it has changed, and fashioned them into a management strategy for rehabilitation of the degraded farmlands and mine spoil surrounding the cave. We have a lease on seventeen acres surrounding the site, now beautifully landscaped with more than six thousand native plants, featuring many species that were predominant in the recent fossil record, including some that are quite rare. Some of the endangered native birds that were prevalent as fossils now once again nest on the property in restored wetlands. College-accredited field school programs train dozens of undergraduates and graduate students annually in paleoecology, archaeology, restoration ecology, and geospatial sciences; a docents program provides free guided tours for the public and educational groups; a self-guiding nature trail runs through the site, built by local Boy Scouts; and for several years now we have hosted a conservation jobs training program for unemployed native Hawaiians. Last year, we had over twelve thousand visitors, who hear a dual message from our site: the past can inform the future, and conservation can really work.

John Pandolfi: When the Great Barrier Reef was up for rezoning in 2004, my 2003 *Science* paper was used by government workers to convince people on the street ("stakeholders") that something had to be done to improve the Great Barrier Reef. The paper included the Great Barrier Reef and showed that reefs everywhere were far from pristine. I was told that this paper was instrumental in the government's ability to convince people of the need to act. As a result the Great Barrier Reef was rezoned in 2004 to increase the amount of "no-take" area from 5% to 33%.

Michael Savarese: For the past ten years, much of my research has been dedicated to the planning and implementation of the Picayune Strand Restoration Project, a Greater Everglades restoration project in southwest Florida. Restoration planning lacked targets for estuarine salinity. A paleoenvironmental reconstruction of oyster reef history throughout the area documented what the pre-alteration salinities were for maximum oyster productivity and the pre-alteration geographic locations of the most productive reefs. These results were then used to model the freshwater flows needed to achieve the necessary spatial and temporal distributions of salinity for optimal estuarine productivity. Ultimately this was used to choose a restoration-engineering plan. The plan is currently underway—the canal and road systems are being repaired in accordance with these salinity targets, and subsequent monitoring of the estuarine oyster reef health will be used to ensure targets are achieved through adaptive management.

6. What has been your biggest disappointment in your efforts to apply the results of your work to the practice of conservation?

Alison Boyer: The lack of communication between conservationists and academics is painfully obvious. We inhabit totally different spheres, and they do not read our literature unless you send it to them. Also, conservation work can take decades to get through the stages of government approval and planning to actual results. We have to be patient.

Mark Brenner: In one case, we did a study on an urban lake and as part of the work we conducted a lake-wide assessment of sediment distribution. The state agency was interested to know if sediment dredging might improve water quality. We determined that, given the ongoing external nutrient loading to the lake, the minor change in lake volume that would be produced by dredging was unlikely to cause any improvement in water quality. We recommended a much cheaper alternative. We felt that much of the public perception about the lake stemmed from the fact that it had a large dump site on its south end, some crumbling bleachers along one shoreline, and a building used for practice by firefighters on another side, all unattractive features. We proposed that the problem was as much aesthetic as biological and that getting rid of the garbage and at least the bleachers would improve the public's image of the water body. We also felt that despite its high productivity, the lake was a wonderful wildlife area, and some information about the wading birds, alligators, and other animals and plants would help the public appreciate the area. Dredging operations were undertaken despite our recommendation, largely in response to intense political pressures. Politics trumped science.

David Burney: The slowness of some bureaucrats, and even some academics, to "get it." Although it is professionally healthy to maintain skepticism in the face of new proposals until the facts are weighed, I often feel that the primary obstacle to using more "new" kinds of information from paleobiology in conservation applications is that most governmental and even university scientists are quite narrowly trained, and anything to do with studies of the past is regarded by some officials and investigators as only about geology and archaeology, not plants, animals, and environments. This is a failing of our educational system but also a limitation of personal curiosity about the world around us that fails to incorporate a sense of the relevance of the effects wrought by the passage of time.

John Pandolfi: Probably that the historical past is forgotten so often and so easily. Even among people who have previously used it or are sympathetic to it, there seems to be this "little brother" attitude about it—welcoming when it is there but often not thinking about it on their own. So it becomes frustrating to constantly remind people that considering the past should be essential to what they are doing. However, I think this is slowly but inexorably changing. More practitioners of conservation paleobiology would assist this along. Also, I think the coral reef community is especially in tune to the value of historical work.

Michael Savarese: While I find conservation paleobiological research rewarding, it is rife with frustration. Restoration projects are readily politicized and tenuously linked to funding. While the Picayune and Caloosahatchee projects were successful and resulted in changes in management practice, the length of time required to effect these changes was enormous. And there are many examples of conservation paleobiology-based reports that ended up neglected or not acted on. The current economic condition of both the State of Florida and the U.S. Federal Government has grossly altered priorities, and environmental restoration is not high on that list.

7. What has been the biggest surprise you have had in your work with agencies and NGOs?

Alison Boyer: I was pleasantly surprised that folks in the ornithology society were so interested in our work and eager to share their hard-earned data.

Mark Brenner: I confess that I had something of a prejudice before I entered the world of applied science—that fundamental, pure science was somehow better. I learned that many of the questions that agencies need answered are interesting and important, and I was particularly impressed that they often gave us the flexibility to broaden the scope of work so we could address additional hypotheses. I also confess that I may have been a bit surprised at the high quality of the scientists who populate our state and federal agencies. After more than twenty-five years of working with agency people, I have come to expect it and it has been a pleasure working with them.

David Burney: I have been pleasantly surprised in recent years to discover that many of the most innovative and adaptive strategies in conservation are now coming not from cash-strapped and sometime motive-conflicted government agencies or even conventional NGOs but more directly from the private sector, through the efforts of large private landowners with an interest in the connections between conservation and paleobiology, such as Ted Turner, Owen Griffiths, and Bette Midler. Some of the best work is likewise being done by grassroots, community-based organizations in which people from all walks of life contribute their time and resources to conservation work they believe in, with little or no role for government agencies or large established NGOs.

John Pandolfi: I am always surprised when I see my work featured, in whatever guise. To be able to communicate the results of my work to government officials and have them take it to heart and ring me about policy and legislation is a very pleasant surprise and gives a whole other dimension to the significance of my work and the gratification that I derive from it.

Michael Savarese: I truly enjoy this type of research. The rewards associated with civic engagement, with helping our region with problems of environmental management, and with having a positive societal impact as a result of my science have been tremendous. I feel like my work is respected and valued.

8. What has been the biggest surprise agencies or NGOs have had in working with you?

Alison Boyer: I'm not really sure yet.

Mark Brenner: That's a tough one, but one of our research findings did cause some surprise at one of the state water management districts. We were conducting a paleolimnology study on small lakes near Tampa, many of which had been "groundwater-augmented" since the 1960s or 1970s. Low water levels in the basins, probably caused by excessive pumping of the aquifer and rerouting of surface water during construction of neighborhoods, prompted local residents to install deep wells to pump water directly into the shrinking water bodies. During ^{210}Pb dating of short sediment cores, we discovered unusually high radium-226 activities in near-surface deposits, a phenomenon we had seen elsewhere but were at a loss to explain. We also found that freshwater mussels in the augmented lakes had high Ra-226 concentrations in their shells, but especially high activities in their soft tissues, 1,000–25,000 x values in the water. We discovered that naturally occurring radium in groundwater was entering the lake via hydrologic augmentation and was being bio-accumulated. It also appeared that mussels and snails that occupy the augmented lakes today did not exist there several decades ago, as waters were too calcium-poor to support such shell-forming biota. Because the water management district now has jurisdiction over groundwater augmentation, this caused a bit of a stir. Further pumping continues to load the systems with radioactivity, but cessation of pumping might lead to drying of the lakes and exposure and drying of ^{226}Ra-rich sediments, which would have to be treated as hazardous waste.

David Burney: People in conservation organizations and government agencies have in several cases funded me to learn more about past characteristics of their sites, with preconceived notions of what I would find, and then had to readjust their position with the new information we were able to collect. For instance, when the picture began to emerge from paleoecological studies that Madagascar's interior and Kaua'i's south shore were typified not by more and bigger examples of surviving native ecosystem remnants (mostly restricted today to very steep or wet terrain) but by no-analogue plant communities before humans, this led to discussions among practitioners as to how to restore such areas and whether it is even possible, and in particular forced the conclusion that many restoration proposals had previously been too conservative in their species list. In other words, if one depends entirely on the shallow time depth of "historical occurrence" of species in many places, relatively few species might be regarded as appropriate. When one incorporates late Holocene prehuman fossil occurrence, all these species are normally included but many more species are shown to have been present before, even though today they may be limited to very small disjunct ranges that do not include the site. Our work has also shown, to the great surprise of some experts, that some species thought to be introduced by humans were in fact natives, and vice versa.

John Pandolfi: I think that agencies or NGOs are often surprised at the heavyweight nature of the work we have done and have had published. I think the major successes we have had in publishing our work in *Science* and other high-profile journals have surprised the NGO community, as they very much take to heart the high profile science in their work and conservation strategies. I have often come across the way folks from these agencies will almost shake their head in resigned disbelief that they really do have to include the past in their work. It's really great because you can see that they are convinced, and they never would have been without the publication of the work and the publicity it attracts.

Michael Savarese: Agencies and NGOs seemed surprised at the value of a geoscience perspective on environmental management and restoration. My involvement with collaborative teams—and our research results—seems to have made a difference.

9. Does your home institution (department, college, or university) value these more applied aspects of your work?

Alison Boyer: It is not obvious that they do. Everyone talks about the importance of interdisciplinary work, but when it comes time to hire someone, it seems they are looking for a disciplinary expert. Students should be aware that research breadth should not come at the expense of depth in your main area of expertise.

Mark Brenner: I think so, but the view may differ among colleges. I am at the land-grant university in the state, and when I was in the "ag school," many faculty members were involved in applied science, which was sometimes viewed disparagingly by those in Liberal Arts and Sciences (LAS). But I think that attitude has changed, and now that I am in a LAS department, I, and many of my colleagues, continue to do such work. I think there is recognition that such work is valid, and we have published many of our findings in the peer-reviewed literature over the years. Such projects also do bring in overhead, and in these lean times, the university is happy to have revenue from all sources. I do recall that when I was a "soft-money" researcher in the "ag school," my strategy was to focus on Florida-centric applied work, because I had the distinct impression that it was valued in that college.

David Burney: Only when it attracts money, frankly. Generally the greatest skepticism comes from colleagues in theoretically oriented subdisciplines who do not work on practical problems.

John Pandolfi: Very much so. In Australia, there is constant pressure to have work that is of social or environmental benefit, so this easily fits in with the whole mission of conservation paleobiology. Even our granting system requires relevance or benefits to Australia, so the more we can show this aspect the better we can acquire outside funding, which of course the university values at all levels.

Michael Savarese: Florida Gulf Coast University does value applied research. Technical reports prepared for agencies, if peer-reviewed, are considered as pieces of scholarship. Environmental work that results in some management or policy change or that is applied to restoration planning or implementation is considered as scholarship of even greater value. At the time promotion portfolios are assembled, the candidate may request letters from agency professionals that attest to the applied value of the research. The university is also committed to civic engagement. Faculty members that incorporate applied environmental projects within their courses are credited for such efforts during annual evaluations and for promotions.

10. Has your work with agencies or NGOs shaped your research directions?

Alison Boyer: Not a lot. I'd like to think that my research was already pretty well-aligned to their interests.

Mark Brenner: Oh yes. I estimate that roughly one-quarter of my publications are related to the applied work that I was involved in with agencies. As mentioned, sometimes we were able to go beyond the scope of the work specified in a contract, but certainly I can say that work for agencies sometimes shaped the research agenda and enabled us to address interesting questions.

David Burney: Yes, it has been very helpful knowing what sorts of information managers can actually use with their chronically limited resources.

John Pandolfi: Yes, it has, but not overly so. I do have major funding right now from the Department of the Environment to study the historical ecology of near-shore reefs of the Great Barrier Reef. This is very much an applied project worth over $1 million. But there is a lot that can be worked on under the guise of European colonization effects of land-use change on coral reefs that goes way beyond applied aspects. It's a very happy marriage that we can work on the history of reef communities for both applied and pure research outcomes. But I am also lucky to be very well funded for pure research as well, so it has never been a case of either/or—or needing the applied funding to keep my research going.

Michael Savarese: Most definitely. The collaborative team approach to restoration has generated new research directions and opportunities for me.

11. Has your work with agencies or NGOs provided you with any new funding opportunities?

Alison Boyer: Not yet.

Mark Brenner: Yes, the work with agencies has been fairly steady for more than twenty years.

David Burney: Yes, whereas my paleoecological research was originally funded primarily by NSF and other governmental science agencies, as well as research-oriented NGOs such as National Geo-

graphic, increasingly in recent years both research and applications have been funded at least in part by the Natural Resource Conservation Service, U.S. Fish and Wildlife Service, and conservation organizations.

John Pandolfi: Yes. This took a while to get going and was small at first, but as the work gained traction it has now blossomed into major research funding.

Michael Savarese: Yes. I have been well funded over the years by a variety of agencies and NGOs. Some of these projects were contracts that fit with my expertise. Others were projects that were self-designed but agreed well with some restoration science need.

12. Has your work with agencies or NGOs provided your students with any new opportunities?

Alison Boyer: Lots of ideas have been discussed, but funding continues to be an issue. Many of these small NGOs are even more cash-strapped than we scientists are.

Mark Brenner: Yes, students have been supported financially on agency-funded projects, and, in one case, a graduate student used the work as the core for her MS thesis.

David Burney: Yes, several of my students have received research funding from the Wildlife Conservation Society, for instance, and a few have subsequently been employed by conservation organizations.

John Pandolfi: We are now very well-funded for our applied work on the Great Barrier Reef, and this now includes five or six PhD students and one postdoc. PhD students on a previous government grant on Great Barrier Reef history got positions at environmental consultant firms and postdocs, so this kind of work can provide training for different career paths.

Michael Savarese: Many of my master's students have completed conservation paleobiology theses that made significant contributions to the Greater Everglades restoration projects. Some of these students were funded as research assistants to conduct the work. Additionally, many of our undergraduates do professional internships with environmental agencies and NGOs. These students, as well as our graduate students, commonly get full-time positions upon graduation.

13. Has your work with agencies or NGOs provided you with any examples that you use in your teaching? If so, in what courses do you use such examples?

Alison Boyer: I have used case studies from our work in New Caledonia in both my strict disciplinary (Ecology) and my interdisciplinary (The Anthropocene) courses.

Mark Brenner: I use both published and unpublished examples in several classes, including Limnology, Paleolimnology, and Perspectives in Florida Lake Management. The latter class is co-taught with a faculty member from another department and we bring in guest speakers from other units on campus, from environmental consulting firms, from NGOs, and from state and federal agencies. Many in the latter group are people I have worked with on projects in the past. We often enlist the help of state agency personnel to run field trips to lakes that are being managed in specific ways. In yet another class, designed specifically for graduate fellows, we visited the South Florida Water Management District and have had several guest speakers from various agencies.

David Burney: I use case studies to inform conservation decisions in lecture courses I have taught, including Natural History of the Hawaiian Islands, Conservation Biology, Environmental Management, Paleobiology, and Ecology.

John Pandolfi: I use results from my research at all levels, from first year introductory biology courses, to second year marine science courses, to third year courses on coral reefs and biological adaptation to climate change.

Michael Savarese: I teach two upper division courses in applied geology: Geobiology and Coastal & Watershed Geology. I use examples from our previous work in both of these courses. Both courses also include a class research project relevant to some aspect of regional environmental management and restoration.

14. How often do you interact with agencies and NGOs?

Alison Boyer: Once or twice a year, and we meet in person when I visit the island to do fieldwork.

Mark Brenner: I interact with agencies and agency personnel on a fairly regular basis, either through projects or through courses. In the fall semester, we host six to eight speakers from agencies as part of the Florida Lake Management class. At other times, we may enlist the help or advice of agency personnel to obtain sampling permits or assist with a project. And I often receive e-mail communications from agency colleagues about questions related to Florida lakes. Throughout the year, such interactions probably occur, on average, every two weeks.

David Burney: I interact daily, and in fact left my post as a tenured professor at Fordham University to work for an NGO full-time, since 2004, as director of conservation at the National Tropical Botanical Garden.

John Pandolfi: From weekly to monthly.

Michael Savarese: When working as a member of a collaborative restoration team, the time spent together can be considerable (meeting monthly, sometimes biweekly, others quarterly). I do interact with agency professionals with greater regularity when we are co-investigators on a project or when we are co-mentoring students.

15. How do you think collaborations between academic researchers such as yourself and agencies and NGOs could be improved?

Alison Boyer: I'd love to see more funding for applied research, and support for reaching out to these organizations. We also need to educate students in ways to effectively communicate with agencies and NGOs.

Mark Brenner: I think we have good relations with a number of agencies, and I have been largely satisfied with the collaborations we have undertaken. Government agencies recognize the fact that academic researchers can deliver more for the money, compared to private consulting firms, because university salaries are largely covered and overhead rates are relatively low. I have had little experience with NGOs, so perhaps that is an avenue worth further exploration.

David Burney: Academic researchers need to make more of an effort to understand and assist the missions of conservation organizations. To this point, the relationship between the disciplines has been highly asymmetric, as conservation practitioners often look for answers in paleobiology, but few paleobiologists have made the effort to find out what conservation managers really need from their expertise or what management questions need further investigation.

John Pandolfi: I think communication is key. One thing researchers don't like to hear is what they should work on and how to do their research by people who are not experts. Yet it is these very same people who often know what is not known and what is needed to be known in order for their lives or the environment to be improved. As scientists we need to be more receptive to the "stakeholders." They should be kept in the loop from early on in the piece. I keep the Great Barrier Reef Marine Park Authority and other key stakeholders abreast of my work on a regular basis and will sound them out at the very beginning of my research projects—often at the project design stage, so that they have "buy-in" from the start. This fosters a sense of ownership in them, and they are more likely to champion your work when the results start pouring in, even if perhaps the results aren't exactly what they want to hear.

Michael Savarese: Because the value of geology and paleontology to environmental management and restoration is not universally appreciated by environmental agencies and NGOs, improving communication is essential. Paleontologists commonly "travel" within our own discipline, attending our own professional conferences and collaborating with other academics. Our expertise and relevance would get greater exposure by our attending, presenting, and organizing sessions at environmental science and restoration conferences. A willingness to adapt your research program to fill specific community needs may better ensure that your work is ultimately used in societally fruitful ways. Finally, changing the culture of valued scholarship within your home institution is essential: research that results in some significant change in environmental management should carry at least as much weight as a peer-reviewed journal publication.

REFLECTIONS ON THE ROUNDTABLE

We were impressed by both the similarities and differences that emerged in the answers to these fifteen questions. Each of our five panelists emphasized the importance of reaching out to agencies and NGOs to let them know how geohistorical data could be relevant to their missions and, perhaps more importantly, to let them know that they want to be actively involved in solving problems. In some cases, such outreach began with a professional article or talk (Pandolfi, Burney). Other contacts occurred through existing institutional relationships (Brenner, Savarese) or through direct initiative (Boyer, Pandolfi). This variety of responses illustrates that there are many ways to become involved, but one essential ingredient may be initiative on the part of the academic scientist. We suspect that simply publishing papers that assert relevance is not an effective way to put conservation paleobiology to work. What seems to work best are in-person meetings, conversations at a professional meeting, phone calls, or even an unsolicited e-mail. While also requiring effort, building on that first connection—networking—can result in additional benefits. Indeed, once a relationship has been established, it need not require the high maintenance of daily contact—unless, of course, all parties want it that way. Much less frequent contact seems to work for most of the panelists.

With regard to completed projects, or those nearing completion, our panelists reported that agencies and NGOs were impressed with the value of geohistorical data. There were no apparent concerns over the quality of the geohistorical data. The worries that many paleobiologists have over the quality of the fossil record is not shared by the end users of such data. Our data can be fit for the purpose—we need not apologize for alleged shortcomings. In addition, the longer temporal perspective proved valuable (Burney, Pandolfi, Savarese) in establishing baselines to both measure impact and provide reference communities for restoration.

We note that all our panelists work in what we call here near time. The value of the deep-time perspective in practical applications remains to be demonstrated. Some studies revealed previously unknown impacts of human activities (Brenner) while in others human agency could be disproven (Brenner, Burney). Our panelists also noted their satisfaction with such applied work, not only for the intangible satisfaction of doing good and having one's contributions valued (Savarese, Pandolfi) but also for the eagerness and innovation of NGO and agency staff (Boyer, Burney), for their willingness to share data (Boyer), and for the high quality of the work being done (Brenner). Panelists also noted that the applied projects provided ample scope for student theses (Savarese, Pandolfi) and for exploring topics of broader or more basic interest (Brenner, Pandolfi).

All had frustrations, perhaps because of a common eagerness to see their efforts make a difference in conservation policy and practice. "We have to be patient," counsels Boyer. Bureaucracies move slowly (Burney, Savarese), institutional memories can be short (Pandolfi), and political decisions can interfere (Savarese, Brenner). And, of course, funding is rarely adequate (all). Such frustrations are surely not unique to conservation paleobiologists.

Success stories based on geohistorical data and analyses continue to accumulate: the integration of geohistorical approaches with management (Brenner); the establishment of a preserve (Burney); the implementation of environmental engineering (Savarese); and the significant expansion of protected areas within the Great Barrier Reef (Pandolfi).

Among the well-established scientists, grants for research projects and contracts for applied work in conservation paleobiology are common. No one is getting rich, but NSF proposals that can cite conservation among the "broader impacts" or Australian Research Council proposals that can speak to social or environmental benefits can be competitive in the scramble for big funding (Burney, Pandolfi). Agencies and NGOs that can outsource their research efforts are another common source of support (Brenner, Burney, Pandolfi, Savarese). Conservation efforts can also be supported by the private sector (Burney), but it always helps to be working on charismatic species or habitats.

All of our panelists are academic scientists who have chosen to seek out applications for their work. Doing so has provided many of them with opportunities for funding (see above) and actual or possible professional opportunities for their students. These scientists' work also provides a rich source of examples that can be used in introductory, advanced, or graduate classes.

Our panelists' perceptions varied with respect to how their home institutions see the value of their efforts. Where institutional criteria for advancement explicitly include applied research (Savarese, Pandolfi, Brenner), our panelists felt valued. A well-earned cynicism, often expressed (Pandolfi, Brenner, Burney), was that the institution was sure to see value in the overhead brought in by funded projects. Concern that some academic scientists see applied or interdisciplinary work as less worthy than pure or basic research was also voiced (Boyer, Brenner, Burney).

As may be common with new fields, all of the panelists became conservation paleobiologists after formal training in some other field. Until very recently, graduate programs in the discipline simply did not exist. Two panelists became conservation paleobiologists through postdoctoral projects that merged the disciplines of biology and geology or limnology (Boyer, Brenner), or entered from established careers in conservation biology (Burney) or paleontology (Pandolfi, Savarese). We suspect that an increasing number of conservation paleobiologists will emerge out of formal training in the field from biology or geology departments.

Suggestions for improving collaborations included better communication with agencies and NGOs to find out their needs (Pandolfi, Burney, Boyer). Academic scientists should not presume to know the missions of agencies and NGOs. Brenner's invitation to agency scientists to present at graduate seminars suggested one avenue for productive next steps. Another suggestion was for academic scientists to travel outside their own circles—to go to the professional meetings that the agency and NGO practitioners attend. These additional efforts do, of course, take both time and money.

In closing, if you are a conservation paleobiologist who wants your geohistorical data to be useful in the practice of conservation, the main lesson you should take away from this roundtable discussion is that intentional steps are needed to develop the skills essential for bridging the gap between science and practice. Doing this takes time and patience, as the personal experiences of our panelists attest—developing management-relevant science requires making connections outside the academic sphere. The effort is worth it.

REFERENCES

ARLETTAZ, R., M. SCHAUB, J. FOURNIER, T. S. REICHLIN, A. SIERRO, J. E. M. WATSON, AND V. BRAUNISCH. 2010. From publications to public actions: When conservation biologists bridge the gap between research and implementation. BioScience, 60:835–842.

BOYER, A. G., AND W. JETZ. 2014. Extinctions and the loss of ecological function in island bird communities. Global Ecology and Biogeography, 23:679–688.

BURNEY, D. A. 2010. Back to the Future in the Caves of Kaua'i: A Scientist's Adventures in the Dark. Yale University Press, New Haven, 198 p.

DIETL, G. P., AND K. W. FLESSA. 2011. Conservation paleobiology: Putting the dead to work. Trends in Ecology and Evolution, 26:30–37.

HULME, P. E. 2014. Bridging the knowing-doing gap: Know-who, know-what, know-why, know-how and know-when. Journal of Applied Ecology, 51:1131–1136.

MUELLER, A. D., G. A. ISLEBE, F. S. ANSELMETTI, D. ARIZTEGUI, M. BRENNER, D. A. HODELL, I. HAJDAS, Y. HAMANN, G. H. HAUG, AND D. J. KENNETT. 2010. Recovery of the forest ecosystem in the tropical lowlands of northern Guatemala after disintegration of Classic Maya polities. Geology, 38:523–526.

NATURE. 2007. The great divide. Nature, 450:135–136.

PANDOLFI, J. M., R. H. BRADBURY, E. SALA, T. P. HUGHES, K. A. BJORNDAL, R. G. COOKE, D. MCARDLE, L. MCCLENACHAN, M. J. H. NEWMAN, G. PAREDES, R. R. WARNER, AND J. B. C. JACKSON. 2003. Global trajectories of the long-term decline of coral reef ecosystems. Science, 301:955–958.

PANDOLFI, J. M., S. R. CONNOLLY, D. J. MARSHALL, AND A. A. COHEN. 2011. Projecting coral reef futures under global warming and ocean acidification. Science, 333:418–422.

SAVARESE, M., AND K. D. PIERCE. 2013. Distribution of oyster reefs and oyster health status in the Cocohatchee River Estuary. Technical report submitted to Estuary Conservation Association, 22 p.

CONSERVATION PALEOBIOLOGY IN THE ANTHROPOCENE

THIS BOOK has covered a lot of paleontological ground. Sections 1 and 2 highlighted some of the ways in which the science of conservation paleobiology could be (and already has been) applied to solving conservation problems. Section 3 followed up by showing how moving from promise to application requires moving beyond the science—into the realm of practice and policy. In this epilogue, we look to the future and at the value of the geohistorical perspective to conservation in the Anthropocene.

The reality of the Anthropocene world we live in is changing how many conservation biologists think and act (Corlett, 2015). In particular, this new normal poses a challenge for traditional approaches to conservation practice, which tend to be reactive and focused on preserving the "natural state" of "healthy" ecosystems or restoring "degraded" ecosystems to some ideal of the past. Critics of traditional approaches to conservation support instead a more proactive and pragmatic conservation biology that may be better suited to deal with change.

For those devoted to traditional approaches to conservation (e.g., protected areas), historical fidelity with species composition and abundance is a primary goal that directs management efforts. But for those resigned to the realities of the Anthropocene, other goals are possible, especially maintaining ecosystem services (Corlett, 2015). This expanding set of conservation goals is necessary because the Anthropocene forces us to admit that we may not be able to put things back the way they were (let alone keep them the way they are now). Once we accept that we can't turn back the clock, we find ourselves having to choose from a complex mix of often conflicting conservation goals (Marris, 2011). Perhaps not surprisingly then, acceptance of an Anthropocene world of rapid change has led to proposals for a broad range of new conservation tools (Corlett, 2015; Holmes, 2015). It has even inspired suggestions for active tinkering with ecosystems to create desirable properties that are beneficial both to humanity and to other organisms (e.g., Hobbs et al., 2011).

Anthropocene-aware conservation thus is about finding ways to design ecosystems or at least assist nature in coping with anthropogenic change. Rewilding is one proposed tool that expands on traditional approaches. The idea—to use surrogates for extinct species in order to restore long-lost ecosystem functions and adaptive potential—is controversial (Svenning et al., 2016). One particularly bold proposal is to move elephants and lions and other large mammals from Africa and Asia to North America (Donlan et al., 2006; Donlan and Greene, 2010). Such deliberate introduction of non-natives can have unforeseen consequences, to say the least. There are also more startling proposals to combine rewilding with synthetic biology through de-extinction (Seddon et al., 2014). For example, ancient DNA found in the remains of extinct woolly mammoths recovered from the frozen tundra holds promise that lost ecosystem productivity could be restored by introducing genetically modified elephants to the Arctic that have a suite of traits better adapted for life in the cold (Shapiro, 2015).

In addition to moving species to recreate recently lost ecosystem functions, another idea is to willfully move species beyond their current ranges through assisted migration (Seddon, 2010). In the future, species that are not able to adapt to climate change by migrating to new habitable areas may require human intervention to move them to new places where they can survive.

If society decides to implement any of these strategies, there is opportunity for conservation paleobiology to inform such decisions. For instance, geohistorical records can help identify recently extinct species and reconstruct aspects of an extinct species' ecology, actions that are central to selecting suitable candidates—either phylogenetically related or ecologically similar—for rewilding projects. Indeed, such work in conservation paleobiology has already begun (e.g., the Makauwahi cave fossil record was used to justify the experimental rewilding of tortoises on the island of Kaua'i to fill the "meso-herbivore" niche that was lost when giant flightless ducks and geese became extinct soon after the arrival of Polynesians at the beginning of the last millennium; Burney et al., 2012).

Recognizing a human-dominated epoch of rapid change also poses a challenge for standards of conservation success (Rudd, 2011). Here, too, conservation paleobiology holds great promise in helping to develop a new set of assessment metrics (Barnosky et al., 2017). For instance, functional—"ecometric" (Polly et al., 2011)—traits of organisms present unparalleled opportunity to monitor ecosystem functions in an Anthropocene world of novel ecosystems. Such traits are a useful bridge between modern and geohistorical contexts because they are scalable and easily quantifiable properties (e.g., leaf shape) of living and fossil organisms that can be related to the functioning of an ecosystem, regardless of the identity of the species involved (Polly et al., 2011). Establishing appropriate baselines of acceptable variation (Wiens et al., 2012) of ecometric traits in the fossil record may help anticipate when and where future environmental change might present the greatest risks to biodiversity and help identify the best opportunities for conservation.

Conservation paleobiologists are also developing other metrics that can inform discussions about the sustainable use of ecosystem services. For example, paleoenvironmental proxies from lake-sediment cores formed the basis for the development of a regional "regulating services index," which captured trends and variation of services as varied as biodiversity, soil stability, air quality, and water quality in China's lower Yangtze basin (Dearing et al., 2012; see also Gillson, 2015).

Given these emerging opportunities, we think that the value of the conservation paleobiology approach will only increase in an Anthropocene future. There will continue to be places where the geohistorical record will reveal the impact of societal activities. There will continue to be times and places where baselines of species composition and abundance derived from the near-time geohistorical record will provide guidance for restoration (Kopf et al., 2015). But in most circumstances, we judge that the greatest value of geohistorical information to conservation lies in providing contextual information for the design of management responses and anticipa-

tion of future changes, with particular concern for the maintenance of ecosystem functions and services.

We hope that this book stands as a snapshot of a young discipline that is rapidly developing an intellectual framework even as it is finding promising applications. The overarching lesson to take away from these pages is that geohistorical records testify to a variety of past species, conditions, and signals of change. Such testimonies may be as relevant to the many concerns, needs, and goals of conservation biology today as they are in navigating an uncertain environmental future. Conservation biology has yet to embrace the full potential of these long-term records (Dietl and Flessa, 2011; Wolkovich et al., 2014; Holmes, 2015), but this situation, too, like the Anthropocene world, is rapidly changing.

REFERENCES

BARNOSKY, A. D., E. A. HADLEY, P. GONZALEZ, J. HEAD, P. D. POLLY, A. M. LAWING, J. T. ERONEN, D. D. ACKERLY, K. ALEX, E. BIBER, J. BLOIS, J. BRASHARES, G. CEBALLOS, E. DAVIS, G. P. DIETL, R. DIRZO, H. DOREMUS, M. FORTELIUS, H. W. GREENE, J. HELLMANN, T. HICKLER, S. T. JACKSON, M. KEMP, P. L. KOCH, C. KREMEN, E. L. LINDSEY, C. LOOY, C. R. MARSHALL, C. MENDENHALL, A. MULCH, A. M. MYCHAJLIW, C. NOWAK, U. RAMAKRISHNAN, J. SCHNITZLER, K. D. SHRESTHA, K. SOLARI, L. STEGNER, M. A. STEGNER, N. C. STENSETH, M. H. WAKE, AND Z. ZHANG. 2017. Merging paleobiology with conservation biology to guide the future of terrestrial ecosystems. Science, 355 (10 February 2017), doi:10.1126/science.aah4787.

BURNEY, D. A., J. O. JUVIK, L. P. BURNEY, AND T. DIAGNE. 2012. Can unwanted suburban tortoises rescue native Hawaiian plants? The Tortoise, 1:104–115.

CORLETT, R. T. 2015. The Anthropocene concept in ecology and conservation. Trends in Ecology and Evolution, 30:36–41.

DEARING, J. A., X. YANG, X. DONG, E. ZHANG, X. CHEN, P. G. LANGDON, K. ZHANG, W. ZHANG, AND T. P. DAWSON. 2012. Extending the timescale and range of ecosystem services through paleoenvironmental analyses, exemplified in the lower Yangtze basin. Proceedings of the National Academy of Sciences of the United States of America, 109:E1111–1120.

DIETL, G. P., AND K. W. FLESSA. 2011. Conservation paleobiology: Putting the dead to work. Trends in Ecology and Evolution, 26:30–37.

DONLAN, C. J., J. BERGER, C. E. BOCK, J. H. BOCK, D. A. BURNEY, J. A. ESTES, D. FOREMAN, P. S. MARTIN, G. W. ROEMER, F. A. SMITH, M. E. SOULE, AND H. W. GREENE. 2006. Pleistocene rewilding: An optimistic agenda for twenty-first century conservation. American Naturalist, 168:660–681.

DONLAN, C. J., AND H. W. GREENE. 2010. NLIMBY: No lions in my backyard, p. 293–305. *In* M. Hall (ed.), Restoration and History: The Search for a Usable Environmental Past. Routledge, New York.

GILLSON, L. 2015. Biodiversity Conservation and Environmental Change: Using Palaeoecology to Manage Dynamic Landscapes in the Anthropocene. Oxford University Press, Oxford, 215 p.

HOBBS, R. J., L. M. HALLETT, P. R. EHRLICH, AND H. A. MOONEY. 2011. Intervention ecology: Applying ecological science in the twenty-first century. Bioscience, 61:442–450.

HOLMES, G. 2015. What do we talk about when we talk about biodiversity conservation in the Anthropocene? Environment and Society: Advances in Research, 6:87–108.

KOPF, R. K., C. M. FINLAYSON, P. HUMPHRIES, N. C. SIMS, AND S. HLADYZ. 2015. Anthropocene baselines: Assessing change and managing biodiversity in human-dominated aquatic ecosystems. Bioscience, 65:798–811.

MARRIS, E. 2011. Rambunctious Garden: Saving Nature in a Post-Wild World. Bloomsbury, New York, 210 p.

POLLY, P. D., J. T. ERONEN, M. FRED, G. P. DIETL, V. MOSBRUGGER, C. SCHEIDEGGER, D. C. FRANK, J. DAMUTH, N. C. STENSETH, AND M. FORTELIUS. 2011. History matters: Ecometrics and integrative climate change biology. Proceedings of the Royal Society of London, Series B: Biological Sciences, 278:1131–1140.

RUDD, M. A. 2011. Scientists' opinions on the global status and management of biological diversity. Conservation Biology, 25:1165–1175.

SEDDON, P. J. 2010. From reintroduction to assisted colonization: Moving along the conservation translocation program. Restoration Ecology, 18:796–802.

SEDDON, P. J., C. J. GRIFFITHS, P. S. SOORAE, AND D. P. ARMSTRONG. 2014. Reversing defaunation: Restoring species in a changing world. Science, 345:406–412.

SHAPIRO, B. 2015. How to Clone a Mammoth: The Science of De-Extinction. Princeton University Press, Princeton, New Jersey, 240 p.

SVENNING, J-C., P. B. M. PEDERSEN, C. J. DONLAN, R. EJRNÆS, S. FAURBY, M. GALETTI, D. M. HANSEN, B. SANDEL, C. J. SANDOM, J. W. TERBORGH, AND F. W. M. VERA. 2016. Science for a wilder Anthropocene: Synthesis and future directions for trophic rewilding research. Proceedings of the National Academy of Sciences of the United States of America, 113:898–906.

WIENS, J. A., H. D. SAFFORD, K. MCGARIGAL, W. H. ROMME, AND M. MANNING. 2012. What is the scope of "history" in historical ecology? Issues of scale in management and conservation, p. 63–75. *In* J. A. Wiens, G. D. Hayward, H. D. Safford, and C. M. Giffen (eds.), Historical Environmental Variation in Conservation and Natural Resource Management. Wiley-Blackwell.

WOLKOVICH, E. M., B. I. COOK, K. K. MCLAUCHLAN, AND T. J. DAVIES. 2014. Temporal ecology in the Anthropocene. Ecology Letters, 17:1365–1379.

Contributors

Warren D. Allmon, Paleontological Research Institution, 1259 Trumansburg Road, Ithaca, NY 14850, USA, and Department of Earth and Atmospheric Sciences, Cornell University, Ithaca, NY 14853, USA. wda1@cornell.edu

Richard B. Aronson, Department of Biological Sciences, Florida Institute of Technology, Melbourne, FL 32901, USA. raronson@fit.edu

Anthony D. Barnosky, Department of Integrative Biology and Museums of Paleontology and Vertebrate Zoology, University of California, Berkeley, Berkeley, CA 94720, USA.
Current affiliation: Jasper Ridge Biological Preserve, Stanford University, Stanford, CA 94305, USA. tonybarnosky@stanford.edu

Alison G. Boyer, Department of Ecology and Evolutionary Biology, University of Tennessee, 569 Dabney Hall, Knoxville, TN 37996, USA. alison.boyer@utk.edu

Mark Brenner, Department of Geological Sciences, University of Florida, Gainesville, FL 32611, USA. brenner@ufl.edu

David A. Burney, National Tropical Botanical Garden, 3530 Papalina Road, Kalaheo, HI 96741, USA. dburney@ntbg.org

Gregory P. Dietl, Paleontological Research Institution, 1259 Trumansburg Road, Ithaca, NY 14850, USA, and Department of Earth and Atmospheric Sciences, Cornell University, Ithaca, NY 14853, USA. gpd3@cornell.edu

Karl W. Flessa, Department of Geosciences, University of Arizona, Tucson, AZ 85721, USA. kflessa@email.arizona.edu

Kena Fox-Dobbs, Department of Geology, University of Puget Sound, Tacoma, WA 98416, USA. kena@pugetsound.edu

Stephen T. Gray, Wyoming Water Resources Data System and Wyoming State Climate Office, Dept. 3943, 1000 E. University Ave., University of Wyoming, Laramie, WY 82072, USA.
Current affiliation: U.S. Geological Survey, Department of the Interior Alaska Climate Science Center, Anchorage, AK 99508, USA. sgray@usgs.gov

Elizabeth A. Hadly, Department of Biology, Stanford University, Stanford, CA 94305, USA. hadly@stanford.edu

Jeremy B. C. Jackson, Center for Marine Biodiversity and Conservation, Scripps Institution of Oceanography, University of California, San Diego, La Jolla, CA 92093, USA, and Center for Tropical Paleoecology and Archaeology, Smithsonian Tropical Research Institute, Box 0843-03092 Panama, Republic of Panama. jbjackson@ucsd.edu

Stephen T. Jackson, Department of Botany, Dept. 3165, 1000 E. University Ave., University of Wyoming, Laramie, WY 82071, USA.
Current affiliation: U.S. Geological Survey, Department of the Interior Southwest Climate Science Center, Tucson, AZ 85721 USA, and Department of Geosciences, University of Arizona, Tucson, AZ 85721 USA. stjackson@usgs.gov

Susan M. Kidwell, Department of Geophysical Sciences, University of Chicago, Chicago, IL 60637, USA. skidwell@uchicago.edu

Paul L. Koch, Department of Earth and Planetary Sciences, University of California, Santa Cruz 95064, USA. pkoch@pmc.ucsc.edu

Michał Kowalewski, Department of Geosciences, Virginia Tech, Blacksburg, VA 24061, USA.
Current affiliation: Florida Museum of Natural History, University of Florida, 1659 Museum Road, Gainesville, FL 32611, USA. kowalewski@ufl.edu

S. Kathleen Lyons, Department of Paleobiology, National Museum of Natural History, Smithsonian Institution [NHB, MRC 121], PO Box 37012, Washington, DC 20013, USA.
Current affiliation: School of Biological Sciences, University of Nebraska Lincoln, Lincoln, NE 68588, USA. katelyons@unl.edu

Loren McClenachan, Center for Marine Biodiversity and Conservation, Scripps Institution of Oceanography, University of California, San Diego, La Jolla, CA 92093, USA.
Current affiliation: Colby College, Environmental Studies Program, 5351 Mayflower Hill Drive, Waterville, ME 04901, USA. lemcclen@colby.edu

Seth D. Newsome, Geophysical Laboratory, Carnegie Institution of Washington, 5251 Broad Branch Road NW, Washington, DC 20015, USA.
Current affiliation: University of New Mexico, Department of Biology, Albuquerque, NM 87131, USA. newsome@unm.edu

John M. Pandolfi, School of Biological Sciences, ARC Centre of Excellence for Coral Reef Studies, The University of Queensland, Brisbane, Queensland 4072, Australia. j.pandolfi@uq.edu.au

Peter D. Roopnarine, Department of Invertebrate Zoology & Geology, California Academy of Sciences, 55 Music Concourse Drive, Golden Gate Park, San Francisco, CA 94118, USA. proopnarine@calacademy.org

Michael Savarese, Department of Marine & Ecological Sciences, Florida Gulf Coast University, Ft. Myers, FL 33965, USA. msavares@fgcu.edu

Bryan N. Shuman, Department of Geology and Geophysics, 1000 E. University Ave., University of Wyoming, Laramie, WY 82071, USA. bshuman@uwyo.edu

John P. Smol, Paleoecological Environmental Assessment and Research Lab (PEARL), Department of Biology, Queen's University, Kingston, Ontario, K7L 3N6 Canada. smolj@queensu.ca

Geerat J. Vermeij, Department of Geology, University of California, Davis, Davis, CA 95616, USA. gjvermeij@ucdavis.edu

Peter J. Wagner, Department of Paleobiology, National Museum of Natural History, Smithsonian Institution [NHB, MRC 121], PO Box 37012, Washington, DC 20013, USA.
Current affiliation: Department of Earth and Atmospheric Sciences, University of Nebraska Lincoln, Lincoln, NE 68588, USA. peterjwagner@unl.edu

Acknowledgments

PROJECTS SUCH as this one come to fruition only with the support and involvement of many people. We owe a deep debt of gratitude to the Paleontological Society for their permission to reprint the proceedings of the short course, which served as the basis for this book. We thank Doug Erwin, Robert Feranec, Linda Ivany, Patricia Kelley, Matthew Kosnik, Charles Marshall, Thomas Olszewski, Roy Plotnick, Scott Starratt, Donna Surge, Marcel van Tuinen, Thompson Webb, Ethan White, Thomas Whitmore, Katherine Willis, and several anonymous reviewers for the insightful reviews that helped us greatly improve the content and focus of this book. We thank our editor Christie Henry, at the University of Chicago Press, for her support and encouragement throughout the long gestation period of this project. Our home institutions also provided the nurturing environment needed for this project's success. Finally, and most obviously, we thank the authors for their thought-provoking contributions. This book would not have been possible without their efforts and expertise.

Index

Page numbers followed by f or t indicate figures or tables respectively.